THIRD EDITION

Science Instruction in the Middle and Secondary Schools

Alfred T. Collette
SYRACUSE UNIVERSITY

Eugene L. Chiappetta
UNIVERSITY OF HOUSTON

Merrill, an imprint of
Macmillan Publishing Company
New York

Maxwell Macmillan Canada
Toronto

Maxwell Macmillan International
New York   Oxford   Singapore   Sydney

Cover photo: Computer-enhanced image of normal human blood cells viewed under scanning electron microscope. Copyright © Fran Heyl Associates.

Editor: Linda James Scharp
Production Editor: Mary Harlan
Art Coordinator: Peter A. Robison
Text Designer: Jill E. Bonar
Cover Designer: Thomas Mack
Production Buyer: Pamela D. Bennett
Illustrations: Precision Graphics

Macmillan Publishing Company
866 Third Avenue
New York, NY 10022

Macmillan Publishing Company is part of the Maxwell Communication Group of Companies.

Maxwell Macmillan Canada, Inc.
1200 Eglinton Avenue East, Suite 200
Don Mills, Ontario M3C 3N1

This book was set in Garamond by Compset, Inc. and was printed and bound by Book Press, Inc., a Quebecor America Book Group Company. The cover was printed by Lehigh Press, Inc.

Photo credits: All photos copyrighted by individuals or companies listed. Jerry Berkowitz: pp. 3, 28, 37, 44, 55, 84, 103, 111, 129, 136, 143, 157, 166, 183, 196, 208, 210, 224, 263, 286, 292, 304, 306, 323, 332, 335, 337, 362, 366, 386, 402, 410, 417, 428, 445, 450, 472, 473, 488; Atomic Industrial Forum, Inc.: p. 21; Authors: pp. 120, 229, 232; National Aeronautics and Space Administration: p. 176; Spring Senior High School, Spring Independent School District, Spring, Texas: p. 230; Shawnee Mission Public Schools: p. 234; Alton Biggs, Allen High School, Allen, Texas, copyright *The American Biology Teacher*: p. 258; The Kinkaid School, Houston, Texas: p. 261; Douglas Elementary School, Houston Independent School District, Houston, Texas: p. 267.

Library of Congress Cataloging-in-Publication Data

Collette, Alfred T.
    Science instruction in the middle and secondary schools / Alfred T. Collette, Eugene L. Chiappetta.—3rd ed.
        p.   cm.
    Includes bibliographical references and index.
    ISBN 0-02-323551-9
    1. Science—Study and teaching (Secondary)—United States.
    I. Chiappetta, Eugene L.   II. Title.
    Q183.3.A1C637   1994
    507'.1273—dc20   93-19065
                                                                    CIP

Printing: 2 3 4 5 6 7 8 9   Year: 6 7 8

To the science educators who made significant contributions to this text through their research, scholarship, and suggestions and to the members of the science teaching profession who are graduates of our respective science education programs

Preface

Science Instruction in the Middle and Secondary Schools is intended to help science teachers prepare students to become scientifically literate. This goal presents a tremendous challenge because science, technology, and their interactions with society are so complex. In addition, the challenge is great because of diversity in students' abilities and cultural backgrounds. These and many other factors have created situations where many students report that science courses are difficult and uninteresting, and where minorities are underrepresented in the science and engineering professions.

We believe that science teachers must consider many facets of science teaching in order to provide science courses that meet students' needs and stimulate them to appreciate science and technology. Teachers must understand the nature of science and technology and possess a strong content background in several of these areas. They must also understand how students learn and must be able to ascertain what knowledge and skills students possess when they enter the classroom. They must be able to use a variety of instructional strategies to help students represent knowledge and find meaning in it. Furthermore, science teachers should be able to

develop portfolio assessment programs that address many learning outcomes in order to determine students' final grades. Finally, science teachers should become active in local, state, and national professional organizations so that they continue to stay up-to-date and grow professionally.

This textbook is meant to be a guide for preservice and inservice middle school and secondary school science teachers. It is a comprehensive methods book that provides important information, practice applications, and resource materials for prospective science teachers, beginning and experienced science teachers, professional science educators, and coordinators and supervisors of science. The book can be used for a full course of study or a short course to improve science teaching practices. In addition, it can be a reference text for other courses, such as those addressing trends and issues and curriculum development in science education.

This textbook is divided into five sections. Section One provides background material for science teaching. The chapters in this section include Directions and Goals of Science Teaching, The Nature of Science, and Cognition and Learning. Section Two deals with teaching strategies and classroom management. These chapters include: Inquiry; Demonstrations; Lecture, Discussion, and Recitation Strategies; Science, Technology, and Society; Laboratory Work; Safety in the Laboratory and Classroom; Science Projects, Science Fairs, and Field Experiences; Computers and Electronic Technology; Using Reading Materials in Science Instruction; and Classroom Management and Discipline. Section Three addresses aspects of planning for instruction. The chapters include: Planning and Teaching Science Lessons; Planning Science Units; and Planning a Science Course. Section Four, which deals with assessment in science teaching, contains two chapters: Evaluation in Science Teaching and Constructing and Administering Science Tests. Section Five includes one chapter, entitled Growing Professionally.

This science methods textbook is the result of more than 40 years of science education experience. During these years many changes have occurred in the profession. Nevertheless, other aspects have remained the same. Science teachers who are knowledgeable and enthusiastic about their work and who make science relevant and interesting seem to produce positive results through their teaching. We have tried to emphasize in this textbook what effective science teachers have always displayed in their teaching as well as to incorporate new research findings.

Acknowledgments

We acknowledge the many science educators and science teachers who contributed to the ideas that have formed the basis of this edition. The contributions of science educators have been invaluable because of the insights they provided regarding many facets of science teaching. Their work has appeared in articles and research reports in *Journal of Research in Science Teaching, Science Education, School Science and Mathematics, The Science Teacher, Science Scope,* and ERIC. Many science educators have contributed significantly to this methods text through their chapter-by-chapter reviews, especially the fine suggestions provided by those who reviewed the third edition of the book: Fred H. Groves, Northeast Louisiana University; Clifford A. Hofwolt, Vanderbilt University; Cheryl L. Mason, San Diego State University; and Herbert H. Stewart, Florida Atlantic University.

We also thank the middle and secondary school science teachers who have supervised our student teachers. Many of these individuals have served as excellent models for our students in training. In addition, the science teachers who have participated in our graduate science teaching programs have influenced our vision of how science teaching should be conducted in the classroom. We have used many of their ideas in the vignettes presented in this textbook. These science teachers and others are a testament that a great deal of exemplary science instruction still takes place in classrooms throughout our nation's schools.

Contents

Background for Science Teaching

BACKGROUND

The directions and goals of science teaching have undergone many changes during the past 150 years, influenced by political, economic, and societal events in the United States. Such events as the Depression of 1873; the American Industrial Revolution, which was a period of dramatic economic and technical change between 1877 and 1919; World War I; the Great Depression of the 1930s; and World War II had significant impact on the status of science teaching during this span of time. More than thirty years ago, the Soviet launching of Sputnik caused the most extensive curriculum reforms in the history of science education in the United States. And today, society is calling for another reform in the teaching of science in our nation's schools. Students are not able to handle problems in our society related to science and technology issues because our science programs are not preparing them to be aware of and understand the contributions of science to personal, social, technological, and cultural improvements. Students and the public in general are not scientifically literate. Science educators, science teachers, scientists, and others are now attempting to force change in the goals and directions of science education in order to develop a more scientifically and technologically literate society.

The most dramatic and significant changes in science in the United States occurred in the late 1950s when the Soviet Union launched the first artificial satellite, Sputnik, on October 4, 1957. This event introduced the space age and triggered an intensive effort to improve science and mathematics education in the elementary and secondary schools of the nation. Scientists, politicians, labor leaders, public figures, and the public in general expressed great concern about the event. Senator Henry Jackson termed the event "a devastating blow to the prestige of the United States as a leader in the scientific and technical world" (Associated Press, Oct. 4, 1987). Wernher von Braun, the scientist who was responsible for the success of the space program

in the United States, said that Sputnik "triggered a period of self-appraisal rarely equaled in modern times" (Associated Press, Oct. 4, 1987); and William O. Baker, who was chairman of Bell Laboratories at the time, recalled on the thirtieth anniversary of Sputnik that "the situation was highly emotional" (Associated Press, Oct. 4, 1987). The event created great public concern and caused President Eisenhower to approve a $1 billion budget to provide direct federal aid to education "to meet the pressing demands of national security for the years ahead" (Associated Press, Oct. 4, 1987). As a consequence, Congress directed a significant sum of money to the National Science Foundation to set up a program to improve science and mathematics education in the United States. By 1962 the foundation's budget skyrocketed to more than $250 million.

These events led to the most dramatic changes in science instruction in the United States. Many pressures forced the U.S. government, through the National Science Foundation, to finance a massive course-content improvement program that produced many new approaches for the teaching of science in elementary and secondary schools. For secondary schools, new approaches were developed for biology, chemistry, physics, earth science, and the physical sciences. Some of the most notable of these curriculum approaches included the Biological Sciences Curriculum Study (BSCS), the Earth Science Curriculum Project (ESCP), the Physical Sciences Study Committee (PSSC), the Chemical Education Materials Study Program (CHEM), and the Chemical Bond Approach (CBA). These courses and others had a tremendous impact on the teaching of science as both a body of knowledge and a research discipline with particular emphasis on scientific inquiry.

During the 1960s the National Science Foundation gave special attention to improving and updating the science backgrounds of practicing science teachers. To do this, the foundation supported science teachers to attend summer institutes, academic year institutes, and in-service institutes offered at universities and colleges throughout the country. The institutes varied—some emphasized content improvement and background, while others trained teachers to use the new curriculum approaches such as BSCS and PSSC. In effect, through these institutes and new curricula, the National Science Foundation had an important part in influencing the directions and goals of science instruction in the United States during the 1960s.

Science education is now undergoing another period of reformation. The National Science Foundation is no longer influencing the goals and directions of the teaching of science. The curriculum materials produced in the 1960s through National Science Foundation funding are now considered obsolete. At present, science teachers, science educators, and scientists, individually and collectively, are attempting to determine the future directions and goals for the teaching of science in the secondary and middle schools.

Many problems in today's world related to science, technology, and society (STS) are forcing a reevaluation of directions and goals of science teaching in the United States. There is a debate among science educators that "centers on the degree to which Science-Technology-Society goals, as opposed to other goals such as reasoning and critical thinking, should be included in the science program" (Bybee, 1987b).

The views of science educators in the United States regarding the teaching of STS have been reported by Bybee (1987b). He surveyed science educators in the United States regarding various issues of STS and found through this study that most science educators believe that it is important to study STS issues in the middle school as well as in high school, college, and adult education. According to Bybee (1987b), "The views of science educators in the United States support the contention that the Science-Technology-Society theme is established in contemporary science education. Because the STS theme is part of science education, it is clear that there are implications for curriculum and instruction, teacher education and developing public support for teaching about Science-Technology-Society in school programs."

The close relationship of science, technology, and society is neither new nor trivial. We often speak of the values and goals of science and society as mutually exclusive. No greater misconception could possibly be the case. Science does not function in a vacuum; furthermore, its values and goals are not at odds with those of society. It is true that science and technology have been used at times to society's detriment (e.g., nuclear weaponry), but such is neither to the detriment of science nor evidence of incongruence between science and society. Let us not forget that science and its methods and technology are products of society and its needs. Consequently, by their very nature, goals and values of science are dictated by society and are pursued for reasons of perceived benefits for all. Certainly, the American Industrial Revolution was fueled by this relationship. The raw materials for the Industrial Revolution were always present, but the scientific knowledge and the subsequent technology for using such materials were not available. Initial advances in scientific knowledge gave rise to such inventions as the cotton gin, automobile, internal combustion engine, steam turbine, and radiography. These inventions led to increased industry and affluence. In short, the standard of living dramatically changed and improved. Society and its industrial sector, having reaped such benefits, strove to achieve more by committing copious financial support for research and development. Research and development led to additional advancements of science and technology that further improved the quality of life. Therefore, the congruence of science, technology, and society is quite evident. Education is also a product of society; thus, the relationships among science, technology, and society can be further shown through an analysis and study of the effects that society has had on the science curriculum.

The brief history of science education in the United States that follows shows how societal pressures, conditions, and technological advances have influenced the goals and directions of science education over the past century and a half.

SCIENCE INSTRUCTION BEFORE 1910

Very little can be stated regarding science instruction in American schools before 1870. It is known that stories and didactic literature emphasizing the study of things and phenomena, as well as religious doctrine, were designed for home use during the middle of the eighteenth century. The materials, based on the philosophies of John Locke and Jean-Jacques Rousseau, gained popularity between 1800 and 1825. Later, when group instruction came into vogue, textbooks based on these materials were developed and used in the schools. Many of the lessons in these textbooks included science content, and the students were required to memorize facts used to support religious doctrine (Underhill, 1941).

Science instruction in the elementary schools between 1860 and 1880 was through "object teaching," a method of teaching based on the precepts of faculty psychology and the writings of the Swiss educator Pestalozzi. The lessons used in object teaching advocated observing, experimenting, and reasoning and denounced the highly verbal and rote memorization techniques that had existed in the classroom. The intellects of children, it was thought, could best be developed if they studied real objects by using all the senses. The training of the mind, and not the learning of facts, was the important objective. Although the pupils were encouraged to study objects and make their own discoveries, object teaching resulted in a great deal of teacher-talk and little or no learning. The method was severely criticized and was banned because it lacked order and direction and involved meaningless memorization. It was also severely criticized by those who wanted to emphasize science

content and deemphasize the personal development aspect of the object lesson (Underhill, 1941; Hurd & Gallagher, 1969).

Elementary science received much attention between 1870 and 1900. During this time, significant changes were occurring in the American rural and urban scenes. The Industrial Revolution initiated massive migration from rural communities to the cities. At the same time, thousands of immigrants were pouring into the United States. Urban industrial societies were being established, which necessitated science programs that were both vocational and utilitarian with an emphasis on content. As a direct response to the social demands of the period, many science curricula appeared that were meant to satisfy the needs of those living in this new, technologically oriented society. A number of science educators, including Col. Francis W. Parker, Wilbur S. Jackman, William T. Harris, G. Stanley Hall, and others, developed elementary science curricula that were practical and utilitarian and stressed the content of science (Underhill, 1941). The nature study movement, under the direction of Liberty Hyde Bailey, also appeared. Nature study grew out of object teaching and was aimed at increasing a child's interest and appreciation of nature, as well as personal development (Underhill, 1941).

The nature study movement, which was started about 1890 by Liberty Hyde Bailey, a professor of nature study at Cornell University, was a direct response to the mass migration from rural into urban areas and was designed to promote individual agriculture in New York State. In 1897 Bailey was given the charge to find ways to stimulate young people's interest in living things and country life. He felt that elementary school pupils could be taught to be interested in and sympathetic with natural habitats of the country. He promoted his philosophy through the publication of the Cornell Rural School leaflets in which he wrote extensively. Pupils learned science by studying common objects such as stones, birds, flowers, and insects found in their environments. They used science processes such as observing, classifying, and measuring to answer questions about things around them. Bailey said that nature study as a process was "seeing the things that one looks at and the drawing of proper conclusions from what one sees. Its purpose is to educate the child in terms of his environment to the end that his life may be fuller and richer" (Bailey, 1904). The nature study program was primarily used in rural elementary schools in New York State and had little effect elsewhere. It is also interesting to note that all the programs that appeared during this time had little effect on what took place in the schools because teachers had difficulty in handling the curriculum materials.

During the late 1920s, Gerald Craig, a science educator of some note, organized an elementary science curriculum that directed elementary science education for more than forty years. Craig developed this curriculum as a result of a study he conducted for his doctoral dissertation, which he completed in 1927. The curriculum consisted of a series of science generalizations intended to answer children's questions. They were written to be utilitarian as well as meaningful to children. His curriculum dominated elementary science instruction until 1960, when inquiry-oriented programs replaced Craig's science readers.

The secondary schools in the United States really began in the academies that started in the middle of the eighteenth century. Before this time the Latin grammar school was responsible for preparing a small number of students for college admission, especially those wanting to become clergymen. The period of the Latin grammar school began in 1635 when the Boston Latin School was established. The schools during this period restricted attendance to affluent, able, male students. The curriculum was chiefly limited to the study of Latin, Greek, and the scriptures, and the students had little or no exposure to scientific and practical fields. The Latin grammar school gradually gave way to the private academies, starting in 1751 with the establishment of the first academy in Philadelphia.

The academies, the forerunners of the secondary schools, thrived for about seventy years. At first, religious instruction dominated the curriculum of the academies, but gradually a more practical curriculum was implemented. The academies had the first organized instruction in science, consisting of courses in natural philosophy (physics), astronomy, chemistry, botany, geology, and physical geography. In addition, courses in navigation, agriculture, and surveying were offered. Lectures and formal recitations were commonly used for instruction. The student population in the academies included both boys and girls, but the schools were considered exclusive because they were private institutions that charged tuition.

The origins of the early high schools can be traced to 1821 when the English classical school was established in Boston. After 1824 this school was called the Boston English High School. At the beginning, courses in practical mathematics, navigation, surveying, mensuration, and astronomical calculations were included in the curriculum of the early high schools. By 1852 courses in astronomy, geology, mechanics, engineering, and higher mathematics were taught. After 1860 many high schools offered meteorology, mineralogy, botany, physiology, and physiography. Biology courses were also included as course offerings and listed as zoology and botany. Botany was offered as a premedical course and had little or no relationship with zoology. The courses were utilitarian, stressing the practical arts and duties of citizenship (Lacey, 1966).

The science courses offered in the early high schools were basically descriptive, with little or no laboratory work connected with them. At first, the teachers who taught science courses were very poorly prepared. Later, teachers with training and experience in science were not uncommon.

Between 1870 and 1910 many changes occurred in the American social structure that influenced the teaching of science in the secondary schools. The Kalamazoo Decision of 1872 paved the way for public, tax-supported high schools. One year later the Great Depression of 1873 occurred. Following the Depression of 1873 the schools were severely criticized for not meeting the needs of a changing society. This was the time when the Industrial Revolution was emerging, causing a significant portion of the population to migrate into urban areas. This was also the period during which thousands of immigrants entered the United States. With increased industrialization, along with related urbanization, the social structure changed. The heterogeneous population led to concerns about individual needs and differences among the student population; industrialization created a strong need for more practical science courses that would better equip individuals to cope with everyday problems encountered in an industrialized society. Science teaching in the secondary schools was subsequently geared to give individuals a greater understanding of science and technology.

The science curriculum was also greatly influenced by the faculties of colleges and universities between 1870 and 1910. It was during the 1870s that universities began to require physics and, later, chemistry for college entrance. Harvard University set the stage in 1872 by establishing physics as an entrance requirement. These requirements had a significant impact on the science offerings in the secondary schools, even to the point of teaching science courses that were shortened versions of college and university courses. Many of the science textbooks that appeared were also abbreviated versions of college science textbooks. Physics became the most important science offering, with chemistry a close second. Biology courses that combined one-year courses in botany, zoology, and physiology were introduced. These courses had little or no integration. Geology was also offered, but its popularity declined toward the beginning of the century.

The period between 1870 and 1910 is often called the period of college domination, which marked the beginning of science as a formal discipline in the secondary schools. With this formalization, the popularity of secondary school science diminished. Courses were established without regard to the students' interests and need to understand their environment. Laboratory instruction became an important part of the science offerings, implemented first in chemistry and later in physics and the biological sciences. The laboratory work was highly specialized and often dull and stereotyped. Science was presented as a body of information in these laboratory courses. Little or no consideration was given to the individual differences and backgrounds of students. The main aims were to develop the abilities to reason, observe, and concentrate.

After 1890 several committees, such as the Committee of Ten in 1892 (Committee on Secondary School Studies, 1893), the Committee of Fifteen (U.S. Office of Education, 1899), and the Committee of College Entrance Requirements, appointed in 1895 (Nightingale, 1899), recommended that science be introduced into the curriculum earlier than the secondary school level. This recommendation, as well as other factors, promoted the reorganization of schools into the 6-3-3 pattern*. In particular, the report of the Committee of Ten stated that all students should be exposed to the same curriculum even if they had no intention of attending college and even suggested the subjects to be taught, as well as the time that was to be devoted to each subject. The academic and intellectual goals were stressed in the report. When the Committee of Ten (1893) set out to standardize high school curricula they emphasized that the secondary schools of the United States did not exist for the sole purpose of preparing boys and girls for college. This

---

*Grades K–6, elementary grades; 7–9, junior high school; 10–12, secondary or senior high school.

report along with others promoted articulation between elementary and secondary school programs. It also caused colleges to have less influence over high school science offerings.

THE PERIOD OF REORGANIZATION (1910 TO 1930)

A number of important changes occurred in the organization of the schools between 1910 and 1930 that greatly influenced the science program. It was during this period that the high school population markedly increased. There was a rapid growth of the 6-3-3 organizational pattern in the schools, and the junior high school movement was at its peak. General science flourished in the curriculum at grade 9; biology became a tenth year subject. Physics and chemistry were taught at either grade 11 or 12.

General science developed very rapidly in the 1920s and was even introduced in the seventh grade in some schools. The unitary aspect of general science influenced the development of many types of integrated courses. This influence caused a deemphasis in the specialized courses such as physics and chemistry. Both biology and general science stressed the study of the environment and the functional aspects of science.

The reorganization of the schools resulted in curriculum innovations, which in turn influenced the objectives of science teaching. Although the objectives of science teaching varied throughout the country, teachers were still primarily interested in presenting scientific information. There was little interest in teaching the overall principles and generalizations of science. Some teachers formulated life- and student-centered objectives, but these appeared to be the exception rather than the rule.

By 1915 broader goals of science instruction in the secondary school deemphasized the requirements for college entrance. A report of the Central Association of Science and Mathematics Teachers Committee on Unified High School Science Courses (Central Association of Science and Mathematics Teachers, 1915) suggested that science should

1. Give students a knowledge of the environment in order to help them with problems of everyday living.
2. Stimulate individuals toward more direct, purposeful activities.
3. Help individuals choose future occupations intelligently.
4. Give students the methods for obtaining accurate information.
5. Enable students to derive greater, clearer, and more intellectual satisfaction from their lives.

In 1918 the National Education Association Commission on the Reorganization of Secondary Education issued a report that stated the seven cardinal principles as the overall goals of education. The goals as stated are (1) health, (2) command of fundamental process, (3) worthy home membership, (4) vocation, (5) citizenship, (6) worthy use of leisure time, and (7) ethical character (Commission on the Reorganization of Secondary Education, 1918). Curricula were to be organized to meet these goals of education. In 1920 a report on the reorganization of science in the secondary schools (Caldwell, 1920) recommended the teaching of science in light of the seven principles of education. The report emphasized incorporating life-centered goals into the science curricula. Subsequent attempts were made to teach general science and biology so that the cardinal principles could be realized. Physics and chemistry continued to be college preparatory courses and were not concerned with meeting social goals.

In 1924 the Committee on the Place of Sciences in Education of the American Association for the Advancement of Science published a report on the problems of science teaching (Caldwell, 1924). The report stressed the importance of scientific thinking as a goal of science instruction and emphasized that scientific observations and experimentation should be used to obtain a better feeling for understanding the scientific enterprise.

Although the mastery of subject matter continued to be a primary aim of the science curricula of the 1920s and 1930s, a great deal of work was done on the formulation of major generalizations that could be used as long-range goals to give continuity to the science program.

BEFORE AND DURING WORLD WAR II (1930 TO 1945)

Between 1930 and 1945 the junior high school movement leveled off, and general science was introduced in grades 7 and 8. Advanced general science courses were proposed for grades 11 and 12 to replace physics and chemistry. Biology became a basic science offering. World War II brought about a great emphasis on the practical aspects of science, and, consequently, courses such as aviation, photography, and electricity were introduced into the curriculum.

Many committees published lists of objectives for teaching science during this period. The most prominent and important of these sets of statements was published in 1932 by the National Society for the Study of Education. A list of thirty-eight generalizations that was designed to give direction to the entire science program from grades 1 through 12 was proposed. One of the major strengths of this list was the balance it provided within the traditional subject matter areas and the interrelationships it promoted among these areas. These objectives represented a broad approach to the problems involved in planning.

The use of major generalizations as objectives, especially in the form outlined in the *Thirty-First Yearbook, Part I* (National Society for the Study of Education, 1932), had its limitations. Science was presented as a body of facts, and students viewed it that way. The other benefits of science instruction derived from problem-solving techniques and the development of certain attitudes and appreciations were neglected.

The Commission on Secondary Curriculum of the Progressive Education Association published a report in 1938 that stressed the need to correlate science with the problems of living and indicated the desirability of selecting content that was useful to boys and girls (Progressive Education Association, 1938). The Commission also regarded reflective thinking and the understanding of economic relationships as important objectives of science teaching.

AFTER WORLD WAR II (1945 TO 1955)

By 1945, about 75 percent of boys and girls of high school age were attending secondary schools. More science courses were introduced to handle increasing enrollment. Courses such as earth science and physical science were offered, and general science was an important offering of the science program. Biology became popular, but physics and chemistry showed marked declines in enrollment. Science fairs and congresses became common during this period.

Science teaching placed notable emphasis on the functional aspects of science. Many organizations such as the National Association of Secondary School Principals, National Education Association, and American Association of School Administrators published statements of objectives for the teaching of science in the secondary schools. The most important statement concerning science instruction was that of the National Society for the Study of Education as presented in its *Forty-Sixth Yearbook, Part I* (National Society for the Study of Education, 1947). The development of objectives in science education between 1932 and 1945 was summarized; during this period, science instruction followed the trends of objectives in general education. Less emphasis was placed on the memorization of information, and more emphasis was directed toward teaching boys and girls the functional aspects of science. The understanding of scientific principles and development of problem-solving abilities were stressed to a greater degree than in the past. However, teaching scientific information was still considered to be the most important aim of science teaching. After World War II, teachers stressed such objectives as developing scientific concepts, principles, skills, and attitudes. Skills in problem solving became a primary aim. An infallible scientific method that could be used as a pattern to gather information was deemed possible. The laboratory took on a new importance as the primary place where this method was used. Skills in gathering and testing data, identifying and solving problems, and examining the validity of conclusions were strongly encouraged as the aims of science education. Nevertheless, the older, well-established aim of teaching scientific information was not abandoned as being unworthy; instead the relative values of objectives were reassessed as the newer goals were added.

The National Society for the Study of Education (1947) published the following categories of objectives, which reflect this type of teaching:

1. Functional information of facts
2. Functional concepts
3. Functional understanding of principles
4. Instrumental skills
5. Problem-solving skills
6. Attitudes
7. Appreciations
8. Interests

The objectives in the *Forty-Sixth Yearbook* (National Society for the Study of Education, 1947) are stated in such broad terms that their implications for science teaching may not be immediately obvious. The professional literature of the late 1940s cites many attempts to break these broad objectives down into usable form. Curriculum groups spent a considerable amount of time on this aspect of planning. Only when a variety of experiences are incorporated into the science curriculum can students begin to advance toward the attainment of these broad objectives. Measuring goals of such scope is difficult, if not impossible.

It is clear that the greatest forces that have influenced changes in science education prior to 1950 have been social, political, or economic. The reforms that have taken place have been direct responses to the demands and needs of society as crises occur and values and social structures change. There seems to be a high correlation between social pressures and the type of curricula that emerge. Bybee (1977) states the following regarding the changes that have occurred in science education prior to 1950:

1. Very few of the changes recommended by individuals or committees have ever been implemented in the science classrooms of America.
2. A model of science teaching in which the primary aim is knowledge has dominated and continues to dominate our science program. Models emphasizing the scientific method, problem solving, personal development, values, attitudes, and aesthetics have had only limited success.
3. The greatest changes have occurred at the elementary level, with fewer changes at the secondary level, and the domain of least change is higher education.
4. Change by academic discipline from most to least would be general/earth and environmental science, biology, chemistry, and finally physics.

THE REVOLUTION IN SCIENCE TEACHING (1955 TO 1970)

The most innovative and drastic changes were proposed for the science program during the period from 1955 to 1970. It was during this time that the student population increased significantly in the secondary schools. The enrollments in most science courses also increased greatly, although physical sciences were not popular at this time. It was also the period when many public critics, particularly scientists, felt that science courses in the public schools were not stimulating students to go into scientific vocations. The courses, they claimed, lacked rigor, were dogmatically taught, were content oriented, lacked conceptual unity, were outdated, and had little bearing on what was really happening in the scientific disciplines. The functional aspects of science teaching were also attacked. Many critics believed that too much emphasis was being placed on life adjustment goals and that there was too much stress on technology. They argued that the laboratory work connected with traditional science courses consisted of exercises that emphasized manipulatory skills and gave students the wrong impression of the scientific enterprise. The traditional science courses were not preparing young people for understanding either the world in which they were living or the future. Science teaching, then, was portrayed as dull, inadequate, and not meeting the demands of the times. The situation called for reform.

The real stimulus for a reform movement came from university scientists and mathematicians. These individuals claimed that the young people entering college were poorly prepared in science and mathematics. These scientists also felt that little was being done to encourage the talented student to go into science and mathematics fields. Because there was a shortage of scientific personnel in the United States, they decided to do something about the situation.

One of the first things done to initiate reform was to examine existing courses of study in science and mathematics. This examination of science textbooks of the 1950s indicated that attempts had been made to keep them up to date by adding bits and fragments of information to existing content. In addition, traditional topics were never dropped. All in all, textbooks were a hodgepodge of information, much of which was outdated and irrelevant to modern science disciplines. The reformers came to the conclusion that existing courses were not salvageable and embarked on writing new courses of study that would be in line with modern science.

The reform movement started gradually even though the situation called for immediate change. However, the Soviet launch of the first space satellite in 1957 sparked a crash program of reform.

The public became very science conscious with the advent of the space age. As Americans we were embarrassed by the progress the Soviets had made in their space program and our lack of progress in our own program. American education was criticized from all points of view. Science and mathematics education received the most severe criticisms because these areas were directly concerned with providing scientific personnel and there was a shortage of this type of manpower.

The reform movement sparked in 1957 was to produce some of the most innovative and spectacular changes in science education ever seen in American public high schools.

The National Science Foundation and other federal agencies came into the act at this time. They were concerned about the critical shortage of scientists and mathematicians. They were also concerned that very few talented individuals were entering scientific vocations. They attributed all this to the deplorable state of science and mathematics teaching in the United States.

These agencies quickly discovered that science courses taught in the public high schools were staffed with unqualified teachers with poor and outdated science backgrounds. The National Science Foundation, in particular, initiated an immediate program to improve the science and mathematics backgrounds of teachers. Colleges and universities set up institute programs sponsored by the National Science Foundation that offered courses in science and mathematics content to update teachers' backgrounds.

At the same time, scientists tried to stimulate change in the science offerings in the public high schools. As a result of their efforts, a number of national curriculum groups composed of scientists and mathematicians were established. Between 1957 and 1965, virtually unlimited funds were available from federal agencies, particularly the National Science Foundation, to bring about changes in the science and mathematics offerings in the public high schools.

Many new approaches were produced by national curriculum groups in the 1960s. All these approaches attempted to lead students through a series of experiences that encouraged the creative process and to bring them to a point where they conceptualized the scientific knowledge they obtained. Students presumably ultimately acquired an understanding and appreciation of the true nature of science—that is, the structure of science and the strategies of science. Courses were organized to permit students to spend a great deal of time in the laboratory performing activities that were investigative in nature. The courses of study focused on the anticipation of things learned and to be learned. Not only were students directly involved with "discovery," but it was hoped that they would also learn about the limitations of science and the scientific enterprise.

A further insight into the type of learning skills that were expected from the programs of the late 1950s and early 1960s is illustrated by the Biological Science Curriculum Study materials. The excerpt that follows is typical of the study group programs developed during the 1960s. The remarks are made to the student in a section called "The Logic of the Research Experience You Have Just Completed."

> In these laboratory exercises we have used a variety of the general approaches employed in modern biological research. We might classify the ones we have used in this series as:
> 1. Observing structures and actions.
> 2. Correlating structural changes with action.
> 3. Analyzing data from experimental measures, particularly for testing hypotheses quantitatively.
> 4. Designing specific experiments to test specific possibilities in specific cases.
> 5. Rationalizing about a structure by comparing it with concepts from other fields, such as engineering.
> 6. Making artificial models that will duplicate certain actions.

> 7. Combining data from anatomy, microanatomy (including electron microscopy), chemistry, and changes during action (function).
> 8. Considering what is evidence, deduction, interpretation, and proof.
> 9. And last, but by no means least, giving our imaginations free play for developing novel concepts and ideas for testing.
> (Richards, 1961, p. 99)

Examination of the experimental courses of the late 1950s and the early 1960s shows the following:

1. The courses were theory and concept oriented. They emphasized the structure of science in that students learned the relationships and relevance of facts, concepts, and principles of the particular science discipline.
2. The courses emphasized breadth and depth of knowledge. The number of concepts to be developed during the course was considered reduced in comparison to traditional-type courses. Students obtained a better understanding of a concept because they were not exposed to an endless number of concepts. Students also obtained breadth in that they studied fewer topics, and these had applications in other situations. (*Breadth* in reference to these courses is not used in the sense of covering a wide *range* of topics.)
3. Courses were less practical than traditional courses. Almost all technology was lacking in these approaches.
4. Courses emphasized the scientific enterprise. Students were permitted to discover for themselves. They were allowed to ask questions, state hypotheses, set up experiments, collect and organize, and interpret data with little or no direction. In this way they obtained a feeling of how scientists work and how they motivate themselves in their work. The student thus took a very active part in the investigatory process and learned to understand it.
5. The courses emphasized that science was changing and that it was not an established

number of facts, concepts, and principles that could not be changed. The open-endedness of science, then, received considerable attention in these approaches.

An attempt to list the goals of science teaching for this period would immediately reveal that precise and concise statements in the usual sense are difficult to make. Instead, a rationale for teaching science is presented. The usual types of statements of goals that are centered around social needs and problems, life needs, individual needs, and life problems are not found. Instead, the objectives of science teaching are derived from the respective science discipline. As Hurd (1969, p. 34) states, "Their logic is simple and straightforward: this is what we know with some degree of reliability; this is how we find out about what we know; and this is how it all fits into a big picture—conceptual schemes."

In summary, the period 1955 to 1970 was a rather active one with respect to curriculum development in science teaching. Some humorously refer to this period as the "Day of the Alphabet Programs" (Fox, 1966) because many new curricula were referred to in this way— BSCS (biology), ESCP (earth science), IPS (physical science), CHEM Study (chemistry), and ISCS (general science). All these programs had several characteristics in common. They were all inquiry oriented, process was stressed, levels of rigor and content were increased, quantification was emphasized, and students were required to gather, record, and analyze data. Such curricula resulted from the public demand for scientifically and technically trained manpower. This demand is easily traced back to the Soviet launching of Sputnik and the advancement of scientific knowledge in the post–World War II period. The pleas during the period were for increased quality, excellence, and rigor in high school science courses. The changes in the science curriculum during 1955 to 1970 were the direct result of the social and political climates of the country.

THE PERIOD OF LITTLE INNOVATION AND DECLINE (1970 TO THE EARLY 1980s)

The National Science Foundation influenced the direction of science teaching between 1955 and 1970. As mentioned, practically unlimited funds were available through this agency for curriculum reform during this period. The directives from the Foundation stipulated that the "new" courses to be produced were to be improved in content and were to cater to the same students as those enrolled in traditional courses. These directives indirectly prevented complete reorganization of the science curriculum. The restrictions imposed by the National Science Foundation also prevented the development of new courses that had no relationship to traditional courses already in existence. Consequently, the courses that emerged had certain shortcomings from the start.

The new curricula developed during the 1960s underwent scrutiny in the late 1960s and early 1970s. Many criticisms were voiced that forced a reevaluation of the prevalent science teaching practices. Nine major criticisms were stated by many science teachers and curriculum developers regarding the approaches that were produced by the national curriculum groups of the 1960s. These criticisms and others influenced the future course of science education in the United States:

1. The courses that had emerged were too discipline centered and specialized, with a professional point of view.
2. The topics covered in these courses were similar to those covered in traditional courses of the early 1950s. In general the topics had a classical orientation. Topics such as human ecology, oceanography, and anthropology— all important for the times and for our future—were either omitted or poorly presented.
3. The courses of study were too highly theory and concept oriented.

4. Science in general education was not emphasized because the courses lacked the humanistic, social, and historical aspects so important to a liberal education.
5. There was a complete disregard for the relationships among science, social studies, the humanities, and technology.
6. The courses were too rigorous for the average student. Although the courses developed were intended for average students who normally took the traditional courses in science, such students were not able to cope with most of the approaches.
7. Most science teachers had difficulty teaching the new approaches. The courses required a special style of teaching that was foreign to most teachers and students. Several years of experience teaching the courses were required before a science teacher could be effective in the approaches.
8. The courses were too long. Even though there was no stipulation that compelled teachers to use the whole course of study, most teachers tried to cover all topics. This, however, was the fault of the teacher rather than the curriculum.
9. The new science programs were developed independently of each other, and the topics from other sciences were used only to give more meaning to a specific topic in a course. The interrelationships of the sciences and the broader meanings of science were not stressed.

There were other criticisms as well, but these were the most outstanding. However, one positive point should be mentioned. The national curriculum approaches emphasized the nature of science and the scientific enterprise—an area that had been neglected in science teaching before 1960.

It is certain that the reform projects discouraged the average student from taking additional science courses that had been developed during the 1960s. The courses were considered too rigorous and highly theoretical and concept oriented, and did not include topics that would prepare students for everyday living. The courses disregarded the social and technological aspects of science teaching, which are important for living in today's world and the world of the future.

Because the national curriculum approaches of the 1960s were regarded as too rigorous and irrelevant for the average student, and because most science teachers found them too difficult to teach, many chose to abandon them. Consequently, science teachers and local curriculum groups began to plan their own courses of study and in many cases made use of some of the ideas and instructional approaches from the national curriculum courses.

New national curriculum courses also appeared in the late 1960s and early 1970s for middle and secondary school science teaching. These courses were considered less rigorous than the national curriculum approaches that dominated science instruction in the 1960s. The approaches were designed to be student centered, stressed individualized instruction, and were flexible, relevant, and of interest to a wide range of students. Among these were the Intermediate Science Curriculum Study (ISCS), Individualized Science Instructional System (ISIS), the Interdisciplinary Approaches to Chemistry (IAC), and the Outdoor Biological Sciences Instructional System (OBSIS). These courses in particular characterized the new developments in the 1970s. Except for these courses and a few others, the 1970s was a period of little activity during which no significant changes were introduced that affected science instruction in the middle and secondary schools in the United States. However, during this period, the role of the National Science Foundation as an active force in influencing the direction of science education ceased because certain citizen groups questioned the content of some curriculum projects sponsored by the National Science Foundation.

A course improvement program that appeared in the mid-1970s sponsored by the National Science Foundation caused a great deal of controversy and finally the demise of NSF-sponsored programs. Man: A Course of Study (MACOS) received very severe criticisms from conservative citizens voicing that the course content was immoral and vulgar. The complaints were brought to the attention of certain conservative members of the U.S. Congress, who managed to stop the appropriation of funds to continue the program. This incident and other reasons created a concern on the part of Congress to the point that federal funds were no longer made available to support NSF curriculum projects. This ended the era of NSF-sponsored curriculum projects. Funds intended to support NSF curriculum projects were routed to support scientific research and projects to improve college science teaching.

Also under attack in the 1970s by certain conservatives was the content of science textbooks used in the public schools. Members of fundamentalist Christian churches and other conservatives scrutinized the content of certain science textbooks, particularly those in biology. The teaching of evolution in the public schools in certain areas of the United States received much attention from Christian fundamentalists. In some states, laws were enacted to control the teaching of evolution as a fact. The debate that occurred was not with the banning of such teaching but rather with the provision of "equal time" for the creationism as an alternative. (See the discussion on evolution versus creationism in Chapter 7.)

In the late 1970s, the National Science Foundation sponsored three research studies on the status of science education in the United States: The Status of Pre-College Science, Mathematics and Social Science Education: 1955–1975; Case Studies in Science Education; and 1977 National Survey of Science, Mathematics and Social Education (Harms & Yager, 1981). The result of these studies prompted the National Science Foundation to fund another study, Project Synthesis. This study stressed the importance of changing the existing goals, programs, and practices of science teaching. Instead of emphasizing academic

preparation for a few students interested in the science professions, it was proposed that major goals should stress the education of all students so they would be able to become knowledgeable about science and technology issues in their everyday lives and be able to deal with science-related issues in today's world.

In summary, science teaching goals changed in the late 1950s and early 1960s because traditional science courses were regarded as obsolete for the times, since they paid little attention to the nature of science and the scientific enterprise. The traditional courses were replaced with new national curriculum courses sponsored by the National Science Foundation, the goals of which emphasized the theoretical, conceptual, and investigative aspects of a science discipline and disregarded the social, humanistic, and technological aspects of the discipline. The scientific enterprise was emphasized in this reform movement.

During the early 1970s new national curriculum courses appeared for middle and secondary school science teaching that were less rigorous and more relevant to average students. In the mid-1970s the National Science Foundation sponsored a course entitled Man: A Course of Study (MACOS). It received so much negative publicity from certain conservative groups that the U.S. Congress stopped funding it. Largely because of this incident, Congress quit supporting further NSF-sponsored curriculum projects.

PERIOD OF CONTROVERSY: THE 1980s

The results of Project Synthesis (Harms & Yager, 1981) indicated that science teaching in the schools in the late 1970s and early 1980s continued to stress academic preparation and the acquisition of scientific knowledge with little or no attention given to application of science to everyday living. The directions and goals of science teaching as we entered the 1980s, according to Harms and Yager (1981), did not change appreciably from those that had been established after the curriculum reform movement of the 1960s. Anderson stated that even though "some of the recommendations of the reformers certainly did not come to fruition, . . . nonetheless it would be difficult to see much difference between the science classrooms of today and those of the mid-sixties" (1983, p. 171). Anderson's statement clearly describes the state of science teaching in the early 1980s. He continued by stating, "the former goals are no longer adequate and . . . a new consensus is needed for a new era. The world has changed since the sixties and schools should be responsive to the society they now serve. It is time for a new consensus on science education goals, a consensus which reflects the 'real world' of today, and problems our students will face tomorrow" (1983, p. 171).

The position that new goals be developed for science teaching to take care of the personal and social needs of students for today and the future has been discussed among science educators since the early 1970s. Hurd in particular and others stressed the importance of redefining the goals of science teaching in the light of science, technology, and society for the 1970s (Hurd, 1972, 1975). Zoller and Watson (1974) advocated a reorientation of science programs toward a science-technology-society emphasis.

The report, *Project Synthesis,* (Harms, 1979) was developed by 23 science educators who presented a set of broad goals intended to meet the needs of young people to live in today's world and the world of the future. The Project Synthesis group developed a set of broad goals divided into the following groups: (1) personal needs, (2) societal issues, (3) academic preparation, and (4) career awareness (Harms & Yager, 1981). The broad goals were further used to define desired goals in biology, physical science, inquiry, science-technology-society, and elementary science.

Following the report of Project Synthesis, the National Science Teachers Association in 1982 published a strong position statement advocating new goals for science teaching that focused on educating citizens to become scientifically literate and recommending that these goals should stress the knowledge, skills, technology, values, and ethics needed to live in a contemporary, scientifically and technologically based society.

Since the publication of *Project Synthesis* and the NSTA position statement, much has been published in the science education literature regarding the future goals of science teaching and the degree to which the science-technology-society theme should be emphasized in middle school and high school science curricula. Many committees were organized in the 1980s to study the state of science education in the United States to come up with proposals that would guide the future directions and goals of science teaching to meet the needs of young people to live in a contemporary society.

The Controversy Regarding the STS Theme

The implementation of the science-technology-society goals continues to be discussed and debated among science educators. Some emphatically believe that STS should be the central focus of science education and that the science curriculum should be organized around this theme (Yager, 1983, 1984, 1985; Hofstein & Yager, 1982).

Other science educators (Kromhout & Good, 1983; Good, Herron, Lawson, & Renner, 1985) argue that science teaching should emphasize the importance of science as a research discipline. They stress the need for students to know what scientists do and how they do it. "We therefore choose to define science education as the discipline devoted to discovering, developing and evaluating improved methods and materials to teach science, i.e., a quest for knowledge as well as the knowledge generated by that quest (Good et al., 1985, p. 140).

Hofstein and Yager (1982) and Yager (1984) define science education differently: "Science education as a discipline is concerned with the interface between science and society." Yager states that the discipline of science education "is concerned with the study of the interaction of science and society, the study of the impact of science upon society, as well as the impact of society upon science" (1984, p. 36).

Although Good et al. (1985) believe that STS issues are important and worth attention in the science curriculum, they do not believe that the interaction of science and society should be confused with science education. "The major concern of science education researchers, however, should be in identifying factors which help people learn science; science as defined by scientists, not by sociologically and politically oriented observers" (Good et al., 1985, p. 140). They continue by saying "to help people understand what science is, it is often necessary to help them understand what it is not. It is not, for example, creationism" (Good et al., 1985, p. 141). Kromhout and Good (1983) disagree with the definition of science education as presented by Hofstein and Yager (1982) and state that "science education should emphasize the coherent structure which is the heart and soul of the scientific method" (p. 648). They do, however, indicate that STS problems could be used as a vehicle for the study of science. "It is to be clear from the outset that we applaud the use of socially relevant problems as motivation for a coherent study of fundamental science" (Kromhout & Good, 1983, p. 648). They further point out that the Hofstein-Yager position is dangerous not only to the goal of scientific literacy but to science and our democratic social structure. They believe that science courses organized around an STS theme open science education to "manipulation and preversion by social activists into a vehicle for reinforcing one-sided or narrow perceptions of social, political, intellectual and religious issues" (Kromhout & Good, 1983, p. 650).

The disagreement among science educators centers around the definition of science education and the extent to which STS should be emphasized in the science curriculum. One position (Hofstein & Yager, 1982; Yager, 1983, 1984,

1985) states that STS should be the central theme of science education with little or no emphasis on the structure and methods of science. The other position (Kromhout & Good, 1983; Good et al., 1985) allows little attention to STS in the science curriculum but stresses the structure and methods of science as the central theme for science today. Bybee believes that both positions give an incomplete picture of the discipline of science education, and says "in their zeal, authors on both sides of the debate have overstated their positions and provided incomplete recognition of the nature of science education as a discipline" (Bybee, 1987a, p. 670).

Bybee presents an extensive clarification of the STS theme in science education. He suggests that the STS theme should be only part of science education: "So, while the STS theme provides an orientation for school programs it is very different from those earlier cited suggestions and criticisms, using specific, narrowly defined issues as the organizational focus for all science teaching" (1987a, p. 674). Bybee also indicates that "there is justification for the STS theme based on an understanding of science education as a social institution, a position only partially recognized in earlier discussions. Recognition of science education as a social institution provides subsequent identification of our general purposes" (1987a, p. 670).

Bybee (1987a) takes the position that science education is a subsystem of education and that several important reports—including *The Cardinal Principles of Secondary Education* (Commission on the Reorganization of Secondary Education, 1918); the report of the Progressive Education Association, *Science in General Education* (1938); *General Education in a Free Society* (Harvard University, 1945); and the 1961 NEA report of the Educational Policies Commission—support the inclusion of personal and social goals. "Schooling in science ought to enhance the personal development of students and contribute to their lives as citizens. Achieving this purpose requires us to reinstate the personal and social goals that were eliminated in the curriculum reform movement of the 1960s and 1970s" (Bybee, 1987a, p. 679).

The Rationale for STS in the Curriculum

Society is becoming increasingly technologically and scientifically oriented with each passing day. We are certainly in the midst of the most rapid advancement of scientific knowledge in the history of humankind. A case in point (and of intimate relevance to the science curriculum) is biomedical research. Rapid progress is being made in such areas as cancer research, auto-immunology, organ transplantation, genetic engineering, neurological dysfunction, and hematology.

The reasons for including STS in the science curriculum are clear. Scientific advances bring problems as well as conveniences. Decisions concerning societal welfare as well as those of personal concern must be made. Recently, new technology has resulted in a machine that can actually "read" the DNA code and assemble parts of genes or complete genes. We are currently on the verge of controlling the traits and health of children before birth. Many questions arise from this new technology. Who will decide the characteristics of new offspring? Should we interfere with the natural process of birth and development?

Of course, the effects of scientific advancement are not restricted to future speculations and predictions. Many people (and society as a whole) have already been forced to reassess and clarify their values and points of view. For example, abortion has always been a controversial issue. The technology for in-vitro fertilization has been basically accepted by those who oppose abortion. Certainly being against termination of life and in favor of the facilitation of life appear to be consistent views. However, public scrutiny of the process revealed that many eggs are fertilized through the test-tube method and only the most viable is implanted in the female. The other eggs are destroyed. An antiabortionist

might have problems with this, because the destroyed fertilized eggs would constitute abortions. Thus, here is a person who does not advocate the ending of life but also feels that the scientific advance of in-vitro fertilization is the miracle of life. This person is now aware that to be consistent, one cannot hold both views. He must reassess his own values. Whether this person will decide to alter his definition of abortion is not important; the point is that scientific advances are coming extremely close to our immediate lives and things we regard as important. Issues concerning the desirability of surrogate mothers, choosing the sex and traits of one's child, and the increase in longevity due to medical advances are very real and relevant issues to individuals and to society as a whole.

These issues are not the same as those that caused an upsurge of concern for ecology and conservation in the United States in the 1960s. The direct and immediate effects of ecological problems were difficult for the general public to grasp. However, today we are suffering the results of our ecological neglect. In 1960, how many would have believed that our energy sources could have become depleted? How many would have believed that we would pay over $1.50 for a gallon of gasoline? These events have become reality; they affect us every day.

And so today we are truly concerned with ecology and conservation. As a matter of fact, in a study by Bybee (1987b), science educators rated the following STS issues concerning ecology and conservation as being among the most important to include in science curricula now and in the future:

1. Water sources
2. Energy sources
3. Land uses
4. Extinction of plants and animals
5. Mineral resources

Another issue that received a high ranking in this study by Bybee (1987b) was nuclear reactors. Would the importance of this issue be so great if it were not for the Three Mile Island incident in 1979 or the Chernobyl incident in 1986?

In short, many of the science, technology, and society issues of today are closer to home than they were in the past. By their very nature, such issues are of the utmost importance and relevance to every future citizen. Such issues will become more pervasive as science and technology further impact our daily lives. According to Bybee, "in general, the picture presented by science educators is that STS problems will be worse by the year 2000 A.D." (1987b, p. 277).

Because the STS theme in the science curriculum stresses scientifically based social issues, it has increasing viability as we enter the 1990s. Science curricula that include STS rely heavily on the link between science, technology, and society, and their goals should include helping students investigate their own values on scientifically related social issues so that they can intelligently deal with them in everyday situations.

The necessity and logical rationale for the reorientation of the science curricula are strong. In contemporary times we witness an ever-increasing number of STS issues. Furthermore, STS issues strike at the very core of the students' interest—their everyday lives. Finally, if we as science educators want to produce informed students who will be our future citizens and leaders, we must provide them with opportunities to think about issues so they can deal with them intelligently. We must provide students with the means to function intellectually in relating science to the present and future concerns of society.

The social dimension of teaching science for present and future living allows students to examine the activities with which they will be involved in the real world. In this way they can be aware of and understand the contributions of science to personal, social, technological, and cultural improvement. This means they must be scientifically and technologically literate to make judgments, resolve problems, and place values on the importance of science in today's society.

Scientific and Technological Literacy

As was mentioned earlier, Project Synthesis examined the state of science education in the United States in the late 1970s and early 1980s (Harms, 1979; Harms & Yager, 1981) and stressed the importance of changing goals, programs, and practices in science education for middle and secondary science. Following this report, the National Science Teachers Association (NSTA) in 1982 took a stand regarding the implementation of major goals to increase scientific and technological literacy among our students. Their statement was adopted to help define the major goals of science education for the 1980s. This position statement is still valid as we approach the year 2000 and can provide guidance for establishing some of the long-term goals for science instruction.

According to the NSTA position (National Science Teachers Association, 1982), scientifically and technologically literate individuals are those who

> understand how Science, Technology, and Society influence one another and are able to use this knowledge in their everyday decision-making. The scientifically literate person has a substantial knowledge base of facts, concepts, conceptual networks and process skills which enable the individual to learn and think logically. The individual both appreciates the value of science and technology, and understands their limitations.

Accordingly then, a scientifically and technologically literate person is one who

1. Uses science concepts, process skills, and values in making responsible, everyday decisions;
2. Understands how society influences science and technology, as well as how science and technology influence society;
3. Understands that society controls science and technology through the allocation of resources;
4. Recognizes the limitations as well as the usefulness of science and technology in advancing human welfare;

Issues and problems associated with the environment have forced a reevaluation of science teaching. The goals for science teaching in the middle and secondary schools are placing more emphasis on producing scientifically literate citizens who understand the major concepts, principles, and theories of science and use this knowledge for making decisions.

5. Knows the major concepts, hypotheses, and theories of science and is able to use them;
6. Appreciates science and technology for the intellectual stimulus they provide;
7. Understands that the generation of scientific knowledge depends upon the inquiry process and upon conceptual theories;
8. Distinguishes between scientific evidence and personal opinion;
9. Recognizes the origin of science and understands that scientific knowledge is tentative, and subject to change as evidence accumulates;
10. Understands the applications of technology and the decisions entailed in the uses of technology;
11. Has sufficient knowledge and experience to appreciate the worthiness of research and technological development;
12. Has a richer and more exciting view of the world as a result of science education;
13. Knows reliable sources of scientific and technological information and uses these sources in the process of decision making. (NSTA, 1982)

The National Science Teachers Association position paper (NSTA, 1982) based on the results of Project Synthesis suggested that the directions and goals of science teaching for the future be developed using the following statements as guidelines:

1. to develop scientific and technological process and inquiry skills.
2. to provide scientific and technical knowledge.
3. to use the skills and knowledge of science and technology as they apply to personal and social decisions.
4. to enhance the development of attitudes, values, and appreciation of science and society.
5. to study the interactions among science, technology, and society in the context of science-related social issues.

The Project Synthesis and NSTA position paper (NSTA, 1982) did not have a significant impact or influence in establishing the future goals and directions of science teaching. In the mid-1980s Bybee (1985) indicated that no progress had been made up to that time in establishing new goals because of our tendency to hold to our outdated programs and practices in science teaching. He stated that we need to restore our confidence in science and technology education in order to move ahead and, therefore, we need to establish new goals that are described by the term scientific literacy (1985, p. 95). By the late 1980s many reports by science educators and others appeared in the science education literature advocating the need for change and reform in science teaching and stating that the science-technology-society theme should be the focus for teaching and learning science.

THE DECADE OF THE 1990s

During the 1980s science educators and others spent much time deliberating about the goals and directions of science education for the 1990s and beyond. Many proposals and reports were generated by committees composed of science educators, scientists, curriculum specialists, science teachers, school administrators, and others that focused on the need for curriculum reform in science education. Most of these reports agreed that (1) students are scientifically illiterate and not prepared to live in a contemporary society; (2) present science instruction consists of presenting concepts, principles, and facts of science with little regard as to how scientific information is generated; (3) students are not exposed to how scientists go about their work, so they know little or nothing about the scientific enterprise; (4) students do not view science as a way of thinking, a way of investigating, and a body of established knowledge; (5) students do not learn how to think and know how to ask important questions in their daily lives; (6) the science textbook is commonly used as the basis for the science curriculum and for science instruction; (7) most students just read science and do not do science; (8) minority students are often discouraged from taking science courses because they regard science as too abstract and too difficult.

Two projects are the most important—Project 2061: Science for All Americans, sponsored by the American Association for the Advancement of Science (AAAS, 1989) and the National Science Teachers Association's Project Scope, Sequence and Coordination (NSTA, 1990). They propose extreme changes in science teaching and may have important implications in guiding the future directions and goals of science teaching. The following statement by Shymansky and Kyle (1991) compares both projects and summarizes well some of their important features:

> Two ambitious projects have the potential to serve as the framework for exploring the critical issues and questions for science education reform. Both projects propose radical changes in the nature, structure, and content of the science curriculum. Both projects recognize that reform: a) must be comprehensive, and b) ought to transpire amid an ambiance of collegiality and collaboration. Further, they agree on several important issues: scientific endeavor ought to be presented as a social enterprise, thus a) emphasis should be placed on human thought, action, and depth of understanding, and the application of science to personal and societal issues; b) learning strategies ought to be based upon a constructivist epistemology; and c) reform must ensure the scientific literacy of virtually all students. (p. 27)

Project 2061: Science for All Americans

Science for All Americans, Project 2061 is sponsored by the American Association for the Advancement of Science and funded by NSF, the Carnegie Corporation of New York, IBM, the Mellon Foundation, and other agencies. The project is directed toward the restructuring of the science curriculum and is based upon scientific literacy for all students. The project is being carried out in three phases over a period of at least 18 years as follows:

Phase 1:   Phase 1 involves the determination of a conceptual base for reform by determining the knowledge, skills, and attitudes that should be acquired by students during their experience from kindergarten through high school. (AAAS, 1989)

Phase 2: During Phase 2, groups of scientists and educators will produce several different alternative curriculum models for grades K-12 for education in science, mathematics, and technology based on recommendations made by Phase 1. The curricula will be field tested and the results evaluated. Phase 2 will also involve the development of blueprints for the preparation of teachers, materials, and technologies necessary for teaching, testing, educational research, educational policies, and the organization of schooling (AAAS, 1989, p. 3). There will be an attempt to increase the pool of experts in science curriculum reform. There will also be an effort to foster public awareness of the need for reform in science, mathematics, and technology education and to have a scientifically literate citizenry (AAAS, 1989, p. 3).

Phase 3: Phase 3 will involve the use of the results of Phase 1 and Phase 2. Many groups involved in educational reform, using the resources of Phases 1 and 2, will attempt to educate the public to become scientifically literate. There will be a nationwide concerted effort to implement reform in science, mathematics, and technology in schools throughout the United States. This phase will require the cooperation of individual schools and school districts and will require more than 10 years to complete (AAAS, 1989, p. 3).

The Scope, Sequence and Coordination Project

The National Science Teachers Association has proposed The Project on Scope, Sequence and Coordination (NSTA, 1990) to affect major reform in science teaching at the secondary level. Like Project 2061, this reform effort is based on scientific literacy. An important aspect of this approach is the spacing of the study of four sciences each year for six years for grades 6 to 12.

Tracking of students will no longer take place and instead students will be exposed to a coordinated science curriculum in which they will study biology, earth and space science, chemistry, and physics each year. The number of topics covered in each science area will be reduced and as a consequence less content will be taught and facts and terms will also be deemphasized. Learning will occur sequentially, starting with direct experiences to build basic concepts and over the years building on these concepts using various forms of sequencing such as "concrete to abstract," "from phenomena to concepts and symbols," or "extracting science from familiar objects, things or situations, or problems from a personal or societal nature."*

The Future Directions and Goals of Science Teaching

Political and societal events in our society often influence change and reform in the teaching of science in the United States. For example, major changes in the teaching of science occurred after the Soviets launched Sputnik in October 1957. Scientists, politicians, and educators were very concerned about the state of science education in the United States because we were not preparing enough scientists to meet the demands of the times. Social conditions during the late 1950s and early 1960s sparked a major effort to improve science education and mathematics education in the nation's schools. As a result, the U.S. government through the National Science Foundation financed an extensive course content improvement program in the various areas of science. The National Science Foundation through these programs influenced the goals and directions of science in the United States for more than a decade.

But since the 1960s major changes in our society have taken place to which schools have not

---

*From a letter by Bill Aldridge, Project Director, to Dr. Marvin Druger, Chairman of Science Teaching Department, Syracuse University, Syracuse, New York, August 31, 1989.

responded. Unfortunately, present goals in science teaching are still based on those espoused in the late 1950s and early 1960s. These outdated goals emphasize academic preparation with no regard for preparing students to live in a contemporary society. Conditions in our society now require that goals be established to prepare students to deal with science-related societal issues, technological devices, and scientific and technical knowledge.

At present, more often than not, science teachers use science textbooks as the basis for their courses, and consequently the science textbook and the teacher more or less dictate the present goals of science teaching. But science textbooks are content oriented and stress academic preparation in various science disciplines, and they exclude topics that reflect goals that will prepare students to deal with science-related problems they will encounter in the future.

Even though the rationale and need to include STS in the science curriculum is obvious, the extent to which STS should be included and emphasized is still being discussed. Certainly, Project Synthesis, Project 2061, and Project Science, Scope and Sequence have made some important recommendations regarding this issue. But the fact remains that we need to produce scientifically literate students through our science programs and other courses in the school curriculum. In formulating the future goals of science education, we cannot disregard the benefits of involving students in the scientific enterprise. Goals must be included that stress science as a way of thinking, as a way of investigating, and as a body of knowledge. We cannot develop scientifically literate citizens until we understand the nature of science and the scientific enterprise and translate this into instruction. To achieve scientific literacy, we must also understand the scientific enterprise and how it relates to science and technology.

It is obvious that a single science teacher, a unit or course of study, or even an entire science program—independently or collectively—cannot produce a scientifically literate society. Courses in mathematics, language arts, and the social sciences must also contribute to meeting this objective. In addition, the experiences that students have in nonschool settings, such as watching television and listening to the radio, can contribute to the development of scientifically literate citizens.

In closing, many science teachers resist change and it will not be easy for them to change from courses that stress academic preparation to courses that are also designed to meet the goals of scientific literacy. Schools and school systems also resist change because change involves financial considerations. The development of new courses will require school systems to devote a significant amount of time and energy in curriculum development and in teacher training to make significant strides in implementing STS instruction. Government and other granting agencies must also provide financial assistance for curriculum development and teacher training programs in order to make significant progress. Science teacher preparation programs offered in colleges and universities must also address this issue. Changes of this magnitude will not take place overnight. It will require many years and the effort of many facets of education to institute radical reform in the teaching of science in our nation's schools.

SUGGESTED ACTIVITIES

1. Produce a checklist that can be used to determine whether a course of study attempts to meet the fundamental goal of scientific literacy. Use the list and guidelines given in this chapter to help develop the checklist. Discuss your checklist with members of your class, as

well as with experienced teachers. Agree on one form of the checklist that can be used to evaluate a course.

2. Select a modern secondary school textbook in physics, chemistry, biology, or earth science and determine, using the checklist you develop from the ideas expressed in this chapter, whether various aspects of scientific literacy are being met by the course of study presented in the textbook. Indicate specifically what aspects of scientific literacy are met. Discuss the findings with others in your class.

3. Interview some science teachers, scientists, and engineers to determine their definition of scientific literacy. Compare their ideas with those that have been presented in this chapter.

4. Study one of the approaches (courses of study) produced by the national curriculum groups, such as Biological Sciences Curriculum Study (BSCS), Earth Science Curriculum Project (ESCP), or Chemical Bond Approach (CBA). Determine whether the criticisms stated in this chapter apply to the course of study selected. Discuss your findings with the members of your class.

5. Select a textbook in biology, chemistry, or physics that is used as a course of study in one of the local high schools and compare this course with the national curriculum approach in the same science area. Determine similarities and differences in concepts and principles stressed, the laboratory work involved, instructional approaches used, as well as other aspects.

6. Examine the contents of Project Synthesis (see Bibliography for reference) and summarize the important points of this report. Discuss your findings with members of your science methods class.

7. Compare Project 2061 with Project Scope, Science and Sequence. What are the main differences between the two approaches? What are their similarities? Discuss the projects with members of your science methods class. What part did Project Synthesis play in developing the two approaches?

BIBLIOGRAPHY

Aldridge, B. G. 1989, January/February. Essential changes in secondary school science: Scope, sequence and coordination. *NSTA Reports.* Washington, DC: National Science Teachers Association.

Aldridge, B. G. 1989, May 12. Essential changes in secondary school science: Scope, sequence and coordination. Washington, DC: National Science Teachers Association (mimeographed statement).

Aldridge, B. G. 1991, May. Basic science or STS: Which is better for science learning? *NSTA Reports.* Washington, DC: National Science Teachers Association.

Aldridge, B. G., and Johnston, K. 1984. Trends and issues in science education. In R. Bybee, J. Carlson, and A. McCormack (Eds.), *Redesigning science and technology education, 1984 NSTA yearbook.* Washington, DC: National Science Teachers Association.

American Association for the Advancement of Science (AAAS). 1989. Science for all Americans, summary, Project 2061. In *2061 Today.* Washington, DC: AAAS.

AAAS. 1991a. Designing blueprints. In *2061 Today,* Summer 1991 (2). Washington, DC: AAAS.

AAAS. 1991b. Curriculum models coming into focus. In *2061 Today,* Fall 1991, 1 (3). Washington, DC: AAAS.

AAAS. 1991c. Project 2061's own classroom. In *2061 Today,* Spring 1991, (1). Washington, DC: AAAS.

Anderson, R. D. 1983. Are yesterday's goals adequate for tomorrow? *Science Education, 67*(2): 171.

Associated Press. 1987. October 4. Lessons learned from Sputnik. *Syracuse Herald American,* p. A10.

Bailey, L. H. 1904. *What is nature study?* Cornell Nature Study Leaflets. Nature Study Bulletin No. 1. State of New York Department of Agriculture. Albany: J. B. Lyon Co.

Brunkhorst, H. K. 1986. A new rationale for science education. *School Science and Mathematics, 86*(5): 364.

Bybee, R. W. 1977. Toward a third century in science education. *American Biology Teacher, 39*(6): 338.

Bybee, R. W. 1985. The restoration of confidence in science and technology education. *School Science and Mathematics, 85*(2): 95.

Bybee, R. W. (Ed.). 1986. *Science-Technology-Society, 1985 NSTA yearbook.* Washington, DC: National Science Teachers Association.

Bybee, R. W. 1987a. Science education and the science-technology-society (STS) theme. *Science Education, 71*(5): 667.

Bybee, R. W. 1987b. Teaching about science-technology-society (STS): Views of science educators in the United States. *School Science and Mathematics, 87*(4): 274.

Bybee, R. W., and Bonnstetter, R. J. 1987. Implementing the science-technology-society theme in science education: Perceptions of science teachers. *School Science and Mathematics, 87*(2): 144.

Caldwell, O. W. 1920. *Reorganization of science in secondary schools.* Commission on Reorganization of Secondary Education, Bulletin No. 26. Washington, DC: U.S. Department of Interior, Bureau of Education.

Caldwell, O. W. 1924. Report of the American Association for the Advancement of Science. Committee on the Place of the Sciences in Education. *Science, 60:* 534.

Central Association of Science and Mathematics Teachers. 1915. Report of the Central Association of Science and Mathematics Teachers Committee on the unified high school science course. *School Science and Mathematics, 15*(4): 334.

Chiappetta, E. L., and Ramsey, J. M. 1991. Curricular models for science programs in Texas. *The Texas Science Teacher, 20*(4).

Commission on the Reorganization of Secondary Education. 1918. *The cardinal principles of secondary education.* U.S. Office of Education Bulletin No. 35. Washington, DC: U.S. Government Printing Office.

Committee on Secondary School Studies. 1893. *Report of the Committee of Ten on secondary school studies.* Washington, DC: National Education Association.

DeBoer, G. E. 1991. *A history of ideas in science education: Implications for practice.* New York: Teachers College Press, Columbia University.

Duschl, R. 1990. *Restructuring Science Education.* New York: Teachers College Press, Columbia University.

Educational Policies Commission. 1961. *The central purpose of American education.* Washington, DC: National Education Association.

Fleming, R. W. 1987. High school graduates' belief about science-technology-society II: The interaction among science technology and society. *Science Education, 71*(2): 163.

Fox, F. 1966. Education and the spirit of science—the new challenge. *The Science Teacher, 33*(8): 58.

Good, R., Herron, J. D., Lawson, A. E., and Renner, J. W. 1985. The domain of science education. *Science Education, 69*(3): 139.

Harms, N. C. 1979. *Project Synthesis.* Boulder: School of Education, University of Colorado.

Harms, N. C., and Yager, R. E. (Eds.). 1981. *What research says to the science teacher* (Vol. 3). Washington, DC: National Science Teachers Association.

Harvard University, Committee on the Objectives of a General Education in a Free Society. 1945. *General education in a free society: Report of the Harvard Committee.* Cambridge, MA: Harvard University Press.

Hofstein, A., and Yager, R. 1982. Societal issues as organizers for science education in the 80's. *School Science and Mathematics, 82*(7).

Hurd, P. D. 1969. *New directions in teaching secondary school science.* Chicago: Rand McNally.

Hurd, P. D. 1972. Energizing perspectives in science teaching for the 1970's. *School Science and Mathematics, 72*(8): 765.

Hurd, P. D. 1975. Science, technology and society: New goals for interdisciplinary science teaching. *The Science Teacher, 42*(2): 27.

Hurd, P. D. 1986. A rationale for a science, technology, and society theme in science education. In R. Bybee (Ed.), *Science technology society, 1985 NSTA yearbook.* Washington, DC: National Science Teachers Association.

Hurd, P. D., and Gallagher, J. J. 1969. *New directions in elementary science teaching.* Belmont, CA: Wadsworth.

Jarcho, I. S. 1985. STS in practice: Five ways to make it work. *Curriculum Review, 24*(3): 17.

Koballa, T., Jr. 1989. "Is there substantial agreement on the goals for science instruction? If so what are they?" in *Research within reach.* Washington, DC: National Science Teachers Association.

Kreiger, J. 1990, June 11. Winds of revolution sweep through science education. *Chemical and Engineering News.*

Kromhout, R., and Good, R. 1983. Beware of societal issues as organizers for science education. *School Science and Mathematics, 83*(8): 647.

Lacey, A. 1966. *A guide to science teaching in the secondary schools.* Belmont, CA: Wadsworth Inc.

National Science Teachers Association (NSTA). 1982. *Science-technology-society: Science education for the 1980's* (a National Science Teachers Association position statement). Washington, DC: NSTA.

NSTA. 1990. *Scope, sequence, and coordination of secondary science: A rationale.* Washington, DC: NSTA.

National Society for the Study of Education. 1932. A program for science teaching. In *Thirty-first yearbook, part I.* Bloomington, IL: Public School Publishing.

National Society for the Study of Education. 1947. *Forty-sixth yearbook, part I.* Chicago: University of Chicago Press.

Nelkin, D. 1977. *Science Textbook Controversies and the Politics of Equal Time.* Cambridge: MIT Press.

Nightingale, A. F. 1899. *Report of the committee on college entrance requirements, addresses and proceedings, 38:* 625. Washington, DC: National Education Association.

Penick, J. E., and Meinard-Pellers, R. (Eds.). 1985. *Focus on excellence: Science/technology/society.* Washington, DC: National Science Teachers Association.

Poogie, A. F., and Yager, R. E. 1987. Citizen groups' perceived importance of the major goals for school science. *Science Education, 71*(2): 221.

Progressive Education Association. 1938. *Science in general education.* New York: Appleton-Century-Crofts.

Richards, G. A. 1961. A laboratory block on interdependence of structure and function. In *Biological Sciences Curriculum Study.* Boulder: University of Colorado.

Rutherford, F. J., and Ahlgren, A. 1990. *Science for all Americans.* New York: Oxford University Press.

Rubba, P. A. 1987. Perspectives on science-technology-society instruction. *School Science and Mathematics, 87*(3): 181.

Shymansky, J. A., and Kyle, W. C., Jr. 1991. Establishing a research agenda: The critical issues of science curriculum reform. *Report of a conference held April 8, 1990.* Atlanta, GA: National Association for Research in Science Teaching.

Staver, J. R., and Small, L. 1990. Toward a clearer representation of the crisis in science education. *Journal of Research in Science Teaching, 27*(1): 79.

Thier, H. D. 1985. Societal issues and concerns: A new emphasis for science education. *Science Education, 69*(2): 155.

Underhill, O. E. 1941. *The origins and development of elementary school science.* New York: Scott Foresman.

U.S. Office of Education. 1899. *Report of the Commissioner, 1893–1894* (Report of the Committee of Fifteen). Washington, DC: U.S. Government Printing Office.

Walberg, H. J. 1991, Spring. Improving school science in advanced and developing countries. *Review of Educational Research, 61*(1): 25.

Yager, R. E. 1983. Defining science education as a discipline. *Journal of Research in Science Teaching, 20*(3): 261.

Yager, R. E. 1984. The major crisis in science education. *School Science and Mathematics, 84*(3): 189.

Yager, R. E. 1985. In defense of defining science education as the science/society interface. *Science Education, 69*(92): 143.

Zoller, V., and Watson, F. 1974. Technology education for nonscience students in the secondary school. *Science Education, 58*(1): 105.

Science should be viewed as a way of thinking, a way of investigating, and a body of knowledge.

CHAPTER 2

The Nature of Science

To educate future citizens of our society to become scientifically and technologically literate must be a major goal for all science teachers at every level of instruction. The educational implication of this goal—science for all—is that understanding the nature of science is useful for living in the world of today and tomorrow, regardless of one's career. This knowledge is essential for making decisions relative to one's health, education, career, leisure, and community responsibilities. Therefore, the education derived through school science programs must go beyond that gained from the traditional study of one or more of the science disciplines. Students must not merely study biology, chemistry, earth science, and physics. They must study many science areas to gain an understanding of the scientific enterprise in general.

Overview

- Present a range of ideas to better understand and explain the nature of science in general.
- Describe three themes related to the nature of science and to scientific literacy.
- Encourage the formulation of a balance of curricular themes, those that are appropriate for middle school and for senior high school science courses.

WHAT IS SCIENCE?

Science is a broad-based human enterprise that can be defined differently from different viewpoints. The layman might define science as a body of scientific information; the scientist might view it as a method by which hypotheses are tested; a philosopher might regard science as a way of questioning the truthfulness of what we know. All these views are valid, but each presents only a partial definition of science; only collectively do they begin to define the comprehensive nature of science. Therefore, science should be viewed *as a way of thinking* in the pursuit of understanding nature, *as a way of investigating* claims about phenomena, and *as a body of knowledge* that has resulted from inquiry. An understanding of these fundamental aspects of science will help science teachers convey to their students a more complete picture of the scientific enterprise.

In one respect, science is the study of nature in an attempt to understand it and to create new knowledge that provides predictive power and application. This is implied in this statement by Edward Teller (1991), an eminent nuclear physicist:

> A scientist has three responsibilities: one is to understand, two is to explain that understanding and three is to apply the results of that understanding. A scientist should have no other limitations. A scientist isn't responsible for that which he has discovered. (pp. 1, 15)

The notion that science is the quest to *understand, explain,* and *apply* offers some utility for teachers. Although some will argue that engineers and technologists apply knowledge to make useful products, many scientists today spend their time applying scientific information to develop useful products. The idea that scientists are not responsible for their work is open to question, however. Science is a human activity and those involved in the scientific enterprise must be responsible for their work. Furthermore, those who participate in this type of inquiry must focus on life and its problems to advance the wisdom of humanity (Maxwell, 1984).

Martin, Kass, and Brouwer (1990) reinforce the point that science is a broad-based discipline with many dimensions. They emphasize that science has many faces, and in order to make school science more authentic science educators must include "the history of science, science, technology, and reflection on the nature of scientific knowledge" (p. 542). Furthermore, they advise that even these aspects of science are incomplete, recommending that science educators and science teachers consider the following to broaden their conceptions of the scientific enterprise:

1. *Methodological fidelity.* Teachers must recognize many methods to gain knowledge, and no one method is accepted as the only method of science. Percy Bridgman's (1950) comment reinforces this idea: "The scientific method, as far as it is a method, is nothing more than doing one's damndest with one's mind, no holds barred. . . ."

2. *Epistemological considerations.* Science must be considered in light of how people acquire and construct knowledge. In essence, the mind forms knowledge by attempting to make sense out of reality—creating ideas from one's personal point of view. The mind does not acquire knowledge as one would fill a bucket with water, but constructs it from personal experience.

3. *Presuppositionalist point of view.* Doing science and creating knowledge are highly passionate activities that are difficult to express in words. Polanyi (1958) indicates that although the contents of science can be articulated and passed on through textbooks and instruction, the art of scientific research cannot. Doing science is best developed through mentor/apprenticeship arrangements. Knowing how, leads to knowing what.

Kuhn (1970) explains that a set of beliefs, methods, and understandings held by a group of scientists provides guidance to those engaging in this line of work. He calls these paradigms and stresses their importance for understanding how science is conducted. Kuhn points out that within each paradigm there are three types of problems that consume scientists: (a) refining the facts, (b) aligning theory with reality, and (c) providing greater articulation of a paradigm theory. Much of science, or normal science as Kuhn refers to it, is "mop-up work" that engages the thinking and problem solving of scientists in these three tasks. Very little normal science addresses the formation of new paradigms and the exploration of novelties. At the time of crises, real science comes into play with the formation of new theories.

4. The falsificationists. The work of Popper (1972) provides a starkly different view of science than that expressed by Kuhn. He stresses that the scientist's ultimate purpose is to attempt to falsify theories. For Popper (1963), confirming a theory is easy, especially if that is what one is attempting to accomplish. Humans possess the propensity to find what they are seeking. But the real test of a useful explanation is to set up conditions in an attempt to discredit it. Popper (1963) states that if we wish to seek truth, we must begin by "criticizing our most cherished beliefs" (p. 6).

5. The "hedonists." Feyerabend (1981) recommends to individuals pursuing knowledge that they do so by producing many ideas that explain nature. The search for knowledge and understanding is one that requires opening up the mind and following one's intuition. He believes that those who conduct science must attempt to find answers and explanations rather than attempting to discredit explanations.

6. Personal science. The private side of science offers a different picture of the scientific enterprise than that which is often presented in textbooks and in research journals. Although the public side of science is often portrayed as objective and precise, the private side is just the opposite. Scientists are passionate about their

ideas and often cling to them regardless of the facts presented (Mitroff, 1974). They search for explanations in idiosyncratic ways, using their own personal language and symbols. The work of the scientists can be learned only by repeatedly doing this type of activity, which is difficult to describe.

7. Historical science. Science has a definite place in the history of humankind, and presenting its evolution can reveal a great deal about its nature. For example, the study of the development of theories over many hundreds of years can bring out themes and principles that are central to understanding important concepts such as the contribution of waves and particles to the nature of matter, and mutations to variations in species. In addition, the history of science illustrates the tentative nature of a theory and the importance of confirmation of ideas to the establishment of scientific knowledge.

8. Societal science. Science is not isolated from society. In fact, it is highly influenced by the religious, ideological, economic, ethical, and cultural values of societies. Early science grew out of myth and religious ideas. Many of the useful discoveries that modern societies enjoy were the result of scientists working together on weapon and defense systems. Both large and small scientific projects are funded by governmental grants, reflecting the values that the government and society place on the acquisition of certain types of knowledge.

One obvious reason for the many faces of science is that the scientific enterprise has evolved over the centuries and continues to evolve—changing constantly. In a developmental sense, science can be viewed at many levels of complexity and abstraction. During the first century B.C., early science was highly descriptive. The Greek philosophers attempted to describe natural phenomena, drawing upon myth and religion for many of their ideas that were used for their purposes. Many of the theories put forth by these early scientists were erroneous (similar to those put forth by today's scientists). Aristotle, for example, believed that objects of unequal size fell to the ground at different rates of speed. Ptolemy believed that the Earth was the center of the solar system.

The sixteenth and seventeenth centuries ushered in modern science in that visible phenomena were described more accurately using logic and mathematics. The movement toward accuracy for explaining nature is evident in the work of Galileo, who presented his ideas in the form of simple formulas to show the distance traveled by a falling body over a given period of time. Similarly, Newton used a simple formula to relate the attractive force between two bodies, which he stated as being directly related to their masses and inversely proportional to the distance separating them.

At the turn of the twentieth century, science probed deeper into the structure of matter, the nature of space and time, and the chemical basis of life. Quantum mechanics attempts to explain matter at the atomic and subatomic levels, while relativity addresses space and time in frames of reference that approach the speed of light and where large gravitational fields exist. Cosmology endeavors to explain the origin of the universe back in time to the blink of an eyelid before its inception. Molecular biology analyzes living systems at the level of the gene.

As the twentieth century draws to a close, a "new science" called chaos is emerging. "Where chaos begins, classical science stops" (Gleick, 1987, p. 3). This new paradigm addresses complex phenomena, which until now have drawn little attention from the scientific community. Chaos concerns itself with phenomena more on the human scale than say quantum mechanics, such as changes in wildlife populations, disorder in the atmosphere, irregularity in heartbeats, the patterns in rising columns of smoke, the clustering of automobiles on expressways, the winding of rivers, and so on. This new direction in science is made possible by computer technology that can present graphic and pictorial images to highlight subtle patterns embedded in seemingly random events.

In some instances, determining what something *is not* helps to better understand what that something *is*. At first thought, this idea appears so useful and simple for distinguishing science from nonscience. Would not most scientists be quick to say that science is much different from astrology or parapsychology, for example? However, identifying a single criterion (or several, for that matter) to distinguish scientific theories from crackpot ideas is a formidable exercise. Most criteria that have been proposed meet with criticism. Morris (1991) presented the example of distinguishing between superstring theory and the claims given by astrologers. He pointed out that if subjecting one's theory to empirical testing is used as the criterion for being scientific, then astrology fares even better than the superstring theorists. Most physicists would agree that it is virtually impossible to observe a superstring because this fundamental entity is hypothesized to be many millions of times smaller than a quark, for whose existence superstrings are theoretically responsible. On the other hand one can set up tests to determine the apparent influences that heavenly bodies exert on human affairs, even if the results may provide little data to confirm these relationships. Morris points out that both scientific and nonscientific theories seem to be the result of thought, imagination, speculation, and problem solving. Both enterprises put forth bizarre ideas. In the final analysis, however, scientific theories generally offer clear, logical, and simple explanations of phenomena, which provide useful insights into nature.

In closing this section, we want to affirm the notion that science is a broad-based enterprise that has changed over the years. To define science as understanding, explanation, and application presents a very simple definition of this discipline. However, in order for science teachers to understand science, they must study the manner in which knowledge is constructed over time, the methods that are used to validate knowledge, the creative and personal side of science, and the place of science in society. They must turn to the history and philosophy of science for this inquiry (Hodson, 1988; Loving, 1991). Therefore, the remainder of this chapter will draw from many sources to assist the practitioner to gain a greater understanding of the nature of science, using three themes as a guide—science as a way of thinking, science as a way of investigating, and science as a body of knowledge.

SCIENCE AS A WAY OF THINKING

Science is a human activity that can be characterized by the thinking that occurs in the minds of people who participate in it. The mental work of scientists illustrates humankind's curiosity and desire to understand phenomena. These individuals possess attitudes, beliefs, and values that motivate them to answer questions and solve problems. Scientists are driven by enormous curiosity, imagination, and reasoning in their quest to figure out and explain natural phenomena. Their work, as viewed by many philosophers of science and cognitive psychologists, is a creative activity whereby ideas and explanations are constructed in the mind. Therefore, the thinking and reasoning of scientists as they go about their work offer important clues regarding the nature of science.

Beliefs

The tendency for scientists to find out seems to be motivated by their belief that a set of laws of nature can be constructed from observation and explained by thought and reasoning. Scientists believe that they, and not some authority, can construct the laws of nature in their minds. These individuals are deeply involved emotionally and personally in their work, constantly striving to better understand the natural and physical world (Roe, 1961). They believe that useful explanations of phenomena can be pieced together by the scientific community. In addition, scientists are sometimes motivated to dispel ideas and beliefs they perceive to be erroneous.

The early scientific thinkers relied on thought and logic to explain what they saw in the skies, for example. They attempted to break away from certain myths and tales to explain objects that changed position in the skies overhead (de Santillana, 1961). Although the beliefs and reasoning of early scientists often seem faulty in retrospect, they provided a foundation for studying important ideas. Aristotle, for example, believed that the planets moved in circular orbits around the earth. His work caused others to examine planetary motion. A later example comes from Paul Ehrlich, a famous bacteriologist:

> Ehrlich's search for substances, which are selectively absorbed by pathogenic organisms, was inspired by his firm belief that drugs cannot act unless fixed to the organism, but today many effective chemotherapeutic drugs are known not to be selectively fixed to the infective agents. (Beveridge, 1957, p. 61)

Curiosity

The belief that nature can be understood is driven by an innate curiosity to find out. Curiosity is most apparent in children who constantly explore their environment and frequently ask the question: Why? Curiosity is characteristic of scientists who often have many interests, even beyond that of unraveling the mysteries of natural phenomena. Nicholas Copernicus, who caused a scientific revolution by placing the sun at the center of our solar system, pursued many vocations. "Copernicus was a churchman, a painter, a poet, a physician, an economist, a statesman, a soldier, and a scientist" (Hummel, 1986, p. 55). Benjamin Franklin likewise demonstrated his many interests as a printer, publisher, inventor, statesman, and scientist.

Imagination

Scientists rely heavily on imagination. The early Greek philosophers and scientists tried to visualize a harmonious universe with heavenly bodies moving in a particular manner. Clerk Maxwell formed mental pictures of the abstract problems that he attempted to solve, especially electromagnetic fields. Many in a college organic chemistry course have listened to the story of how August Kekule came to visualize the benzene ring during a dream:

> . . . The atoms flitted before my eyes. Long rows, variously, more closely, united; all in movement wriggling and turning like snakes. And see, what was that? One of the snakes seized its own tail and the image whirled scornfully before my eyes. . . . (in Beveridge, 1957, p. 76).

Reasoning

Associated with imagination is reasoning. The history of science provides many examples of how those who participate in the scientific enterprise study phenomena with considerable reliance on their own thinking as well as that of others. According to Albert Einstein, science is a refinement of everyday thinking, a belief which becomes evident when one studies the work of scientists in their attempt to construct ideas that explain how nature works. Science teaching can benefit greatly from the inclusion of narratives about the development of the major theories of the natural and physical world, illustrating how these explanations evolved and changed over time (Duschl, 1990). Examples of science courses that focus on the nature of scientific thinking can be found among the national curriculum projects produced during the 1960s. Harvard Project Physics, in particular, centered its textbook and supplemental reading on historical accounts of how scientific principles and theories were studied by various individuals. A classic example of how a scientist used a thought experiment to illustrate correct and incorrect reasoning about the motion of celestial bodies is given in Galileo's *Two New Sciences,* where he presents a conversation among three characters: Simplicio, who represents the Aristotelian view of mechanics; Salviati, who represents the new view of Galileo; and Sagredo, who represents a

man with an open mind who is eager to learn. This short excerpt taken from a dialogue illustrates how Galileo attempted to reason that two bodies, regardless of their mass, will fall to the ground at the same rate.

Simplicio: There can be no doubt that one and the same body moving in a single medium has a fixed velocity which is determined by nature and which cannot be increased except by addition of impetus or diminished except by resistance which retards it.

Salviati: If we take two bodies whose natural speeds are different, it is clear that on uniting the two, the more rapid one will be partly retarded by the slower, and the slower will be somewhat hastened by the swifter. Do you not agree with me in this opinion?

Simplicio: You are unquestionably right.

Salviati: But if this is true and if a larger stone moves with a speed of, say, eight, while a smaller moves with a speed of four, then when they are united, the system will move with a speed of less than eight; but the two stones when tied together make a stone larger than that which before moved with a speed of eight. Hence the heavier body moves with less speed than the lighter one; an effect which is contrary to your supposition. Thus you see how, from your assumption that the heavier body moves more rapidly than the lighter one, I infer that the heavier body moves more slowly.

Simplicio: I am all at sea. . . . This is, indeed, quite beyond my comprehension. . . .

Simplicio retreats in confusion when Salviati shows that the Aristotelian theory of fall contradicts itself. But while Simplicio cannot refute Galileo's logic, his own eyes tell him that a heavy object does fall faster than a light object. (Project Physics Course, 1975; pp. 43–45)

If educators turn to the history of science, they will find many examples of thought experiments that illustrate creativity, imagination, and reasoning. The writings of Ernst Mach contain numerous examples of puzzles that were contrived to cause one to think deeply about physical phenomena. Mach (1838–1916), a great scientist and philosopher, presented thought experiments for his readers to perform in each edition of his Zeitschrift. One such mind teaser asks, "What happens when a stoppered bottle with a fly on its base is in equilibrium on a balance and the fly takes off" (cited in Matthews, 1991, p. 15).

Scientists engage in a variety of reasoning behaviors to elucidate patterns in nature. Sometimes they use inductive thinking, while at other times they use deductive thinking. Through inductive reasoning, one arrives at explanations and theories by piecing together facts and principles. Although Sir Francis Bacon did not originate the inductive method, he did a great deal to promote this approach. Bacon argued strongly that the laws of science are formed from data collected through observation and experimentation (Bruno, 1989). In doing so, he stressed empiricism over deductive logic. Nevertheless, deductive thinking is no less important in the thinking process.

Deductive thinking involves the application of general principles to other instances. The deductive process makes inferences about specific situations from known or tentative generalizations. This form of reasoning is often used to test hypotheses, either confirming or disproving them. Deduction is frequently used in astronomy to predict events from existing theories. It has also been used extensively in the area of theoretical physics to predict the existence of certain subatomic particles, many of which have been discovered years after their announcements. The hypothetical-deductive approach is also a suggested strategy to use for instructing students in the biological sciences to help them better understand this discipline (Moore, 1984).

Self-Examination

Scientific thinking is more than an attempt to understand nature; it is also an attempt to understand itself, to look into the ways by which people arrive at their conclusions about nature. To paraphrase Poincare, an intelligent person cannot believe everything or believe nothing.

What then can an intelligent person believe and what must he or she reject, and what can he or she accept with varying degrees of reservation?

Scientists have always concerned themselves with these questions; indeed physics was once called "natural philosophy." But never more than today have scientists spent more effort in examining their processes of reasoning. Investigations into the particulate nature of matter, quantum mechanics, and acceptance of the principle of indeterminancy have undermined notions of nature that once were never questioned; even the more reluctant scientist has been forced to look closely at his or her ways of thinking.

Addressing the 1964 National Science Teachers Association, Cohen outlined a few of the problems that face scientists (Cohen, 1964, p. 27):

1. What is an explanation? Are the laws of natural phenomena descriptions or perhaps conceptual umbrellas that cover distinct types of phenomena? Are they causal?

2. What is a causal explanation? Is cause a metaphor derived from the daily life of actions and effects? If so, where is the active agent and the causal connection? Where in the difficult indeterminacy of microphysics do we locate causal understanding?

3. Is it true that reversals of time may occur so that effects may occur at a moment before their sources? If so, how shall we understand our deeply intuitive feeling for past and future?

4. To what extent do successful predictions validate a theory? To what extent do negative outcomes falsify our theories? How much are present theories "stretched" to cover uncomfortable facts?

5. Can we understand the twin characteristics of every experiment—accuracy and vagueness—and can we pinpoint by experiment simple properties in a world that is so grossly complicated?

6. We work with ideally isolated factors and parts; we never seek the whole. So we must ask is the universe the simple additive sum of the isolates? Is it to be understood as a superposition of externally described forces and effects

and entities? How shall we best express the old truth of biology that function depends on the position within the whole and that individuals run in accordance with the whole of which they form a part?

7. How has the experimenter distorted the object of his or her search? Do we know only what we observe, and do we observe an object only in its relatedness to our observing selves? Is there no object other than the observer?

These and a dozen more similar problems bother scientists who look beyond the test tube and the microscope. The basic problems of science are not the discovery of another nuclear particle or another item in the DNA code, but the ways to bring our ideas about the world into agreement with what we see.

Beyond all the above, many other ideas can be assigned to science as a way of thinking. For example, objectivity, open-mindedness, and skepticism are often ascribed to the scientific attitude. Although these images of scientists are put forth in textbooks, they may not necessarily represent the manner in which scientists conduct their work. Holton (1952) reminds us about the distinction between public and private science, which challenges the empiricist stereotype of the detached, objective researcher, who is what the public sees in the final edited version of the scientist's works. Gauld (1982) cautions science educators about their beliefs regarding the empiricist conception of the scientific attitude and suggests perhaps eliminating it from science education. "Teaching that scientists possess these characteristics is bad enough but it is abhorrent that science educators should actually attempt to mold children in the same false image" (Gauld, 1982, p. 118).

SCIENCE AS A WAY OF INVESTIGATING

Those who desire to understand nature and to discover its laws must study objects and events. The clues to nature's "clockwork" are embedded in reality. They must be uncovered through experimentation and observation, and reasoned out through the human thought process. Some go so far as to say that "Understanding the process—the way scientific information is obtained, tested, and validated—is more important than knowledge of any specific product of science" (Hetherington, Miller, Sperling, and Reich, 1989).

Science as a way of investigating illustrates many approaches to constructing knowledge. Science has many methods, which demonstrates humankind's inventiveness for seeking solutions to problems. Some of the approaches used by scientists rely heavily on observation and prediction as in astronomy and ecology. Other approaches rely on laboratory experiments that focus on cause and effect relationships such as those used in microbiology. Among the many processes often associated with science and inquiry are observing, inferring, hypothesizing, predicting, measuring, manipulating variables, experimenting, and calculating. A realistic idea of the many aspects that are related to inquiry and research is offered by Franz (1990):

- experimentation
- reason
- chance
- observation
- hypotheses
- strategy
- intuition
- overcoming difficulties
- serendipity

Hypothesis

A hypothesis is an investigative tool that helps the inquirer to clarify ideas and state relationships so they can be tested. A hypothesis is a concise statement that attempts to explain a pattern or predict an outcome. Hypotheses stem from questions that a scientist asks concerning a problem under study. They are tentative statements and must be tested by additional observation, experimentation, or prediction. Hypotheses set the stage for challenging an idea to determine if it merits at least a temporary place in the fabric of scientific knowledge. All descriptions of scientific methods agree on the need for good research hypotheses to carry out meaningful investigations.

Although setting up a situation that could disprove a hypothesis, contrary evidence may not necessarily lead to the complete rejection of a hypothesis.

> If the generalization has any reasonable body of supporting data, the finding of new facts which do not fit usually leads to the refinement or elaboration of the original hypothesis. . . . On the other hand, if the original basis for the hypothesis was slender, the unfavorable instances may so outweigh the favorable ones as to make it reasonable to believe that the earlier agreement was a matter of pure chance. Also a new hypothesis may be developed which fits the original data as well. (Wilson, 1952, p. 28)

People who engage in experimentation probably do so from a more optimistic point of view than attempting to discredit their ideas, as suggested by Popper (1963). In reality, they are betting on their hypotheses, hoping their experiments will confirm their predictions. However, what generally occurs is an unexpected result, which they have difficulty explaining. Chamberlin (1965) urges researchers to propose not just one hypothesis but as many as the mind can invent, thus freeing the mind from bias that might result from excessive love of one's intellectual child. With many hypotheses, the researcher can be somewhat more objective and increase the probability for identifying several causes, knowing that some hypotheses will not survive. Nature is complex; attempting to explain it by limiting ourselves to one hypothesis is too narrow and self-defeating.

Observation

Observation is certainly one cornerstone of science. Although most observations are carried out through the sense of sight, other senses like hearing, feeling, smelling, and tasting can also be part of the observation process. Through observation, data and information are gathered and

organized in order to make sense out of reality. This is how facts are established so that scientists can arrive at theories. The body of knowledge formed by scientists is the result of extensive observations that eventually coalesce into concepts, principles, and theories.

The list of individuals who have made significant contributions to science, as the result of their keen observations, is long. Among those worth mentioning is Tycho Brahe, who amassed the most important observations regarding the motion of stars and planets during the Middle Ages. These observations, in turn, were used by Kepler to produce the now accepted idea that planets move in elliptical orbits around the sun. Tycho used his eyes as well as a few simple devices to study the movement of celestial bodies.

Accurate and relevant observations are essential in order to test hypotheses.

Tycho made his observations with scrupulous regularity, repeating them, combining them, and trying to allow for the imperfection of his instruments. As a result he reduced his margin of error to a fraction of a minute of an arc, and provided the sharpest precision achieved by anyone before the telescope. (Boorstin, 1985, p. 307)

Although observation plays a central role in scientific investigation, this skill is tied closely to the knowledge, thinking, and motivation of the observer. As Goethe said: "We see only what we know." Thus, the scientist as well as the layman sees with the mind. Beveridge (1957) points out that what people observe in a situation depends upon their interests. Furthermore, he claims that false observations occur when the senses provide the wrong information to the mind or where the mind plays tricks on the observer by filling in information from past experiences. Often the observer focuses attention on what is expected, thus missing unexpected occurrences and valuable facts. Although observation seems to be the most basic and fundamental of the inquiry skills, it is a complex activity that merits careful study in and of itself.

Experimentation

Along with observation, experimentation is central to the scientific enterprise and the development of modern science. Through experimentation, valid knowledge can be formed and in some instances erroneous beliefs that have been passed down by authority can be discarded. Through experimentation, one can probe nature's secrets, which seem to be tightly guarded and often disguised. Experimental activities range from using a match to burn a peanut to using a particle accelerator to smash a proton. Modern science came into being in the sixteenth century when individuals like Galileo and Copernicus turned away from Greek science, which was dominated by philosophers who distrusted experimentation. Many Greek philosophers relied on reason alone to develop their ideas. Some of these individuals placed little value upon practi-

cal activities that could reveal the obvious. They shunned the technologies of the craftsman, such as mechanical and optical devices that could provide accurate observations (Aicken, 1984).

When one experiments, events are introduced producing relationships that can be studied carefully. The conditions of these interventions are known and controlled. In this manner, the procedure is documented and thus can be reproduced by others. The controlled experiment is often used to test a hypothesis concerning cause and effect relationships. Condition A, for example, is altered to determine its effects on condition B. The situation that is manipulated by the researcher is called the independent variable while the resultant variable is called the dependent variable. Other variables are held constant and their conditions noted. In situations where living organisms are under study, randomization of organisms or subjects into experimental and control groups is most desirable.

We must guard against placing ultimate trust in experimentation for evidence. Beveridge (1957) warns not to put excessive faith in experimentation because the possibility of error in technique always exists, which can result in misleading outcomes.

> It is not at all rare for scientists in different parts of the world to obtain contradictory results with similar biological material. Sometimes these can be traced to unsuspected factors, for instance, a great difference in the reactions of guinea pigs to diphtheria toxin was traced to a difference in diets of the animals. In other instances it has not been possible to discover the cause of the disagreement despite a thorough investigation (p. 34).

In spite of what is sometimes reported about scientific investigations, especially in secondary school science textbooks, inquiry is often idiosyncratic. Furthermore, there is no "scientific method." *The notion that there is a scientific method should be dispelled from science teaching.* For example, the work of Karl Pearson (1937) and others who believed that they could capture the scientific method in the following steps is still taught in some science courses and it often appears in the introductory chapter of many middle and senior high school science textbooks.

1. Observing
2. Collecting data
3. Developing a hypothesis
4. Experimenting
5. Concluding

Researchers do not necessarily follow this set of procedures or any other set, in spite of the fact that these steps might be evident in the polished, edited, research reports that appear in scientific journals.

Mathematics

Roger Bacon expressed it well when he said: "Mathematics is the door and the key to science." His quote suggests that mathematics is necessary to the understanding of nature's clockworks and without it, we cannot get inside to find out what is taking place. The formulas and symbols assist us to identify relationships that represent laws and patterns in nature. At a very deep level, mathematics helps us to express models of situations that we cannot possibly see. The atom, for example, is so small that we cannot see it or the electrons and protons that form its structure. The solar-system model that we often see in textbooks or the packet-of-pulsating-waves model may be inaccurate visualizations of the atom. According to scientists who study atoms and subatomic particles, the atom may be an entity that we cannot visualize. Nevertheless, mathematical models have been devised that predict with great accuracy the behavior of groups of atoms and subatomic particles. Mathematics has been used successfully to provide useful representations of nature that are out of our perceptual reach.

The power of numbers and the intimate association between mathematics and the evolution of science is evident throughout recorded history (de Santillana, 1961). For example, Pytha-goras mingled astronomy, geometry, music, and arithmetic in his studies of nature. He indicated that numbers were special and that the universe produced melody and is put together with harmony. He described the motion of stars in terms of rhythm and melody. Pythagoras's one physical discovery had to do with the patterns of sound produced by changing the length of a plucked string. He noted that moving a bridge to different locations or intervals changed the sound. The ratios 1 : 2, 4 : 3, and 3 : 4 were important numerical relationships that resulted in distinct sounds. He further discovered that the numbers 1, 2, 3, and 4, which formed these ratios, added up to 10, the "perfect number."

Galileo, it is often said, made significant contributions to modern science because he attempted to explain phenomena with evidence, using mathematical relationships. Many connect Galileo with the falling stone controversy. Those who subscribed to Aristotle's notion of falling bodies held that the heavier object falls at a faster rate than the lighter object. Galileo said, wait a minute, if you try this out you will find that they both fall at the same rate, provided that air resistance is ignored. His work on this problem resulted in the celebrated formula

$$s = 16t^2$$

which states that a falling body has traveled a certain distance s after falling t seconds. Galileo also showed that the trajectory of a ball thrown in the air follows a parabolic path.

SCIENCE AS A BODY OF KNOWLEDGE

The body of knowledge produced from the scientific disciplines represents the creative products of human invention that have occurred over the centuries. The enormous collection of ideas pertaining to the natural and physical world is organized into astronomy, biology, chemistry, geology, physics, and so on. The result is a compilation of carefully catalogued information, containing many types of knowledge, each of which makes its own unique contribution to science. The facts, concepts, principles, laws, hypotheses, theories, and models form the content of science. These ideas possess their own specific meaning, which cannot be understood apart from the processes of inquiry that produced them. Therefore, the content and methods of science are tied together, and teaching one without the other distorts the learners' conception and appreciation for the nature of science.

Facts

The facts of science serve as the foundation for concepts, principles, and theories. A fact is a truth and the state of things. Facts represent what we can perceive. Often two criteria are used to identify a scientific fact: (a) it is directly observable, and (b) it can be demonstrated at any time. Consequently, facts are open to all who wish to observe them. We must remember, however, that these criteria do not always hold since factual information regarding a one-time event, such as a volcanic eruption, may not be repeatable. In addition, there is uncertainty and limitation in measurement that accompanies facts. The presentation of the facts in a science course is not enough because the receiver of the information should know how the facts were established.

Concepts

Facts have little meaning by themselves. They are raw material in a sense and must be examined to form meaningful ideas and relationships. Thinking and reasoning are required to identify patterns and make connections among the data, thus forming relationships we call concepts. A concept is an abstraction of events, objects, or phenomena that seem to have certain properties or attributes in common. Fish, for example, possess certain characteristics that set them apart from reptiles and mammals. Most bony fish have scales, fins, and gills. According to Bruner, Good-

now, and Austin (1956), a concept has five important elements: (1) name, (2) definition, (3) attributes, (4) values, and (5) examples. The process of concept formation and attainment is an active one and requires more than just conveying these elements to the learner. The student must establish some of the attributes and discover some of the patterns among data if the concepts are to become linked to other meaningful ideas in the mind. In addition, concepts can be affective in nature as well as cognitive.

Principles and Laws

Principles and laws also fall into the general category of a concept. Although they can be considered broader than a simple concept, principles and laws are often used synonymously. These higher order ideas are used to describe what exists. They are often accepted as facts; nevertheless their distinction and empirical basis must be remembered. Principles and laws are composed of concepts and facts. They are more general than facts but are also related to observable phenomena. Gas laws and laws of motion, for example, specify what can be observed under certain conditions. The principles that regulate growth and reproduction provide reliable information regarding changes that take place in living systems.

Theories

Science goes beyond the classification and description of phenomena. Scientists use theories to explain patterns and forces. Theories are ambitious intellectual endeavors because they deal with the complexities of reality—that which is obscure and hidden from direct observation. This idea becomes evident when one considers the theory of the atom, which states that all matter is made up of tiny particles called atoms, many millions of which would be required to cover the period at the end of this sentence. This visual conception becomes even harder to grasp when we consider the aspect of the theory that suggests an atom is mostly empty space with a small dense center and charged particles moving in certain regions of space far out from the center.

Theories have a different purpose than facts, concepts, and laws, but scientists use these types of knowledge to present their explanations of why phenomena occur as they do. Theories are of a different nature, and they never become fact or law, but remain tentative until disproven or revised:

> Any physical theory is always provisional, in the sense that it is only a hypothesis: you can never prove it. No matter how many times the results of experiments agree with some theory, you can never be sure that the next time the results will not contradict the theory. On the other hand, you can disprove a theory by finding even a single observation that disagrees with the predictions of the theory. As philosopher of science Karl Popper has emphasized, a good theory is characterized by the fact that it makes a number of predictions that could be disproved or falsified by observation. Each time new experiments are observed to agree with the predictions the theory survives, and our confidence in it is increased; but if ever new observation is found to disagree, we have to abandon or modify the theory. At least that is what is supposed to happen, but you can always question the competence of the person who carried out the observation. (Hawking, 1988, p. 10)

Models

The term model is often used in the scientific literature. A scientific model is the representation of something that we cannot see. These models become mental images that are used to represent phenomena and other abstract ideas. They include the most salient features of the idea or theory that the scientist is attempting to make comprehensible and to explain. The Bohr model of the atom, the planetary model of the solar system, the wave and particle models of light, and the double helix model of DNA are all concrete representations of phenomena that we cannot perceive directly. Generally, models are deduced from abstract ideas, and sometimes there are no

sharp distinctions among models, hypotheses, and theories. Textbooks are the major referent for most of our notions about scientific models. They are useful to help us to become familiar with important ideas. Unfortunately, many come to believe that the models presented in science textbooks are the real thing, forgetting that a model is used only to help one conceptualize the salient features of a principle or theory, and that the mental picture is not what exists in reality.

IMPLICATIONS FOR TEACHING SCIENCE

A major question regarding the nature of science is this: To what degree is an authentic view of science incorporated into school science programs? When one observes the activities taking place in science classrooms, is it evident that students are thinking, finding out, and explaining? Or does teacher talk predominate along with the excessive use of worksheets? Students must be engaged in more than passive learning, and science course experiences must go further than conveying information. Students must acquire a broad and authentic sense of science, and science courses must provide them with useful attitudes, skills, and information to use in a modern world.

If one were to examine the written materials for a middle school or senior high school science curriculum, what could be concluded regarding the potential to produce learning outcomes reflective of the three themes discussed above? Furthermore, would the balance of themes be appropriate for the intended audience? Course outlines, lesson plans, textbooks, laboratory experiences, and tests must reflect a balance of themes appropriate for the unique cognitive and affective characteristics of any given group of learners.

The curriculum component that drives the curriculum in most school districts is the assessment. If science tests assess information, especially at the lower levels of thinking, then science courses will emphasize teaching and testing for science as a body of knowledge with no concern for how this knowledge was acquired. Secondary schools exhibit a long tradition of using the college instructional mode of science instruction. The lecture/laboratory format is evident in middle as well as in senior high schools. This approach does not reflect how science is conducted nor does it reflect effective instructional practices. If science as a way of thinking and as a way of investigating are important, then these themes must be evident in the assessment procedures used to measure learning outcomes.

SUGGESTED ACTIVITIES

1. Ask scientists, technicians, engineers, business people, science teachers, students, and so on for their definitions of science. Compare these responses with those given in this chapter.

2. Read Kuhn's book *The Structure of Scientific Revolutions* and discuss the main ideas with others in your science methods class. Do the same with some of the other books listed in the bibliography.

3. Working in small groups with those in your science methods class or with colleagues, present a variety of models for scientific literacy that are appropriate for middle school and senior high school science courses. Indicate the balance of the three themes described in Chapter 2 that you would recommend for a given level and course, and give reasons for these choices. In your deliberations, consider the ethnic mix of students, the academic orientation of the science program, and whether the school administration will support a traditional science curriculum or a more innovative curriculum.

4. Develop a checklist for the three themes of scientific literacy described in this chapter that you would use to assess science teaching. Use the checklist to evaluate the extent to which a science course's lecture/discussions, laboratory exercises, textbook chapters, and so forth provide an appropriate balance of the three themes.

5. Select one chapter in a science textbook that is related to the science discipline of your choice. Categorize each major scientific term in that chapter as either a fact, concept, principle or law, or theory. Evaluate the development of the major ideas. In addition, determine how well the authors present the discovery of scientific knowledge through the thoughts and actions of scientists.

6. Identify exercises from laboratory manuals. Determine the extent to which each exercise promotes science as a way of thinking and investigating.

BIBLIOGRAPHY

Aicken, F. 1984. *The nature of science.* London: Heinemann Educational Books.

Beveridge, W. I. B. 1957. *The art of scientific investigation.* New York: Vintage Books.

Boorstin, D. J. 1985. *The discoverers.* New York: Vintage Books.

Bridgman, P. W. 1950. *The reflections of a physicist.* New York: Philosophical Library.

Bruner, J. S., Goodnow, J. J., and Austin, G. A. 1956. *A study of thinking.* New York: John Wiley.

Bruno, L. C. 1989. *The landmarks of science.* New York: Facts on File.

Chamberlin, T. C. 1965. The method of multiple working hypotheses. *Science, 148:* 754.

Cohen, R. S. 1964, May. Individuality and the common purpose: The philosophy of science. *The Science Teacher, 31:* 27–28.

de Santillana, G. 1961. *The origins of scientific thought.* New York: New American Library of World Literature.

Duschl, R. A. 1990. *Restructuring science education.* New York: Teachers College Press.

Feyerabend, P. K. 1981. *Philosophical papers.* Cambridge: Cambridge University Press.

Franz, J. E. 1990. The art of research. *Chem Tech, 20*(3): 133–135.

Gauld, C. 1982. The scientific attitude and science education: A critical reappraisal. *Science Education, 66:* 109–121.

Gleick, J. 1987. *Chaos.* New York: Penguin Books.

Hawking, S. W. 1988. *A brief history of time.* New York: Bantam Books.

Hempel, C. 1966. *Philosophy of natural science.* Englewood Cliffs, NJ: Prentice Hall.

Hetherington, N., Miller, M., Sperling, N., and Reich, P. 1989. Liberal education and the science. *Journal of College Science Teaching, 19*(2): 91.

Holton, G. 1952. *Introduction to concepts and theories in physical science.* Reading, MA: Addison-Wesley.

Hodson, D. 1988. Toward a philosophically more valid science curriculum. *Science Education, 77:* 19–40.

Hummel, C. E. 1986. *The Galileo connection.* Downers Grove, IL: InterVarsity Press.

Kuhn, T. S. 1970. *The structure of scientific revolutions.* Chicago: The University of Chicago Press.

Loving, C. C. 1991. The scientific theory profile: A philosophy of models for science teachers. *Journal of Research in Science Teaching, 28:* 823–838.

Martin, B., Kass, H., and Brouwer, W. 1990. Authentic science: A diversity of meanings. *Science Education, 74*(5): 541–554.

Martin, M. 1985. *Concepts of science education.* Lanham, MD: University Press of America.

Matthews, M. R. 1991. Ernst Mach and contemporary science education reforms. In M. R. Matthews (Ed.), *History, philosophy, and science teaching* (pp. 9–18). New York: Teachers College Press.

Maxwell, N. 1984. *From knowledge to wisdom: A revolution in the aims and methods of science.* Oxford, England: Basil Blackwell.

Mitroff, I. 1974. *The subjective side of science.* Amsterdam: The Elsevier Scientific Publishing Company.

Moore, J. A. 1984. Science as a way of knowing. *American Zoologist, 24:* 467–534.

Morris, R. 1991. How to tell what is science from what isn't. In J. Brockman (Ed.), *Doing Science.* New York: Prentice Hall.

Pearson, K. 1937. *The grammar of science.* London: Dutton.

Polanyi, M. 1958. *Personal knowledge.* Chicago: University of Chicago Press.

Popper, K. 1963. *Conjectures and refutations: The growth of scientific knowledge.* New York: Harper & Row.

Popper, K. 1972. *The logic of scientific discovery.* London: Hutchinson of London.

Project Physics Course. 1975. *Project physics.* New York: Holt, Rinehart and Winston.

Robinson, J. T. 1968. *The nature of science and science teaching.* Belmont, CA: Wadsworth Publishing Co.

Roe, A. 1961. The psychology of the scientists. *Science, 134:* 456–459.

Teller, E. 1991. Teller talks. *The Daily Cougar, 57*(83): 1, 15. Houston: University of Houston.

Wilson, E. B., Jr. 1952. *An introduction to scientific research.* New York: McGraw-Hill.

Science teachers should use instructional strategies to help students represent ideas and explain them.

CHAPTER 3

Cognition and Learning

Adolescents who participate in middle and senior high school science programs must acquire knowledge, skills, and attitudes that help them to make important decisions throughout their lives. This goal requires students to learn useful information, understand its significance, and apply it to a variety of situations. To achieve these ends, new programs and approaches to science teaching must be implemented in our nation's schools. These curricula must focus upon what students know and how they reason, utilizing the findings and recommendations from the field of cognitive psychology and behavioral psychology as well as from the extensive work of science educational researchers.

Overview

- Address some of the major themes that cognitive psychologists believe are important, for example, thinking, constructing knowledge, representing ideas, and finding meaning from one's studies.
- Describe cognitive development and its implications for teaching science.
- Present teaching strategies that can help students construct knowledge and find meaning in their studies such as concept learning, concept mapping, conceptual change strategies, and problem solving.
- Discuss the importance of students' attitudes toward science and their interest in this subject, and how these affective dimensions of learning impact science education.

MAJOR THEMES CENTRAL TO COGNITION AND LEARNING

Knowing one's subject is necessary but not sufficient to bring about meaningful learning in the teaching of science. Science teachers must understand what goes on in the minds of their students and possess a sense of what students know when they enter the classroom. They must be clear regarding the extent to which students are able to reason and think. They must also realize the limitations that students have for processing certain types of information. Furthermore, science teachers should be sensitive to students' feelings toward science and how their interests can aid or interfere with learning this subject. For greater insight into these matters, science teachers can turn to the fields of cognition and learning regarding the processes that take place in their students' minds and the behaviors that result.

An examination of cognition and learning is critical for those serious about the education of scientifically literate individuals who can acquire useful information and apply it to solve problems and make decisions. Achieving these ends is no trivial matter; it is an enormous challenge. Considering these important cognitive and affective goals, how should science teachers organize their teaching to maximize learning? How should science teachers plan courses and instruction to help adolescents acquire useful reasoning skills, desired content knowledge, and positive attitudes toward science?

Cognitive psychologists have put forth numerous theories to explain the mental processes that constitute knowledge and thinking. Their work has centered around topics such as the mental representation of ideas, the manipulation of symbols, the processing of information, problem solving, learning to learn, the construction of knowledge, computational skills, forming images, language, and memory. Cognitive psychologists have studied the development of many types of knowledge, and their ideas have been influenced by those working in many disciplines such as mathematics, computer science, linguistics, anthropology, and neuropsychology. Most important, cognitive psychologists have gone beyond the work of behavioral psychologists, who have focused their inquiry primarily on observable behavior.

Thinking is a central topic in cognitive psychology, and some in that field advise educators to help students develop a disposition for critical thinking within a variety of contexts (Resnick, 1987). They recommend the identification of learning outcomes that are of great social value and that are useful to students when they go out into the real world. However, in order to achieve these ends, cognitive psychologists advise implementing instructional programs that help students to construct mental representations of ideas and to relate them to their prior experiences in a manner that makes sense (Resnick, 1987).

In addition to thinking, the *construction of knowledge* is a central theme of much of the work taking place in cognitive studies. This theme is based upon the notion that ideas, images, and skills are formed in the mind through mental activity directed toward the physical and social world. Some view this process as the integration of new ideas from outside of the mind into existing ideas within the mind. Others view the process as the construction of ideas inside of the mind to fit what exists in the physical and social world. Still others believe this is a two-way process whereby ideas can be assimilated into existing structure and existing structures can be altered to fit that which exists outside of the mind.

The *representation of ideas* to be learned has also been a topic of considerable inquiry. Some of this work has centered around concept learning where the definition and components of the ideas are delineated. Other aspects of this work have addressed the importance of ascertaining what the learner knows, especially misconceptions. Additional areas of investigation have considered techniques for presenting the relationships among the learning outcomes using learning hierarchies and concepts maps.

Another important dimension of cognition relates to the *meaning* that students find in their studies. Obviously, students must find meaning in the curriculum in order for it to have worth in their lives. Thus students' attitudes toward the subject matter, the teacher, other students, and instructional activities are critical areas of concern and study by science educators. In addition to attitudes, students' interests deserve attention because of their potential to be used in the assessment of various contexts for their motivational effects on learning.

If meaningful learning is to take place in science programs, students must assume a great deal of control over their own learning process. They must become aware of their own learning, a concept referred to as *metacognition*. Metacognition is the knowledge and awareness of one's own learning. When students shift into this mode of learning, they can consciously control what they are doing. Therefore, the learning environment that science teachers manage must make purposeful inquiry one of its major goals.

A question often arises regarding cognition: How does learning proceed? Do people piece small ideas together to form larger ones, or do they begin with the big picture and use general strategies to gain greater insight into smaller ideas that serve to bring the whole picture into greater focus? The Gestalt psychologists would say that people view the big picture, using their existing schemas and strategies to incorporate new ideas into their thinking. Contemporary information processing and behavioral psychologists might suggest that people process smaller units of information that form larger ideas. The whole/part, top/down, or big/small metaphor can be used to organize science curricula by visualizing the direction of the learning process. Although there is no definite answer as to which way learning should take place in science, the discussions that follow can be used to consider these ideas and the many other matters regarding cognition and learning. Figure 3–1 presents a diagram of the themes and topics that are important to learning and science instruction.

COGNITIVE DEVELOPMENT AND TEACHING SCIENCE

Jean Piaget provided educators with a theory of cognitive development that focused on the construction of knowledge as opposed to what he perceived to be the conventional view of learning. He emphasized (Piaget, 1964) that learning is a specific process that is provoked by situations similar to those experienced inside and outside of school, whereas he viewed development as a general process that occurs spontaneously as the result of physical and social interactions, neural maturation, and the progressive differentiation of mental functions. Therefore, for Piaget, development provided the underlying foundation for learning that helped to explain this process. For these reasons, and because of the results of some of the research that has been conducted on the topic, cognitive development can be used to evaluate the appropriateness of science curricula for students and to make instructional decisions by science teachers.

Operative Thinking

For Piaget, knowledge was a dynamic entity, constructed through mental actions that changed over time as the result of many experiences. Central to his definition of knowledge is the idea of an *operation*. An operation is a mental schema that forms knowledge and is integrally connected with it. As Piaget stated:

> Knowledge is not a copy of reality. To know an object, to know an event, is not simply to look at it and make a mental copy or image of it. To know an object is to act on it. To know is to modify, to transform the object, and to understand the process of this transformation, and as a conse-

FIGURE 3–1   Themes and topics that are central to cognition and learning science

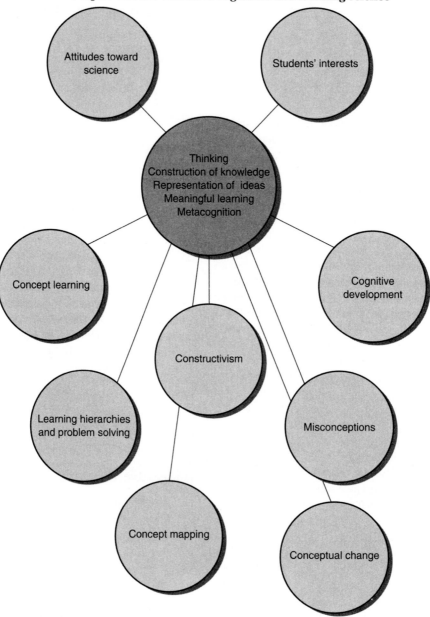

quence to understand the way the object is constructed. An operation is the essence of knowledge; it is an interiorized action which modifies the object of knowledge. (Piaget, 1964, p. 176)

Some of the thinking skills associated with Piagetian operative knowledge are ordering, classifying, adding, subtracting, hypothetical reasoning, proportional thinking, and controlling variables. These operations, schemas, or thinking skills are tied to more fundamental cognitive structures, which Piaget described in logical-mathematical terms. These structures are more easily understood in terms of mental blueprints (Lawson & Renner, 1975) associated with sensorimotor, preoperational, concrete operational, and formal operational reasoning.

Lawson (1982) helps to clarify Piaget's conception of operative knowledge with a comparison of two types of knowledge that are familiar to science teachers. He points out that knowledge, from a Piagetian perspective, must be separated into at least two domains, operative and figurative. Operative knowledge is the most active form of thinking. It relates to "knowing how" to carry out tasks, be they psychomotor or cognitive in nature. This type of thinking is illustrated by the student who: (a) observes fish in an aquarium in order to determine their behavior, (b) connects an electric meter to measure the current in a circuit, or (c) reasons out many ways to compare the density of objects that vary in size, shape, and material. Operative knowledge is related to science process skills such as classifying, inferring, predicting, hypothesizing, and manipulating variables. This action knowledge is synonymous with the term procedural knowledge used by many cognitive psychologists.

In contrast to operative knowledge, there is figurative knowledge (Lawson, 1982). Figurative knowledge is "knowing what" something is. This type of knowledge is illustrated by the student who knows: (a) the name of the fish in an aquarium, (b) the name of the meter used to measure the current in a circuit, or (c) the definition for mass as a property of objects. Figurative knowledge relates to the discipline-specific concepts that comprise the subject matter content taught in science courses. It is also synonymous with the term declarative knowledge used by many cognitive psychologists.

Lawson reinforces Piaget's claim that figurative knowledge is subordinate to operative knowledge. He emphasizes that the general developmental level of students' operative thinking ability is directly related to their degree of understanding of science subject matter. Therefore, learning science must be a process of active inquiry where students are given many opportunities to act on objects and events, and to test out their ideas.

Stages of Cognitive Development

Along with mental operations, stages of cognitive development are another important aspect of Piagetian theory. These four stages of development are commonly known as the sensory-motor, preoperational, concrete operational, and formal operational stages. They represent phases of intellectual growth from infancy to adolescence that individuals pass through. The operations associated with these stages and their underlying structures become more complex as development ensues, permitting the individual greater range of thought and depth of understanding. Furthermore, these stages of development are invariant and qualitatively different.

Sensory-Motor Stage. The first developmental period is called the sensory-motor stage, which begins at birth and continues for approximately two years. Infants are primarily sensory and physical in their encounters with the environment. Arm, leg, head, lip, and eye movement is conspicuous and a dominant form of action during this period of development. Infant's behavior can be described in the following terms: nonverbal, reflex actions, play, imitation, and object permanence (table 3–1).

TABLE 3–1   Intellectual characteristics at different stages of cognitive development

Stages	Characteristics
Sensory motor	Nonverbal
	Reflex actions: sucking, crying, and grasping
	Play
	Imitation
	Object permanence
Preoperational	Language development
	Egocentrism
	Irreversibility
	Centration
Concrete operational	Reversibility
	Seriation
	Classification
	Addition, subtraction, multiplication, and division
	Conservation: number, substance, area, weight, and volume
Formal operational	Combinatorial reasoning
	Proportional reasoning
	Identification and control of variables
	Hypothetical-deductive reasoning

Infants primarily display a nonverbal type of mental activity. They lack the ability to label objects and to communicate with words, yet they are interested in playing with objects in order to derive enjoyment. Children practice imitation, and establish their own idiosyncratic symbol system upon which they later attach words. For example, moving arms and legs frantically may be an attempt to represent the movement of a mobile that hangs over the infant's crib. Reflex actions such as sucking, grasping, and crying occur frequently in this stage. These actions are adaptive behaviors that assist in the infant's survival. Knowledge and interest in the whereabouts of objects change throughout this stage. In the early months of life, if an object is hidden from view, the infant forgets about it. However, as the child matures, she will pursue an object that is hidden from view. Here we observe the formation of memory and object permanence.

Piaget noted the beginning of intelligent behavior in infants during the sensory motor stage. For him these primitive behaviors or schemas indicated the development of deeper cognitive structures that would become more numerous and complex as the child matures toward adulthood.

Preoperational Stage.   The second period of cognitive development extends from approximately the age of 2 to 6 years. The preoperational stage is marked by rapid language development, egocentrism, irreversibility, and centration. Preoperational children are talkers and thinkers. However, their reasoning is not always logical from an adult's point of view, and their verbalization is often misleading in that they may not know as much as their words lead one to believe.

Language development appears to dominate this stage, with children demonstrating the ability to attach words to objects and events. However, language at this stage is egocentric in that it is not used primarily to communicate in a meaningful manner with others, but it is used to vocalize how children perceive the world. At this stage children have a tendency to center their attention on specific attributes and events, seemingly unable to reverse their thinking or to address many attributes associated with a situation. Children at this stage are perceptually bound and are lacking in logical reasoning.

The tendency to center on perceptual attributes and the inability to reverse one's thinking is demonstrated with the conservation of continuous quantity task. The child is presented with two identical containers with equal amounts of water (figure 3–2). After inspection, the child establishes that the containers A and B have equal amounts of water. The liquid from container B is poured into container C, which is taller and thinner. Upon being asked if C contains more water than A, the preoperational child answers, "Yes." From this response, it is evident that perception

FIGURE 3–2 A setup used to assess for conservation of continuous quantity. After establishing that containers A and B have an equal amount of liquid, the liquid in container B is emptied into C. The subject is asked to compare the amount of liquid in C with that in A.

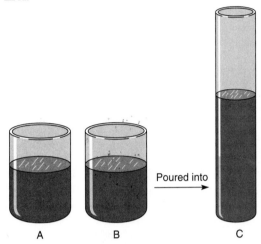

Poured into →

A   B   C

dominates thinking. The child centers on the height of the water in C, but is unable to think back and recall that the water came from B, which had the same amount of water as A.

Concrete Operational Stage. The concrete operational stage generally begins around the age of 6 years and extends into the teens and even into the adult years for some individuals. This stage marks the onset of the ability to think logically and to perform mental operations that are tied to concrete objects and events. During this stage, individuals acquire the ability to order, classify, add, subtract, multiply, and divide. They acquire the ability to determine the cause of events and to represent spatial relationships. They also develop the ability to conserve, which helps them to understand mechanical transformations. Concrete operational reasoning is necessary for learning a great deal of science course subject matter.

Children attain many concepts as they develop during the concrete operational stage. Early in this period of development, they are able to classify plants, animals, and common objects based on perceptible attributes. As this ability improves, children can arrange concepts hierarchically, forming classes and subclasses. Forming classes and understanding their relationships facilitate learning science subject matter, which is a large body of organized knowledge.

Conservation is another important Piagetian concept that individuals develop who are going through the concrete operational stage. Conservation is the ability to understand mechanical transformations, that is if we alter one variable in certain systems, another variable remains constant. This ability develops in different areas at different times during intellectual growth. First conservation of number is attained, followed by substance, then area, weight, and finally toward the end of the concrete operational stage volume is attained. If you refer back to the conservation of substance example given above for the preoperational child, you can contrast the two stages. The concrete thinker can follow the transformation that occurs when a given amount of liquid is poured from the short, wide container into the tall, narrow container. She can reason that change in the liquid height does not change its amount. However, an increase in liquid height means an increase in liquid amount to the preoperational thinker.

Formal Operational Stage. The final stage of cognitive development is the formal operational stage, which begins about the age of 12 and is attained during early adolescence by some, and by late adolescence and early adulthood by others. Individuals at this stage can think more abstractly and deal with many more variables than the concrete reasoner. They can make assumptions about hypothetical situations and reason logically about these assertions. Formal operational thinkers have a greater facility to deal with cause and effect in a multiple variable system, to formulate hypotheses, to use combinatorial thinking, and to use proportional reasoning than concrete operational thinkers. The formal operational reasoner has the capacity to use mental operations on abstract symbols with speed and accuracy. Individuals at this level of cognition have the potential to understand abstract and complex ideas through oral and written discourse.

Consider proportional reasoning as an example to illustrate the cognitive ability of the individual who has attained the formal level. Proportional reasoning represents the ability to understand part-whole relationships, such as $a/b = c/d$. When the part-whole relationships are simple ($1/2 = 2/4$), they can be understood by concrete operational subjects. However, when the relationships are more complex ($36/54 = x/72$), they require formal reasoning patterns for their understanding.

Piaget (Inhelder & Piaget, 1958) organized proportional thinking into four levels: low concrete (IIA), high concrete (IIB), low formal (IIIA), and high formal (IIIB). The first two levels are designated as qualitative. Individuals who are low concrete proportional reasoners can balance equal weights on the balance arm (figure 3–3), but experience difficulty balancing unequal weights. The IIA individuals cannot balance a 5 g and a 10 g weight, for example. They do, however, realize that the lighter weight should be placed further out from the fulcrum than the heavier weight. High concrete proportional reasoners (IIB) can balance both equal and unequal weights, realizing that the lighter weight must be placed a greater distance from the fulcrum than the heavier weight. However, they achieve equilibrium on the balance with unequal weights "by eye" or through trial and error. They do not know exactly where to place the two unequal weights on the balance arm to make it level. These individuals are not metric or quantitative in their thinking. However, high concrete operational reasoners can accurately reason with weights that have a simple 2 : 1 or 3 : 1 ratio. In these instances, they are qualitative in their thinking.

Low formal operational thinkers can balance weights of unequal mass on the balance. They accomplish this metrically. That is, if a 6 g weight is placed 8 units out from the one side of the fulcrum, they can place a 12 g weight 4 units out from the other side of the fulcrum. They can predict where unequal weights must be positioned to balance each other. The IIIA individuals can state that the 12 g weight is placed one-half the distance from the fulcrum from which the 6 g weight is placed, or that a 5 g weight should be placed three units from the fulcrum and a 3 g weight must be placed 5 units from the fulcrum. However, they cannot give a general rule to explain how equilibrium is achieved on a balance. Only the high formal IIIB individuals, with respect to proportional reasoning, can state a general rule that the weight of one object times its distance from the fulcrum must equal the weight of the second object times its distance from the fulcrum. The IIIB individual can figure out quantitatively exactly where weights of unequal mass should be placed, and give a general rule to explain this law such as $W \times D = W' \times D'$.

Hypothetical-deductive reasoning is another ability that distinguishes concrete from formal operational thinkers. Concrete operational individuals can deal with the real; they focus on what is, that is, data and events that are observable (Flavell, 1963). Formal operational individuals

FIGURE 3–3 An equal-arm balance is used to assess proportional reasoning. Subjects are asked to balance weights of different masses at various points along the balance arm.

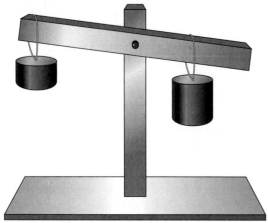

are inclined to deal with the possible; they focus on what could be. Formal thinkers envisage many possibilities, what could be and what is. Hypothetical-deductive reasoning is a strategy that determines reality (what is) from the possible (what might be). They seem determined to state what is possible as a set of hypotheses, then attempt to confirm or reject these hypotheses. Thus, hypothetical refers to proposing what might be and deductive refers to determining what is.

Although Piaget is credited with making an enormous impact upon the field of cognitive psychology, aspects of his developmental theory have been rejected by researchers in this field. Carey (1990) stated that the contribution of domain-general structures is not as great as Piaget believed. She posits that domain-specific structures make a greater contribution to children's success on various Piagetian tasks than domain-general structures. In fact, Carey (1990, p. 162) states: "It seems that cognitive development is mainly the result of acquiring knowledge in particular content domains." She points out that even very young children, such as preschoolers, have many of the representational and computational capacities that Piaget believed they lacked. When researchers have controlled or provided for knowledge of certain concepts, some children demonstrate performance similar to adults. The work conducted by many cognitive psychologists acknowledges the importance of the interview technique that Piaget used and his study of knowledge development. However, they provide different explanations for the thinking and intellectual competence, which seem tied more to experiences with figurative knowledge rather than operative knowledge.

Developmental Level and Science Instruction

The early literature reporting Piaget's assessment of stages of cognitive development might lead educators to believe that most students are formal operational reasoners by the ages of 15 or 16. This was true for the students Piaget studied in Geneva because these students were the more able students in that educational system. This is not the situation in the United States, where most of the vocational and academic track students are educated in the same middle and senior high schools. Therefore, there is a large range of developmental levels of students in our secondary schools. It is possible that as many as 75% of the middle school students are concrete operational thinkers and that at least 50% of the high school students are concrete operational (Chiappetta, 1976). These data have implications for the cognitive demands of school science programs.

Many curriculum writers and educators have used the Piagetian model of development in an attempt to provide appropriate learning experiences for students. Where there is a good match between students' developmental stage and cognitive complexity of the instructional material, students have a greater chance to achieve the desired learning outcomes. However, when the material is too abstract and complex, many students may fail to comprehend the subject matter.

Shayer and Adey (1981) used Piaget's model to assess the suitability of science curricula for students in Britain. The researchers evaluated the cognitive complexity of the subject matter to determine its suitability for a given population of students. They concluded the following regarding assessments made in their nation: "there is a massive mismatch in secondary schools between the expectations institutionalized in courses, textbooks, and examinations and the ability of children to assimilate the experiences they are given" (Shayer & Adey, 1981, vi). Although their study took place more than a decade ago with the Nuffield O level curriculum materials intended to teach fundamental principles of biology, chemistry, and physics, some of this mismatch continues to exist in Britain and in the United States as well.

If science teachers and curriculum writers wish to identify learning outcomes that students can attain, they must be aware of the cognitive operations, schema, or reasoning patterns required to learn a given amount of material. They must ask whether the reasoning patterns possessed by the students are compatible with the cognitive complexity of the subject matter. Shayer and Adey (1981) advise that chemistry is perhaps the most difficult of the basic sciences for students to assimilate because it involves so many abstract ideas that are imperceptible. Also this discipline requires the use of abstract models in order to convey basic ideas to the learner. Biology and physics seem to make less of a conceptual demand on the learner than chemistry. However, the great integrating ideas of biology and physics are no less demanding than those of chemistry.

An examination of Lawson's extensive line of inquiry in this area of developmental psychology consistently reveals a close relationship between operative thinking and comprehension of science course subject matter. One illustration is presented in a study of seventh graders who participated in a unit on genetics and natural selection (Lawson & Thompson, 1988). The unit consisted of lecture/discussion, reading from the textbook and answering questions at the end of the chapter, and an occasional student activity. Students participated in this instructional plan for one month, during which they studied the theory of evolution, dominant and recessive traits, gene theory, sex determination, mutation, and natural selection, among others. At the end of the unit, a large percentage of concrete operational students failed to understand the basic concepts of genetics and inheritance. They continued to hold onto the naive conception that acquired traits can be passed on to offspring. Only a very small percentage of formal operational students held to this misconception about acquired traits; they seemed to understand that a combination of parental traits determine the characteristics of offspring.

Lawson's message to science educators is clear. He believes that the purpose of education is to promote thinking, and a major goal of science teaching is to help students attain formal reasoning. The ability to use formal reasoning patterns demonstrates scientific thinking. Furthermore, the general nature of formal reasoning has been shown to be closely related to achievement in many school subjects as well as science (Lawson, 1985). Lawson believes that the percentage of formal reasoners can be increased as the result of school science programs, if the appropriate learning environments and instruction are provided.

Science teachers should not assume that the developmental level of students is the sole determinant of their achievement and that it is the only reason for the low achievement and poor interest that have been reported about school science. There are many reasons for these outcomes. For example, if too much material is taught and covered too fast, especially through expository means, concrete operational students will probably do poorly in middle and secondary school science courses. However, if the subject matter is carefully selected and courses planned so that there is a high degree of interaction between the students and the learning environment, concrete operational thinkers can experience success. These individuals can reason logically, attain concepts, use abstract symbols, and deal with a small number of variables. They have many thought processes and means to mentally represent knowledge, which can facilitate their concept attainment. Furthermore, instruction aimed at developing discipline-specific concepts can produce positive results when educators consider the knowledge of the learner, provide for meaningful experiences, and permit individuals to construct their knowledge.

CONSTRUCTIVISM

Jean Piaget's efforts to understand the development of knowledge has given educators a constructivistic basis upon which to view learning and teaching. Piaget stressed that learning is an active process and knowledge is constructed in the mind of the learner. Therefore, learning is a creative act where concepts and images are formed by thinking about objects and events and acting upon them. From a constructivistic point of view, learning is an adaptive process where cognitive activity organizes experiences with mental structures that exist in the mind. This is purposeful activity in which the learner strives to *fit* mental ideas to reality. In a sense, cognition begins and ends in the mind. To learn and to know is not to form a picture of objects and events or to receive information, since learning does not result from a copy of objects and events. To know, from a constructivistic view, is the result of conceptual activity to make sense out of experience (von Glasersfeld, 1990).

There is an important difference between the traditionalists' beliefs regarding learning and the constructivists' views. The traditionalists or realists say that there is an objective world, which is the standard. People form copies of this standard reality, and build replicas of it in their minds. Individuals seek truth by *matching* their knowledge with reality. Therefore, the truthfulness of one's knowledge is gauged by the degree to which it matches the realities of the real world (Bodner, 1986).

In contrast to the traditionalist point of view, the constructivists stress that humans construct the objects and relationships that they perceive to exist in their world. They perform mental acts to the extent that their conceptions fit the environment (von Glasersfeld, 1984). Knowledge is valid if it has utility. Knowledge is useful if it helps individuals adapt to their world. Truth is of no concern, because: What is truth?

From these descriptions of the traditional and constructivistic views of learning, it is evident in which direction U.S. education has evolved. Apparently, it is easier to specify learning outcomes and organize instructional activities in a traditional school setting. Educators accomplish their ends by specifying the terms and ideas to be learned, putting the ideas into print, and establishing them as truth. They then take a direct

The learning outcomes of science courses should relate to the development of scientifically literate students. In the process of becoming scientifically literate, students must construct some of the knowledge they acquire through scientific investigations.

path to getting these ideas into the minds of the learners, and there you have the curriculum.

In a constructivist-based curriculum, the situation is not as simple. Learning outcomes must be based upon shared experiences between the teacher and individual students. They must fit what the students can arrive at when they interact with the instructional environment. The learning outcomes must be meaningful and compatible with thinking patterns that exist in the minds of students. Instruction must be structured to allow time for students to organize their thinking, represent their knowledge, and test their hypotheses. Furthermore, knowledge must serve some useful purpose. Obviously, this type of interactive curriculum requires a different philosophy of schooling than that which now exists, and more time and energy to implement.

FIGURE 3–7   A concept map for learning about water. From *Learning How to Learn* (p. 16) by J. D. Novak and D. B. Gowin, 1984, New York: Cambridge University Press. Reprinted by permission.

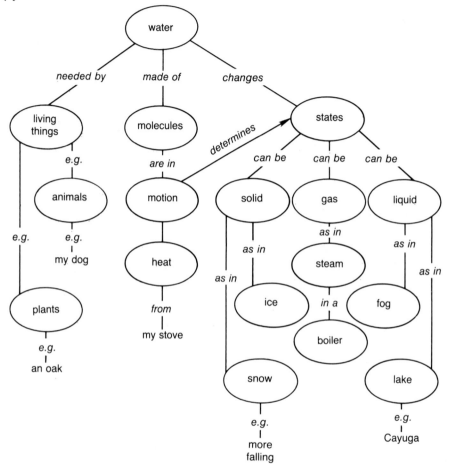

Concept mapping is used in other fields besides science instruction. For example, it is used under the name of semantic mapping for language arts instruction.

Concept Circles

Wandersee (1990) has extended the work of Novak and Gowin, providing science teachers with more explanations and techniques for graphically representing scientific knowledge. He points out that mapping and cartography are creative processes that involve many cognitive activities such as encoding, visualization, problem solving, and decision making. Concept mapping is a metalearning strategy; therefore, it is a method to get students to learn how to learn. Wandersee warns that a concept map appears deceptively simple and perhaps similar to a flow chart. Nevertheless, a flow chart is linear and a concept map is branched and better reflects human learning, which is webbed and hierarchical.

Concept maps, with their circles and lines, take time for students to develop, sometimes taking them 8 to 10 weeks to grasp the technique.

Wandersee has developed what he calls concept circles. In the following list, he suggests how to construct these learning diagrams:

1. Let a circle represent any science concept.
2. Print the name of that concept (e.g., plant) inside the circle using lowercase letters.
3. To show that one concept is included within another (e.g., all birds are vertebrates), draw a smaller circle within the larger circle. Label the smaller circle by printing the name of the narrower, more specific concept within it.
4. When you want to show that some instances of one concept are also part of another concept (e.g., some water contains minerals), draw partially overlapping circles and label appropriately.
5. When you want to show that two concepts are mutually exclusive, draw and label separate circles.
6. You may use up to five concepts in a single diagram, no more. This is due to limitations of the capacity of working memory in humans. Circles can be separate, overlapping, included, or superimposed. All circles must be labeled.
7. The relative sizes of the circles used in a diagram can show the level of specificity for each concept. Large circles stand for more general concepts.
8. The areas of the circles in a concept circle diagram can be used to represent relative amounts or numbers of instances of that concept.
9. Time relationships (such as those found in the history of biology) can be represented by drawing nested (or concentric) circles with the oldest concept being the center one. If chronological relationships are being shown, a "t" should be within the diagram near the central concepts label and it should be enclosed by parentheses (t).
10. One concept circle diagram may be connected to another by "telescoping" graphics. Telescoped diagrams should be made to read from left to right. Several stages of telescoping can be used if a large, scroll-like piece of paper is available.
11. All concept labels should be written horizontally. An exception should only be made for the largest circle, where a lengthy label may replace the upper curve of the circle.
12. Most concept circle diagrams can be improved if redrawn to leave sufficient space around the labels to give the diagram an uncluttered look.
13. Colored pens, markers, pencils, or highlighters may be used to add color to a concept circle diagram with the goal of making the relationships between concepts easier to visualize, understand, and recall.
14. Empty space (white space) around included concepts is used to imply that there are concepts that are not shown. A shaded or colored area surrounding included concepts is used to indicate that no important concepts have been omitted.
15. When the concept circle diagram is finished, a title that describes what the diagram is about should be written in the upper left-hand quadrant of the page and an explanatory sentence should be written in the area directly beneath the diagram. (Wandersee, 1987)

Figure 3–8 presents a concept circle diagram for agents that cause diseases. The picture represents a unit that focuses upon viral diseases and for this reason telescoping is used to project the virus circle. Note that the worms circle is smaller than the others because of its relative unimportance in diseases, especially among humans.

LEARNING HIERARCHIES AND PROBLEM SOLVING

For Robert Gagné, the primary objective for educators is to identify the major learning outcomes for students and state them in behavioral terms. Then it is possible to identify a set of learning tasks, which will increase the probability that the primary outcome will be attained by

FIGURE 3–8   An example of concept circle diagrams illustrating telescoping. Adapted with permission from "Concept Circles" (workshop handout) (p. 4), by J. Wandersee, 1987.

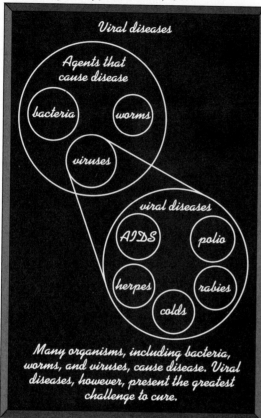

FIGURE 3–9   A learning hierarchy of a hypothetical task analysis

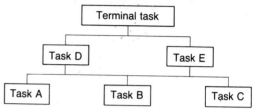

that would precede and lead up to the terminal task. The subordinate learning tasks would also be analyzed, and the process continued until a hierarchy of learning tasks is constructed, which serves as a road map for instruction. In figure 3–9, students must be able to perform task D and E before they can perform the terminal task, and they must be able to perform tasks A, B, and C before they can perform tasks D and E. So the pattern continues.

Once the learning hierarchy has been developed and validated, students can be pretested to determine at which point on the hierarchy they are ready to begin instruction. In an individualized program, students can be given instruction at the appropriate level. In group-paced programs, all students will probably begin instruction at the bottom of the hierarchy and work up through the learning tasks. This will probably provide the least capable students with the instruction they need and the most capable students with a review of what they already know.

Gagné (1970) has organized eight types of learning into a hierarchy (table 3–2). The most sophisticated type of learning is *problem solving.* According to Gagné, problem solving consists of

students. Gagné emphasizes that the primary objective be placed at the top of the *learning hierarchy* with the other objectives below it in an arrangement that leads to the attainment of the terminal behavior (figure 3–9).

In the development of a learning hierarchy, Gagné would ask, "What do you want the learner to know or do at the end of the instruction?" The answer to this question would identify the terminal task, which would occupy the top of the hierarchy. Then he would ask, "What must the learner know or be able to do in order to reach this end?" The answer to this question would identify a series of subordinate learning tasks

TABLE 3–2   Gagné's eight types of learning

Type 8: Problem solving
Type 7: Rule learning
Type 6: Concept learning
Type 5: Discrimination learning
Type 4: Verbal learning
Type 3: Chaining
Type 2: Stimulus-response learning
Type 1: Signal learning

using principles or rules to achieve some worthwhile goal. What is learned in problem solving is a higher order principle or rule. Before the learner can solve a problem, however, she must understand the rule. Hence, the principle or rule is the second most sophisticated type of learning. Gagné points out that principle learning consists of relating two or more concepts. Science concepts must first be learned before they are related. Concept learning becomes the third most sophisticated form of learning. Gagné continues the breakdown of learning types to the most simple form—signal learning. Whether educators wish to think of problem solving as a bottom-up or top-down approach to learning, this type of learning occupies an important place among the priorities of psychologists and science educators.

All normal people engage in problem solving, which generally involves a great deal of cognitive activity. Problem solving can result in many important learning outcomes from simple stimulus-response to the application of rules. These activities can affect students' attitudes and interest. Obviously, problem solving is an active form of mental activity that facilitates scientific thinking.

John Dewey (1938) was an early proponent of problem solving. He defined a problem as anything that gives rise to doubt and uncertainty and stressed that problems to be solved should be important to society and relevant to students. Therefore, problem solving should not be confused with substituting numbers into a formula and computing an answer. Gable (1989) reinforces this idea, emphasizing that problem solving should not be a means of obtaining the correct solution but a means to promote understanding. Watts (1991) stresses that creative elements are involved in problem solving and that many approaches can be taken to arrive at a solution.

Gagné (1977) has some recommendations for teachers who wish to promote problem solving through their instruction. First, one should develop problem-solving tasks around novel ideas

and get away from associating them with routine drill exercises. Second, one should analyze the problem-solving task to determine the prerequisite knowledge and skills necessary to arrive at the solution. One must determine if the students can recall relevant rules and information to solve the problem. Merely asking the students questions can determine this. Third, one should be certain that the students understand the nature of the problem; asking the students to state and restate the problem will help achieve this understanding of the problem. Finally, one must be careful not to give away the solution to the problem. This may occur when the teacher is trying to determine if the students have the prerequisite knowledge. Teachers should remember that learners must discover the higher order rule themselves if they are to exhibit problem solving. This strategy will be taken up in Chapter 4.

ATTITUDES TOWARD SCIENCE

Attitude forms one of the most important affective concepts in science education. Students' attitudes toward science are critical indicators of the worth they place on this subject and how they view the scientific enterprise. These attitudes are learned and can be modified in school science programs. Although attitude has occupied an important place in research on science education, it has been poorly understood and measured (Munby, 1983; Shrigley, 1983; Shrigley, Koballa, & Simpson, 1988). Fortunately, this situation is improving.

Attitude reflects one's disposition toward an idea. This construct represents a favorable or unfavorable *feeling* toward something (Koballa, 1989), and therefore attitude is primarily an affective construct that centers upon the evaluation of an idea. The word attitude is often confused with belief, interest, opinion, and value, which although related are somewhat different. Beliefs are the cognitive basis for attitudes and are the more informational and factual. Although the term interest is thought of as synonymous with attitude, it has a slightly different meaning and has been addressed in a different line of research. Interest addresses someone's willingness to respond to something. Opinion is often used interchangeably with attitude and interest, but it is seldom used as the focus of inquiry in affective studies in science education. Values relate to moral and ethical issues of right and wrong, and are much broader than attitudes.

Shrigley (1991) states that the attitudes or feelings that affect science teaching are manifested on four fronts: "(a) those likes and dislikes of students toward science-related objects and ideas, (b) science anxiety, (c) the historical fear-trust of science rendered by scientific invention, and (d) scientific attitudes and related beliefs about the nature of science modeled by scientists" (Shrigley, 1991, p. 144). The relationships of these ideas are shown in figure 3–10.

The likes and dislikes of science are those feelings that students as well as the general public have toward science-related objects. Some people have formed positive feelings toward science as the result of taking field trips to science museums, building useful devices, or conducting investigations on topics of personal interest. Conversely, others have formed negative feelings about this subject from activities carried out in their science courses, such as memorizing long lists of vocabulary terms, using formulas to solve word problems, or studying ideas that are remote from their lives.

Science anxiety is a state of mind that exists among many individuals in our society. For these people, the thought of taking a high school or college physics course produces considerable distress. A large percentage of individuals become frustrated when they have to apply formulas and use mathematics to solve word problems in physics. Mathematically based science and engineering courses are avoided by most high school and college students. Math and science anxiety is widespread.

The trust-fear cycle has considerable influence on attitudes toward science. We can look back over the past century for examples of many

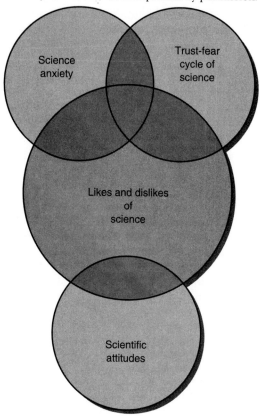

FIGURE 3–10   The attitudes that affect science education are manifested on four fronts. From "The Role of Attitudes in School Science Instruction" (p. 145), by R. L. Shrigley. In S. K. Majumdar, L. M. Rosenfeld, P. A. Rubba, E. W. Miller, and R. F. Schmalz (Eds.), *Science Education in the United States: Issues, Crises and Priorities,* Easton, PA: The Pennsylvania Academy of Science, 1991. Reprinted by permission.

scientific discoveries that have contributed to these cycles. The making of the atomic bomb was a monumental scientific and engineering feat. In the 1940s, the efforts of hundreds of scientists and engineers working together produced nuclear fission, which resulted in the ability to control nuclear reactions and to produce an atom bomb. These individuals had to unlock one of nature's most tightly held secrets—that the center of the atom contains enormous energy, some of which can be released when this tiny mass is disturbed. However, the proliferation of nuclear arms throughout the world and the release of radiation into the atmosphere by nuclear power plant accidents have placed considerable doubt in the minds of many regarding the value of nuclear power. Other products that have contributed to the trust-fear cycle are plastics, pesticides, genetically engineered genes, internal combustion engines, and mind-altering drugs.

Shrigley further points out that scientific attitudes are related but peripheral to the notion of attitudes toward science that dominates recent studies in this area. Scientific attitudes are more cognitive than science attitudes. They pertain to the nature of science and philosophical beliefs thought to be held by scientists. For example, some of the attributes used to portray men and women of science are curiosity, open-mindedness, objectivity, skepticism, self-examination, and logical reasoning. Students form these ideals from what they hear from science teachers, read in textbooks, see at movie theaters, and view on TV. Unfortunately, these ideals are stereotypes of scientists, which are not all together true.

The science education literature in the early part of the twentieth century mainly addressed scientific attitudes as opposed to students' attitudes toward school science. This is evidenced in the statement: "The attitudes of science are those of respect for tested truth and the methods by which it is revealed" (Powers, 1932). Investigations conducted during the first part of the century were primarily aimed at determining the extent to which scientific attitudes were understood by students and teachers, and whether these ideas were reflected in science textbooks (Curtis, 1926; Crowell, 1939).

Investigations conducted over the past 20 years, especially the national assessments in science, indicate that students' attitudes toward science are positive in elementary school, but begin to decline as they enter middle school and continue into high school. By the end of high school, a large percentage of students have formed negative attitudes toward science, especially toward the physical sciences. Only a small percentage of students take high school chemistry and physics. Many students find science difficult and not very useful in their personal lives. The relationship between students' attitudes toward science and their achievement in this subject is interesting, but not what one might expect.

One might hypothesize that there is a high correlation between students' attitudes toward science and their achievement in science. Students who exhibit a positive attitude toward science should do well in science courses. Surprisingly, this does not seem to be the case, and it has been reported that students' attitude toward science is not a significant predictor of science achievement (Keeves, 1975; German, 1988). In fact, one study found that middle school students who dislike science do well and those who like it do poorly (Baker, 1985).

If there is no apparent relationship between student attitude toward science and achievement in this subject, to what is this construct related? Talton and Simpson (1987) provide a different orientation to this line of inquiry. They directed their attention to the classroom environment, which is made up of its emotional climate, various components of the curriculum, the physical environment, the teacher, other students, and friends. They reported a high correlation between students' attitudes toward science and the classroom environment variables, and a very low correlation between attitudes toward science and achievement. These findings reinforce the idea that students feel positive about science in classrooms that have supportive learning environments, with fun activities and friends who like science.

Classroom climate provides important clues regarding students' feelings toward science. The interpersonal interactions that take place between the science teacher and students affect their beliefs and emotional responses toward science. A learning environment guided by a teacher who has high regard for all of the students in a class may tip the attitude scale toward the positive end. We emphasize "all students" because students are greatly affected by what their friends say about teachers and courses. If a science teacher establishes positive rapport with some students but not others, the chances increase that one or more of a given student's friends will become discontent with science class. Discontent spreads to others, and is evidenced by statements such as: "Science is hard." "This class is dumb." "Why do we have to take this?" "This is stupid, I will never use it."

Frasier (1987) provides science teachers with useful insights into the dynamics of classroom climate from studies conducted in many science classrooms. He highlights these results best in comparisons between exemplary and other science teachers' effects on students' attitudes regarding various psychosocial dimensions. In general, students in the classrooms of exemplary teachers view their educational experiences more favorably. They seem to derive more satisfaction from these learning environments and perceive them to be more cohesive. They also report that their classes are less competitive and difficult. Frasier pointed out that outstanding science teachers include in their curriculum the same complex material as the other science teachers, but they make it seem easier. Other dimensions upon which exemplary science teachers receive higher ratings are student involvement, class affiliation, and organization.

As just indicated, attitude does not appear to be a good predictor of achievement in science, but the reverse may be different—*science achievement may influence science attitude.* There is some evidence that science achievement in grades 6 through 10 is a predictor of science attitude. Although this relationship is small, 0.2 to 0.3, it is consistent over many studies (Wilson, 1983). Perhaps success in learning science at the middle school level will have a positive impact on students' later attitudes toward and interest in science. A learning environment that helps students construct and reform their knowledge through many experiences may affect their beliefs about science and feeling toward this subject.

If achievement holds a clue to students' attitudes, and knowledge and beliefs lead to attitudes, then perhaps instruction can modify students' feelings toward science. One approach to developing positive attitudes toward science is to include the positive aspects of science and technology in the curriculum. Science and technology have contributed many products to society that benefit everyone. Most of the products we use can be traced back to the efforts of men and women engaging in science and technology—from the packaging of food products, to the distribution of natural gas, to the purification of water, to the manufacture of light bulbs.

Science teachers must not only address the positive aspects of science and technology but they must do so in a manner that replaces negative ideas with more positive beliefs. Negative attitudes must be replaced with new ideas and images that cause the learner to abandon those bad feelings (White, 1988). For example, many people with negative feelings toward science are influenced by products and techniques that are harmful to humans and animals, such as atomic bombs, pesticides, and animal research. These feelings can be replaced only over time through carefully designed instruction that helps students construct new knowledge that casts scientific understanding and technological products in a new light.

STUDENTS' INTEREST

Interest, like attitude, is a critical concept in science education. This affective construct reflects one's willingness to engage in an activity, rather than one's evaluation of it. Interest is defined as curiosity or fascination for an idea or event that engages attention. This psychological concept helps to explain some aspects of motivation, which are essential to meaningful learning. Many successful science teachers have come to realize the positive effects that can occur when students' interests are addressed in the curriculum. These educators are sensitive to boredom, and work hard to make their instruction appealing. They conduct activities, present information, and include topics that motivate students to attend to and participate in the learning environment.

The work of many psychologists and educators has contributed to our understanding of interest and its importance to learning. Dewey (1913) was an early proponent of the effects of interest on school learning. He advised teachers to determine what interested their students and to consider this in their teaching. He advised educators that when students are interested in a topic or activity, they are engaged "in a whole-hearted way" (p. 65). However, Dewey was very insightful regarding how teachers should go about the task of making instruction appealing to students. He warned that students will attend to the attractive wrappings that educators place around important learning outcomes to add appeal, which will cause students to miss the essential facts. Therefore, science teachers must select the most interesting concepts and principles that relate directly to the subject matter under study if they wish to generate interest that affects understanding. Furthermore, back around the turn of the twentieth century, Dewey noted the problems that students from lower socioeconomic backgrounds were experiencing in school when he stated:

> The pressure of poverty does not seem to be so great an influence on the elimination of pupils as the attitude of children and parents which doubts the worth of further schooling. And we find many children, whom we have considered backward or perverse, are merely bored by the unappealing tasks and formalities of school life. The major difficulty with our schools is that they have not adequately enlisted the interest and energies of children in school work. (Dewey, 1913, p. vii)

Bartlett (1932) in his pioneering work on memory showed that interest played a major role in memorization and recall. Even Piaget (1981), in the later part of his career, pointed out the useful effects that affective factors exert on learning. Jersild and Tasch (1949) found that children of all ages are preoccupied with people and personal relations. They also found that with age there is an increase in student interest in

various forms of *self*-improvement, vocational fitness, educational opportunity, and understanding self and others. These researchers produced a one-page interest finder. Some of the items included in this inventory are:

- My three wishes.
- What I would like to learn more about at school.
- What I don't care to study about.
- What I like best outside of school.
- What I like least at school.
- What I like least outside of school.
- What I want to be or do when I grow up.
- The most interesting thing I have done at school during the past week or so.
- One of the places I especially like to go.
- One of the happiest days in my life.

More recently, the investigations carried out by researchers on interest in textbooks and other reading material have extended our understanding of this idea. Hidi (1990), in a review of research on interest, stated: "Interest-based activities (whether playing with a toy or reading on a topic of interest) are seen as highly motivating and involve attention, concentration, persistence, increased knowledge and value" (p. 554). The work of Hidi and others has classified written material into categories of interest and importance. Interesting materials are those whose ideas and salient details attract the readers' attention, such as personal information about persons or life/death situations, novel or unusual information, life themes, high activity, or characters with which one can identify (Anderson et al., 1987). Important or meaningful material is classified as an idea that relates directly to the essential learning outcomes of the curriculum—the science concepts and principles that we want students to internalize and understand.

The literature is clear on the effects of interest in text. Children and adults understand and remember information from reading topics of high interest. However, punching up school science texts with interesting but unimportant details can result in unwanted outcomes whereby the interesting details are recalled but not important generalizations (Garner et al., 1991). For teaching science, students must learn important principles and theories. These ideas are necessary in order to understand natural phenomena, technological devices, and science/societal issues. On the one hand, writers and teachers must avoid presenting material in a dry unappealing manner that turns students off. On the other hand, they must avoid spiking up presentations to the point where students remember only the spicy contextual details and not the important material.

Garner et al. (1991) illustrate the effects of making science texts more attractive to the reader. They developed many reading passages about the eminent physicist Steven Hawking, who holds Newton's chair at Cambridge University. Hawking is a brilliant theoretical physicist who has been compared to Einstein. He is well known in the scientific community for his theories about black holes and his attempt to link relativity with quantum mechanics in what is known as the Grand Unified Theory. The general public recognizes Hawking from pictures in magazines where he is shown confined to a wheelchair with the paralyzing condition known as Lou Gehrig's disease. Garner and colleagues constructed many passages about Hawking so that interesting detail was: (a) presented as an aside in a passage that was generally interesting, (b) presented as an aside in an uninteresting passage, (c) embedded in a passage that was generally interesting, and (d) embedded in a passage that was generally uninteresting. The results were as expected. The interesting details about a bet Hawking made with a colleague concerning black holes were recalled better than the important scientific generalization about the Grand Unified Theory.

The effects of tapping into students' interests in a manner that closely aligns these interests with the subject matter under study is demonstrated by the work of Daponte (1992), who has instructed minority students in high school physical science. Daponte noticed that his stu-

dents had little familiarity with the objects used in the assigned textbook to illustrate Newton's laws of motion. For example, most of these students had never ridden in a motorboat, which was used to explain Newton's third law. However, all of the students were familiar with and interested in sports. Therefore, Daponte devised enrichment activities for the students to read and questions to answer that centered around the objects used to play baseball, basketball, and football. The results were dramatic in that the students who engaged in the enrichment activities centered around sports did much better on tests that measured their knowledge and understanding of Newton's three laws of motion than did students who engaged in enrichment activities centered around nondescript and less interesting objects.

Some science teachers are successful at improving achievement and student interest in science. These educators seem to possess a talent for providing a context in which meaningful learning can take place. They alter the learning conditions in the classroom so that students find the instructional process supportive and comfortable. These teachers modify the context within which the subject matter is taught so that it connects with what students know and perceive to be relevant, yet in a manner that relates to the curriculum. They become familiar with their students' background, culture, and activities outside of school, and use this information to weave a context within which students construct meaning and understanding.

For years science educators have observed that interest in science drops about the time youngsters enter the middle school. One of many possible causes for this phenomenon is that elementary teachers focus more on students, while secondary teachers focus more on subject matter. When students enter the sixth and seventh grade, they have a full period of science where they receive large doses of information. Frequently, this subject matter emphasis is centered on teaching an organized body of scientific content, which can be dry and abstract. As children make the transition into middle and junior high school, science teachers must consider their interests and concerns, and bring the curriculum into better alignment with these factors. High school science programs must also focus upon many areas of students' interests, such as jobs, self improvement, leisure, sports, fitness, and health.

IMPLICATIONS FOR TEACHING SCIENCE

The students who participate in science courses vary dramatically in their cognitive abilities and cultural backgrounds. Within the same classroom there can exist concrete and formal operational thinkers. Some students can barely utter a few words in English, while others demonstrate great facility with the language. A large percentage of our youth demonstrate modest reading and writing skills. Some have never been to a museum. With all of this variation in knowledge, skills, and background, considerable attention must be given to cognition and learning in order to provide a better balance of behavioristic and cognitivistic strategies to improve science education.

Science teachers may experience problems in implementing into their curricula some of the ideas discussed in this chapter that are associated with cognitive psychology. Most of the middle and senior high school science programs are centered in the behaviorist tradition. These curricula have specified learning outcomes, assigned textbooks, and prepared tests. The curriculum is set and nonnegotiable. Often teachers are expected to cover a large amount of subject matter and to prepare students for a departmental or district test at prescribed intervals. As a consequence, new and even experienced teachers often express frustration with this situation; they also experience difficulty in getting a large percentage of their students to demonstrate reasonable levels of achievement. They also express frustration in their attempts to alter the system so that less information is covered and greater understanding is achieved.

Science teachers who wish to help students develop a deeper understanding of science concepts and to find greater meaning in what they study in science can begin by attempting to modify the behavioral-based curricula by infusing into them some of the strategies associated with cognitive psychology. For example, teachers can cover less subject matter and provide a greater variety of activities for the most important ideas to be studied. They can identify learning outcomes that are compatible with the developmental levels of students. In addition, science teachers can select topics that students find relevant and interesting, which are directly related to important science concepts and principles under study. In some instances, students themselves can be involved in the process of identifying learning outcomes they wish to achieve. Most importantly, science teachers must begin the instructional process by finding out what students know and believe regarding the subject under study. Then they can provide a variety of learning experiences, giving students numerous opportunities to gather information, present their knowledge, and test out their ideas within contexts that are relevant to their lives.

SUGGESTED ACTIVITIES

1. Select a middle school or high school science curriculum guide and accompanying textbook. Identify several major concepts, principles, and theories. Evaluate these discipline-related ideas in terms of their appropriateness for the concrete and formal operational reasoners for whom they are intended.

2. Select a middle school or high school science textbook. Identify several major concepts, principles, and theories. Analyze how the text

presents these ideas using the five major components of a concept discussed in this chapter: (1) name, (2) definition, (3) attributes, (4) values, and (5) examples. Then assess the extent to which the reading material (a) explains these ideas, (b) provides sufficient examples to promote understanding of them, and (c) develops these ideas within a context that is relevant to students.

3. Identify a major concept, principle, law, or theory that students are expected to learn in a middle school or high school science course. Prepare some questions that can be used to assess students' understanding of this content knowledge. Then interview a small number of students in that course to ascertain their understanding of these ideas and their misconceptions.

4. Find a unit of study that a middle school or high science teacher is using to teach a topic.

Examine the instruction to determine the potential the unit holds for promoting conceptual change among the students who will participate in its instruction. From this exercise, consider the conceptual change models discussed in this chapter and their focus on (a) eliciting ideas, (b) clarifying knowledge, (c) examining conflicting views, (d) constructing new ideas, (e) applying new knowledge, and (f) reviewing the knowledge restructuring that took place during the unit.

5. Identify a class of students participating in a middle school or high school science course. Prepare a list of questions to assess their attitudes toward the science they are studying. Also examine the classroom environment and the instructional materials and evaluate how much interest they hold for the students.

BIBLIOGRAPHY

Anderson, R. C., Shirey, L. L., Wilson, P. T., and Fielding, L. G. 1987. Interestingness of children's reading material. In R. E. Snow and M. J. Farr (Eds.), *Aptitude, learning, and instruction: Vol. III. Cognitive and affective analyses* (pp. 287–299). Hillsdale, NJ: Lawrence Erlbaum Associates.

Ausubel, D. P. 1963. *The psychology of meaningful verbal learning.* New York: Grune & Stratton.

Baker, D. R. 1985. Predictive value of attitude, cognitive ability, and personality to science achievement in the middle school. *Journal of Research in Science Teaching, 22:* 103–113.

Bartlett, F. C. 1932. *Remembering: A study in experimental and social psychology.* New York: Cambridge University Press.

Bodner, G. M. 1986. Constructivism: A theory of knowledge. *Journal of Chemical Education, 63:* 873–878.

BouJaoude, S. 1991. A study of the nature of students' understandings about the concept of burning. *Journal of Research in Science Teaching, 28:* 689–704.

Bourne, L. E., Jr. 1966. *Human conceptual behavior.* Boston: Allyn & Bacon.

Brook, A., and Driver, R. 1984. *Aspects of secondary students' understanding of energy: Summary report,* Children's Learning in Science Project, Centre for Studies in Science and Mathematics Education. Leeds, England: The University of Leeds.

Brumby, M. N. 1984. Misconceptions about the concept of natural selection by medical biology students. *Science Education, 68:* 493–503.

Bruner, J. S., Goodnow, J. J., and Austin, G. A. 1956. *A study of thinking.* New York: John Wiley & Sons.

Carey, S. 1990. Cognitive development. In D. N. Osherson and E. S. Smith (Eds.), *Thinking: An invitation to cognitive science.* Cambridge, MA: The MIT Press.

Carroll, J. B. 1964. Words, meanings and concepts. *Harvard Educational Review, 34:* 178–202.

Case, R. 1978. The developmentally based theory and technology of instruction. *Review of Educational Research, 48*(3): 439–463.

Chiappetta, E. L. 1976. A review of Piagetian studies relevant to science instruction at the secondary and college level. *Science Education, 60:* 253–262.

Claxton, G. L. 1984. Teaching and acquiring scientific knowledge. In T. Keem and M. Pope (Eds.), *Kelly in the classroom: Educational applications of personal construct psychology.* Montreal: Cybersystems.

Crowell, V. L. 1939. Attitudes and skills essential to the scientific method, and their treatment in general-science and elementary-biology textbooks. In F. D. Curtis (Ed.), *Third digest of investigations in the teaching of science.* Philadelphia: P. Blakiston's Son & Co.

Curtis, F. D. 1926. An experiment in developing scientific attitudes in eighth and ninth grade pupils. In F. D. Curtis (Ed.), *Third digest of investigations in the teaching of science.* Philadelphia: P. Blakiston's Son & Co.

Daponte, T. 1992. *Investigating student's understanding of Newton's laws of motion through schema theory and sporting activities.* Unpublished doctoral dissertation, University of Houston.

Dewey, J. 1913. *Interest and effort in education.* New York: Houghton Mifflin.

Dewey, J. 1916. *Democracy and education: An introduction to the philosophy of education.* New York: Macmillan.

Dewey, J. 1938. *Experience and education.* New York: Macmillan.

Dressel, P. L. 1960. How the individual learns science: Rethinking science education. In N. B. Henry (Ed.), *Fifty-ninth yearbook of the National Society for the Study of Education.* Chicago: University of Chicago Press.

Driver, R. 1988. Theory into practice II: A constructivist approach to curriculum development. In P. Fensham (Ed.), *Development and dilemmas in science education* (pp. 133–149). Philadelphia: Falmer Press.

Flavell, J. H. 1963. *The developmental psychology of Jean Piaget.* Princeton, NJ: Van Nostrand Reinhold.

Frasier, B. J. 1987. Psychosocial environments in classrooms of exemplary teachers. In K. Tobin and B. J. Frasier (Eds.), *Exemplary practices in science and mathematics education.* Perth, Western Australia: Science and Mathematics Centre, Curtin University of Technology.

Gabel, D. 1989. Introduction. In D. Gabel (Ed.), *What research says to the science teacher, Vol. 5: Problem solving.* Washington, DC: National Science Teachers Association.

Gagné, R. M. 1970. *The conditions of learning.* New York: Holt Rinehart and Winston.

Gagné, R. M. 1977. *The conditions of learning.* New York: Holt Rinehart and Winston.

Gardner, H. 1985. *The mind's new science.* New York: Basic Books.

Garner, R., Alexander, P. A., Gillingham, M. G., Kulikowich, J. A., and Brown, R. 1991. Interest and learning from text. *American Educational Research Journal, 28:* 643–659.

German, P. J. 1988. Development of the attitude toward science in school assessment and its use to investigate the relationship between science achievement and attitude toward science in school. *Journal of Research in Science Teaching, 25:* 689–703.

Herron, D. J., Canter, L. L., Ward, R., and Srinivasan, V. 1977. Problems associated with concept analysis. *Science Education, 61:* 185–199.

Hidi, S. 1990. Interest and its contribution as a mental resource for learning. *Review of Educational Research, 60:* 549–571.

Inhelder, B., and Piaget, J. 1958. *The growth of logical thinking from childhood to adolescence.* New York: Basic Books.

Jersild, A. T., and Tasch, R. J. 1949. *Children's interests.* New York: Teachers College, Columbia University.

Keeves, J. P. 1975. The home, the school, and achievement in math and science. *Science Education, 59:* 439–460.

Koballa, T. R., Jr. 1989. Changing and measuring attitudes in the science classroom. National Association for Research in Science Teaching, *Research Matters . . . To the Science Teacher, No. 19.*

Koballa, T. R., Jr., and Crawley, F. E. 1985. The influence of attitude on science teaching and learning. *School Science and Mathematics, 85:* 222–232.

Kuhn, T. S. 1970. *The structure of scientific revolutions.* Chicago: University of Chicago Press.

Lawson, A. E. 1982. The reality of general cognitive operations. *Science Education, 66:* 229–241.

Lawson, A. E. 1985. A review of research on formal reasoning and science teaching. *Journal of Research in Science Teaching, 22:* 569–617.

Lawson, A. E., and Renner, J. W. 1975. Piagetian theory and biology teaching. *The American Biology Teacher, 37:* 336–343.

Lawson, A. E., and Thompson, L. D. 1988. Formal reasoning ability and misconceptions concerning genetics and natural selection. *Journal of Research in Science Teaching, 25:* 733–746.

Magoon, J. A. 1977. Constructivist approach in educational research. *Review of Educational Research, 47:* 651–693.

Munby, H. 1983. Thirty studies involving the "scientific attitude inventory." What confidence can we have in this instrument? *Journal of Research in Science Teaching, 20:* 141–162.

Needham, R. 1987. *Teaching strategies for developing understanding in science,* Children's Learning in Science Project, Centre for Studies in Science and Mathematics Education. Leeds, England: The University of Leeds.

Novak, J. D., and Gowin, D. B. 1984. *Learning how to learn.* New York: Cambridge University Press.

Osborne, R. J. 1984. Children's dynamics. *The Physics Teacher, 22:* 504–508.

Osborne, R. J., and Freyberg, P. 1985. *Learning in science.* London: Heinemann.

Osborne, R. J., and Wittrock, M. C. 1983. Learning science: A generative process. *Science Education, 67:* 489–508.

Piaget, J. 1964. Cognitive development in children: Development and learning. *Journal of Research in Science Teaching, 2:* 176–186.

Piaget, J. (Ed. and Trans.). 1981. Intelligence and affectivity: Their relationship during child development. *Annual Reviews Monograph.* Palo Alto, CA: Annual Reviews.

Posner, G. J., Strike, K. A., Hewson, P. W., and Gertkzog, W. A. 1982. Accommodation of a scientific conception: Toward a theory of conceptual change. *Science Education, 66:* 211–228.

Powers, S. R. 1932. What are some of the contributions of science to liberal education? In *National Society for the Study of Education: The thirty-first yearbook, Part 1, A program for teaching science.* Chicago: University of Chicago Press.

Resnick, L. B. 1987. *Education and learning to think.* Washington, DC: National Academy Press.

Roth, K., and Anderson, C. W. 1990. A conceptual change model of science instruction. An instructional handout. East Lansing, MI: Michigan State University, Institute for Research on Teaching.

Rowe, M. B. (Ed.) 1990. *The process of knowing: What research says to the science teacher, volume six.* Washington, DC: The National Science Teachers Association.

Sequeira, M., and Leite, L. 1991. Alternative conceptions and history of science in physics teacher education. *Science Education, 75:* 45–56.

Shayer, M., and Adey, P. 1981. *Towards a science of science teaching.* London: Heinemann Educational Books.

Shrigley, R. L. 1983. The attitude concept and science teaching. *Science Education, 67:* 425–442.

Shrigley, R. L. 1991. The role of attitudes in school science instruction. In S. K. Majumdar, L. M. Rosenfeld, P. A. Rubba, E. W. Miller, and R. F. Schmalz (Eds.), *Science education in the United States: Issues, crises and priorities.* Easton, PA: The Pennsylvania Academy of Science.

Shrigley, R. L., Koballa, T. R., and Simpson, R. D. 1988. Defining attitude for science educators. *Journal of Research in Science Teaching, 25:* 659–678.

Smith, E. L. 1990. Implications of teachers' conceptions of science teaching. In M. B. Rowe (Ed.), *The process of knowing: What research says to the science teacher, volume six.* Washington, DC: The National Science Teachers Association.

Talton, E. L., and Simpson, R. D. 1987. Relationships of attitude toward classroom environment with attitude toward and achievement in science among tenth grade biology students. *Journal of Research in Science Teaching, 24:* 507–525.

Tennyson, R. D., and Cocchiarella, M. J. 1986. An empirically based instructional design theory for teaching concepts. *Review of Educational Research, 56:* 40–71.

Thorndike, E. L. 1935. *Adult interests.* New York: Macmillan.

Treagust, D. F., and Smith, C. L. 1989. Secondary students' understanding of gravity and the motion of planets. *School Science and Mathematics, 89:* 380–391.

von Glasersfeld, E. 1984. A radical constructivist model of knowledge. In P. Waltzlawick (Ed.), *The invented reality: How do we know what we believe we know?* New York: Norton.

von Glasersfeld, E. 1990. An exposition of constructivism: Why some like it radical. *Journal for Research in Mathematics Education,* Monograph No. 4, 19–29.

Wandersee, J. H. 1987. Concept circles. Ithaca, NY: Cornell University.

Wandersee, J. H. 1990. Concept mapping and the cartography of cognition. *Journal of Research in Science Teaching, 27:* 923–936.

Watts, M. 1991. *The science of problem-solving.* Portsmouth, NH: Heinemann.

White, R. T. 1988. *Learning science.* New York: Basil Blackwell.

Wilson, V. L. 1983. A meta-analysis of the relationship between science achievement and science attitude: Kindergarten through college. *Journal of Research in Science Teaching, 20:* 839–850.

Teaching Strategies and Classroom Management

TABLE 4–2   Basic and integrated science process skills

Process skill	Definition
Basic skills	
Observing	Noting the properties of objects and situations using the five senses
Classifying	Relating objects and events according to their properties or attributes (This involves classifying places, objects, ideas, or events into categories based on their similarities.)
Space/time relations	Visualizing and manipulating objects and events, dealing with shapes, time, distance, and speed
Using numbers	Using quantitative relationships, e.g., scientific notation, error, significant numbers, precision, ratios, and proportions
Measuring	Expressing the amount of an object or substance in quantitative terms, such as meters, liters, grams, and newtons
Inferring	Giving an explanation for a particular object or event
Predicting	Forecasting a future occurrence based on past observation or the extension of data
Integrated skills	
Defining operationally	Developing statements that present a concrete description of an object or event by telling one what to do or observe
Formulating models	Constructing images, objects, or mathematical formulas to explain ideas
Controlling variables	Manipulating and controlling properties that relate to situations or events for the purpose of determining causation
Interpreting data	Arriving at explanations, inferences, or hypotheses from data that have been graphed or placed in a table (this frequently involves concepts such as mean, mode, median, range, frequency distribution, t-test, and chi-square test)
Hypothesizing	Stating a tentative generalization of observations or inferences that may be used to explain a relatively larger number of events but that is subject to immediate or eventual testing by one or more experiments
Experimenting	Testing a hypothesis through the manipulation and control of independent variables and noting the effects on a dependent variable; interpreting and presenting results in the form of a report that others can follow to replicate the experiment

Note: Data compiled from American Association for the Advancement of Science, *Science: A Process Approach. Commentary for teachers,* Washington, DC, 1965.

students in an extensive laboratory investigation, centering upon the "string telephone." Mr. Roosevelt initiates this activity by demonstrating how two metal soup cans, connected with a wire, can transmit sounds between two people. He challenges the students to construct many different phones to determine which pair will transmit the clearest voice messages. He also asks them to predict which phones will work best and to explain why. Working in groups, the students bring to class many cans and containers of various sizes and compositions—from small metal cans to large coffee cans, from small paper cups to giant soft drink containers, and from plastic foam cups to dairy containers. The students also bring to class a variety of lines to connect the phones, such as thread, string, monofilament fishing line, and wire.

After the students have tested many combinations of phones and lines, they select a set of phones that produces very clear voice sounds. For example, a pair of paper cups connected by carpet thread produces amazing results. Mr. Roosevelt asks his students to identify the variables that seem to produce good sounds through these simple devices. He guides students' thinking so that they realize that they were conducting an experiment and controlling variables. The students not only come to realize that the size and composition of the phones are important but that these variables affect the vibration of the transmitting material. Even though the students lack the scientific terminology to explain the effects of elasticity on vibration, they feel good about conducting an investigation that seems scientific.   ▲

Teacher 2

▼ Ms. Suki is a high school physics teacher who engages her students in the careful examination of variables that affect the transmission of sound in different media, for the purpose of gaining conceptual knowledge. She begins the study of sound with a laboratory exercise in which she demonstrates the difference in clarity of voice in two "string telephone" systems. One pair of phones is made from paper drinking cups, the other pair from heavy plastic containers in which metal parts are packed for shipping. From the demonstration it is obvious that the paper cups are superior to the plastic containers in transmitting a conversation between two classmates. After this fact has been established, the students are given the charge to determine why the paper does so well in voice transmission. Ms. Suki asks: "What are the variables that affect the transmission of sound?" However, before the students proceed, the teacher requires them to explain in writing what factors in a medium promote sound transmission and why this is true.

In a large box, Ms. Suki has many pairs of phones, all made from materials different in composition, thickness, and flexibility. It does not take long for students to realize that the transmission of sound is directly related to the elasticity of the medium and inversely to its density or inertia. This is reinforced when students find a chart in their physics textbook showing that sound travels much faster in steel (5,200 m/sec) than in lead (1,200 m/sec). ▲

In the examples given above, the two teachers differ in their purpose for using the process approach for science instruction. The middle school teacher uses experimentation and controlling variables to promote the spirit of inquiry through the study of sound. This teacher desires that students gain only some familiarity with basic concepts of sound transmission. The high school physics teacher uses this approach to help students clarify any misconceptions they may have about the transmission of sound and to help them develop correct ideas about this phenomenon. For the physics teacher, the primary outcome is conceptual development, and

she uses the identification and examination of variables to assist in achieving this end.

Discrepant Events

A stimulating approach to initiate inquiry is to use discrepant events. A discrepant event is one that puzzles the observer. It causes the observer to wonder why an event occurred as it did, and it leaves the observer at a loss for an explanation. An inquiry session initiated through a discrepant event usually begins with a demonstration or film, preceded by some directions to focus students' attention on what they are to observe.

In an inquiry session that centers on Bernoulli's principle, for example, the teacher might say, "I want you to study this situation. Watch this demonstration and tell me what you observe happening to the Ping-Pong ball."

A Ping-Pong ball attached to a string is brought near a stream of water rushing out of a faucet, as shown in figure 4–1. The students observe that the Ping-Pong ball "adheres" to the

FIGURE 4–1 The Ping-Pong ball "adheres" to the stream of water because of the reduced pressure of the air created by the rapid motion of the fluid.

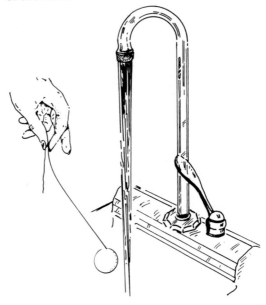

stream of water. The teacher repeats the procedure of pulling the ball away from the stream of water and then bringing it back to the stream, allowing the students to see the ball "attracted" to the stream of fast-flowing water.

While this demonstration is taking place, the teacher asks the students to describe what they observe. When students are called on to respond, most will indicate that the Ping-Pong ball clings to the stream of water. A follow-up question might be: "What did you expect the Ping-Pong ball to do when it was brought near the stream of water?" Some students will respond with the comment: "I thought that the stream of water would push the Ping-Pong ball away." This leads to further inquiry and to the principle under study.

What is it that students should discover in this discrepant event demonstration? Primarily, the students are to discover what happens to an object when a fluid flows past that object. The explanation for this phenomenon is not the objective at this point because it would probably dampen further inquiry. Instead the teacher can continue with this demonstration by asking students to predict what will happen to the attrac-

tion between the Ping-Pong ball and the stream of water if the faucet is turned on full force. This will stimulate more thought regarding the discrepant event.

There are many other demonstrations that the teacher can present to give the students additional examples of Bernoulli's principle. With additional demonstrations already prepared, the teacher might first ask the students if they can give an example of this principle in everyday life. If a student responds by mentioning the lift of an airplane that occurs when air flows over the wing of the plane, this can be demonstrated in class. Here the principle can be shown by giving students sheets of paper and asking them to blow over their surface. They will observe that the paper rises because the pressure over the upper surface of the paper decreases due to Bernoulli's principle (Figure 4–2).

Up until this point in the inquiry session, very little has been said about Bernoulli's principle and its mechanism. This is as it should be, because there is usually a tendency to present a demonstration and allow for a brief question-and-answer session, and then explain the idea at hand. It seems to be difficult for many science

FIGURE 4–2   The sheet of paper at the left acts like an airfoil. When one blows over the top of the paper, it rises up, demonstrating that the air rushing over it creates a lift effect due to reduced air pressure.

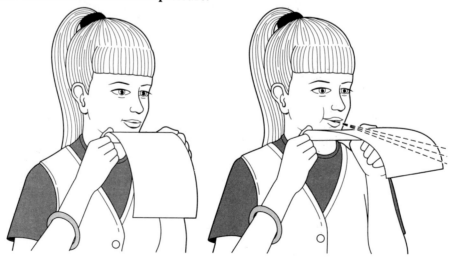

teachers to continue the question-asking process and to maintain student inquiry for very long. How do the students learn to explain Bernoulli's principle?

There are several ways to explain the principle. One way is to ask the students for their explanations. During this part of the inquiry session the teacher can respond with yes or no to the students' responses, thus not giving them the explanation at this point. The students should be encouraged to phrase their explanations in the form of a question so that the teacher can respond only with a yes or no. Another way to approach this matter is to direct the students to their textbooks or other sources to find an explanation of the principle. Finally, the teacher can help to explain the principle.

This dialog can be followed by presenting another discrepant event such as the one shown in figure 4–3. In this demonstration, students are asked to (1) push a straight pin through a small piece of paper, (2) hold the paper in one hand and a spool in another, with the pin extending up into the spool hole, and (3) blow down through the hole. The question posed: "What will happen when you let go of the paper while blowing through the spool hole?" Many students will predict that the paper will be violently blown away from the spool and fall to the ground. The discrepant events presented above are just three of many that can be used to pique students' curiosity and to develop their knowledge about Bernoulli's principle.

Discrepant events can be used to stimulate inquiry about numerous concepts and principles. Discrepant event demonstrations of the laws of motion, center of gravity, Pascal's principle, density, and vacuum, among others, can be used to initiate inquiry sessions. The discrepant event approach receives support from learning psychologists as a valid instructional method. Discrepant events influence equilibration and the self-regulatory process, according to the Piagetian theory of intellectual development. Situations that are contrary to what a person expects cause him to wonder what is taking place. With

FIGURE 4–3 When one blows vigorously down through the hole in the spool, the paper remains suspended in the air up against the base of the spool. The straight pin keeps the paper centered under the hole.

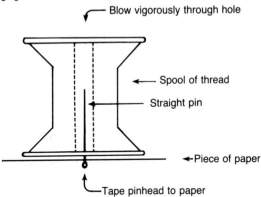

proper guidance the individual will try to figure out the discrepancy and attempt to find a suitable explanation for this "curious event." When a student finds an acceptable explanation, he rests temporarily at a new cognitive level. The individual is then better equipped mentally to attack new situations that cause curiosity and puzzlement (Piaget, 1971).

Science teachers must exercise care in selecting events and problems that directly influence self-regulation and inquiry. They must construct situations that present a challenge to students but that can be understood by them given proper guidance and ample time. Some researchers have argued that a discrepant event approach can even be used with homework problems, providing the following two factors are present:

Problems must be chosen so that the student can partially but not completely understand them in terms of old ideas . . . and sufficient time must be allowed for the student to grapple with the situation, possibly with appropriate hints to direct this thinking, but allowing him to put the ideas together himself. (Lawson & Wollman, 1975, pp. 470–471)

These same two factors or guidelines should be considered for discrepant events, whether they be in the form of live demonstrations, filmed demonstrations, or word problems.

Inductive Activities

The inductive approach provides students with a learning situation in which they can discover a concept or principle. In this approach the attributes and instances of an idea are encountered first by the learner, followed by naming and discussing the idea. This is, in a sense, an "experience before vocabulary" approach to learning. This strategy is the opposite of the deductive approach, in which the name and explanation of an idea are given first, followed by experiences that illustrate attributes and instances of the idea.

An example of an inductive approach that can be used in biology is one in which students are introduced to the theory of natural selection before it is mentioned or discussed in the classroom. They can form the notion that evolution can occur over time through the selection process. This is illustrated through an outdoor activity explained below.

▼ Students are told that they will play the role of predators and will have an opportunity to go outside and feed on insects. For this exercise, the teacher scatters 200 red and 200 green small objects over a grassy lawn on the school grounds. (Soup macaroni, popcorn dyed with food coloring, or colored beads are often used in this exercise.) The students are given signals to start and stop "feeding," each usually a few minutes in duration.

They are instructed to walk upright, only bending to pick up one object at a time. Students must take two steps before picking up another object. A few time intervals of approximately one minute in duration can be used in this exercise.

After the students return to the classroom, the data are entered on a table on the chalkboard. The total number of red and green "insects eaten" is entered in the table. Inspection of the data will show the number of "insects eaten" during each interval, the total number of "insects eaten," and the percentage of "insects" remaining (table 4–3).

The question that can be asked is: "What happened to the 'insect' population?" After inspecting the data, the students may respond that the red "insects" were consumed at a higher rate (toward extinction) than the green "insects." They will realize that this situation is due to the colors of the "insects" and the grass. Because the red "insects" were easier to see, they were consumed faster than the green "insects."

As a result of this inductive activity, some students now have an idea of one aspect of natural selection—how color can serve to camouflage some species while exposing others to predators. For other students this activity merely broadened their knowledge of natural selection.   ▲

The inquiry-oriented teacher will extend this activity so that students can expand their knowledge by discovering additional examples of natural selection. The teacher might present the following questions:

■ What traits have helped certain species of birds to survive?
■ What traits or behavior patterns are causing some species of birds to become scarce?

TABLE 4–3 Consumption of insects

Species	Number "eaten" in each interval*		Total number "eaten"	Percentage remaining
	1	2		
Red	63	88	151	24
Green	21	34	55	72

*Initially there were 200 red and 200 green "insects."

- What traits have assisted sharks to survive over thousands of years with very little change?
- Name some animals that have become extinct. What might have caused their extinction?
- Identify some animals or plants in your community that may soon become extinct. Why might this occur?
- Identify some animals or plants in your community that might thrive in the near future. Why might this occur?

Students should be encouraged to select one of these questions and to find information to answer their question from books, magazines, television programs, newspapers, people, and personal observations. At a designated time the students can give a short report to the class concerning their findings.

The Learning Cycle. The inductive approach is illustrated in an instructional strategy called the Learning Cycle. It was developed as the primary teaching method for an elementary science program, Science Curriculum Improvement Study (SCIS), produced in the 1960s under the direction of Robert Karplus (SCIS, 1974). In addition to the elementary level, the strategy has been used at the junior high, senior high, and college levels to promote inquiry and intellectual development (Fuller, 1980; Renner et al., 1985). The Learning Cycle has three phases—exploration, invention, and application.

Exploration. The exploration phase allows students to experience objects and events in order to stimulate their thinking about a concept or principle and to permit them to discover patterns and relationships. During this phase, students are given some guidance to keep them with the learning task. Questions are posed and cues are given to channel thinking. However, students are not given answers or labels. This phase is illustrated in the natural selection example wherein the students are directed to pretend they are birds and should "eat" the "insects" that have been scattered over the lawn.

Invention. The invention phase allows students to determine relationships among objects and events that they have experienced. Initially, the teacher serves as a guide to channel thinking, encouraging students to construct appropriate labels for the relationships they have just discovered. However, at some point it is appropriate for the teacher to provide labels and terms that are given in the curriculum guide and the textbook. This phase is illustrated in the natural selection example during the inspection of the data in table 4–3. The students are asked to explain the consumption pattern of the insects. Why were more red insects eaten than green insects? Here the teacher discusses the terms natural selection and evolution.

Application. The application phase allows the students to apply their knowledge of a given concept to other situations. The teacher encourages the students to find (discover or inquire into) examples to illustrate the concept they have just experienced. This phase permits students to generalize their learning, thus reinforcing newly acquired knowledge and attaining concepts. In the natural selection vignette, for example, the students are encouraged to find additional examples of the theory in textbooks, magazines, television programs, first-hand observations, and so on, and to share this material with other class members.

The labels for the three phases of the Learning Cycle have been modified by different educators. Renner, Abraham, and Birne (1985) used the terms exploration, conceptual invention, and expansion of the idea. Rakow (1986) used exploration phase, concept introduction, and concept application. The terms used for these phases are not so important as giving students the opportunity to experience the sequence of instruction, probably causing them to enjoy it and to believe they are learning more in science class (Renner et al., 1985).

Lawson, Abraham, and Renner (1989) have researched and written extensively about the Learning Cycle. They discuss three types of

Learning Cycles in their work, which they term descriptive, empirical-abductive, and hypothetical-deductive. These strategies differ according to what the learner is expected to do with the data during the invention phase.

In the descriptive learning cycle, students discover empirical patterns during the exploration phase. The students are basically describing perceptible phenomena and telling "what" they have found out. With the empirical-abductive learning cycle, students also attempt to discover and describe empirical patterns. However, they go one step further by attempting to determine the cause of these patterns. The term abductive is used here to specify an activity where the learner studies data and patterns and attempts to explain "why." In the hypothetical-deductive learning cycle, students are asked to generate alternative explanations for phenomena and to design experiments to test them out. This approach certainly engages the students in more reflective and deductive thinking than the other two approaches.

Lawson, Abraham, and Renner (1989) point out that the three phases and their instructional sequences are important. They recommend beginning with exploration and moving to term introduction (Invention), which is the inductive aspect of this approach. This sequence may be more effective with average students, while a more deductive approach may be better suited to the more able students. Furthermore, the inductive approach may be more effective when studying complex concepts that make a high intellectual demand on the learner.

Koran, Koran, and Baker (1980) warn that inductive approaches must be carefully planned so that students, especially those of low ability, recognize the relevant stimuli that constitute the concepts and principles being taught. The low ability students need cues to direct their attention and to assist them in processing the information. These researchers reported success with an inductive cueing approach to teach the biological topic of monocots to ninth- and tenth-grade students. They used a slide-tape program

and preceded each group of slides with one of the following cues:

(a) "Pay attention to the appearance of the leaf views,"
(b) "Pay attention to the petal number," or
(c) "Pay attention to the fibrovascular bundle organization." (Koran, Koran, and Baker, 1980, p. 167)

The inductive or discovery method can increase student interest and motivation in science, but the right conditions must prevail. First, students must be able to discover the concept or principle under investigation. Then, they must be able to achieve this end with a modest amount of searching for relevant information. If the concept is too difficult, or if there is a great deal of irrelevant information, the discovery approach should give way to the expository method.

5 E Instructional Model.   The 5 E instructional model was developed for the elementary school science program Science for Life and Living by the Biological Sciences Curriculum Study organization (BSCS, 1990). This approach to teaching science through inquiry is an elaboration of the Learning Cycle and has application at the middle and senior high school, as is recommended by BSCS in their curriculum framework for teaching about the history and nature of science and technology (BSCS & SSEC, 1992). As the name implies, the 5 E model has five phases, which are described below.

Engagement.   The first phase of the model serves to capture students' attention and to engage them in the study of a concept, principle, issue, or problem. These activities can be in the form of a question, discrepant event, problem, puzzle, or any other strategy used to focus thinking. The engagement phase introduces students to the investigation that they will undertake and the manner in which they will go about their inquiry. During this phase the teacher guides students to make connections between what they know and what they are about to study.

Exploration.   During the second phase of the model, students carry out investigations. Students gather information, test out ideas, record observations, experiment, etc. This phase provides concrete, hands-on experiences that bring students into contact with the phenomena or situations they are studying. As a result, students should realize patterns and relationships, and raise questions. The teacher serves as a facilitator of learning, helping students to find out and discover ideas.

Explanation.   The third phase permits students to make sense from their explorations. They are encouraged to find patterns, relationships, and answers to questions. Students are urged to explain their findings and to demonstrate their understanding. The teacher uses questions to guide thinking and reasoning. In this phase, it is important for students to use their own words to explain ideas, which may be followed by more precise terminology put forth by the teacher.

Elaboration.   This phase in the instructional sequence gives students opportunities to apply to different situations the concepts and skills that they have acquired. The application of newly acquired information and skills is reinforced in new contexts, making learning more meaningful. Misconceptions are addressed. Cooperative group work and individual study are appropriate strategies to use at this point in the teaching sequence. The elaboration phase can take place over many periods of instruction and serve to enhance student cognition.

Evaluation.   The last phase of the 5 E instructional model calls upon students to demonstrate their knowledge and understanding of the ideas under study. These activities also serve to reinforce important learning outcomes. This assessment can be based upon the measurement of many types of cognitive and psychomotor behaviors such as producing written works, demonstrating laboratory skills and procedures, and completing projects.

The 5 E instructional model is similar to the conceptual change teaching sequence discussed in Chapter 3. The model offers science teachers a psychologically and scientifically based approach to instruction—one that holds the potential to promote inquiry as well as the attainment of science concepts.

Deductive Activities

The most extensively used method to teach science is the deductive approach. With this strategy, ideas are discussed using appropriate labels and terms, followed by experiences to illustrate the ideas. The deductive approach is a "vocabulary before experience" model of teaching or a lecture/laboratory approach. It is the opposite of the inductive approach. However, just because students do not have to discover patterns and relationships initially through exploration does not mean that they do not have to figure things out for themselves.

The deductive approach can be used to promote inquiry sessions and to construct knowledge. The first phase presents the generalizations and rules about the concept or principle under study, and the second phase requires students to find examples of the concepts or principles. The following example illustrates an inquiry session in teaching acid-base chemistry by beginning deductively.

Phase One.   Students are presented with definitions and properties of acids and bases. This presentation can begin with the definitions of acids and bases that are based on three theories. The first is the Arrhenius theory, which explains acids and bases in terms of ionization. The second is the Brønsted-Lowry theory, which defines these disjunctive concepts in terms of proton acceptors and proton donors. The third definition is based on the Lewis theory, which defines acids and bases in terms of electron-pair acceptors and donors.

The discussion can also include the attributes or properties of acids and bases that distinguish them from other chemical compounds in aqueous solution. For example, acids turn the litmus indicator red, taste sour, and neutralize bases; bases turn litmus blue, taste bitter, and neutralize acids.

The notion of pH might be introduced at this time to develop a quantitative idea of the acidity or basicity of a solution. Here, the pH scale would be discussed, stating that a pH of 7 indicates a neutral solution, a pH of 1 is a strong acid, and a pH of 14 is a strong base.

Phase Two.   After a lecture/discussion regarding the definitions and properties of acids and bases, students are ready to gather information about these concepts through laboratory activities.

▼ One of Mr. Beard's favorite laboratory experiments is on acids and bases. He introduces the laboratory experience by saying, "Today you are going to become experts on identifying acids and bases. On the laboratory bench to your right are many solutions, some of which are commonly found in a chemistry lab, some of which are commonly found in your home, and some of which are commonly found in industrial plants and factories. I want you to classify these solutions as either acids or bases and give their pH. You can use these acid-base indicators—methyl orange, litmus, indigo carmine—to determine which is an acid and which is a base and to determine their relative strengths. Analyze data and order your chemicals from the strongest acid to the strongest base. Then check the pH of these solutions with the pH meter that is on the front demonstration table and reorder the solutions if necessary." ▲

This method of teaching science through inquiry—a lecture/discussion phase followed by a firsthand experience phase—is a popular and common practice among science teachers. Nevertheless, the approach should be carried out with the spirit of "let's find out." It is an effective way to teach difficult content (Herman & Hincksman, 1978). With difficult subject matter, all students may not induce the rule or principle through an inductive approach; thus, a deductive approach may be more effective with the majority of students.

The traditional, lecture/lab, deductive approach is often overused in secondary school science teaching. Consequently, this method can become too routine and is often used when an inductive approach might be more effective. Furthermore, this approach loses some of its effectiveness when too much information is presented during the lecture phase. Perhaps elements of both the deductive and inductive strategies should be combined and used with the conceptual change strategy that was discussed in Chapter 3. Remember, the conceptual change strategy emphasizes that the teacher initiates instruction with brief *orientation* to introduce students to what they will be learning. Then the approach specifies that the teacher *elicit* from students what they know about the topic under study.

Gathering Information

Scientific inquiry includes more than constructing knowledge through hands-on activities. Laboratory work and experimentation are only some of the ways in which we find out about reality. A great deal of the inquiry that scientists and engineers carry out involves reading and conversing with others. Many of these professionals probably spend more time gathering ideas and information from literature sources and other people than they spend in the laboratory. Many laymen are also good inquirers. For example, successful business people are constantly looking for new products to make their businesses thrive. Thus, they converse with many individuals regarding products, marketing, and sales. Thrifty homemakers are constantly seeking food and clothing bargains to make sure their families have what they need—within the family budget.

Science teachers must encourage students to obtain information from a variety of sources at many points during the inquiry process. Information gathering can occur during the application phase of the learning cycle, for example, when sending students to read about a topic. Reading articles and reading the textbook may be appropriate at this point because the students have had firsthand experiences from which to relate. In other instances, the teacher may ask students to bring in newspaper clippings on a topic just as the class begins to study it.

Reading Printed Material.   Newspapers and magazines are rich sources of information for students to improve their scientific knowledge. One science course requirement can request students to cut out or photocopy articles and organize them into a notebook. Another technique to improve knowledge and understanding of a given topic is to require a short written report. These reports can be made from a single source or a few sources. When long reports are desired, students should be required to use a variety of sources such as newspapers, textbooks, magazines, encyclopedias, and journals. Students can research information in their homes, the school library, and the public library. They should be taught how to cite information sources in their reports. Effective science teachers require information gathering throughout the school year but are careful not to burden students with this type of work.

Assignments of this nature are not always successful because student motivation and competence are involved. Therefore, some science teachers identify topics that students can investigate before giving these assignments. They arrange topics on three-by-five index cards, and along with a topic title they list appropriate literature sources that are readily available to the students. Also, there may be several index cards with the same topic title, because there may exist a limited number of topics that can be researched on a given subject for a class of thirty students. Using this guided approach, students are led by the hand during their information gathering experiences. However, as students gain more experience and competence with this procedure, they require less direction and guidance.

Seeking Information from Individuals. Seeking information from individuals is an important aspect of knowledge acquisition and the scientific enterprise. People are a rich source of information and ideas. They can explain concepts to teenagers and improve their understanding of these ideas, often better than a textbook or science classroom explanations. Older siblings, parents, aunts, uncles, pharmacists, lawyers, nurses, doctors, fire fighters, engineers, construction workers, electricians, bakers, mechanics, coaches, musicians, and florists are among those with whom science students can interact to learn more about a topic. These people are often willing to spend time with young people to explain how something works or to clarify ideas and concepts.

Inquiry techniques that engage students in gathering the opinions of others is an excellent way for students to find out what others believe about issues and problems. This approach can also teach students how to develop questionnaires and survey instruments. Furthermore, this is an excellent way to make science relevant, and to illustrate its relationship to society.

▼   Ms. Gelespie encourages her life science students to find out what others think about various issues that relate to living organisms. She has found that when the students gather information, they feel as though they are conducting scientific inquiry. Therefore during the course, Ms. Gelespie requires her students to survey the opinions of others on at least four topics: nutrition, population, drugs, and chemical waste. She organizes students into groups to investigate these topics. Each group is assigned a particular aspect of the topic to investigate and they are encouraged to collect data from a wide variety of individuals such as peers, parents, and professionals in a given field. Students conduct interviews face to face, over the telephone, or with a survey form. For these data collection approaches, the teacher has the students develop a questionnaire to guide the interview process. Ms. Gelespie reports that her students enjoy this type of inquiry, especially using the telephone, which is one way to tap into what they do well—talk on the phone.   ▲

Science courses that require long-term investigation and problem solving encourage students to interact with a variety of contexts and people, causing them to become deeply involved in their science course work. This involvement creates a great deal of ownership of what they are studying because students invest a great deal of thinking, creativity, and time in their inquiries. At this level of involvement, most students demonstrate an intrinsic motivation to "find out and tell why."

Problem Solving

The problem solving approach to science instruction has the potential to engage students in authentic investigations and to develop their inquiry skills. This strategy can give students a feeling for science and engineering, and how professionals in these fields go about their work. Problem solving can also make science course learning more meaningful and relevant for teenagers.

Problem solving is often used synonymously with inquiry and science process skill reasoning (Helgeson, 1989). As such, this concept is associated with the nature of scientific inquiry as well as instructional methodology. The term problem solving will be used in this chapter to represent higher level thinking. Gagné (1977), for example, placed problem solving at the top of his learning hierarchy to illustrate how the learner must go beyond merely applying a rule to solve a problem (see Chapter 3). He pointed out that the end result of problem solving is when the learner actually discovers a higher order rule or generalization and constructs a new relationship and meaning for a concept under investigation. In addition, the term problem solving also will be defined as situations arise that are relevant to students' lives and which raise their doubt or uncertainty (Dewey, 1938). Problem solving will also refer to inquiry that engages students in investigations where they raise questions, plan procedures, collect information, and form conclusions. Problem solving can be short in duration or long, taking up to several months to complete. Furthermore, problem solving can focus upon many learning outcomes that are central to science and technology, such as those presented in figure 4–4. This strategy is not the sort of activity that directs students to answer questions at the end of the textbook chapter or to substitute numbers into a formula and compute the answer.

After studying problem solving, Watts (1991, p. 3) pointed out: "Problems come in all shapes and sizes, and from all areas of life.... They may be problems, or puzzles that need an explanation or simple 'itches' that may be the first steps in an investigation to be followed by a few well-thought-out tests to reach an answer." Watts also stressed that students who engage in problem solving take ownership for their learning and are motivated to find out.

In some science programs, problem solving can be used as a primary learning outcome as well as a major instructional strategy. In these programs of study, fundamental science subject matter is learned as the investigation ensues. Problem-solving-based science courses can sometimes be found in the middle school where the curriculum is more flexible and open-ended than in the high school.

▼ Ms. Janis teaches seventh-grade life science in a neighborhood where many blue collar families live. Part of the curriculum includes topics on nutrition and exercise. Ms. Janis begins the study of nutrition and exercise by referring to the bulletin board at the back of the room where the pictures of many teenage models, movie stars, and athletes are shown. The students are attracted to these pictures and like to identify with the people in them. Ms. Janis poses the following question: "How do these people get to be where they are in life?" This question brings responses from everyone in the class. Students have many opinions regarding fame and success, and are willing to express these ideas freely. During the discussion, invariably someone will mention a good diet and the advantages of exercise. The teacher then probes to find out what the students believe regarding a good diet and the diets of these famous people. Ms. Janis asks: "Do

FIGURE 4–4   Many types of problem solving investigations can be conducted in a science course; some focus on a small number of learning outcomes and may be short in duration, while others focus upon many learning outcomes and may require a long time to complete.

Short-term investigations take from part of a class period to a few periods to complete.

⟶

Science process skills

Science concepts   Science phenomena

Science topics   Technological devices   Design projects

Science/technology based services   Consumer projects

Science/technology/society issues and problems

⟶

Long-term investigations take many class periods to complete and may extend over many months.

you have the same diet as some of these famous people?"

This question leads to a great deal of uncertainty and many different opinions regarding what fashion models and body builders eat. Consequently, one of the first investigations that the students undertake is to keep a record of what they eat. They arrange this information so that the number of calories consumed and the percentages of fat, proteins, and carbohydrates in their diet are clearly displayed. A food diary is kept for at least three weeks by every student.

While the students are keeping their own nutritional diary, they are asked to inquire about the dietary needs and recommendations for young people and adults. The students use the school and city library to gather this information. They also ask people in the community for information regarding the topic. Some students interview nurses, dietitians, physical therapists, coaches, body builders, and doctors. The students are required to present the information and their sources. Tables and graphs are used in the reports.

The results are always startling to the students when they compare the nutritional balance in their diet with what professionals recommend. The students eat a great deal more fat than is recommended. Some students' diets consist of approximately 60% fat. These young people find it a bit unsettling to learn that their diets are so high in fat, especially when the recommended amount is

around 35%. Many of the boys in the class are shocked to learn about the diets of body builders who consume many calories; however, most of the calories are in the form of carbohydrates, with a relatively small percentage of fat and very little sugar. These activities lead many of the students to reconsider what they are eating and change their eating habits. As a result of these discussions, the food diaries of many students begin to show a different pattern of eating than at the beginning of the unit.   ▲

In contrast to the above example where a middle school teacher implements a problem-solving-based curriculum, let us examine a senior high school teacher's instruction where problem solving is used only occasionally. Consider the following example taken from Mr. Morton's high school chemistry class.

▼   After a short period of time into the study of water and aqueous solutions, one student brought up the question of the quality of the drinking water in the community. The student made statements to the effect that the tap water was contaminated and that it was unsafe to drink. Students reacted to this assertion, some seeming to agree with it, others thinking it was crazy and simply not true. This situation resulted in a heated debate and challenges that led the class to investigate the quality of the city's drinking water.

Mr. Morton helped the students turn their concerns about the quality of the drinking water into a research question: Is the city's water supply safe to drink? With this, students formed research groups to work on the question. Mr. Morton asked each group to design an investigation that they could carry out to answer the question. He wanted students to identify as many sources as possible to gather data regarding water quality. The students proved quite ingenious in this aspect of their inquiry. Some thought of going to the Water Department to find out how they analyze water to determine its purity. Some suggested that they take samples of their water and test it in the school's chemistry laboratory. One group thought of obtaining information from local environmental activists who have made claims that industry is contaminating the drinking water. Someone brought up the idea of looking into the city's health records for any problems that might be related to contamination of the drinking water. During the water quality investigation, which was conducted mainly after school, the students continued to study the textbook unit on water and aqueous solutions presented in the textbook.

When the students completed their investigations about the city water, they were given class time to present their findings and conclusions. As a result of the investigation, students seemed convinced that the water was safe to drink. Although they found that the water was not pure, the percentage of contaminants was small and did not seem to affect the health of the public. One group did find a contamination problem that occurred several years ago in a rural area where families were drinking from well water that contained a relatively high percentage of fecal matter. In all, the students became excited about their findings and admitted that they learned a great deal more about the chemistry of water as a result of this problem solving activity. ▲

GROUPING AND COOPERATIVE LEARNING

Effective science teachers often group students and assign them tasks in order to facilitate inquiry in the science classroom. They organize academic work because this approach seems to increase student involvement in the learning environment. When students have a specific task to carry out, they seem to have more direction and interest in their own learning. Grouping and role techniques have also been found to be a useful management strategy, because the teacher's role changes from a dispenser of information to a manager of student-directed learning where students tend to be more productive with fewer behavioral problems.

Grouping and assigning roles are a natural way to promote cooperative learning, all of which hold the potential to produce many important educational outcomes. For example, permitting students to work in groups to solve problems can promote scientific inquiry and develop in students a feeling for "doing" science. Cooperative learning can improve achievement and mastery of content (Slavin, 1989/1990). This approach can also develop team-building and a positive classroom environment (Kagan, 1989/1990). Cooperative learning, as its name implies, gets students to work together, eliminating some of the competitiveness and isolation that can exist in most academic environments. Hassard (1990b, p. 5) pointed out:

> Cooperative learning is a powerful learning model in which groups of students work together to solve problems and complete learning assignments. Cooperative learning is a deliberate attempt to influence the culture of the classroom by encouraging cooperative actions among students. Cooperative learning is a natural approach for the teaching and learning of science and is one that will be shown to be effective with a wide range of students.

Some of the critical elements of cooperative learning have been summarized by Watson (1992). He pointed out that cooperative groups have from two to seven students working together. Either the students can be assigned different tasks or each student can study the same body of information. Groups are generally heterogeneous with respect to academic ability, gender, or racial background. Watson advised

Grouping students and assigning roles to them should be used to facilitate inquiry teaching. When students have specific tasks to carry out, they seem to have more direction and interest in learning.

that incentives are necessary so that individuals and groups of students are rewarded for success. This reinforces learning and cooperation. However, students must be accountable for their personal achievements and contribute to the group. Experienced teachers know well that it is easy for some students to ride along on the efforts of others during group work; therefore, teachers who are just beginning to use this strategy should be on guard against this occurrence.

Cooperative instruction can take many forms, having no set number of steps to follow. Nevertheless, the following steps are discussed to highlight important aspects of this strategy.

Step 1:   Organize students into groups, using criteria to make decisions regarding this process. Determine the desired cognitive and affective outcomes for the investigation to be undertaken, then place students into groups accordingly. For example, if you wish to assign tasks to individual students, identify students in each group who can carry out the task and who also work well together with others in the group.

Step 2:   Identify ideas or topics that will motivate student inquiry. Some science teachers provide a preliminary list of ideas for their students that relates to the course or unit under study, focusing their thinking process. However, this approach should encourage brainstorming in order to identify additional ideas for student investigation. Problem solving can focus upon many ideas such as those listed earlier in figure 4–4 such as science concepts or principles, science topics, STS issues and problems, consumer products, and technological devices.

Step 3:   Ask each group to provide a preliminary outline of their project or study. This step immediately places students on a productive path. When you examine the outline, provide suggestions and guidance. Be sure each student in the group knows exactly what to do.

Step 4:   Monitor the investigations. You should have a good idea where each group is while the investigations are carried out. Some inquiries and projects will be conducted during class time, making them easy to monitor. Other investigations will take place after school and on weekends. For this type of work, take some time during class to ask for information to determine how groups as well as individuals are progressing.

Step 5: Help students to prepare their final reports so that they do well and feel good about their work. Help students form an outline for these reports and designate who will do each part of the write-up. The report is an opportunity for students to demonstrate their science process skill reasoning through the questions they attempt to answer, the inferences and hypotheses they form, and tables and graphs they construct to communicate their findings. This phase of the work is ideal for helping students represent knowledge, visualize models, give explanations, and demonstrate understanding.

Step 6: Assist each group to identify several, if not all, students to take part in presenting their report. This aspect of cooperative group work develops presentation skills and confidence in speaking before others. Try to avoid the same students making all of the presentations.

Step 7: Evaluate the investigations and projects. This often takes the form of assigning points to groups and individual students and entering them into the grade book. Generally, students put a great deal of effort into the activities so they should be rewarded accordingly.

To illustrate the cooperative group role assignment approach, consider the investigation of a consumer product Mrs. Cox proposed to her sixth-grade students.

▼ Mrs. Cox posed this question to the class: "These are five brands of paper towels I purchased in a supermarket. Which brand is the best?" She then divided the class into five groups of four students each. Mrs. Cox asked each group to brainstorm the question and come up with a list of characteristics that they would use to determine the quality of paper towels. She then asked each group to identify two characteristics they wanted to test to determine quality. The characteristics identified included softness, tensile strength, water absorption, and absorption of grease and oil. Mrs. Cox gave the following directions to one of the groups and similar directions but different assignments to the others:

- *Roger:* Organize the research project and be sure that everyone does his or her job.
- *Cindy:* Write up the report of the investigation and assign each person in your group a part in making the presentation to the class.
- *Barbara:* Gather the brands of paper towels to be tested and help Brian investigate the criteria established to determine the quality of paper towels.
- *Brian:* Work with Barbara to test the criteria used to determine paper towel quality. Gather materials necessary to the investigation. Clean up after the task has been completed.

The students obtained samples of the five brands of paper towels to test. They measured the absorption potential of the paper towels by rolling up a sample of each type and placing it into a beaker containing 100 mL of water. Each sample was treated in the same way. Students allowed their paper samples to absorb water for a 4-second interval. The amount of water absorbed by each sample was determined by finding the difference between the amount of water in the beaker before the towel was immersed and after it was removed (table 4–4).

Mrs. Cox used this problem and others that center around the claims made in advertisements and commercials to improve her students' critical thinking and problem solving skills. She found that this approach prompted many students to feel and think like scientists. When she used commercials and advertisements as the focus of the investigations, the students seemed motivated and interested in their work. Of course, after the students felt comfortable with the group investigative approach, there was much more stress on specific course content in subsequent inquiry sessions. ▲

Many years ago Seymour and Padberg (1975) emphasized the positive effects of group work in the science classroom in their research on the role approach. Their findings support those of other researchers who conclude that this approach improves communications, increases insight into scientific phenomena, develops cooperation and responsibility, and improves problem solving and task completion.

TABLE 4–4 Paper towel absorption test

Brand	Amount of water absorbed in mL			
	Trial 1	Trial 2	Trial 3	Mean
A	10.1	14.5	13.3	12.6
B	42.5	44.6	50.2	45.8
C	65.9	60.8	58.0	61.6
D	20.2	31.5	26.8	26.2
E	44.5	48.7	45.4	46.2

The Inquiry Role Approach (IRA) is a technique developed for secondary school biology courses at the Mid-Continental Regional Educational Laboratory in the early 1970s. It stresses inquiry in the laboratory by assigning students to specific roles. Biology content, scientific attitude, and inquiry skills are developed through group problem solving (Seymour, Padberg, Bingman, & Koutnik, 1974).

The IRA technique organizes students into teams of four. Each student is given one of the following roles: coordinator, technical advisor, data recorder, and process evaluator. Students change roles as they work on different investigations. In this manner they practice and acquire a variety of group problem solving skills. The IRA approach begins with a great deal of structure and teacher direction and moves toward more student-directed activities during the course, with the teacher playing more of an advisory role (General Learning Corp., 1974).

The Research Team Approach to Learning (ReTAL) is another group/role approach that has been used successfully with elementary, junior high, senior high, and college students. According to its developer, Bette DelGiorno (Del-Giorno and Tissair, 1970), ReTAL places emphasis on team learning and stresses an open-ended approach to science instruction. This technique is structured to fit into existing courses, extending science content and processes beyond a given textbook and traditional classroom experiences.

ReTAL consists of a four-phase process. The technique provides another method of facilitating inquiry through group procedures (DelGiorno, 1969). In addition, this is a deductive approach to inquiry that many science teachers may feel comfortable using. The phases of this technique are as follows:

1. *Lecture phase or the deductive process.* This is the telling phase where the teacher introduces a given science topic to the whole class. Key terms, concepts, and relationships are presented to the students.
2. *Independent study phase or the confrontation process.* The students are urged to find resource people, books, and films to extend their knowledge of the topic under investigation. The library, field trips, brainstorming sessions, and consultation with experts in the field are stressed.
3. *Experimental phase or the inductive process.* Students design and conduct investigations or create models and products. In general, they try out their ideas.
4. *Discussion phase or the evaluation process.* At this point, individual teams meet to discuss their work and to determine if further research is needed for their topic. When the team determines that their investigation is complete, they write a research report. The team reports can be given to the entire class in a final discussion of topic(s) in a large group setting.

The number of investigations that can be pursued using ReTAL in a science course is practically limitless. For example, here is a set of questions given in the ReTAL *Teacher's Guide* (DelGiorno, 1969) on the study of the earthworm:

How does the earthworm behave?
Response to light and darkness?
Response to hot and cold?
Response to wet and dry?
Response to seawater, fresh water?
Response to colors?
Response to sound?

Response to vibration?

Response to pressure?

Response to frustration?

Response to maze tasks?

CONSIDERATIONS FOR IMPLEMENTING INQUIRY INSTRUCTION

Some science teachers claim that they experience problems implementing the inquiry approach. Although they believe inquiry is important, they cite many obstacles in their attempts to use this form of instruction in the classroom. Some claim that this method is too time consuming and that it does not permit them to cover the required amount of subject matter. Furthermore, these teachers believe that they must teach many terms and definitions so that students will be successful on tests prepared by their science department, school district, or state education agency. Assessment is an important aspect of schooling, and it places a great deal of pressure on science teachers to teach toward the test, thus deemphasizing inquiry.

Some parents complain when science teachers attempt to use inquiry strategies and science programs that stress investigation, thinking, and discovery, especially where hands-on instructional materials are used in place of textbook-based courses. These parents want textbooks to be used and for their children to be drilled on the vocabulary words that relate to the topic under study. They also want a textbook that their children can take home to study. Many parents want their children to go to college and recognize only direct instruction of subject matter content as the most appropriate form of instruction for this preparation.

Materials and equipment are another aspect of teaching science through inquiry that can be problematic (Stake & Easley, 1978). Teachers who engage students in firsthand investigation must acquire and maintain materials. Funds are needed to buy equipment, which is often relatively expensive when purchased from science supply companies. Expendable materials are also needed throughout the school year. Middle school science programs often suffer from the lack of equipment to conduct a variety of science experiments. Furthermore, the facilities in many middle and junior high schools are not designed for laboratory activities to be carried on in all rooms where science is scheduled.

Some science teachers have found that their attempts to use inquiry methods have not been successful. They report that many students seem to be confused and do not learn intended outcomes. The opportunity to explore and find out in the traditional classroom environment can also lead to discipline problems because only the brightest students seem capable of handling investigative activities. Exploration and openness tend to remove control of the classroom environment from the teacher, allowing active and unruly students to be disruptive (Helgeson, Blosser, & Howe, 1977).

Throughout the years it has become obvious that many school systems are not organized to conduct programs that foster curiosity, discovery, and thinking. These systems often pay lip service to promoting creativity and learning how to learn, but they do not support their claims with an instructional program that fulfills these ideals. Their curricula are impoverished, and teaching for understanding is practically nonexistent. Many of the science teachers in these schools have little understanding of inquiry methods and how to use them. They are inadequately trained to teach science. Some science teachers have not participated in scientific research, and the science courses they took in college did not model scientific inquiry.

Learning psychologists have identified at least two inherent problems with the discovery process that must be considered by those who desire to promote inquiry teaching. These problems center around errors and time (Shulman & Keislar, 1968). Students who are left to discover ideas for themselves often make mistakes in their thinking (Glaser, 1968). They go up blind alleys in their search for answers to questions. Glaser cautioned educators about situations that involve too much exploration and error in learning. He emphasized that the teacher should maintain control over the learning environment to reduce errors and minimize student frustration, both of which can interfere with learning. The constructivist and conceptual change approaches that are recommended by cognitive psychologists and science educators support this view (see Chapter 3). These professionals emphasize that instruction must begin with what students know by providing them with opportunities to discuss what they believe to be true about a given situation, before students attempt to test out their ideas. Carefully prepared questions can guide student thinking. The introduction of terms at appropriate places in the instructional process can facilitate concept development. Therefore, teaching science through inquiry should be a guided approach to discovery rather than an unstructured search for answers (Wittrock, 1968). Furthermore, students should not be expected to discover all learning outcomes in the curriculum, because certain facts and concepts can be given to them at the proper time in the instructional process (Gagné, 1968).

SUGGESTED ACTIVITIES

1. Construct a file of activities that can be used to teach science through inquiry and identify them in your teaching area (i.e., life science, earth science, physics). These activities can be found in science textbooks, paperback books, and laboratory manuals. Some of the best inquiry-oriented activities will come to mind when you reflect upon what takes place in your everyday surroundings.

2. Obtain for your own professional library textbooks, paperback books, magazines, journals, and manuals that can provide ideas for inquiry sessions. These materials are good sources for discrepant events, problem solving, and inductive activities.

3. Plan and teach an inquiry lesson to your peers, or to middle school or senior high school students, that emphasizes one or more ways to initiate and carry out inquiry instruc-

tion: process skills, discrepant events, inductive activities, deductive activities, and problem solving. Participate in the critique and feedback of these teaching sessions with others.

4. Develop an inquiry activity that uses grouping, role assignments, and cooperative learning. Try out your activity with peers or middle or high school students to determine the effectiveness of this technique.

5. Conduct an analysis of laboratory exercises and other science instructional materials in order to determine the type of inquiry strategies emphasized and the extent to which they engage students in this method. Refer to Schwab's continuum to assist you in these assessments, as well as to the figures and tables in this chapter that explain inquiry.

BIBLIOGRAPHY

Bateman, W. L. 1990. *Open to question: The art of teaching and learning by inquiry.* San Francisco: Jossey-Bass.

Biological Sciences Curriculum Study. 1978. *Biology teacher's handbook.* New York: John Wiley & Sons.

Biological Sciences Curriculum Study. 1990. *Science for life and living: Integrating science, technology, and health, Grades K-6.* Dubuque, IA: Kendall/Hunt Publishing Co.

Biological Sciences Curriculum Study and Social Science Education Consortium. 1992. *Teaching about the history and nature of science and technology: A curriculum framework.* Colorado Springs, CO: The Colorado College.

Bruner, J. 1961. The act of discovery. *Harvard Educational Review, 31*(1): 21.

Bruner, J. S. 1966. *Toward a theory of instruction.* Cambridge, MA: Harvard University Press.

Champagne, A. B., and Klopfer, L. E. 1981. Problem solving as outcome and method in science teaching: Insights after 60 years' experience. *School Science and Mathematics, 81*(1): 3.

Chiappetta, E. L., and Russell, J. M. 1982. The relationship among logical thinking, problem solving instruction, and knowledge and application of earth science subject matter. *Science Education, 66*(1): 85.

Cothron, J. H., Giese, R. N., and Rezba, R. J. 1989. *Student research: Practical strategies for science classrooms and competition.* Dubuque, IA: Kendall/Hunt Publishing Co.

DelGiorno, B. J. 1969. *The research team approach (ReTAL): A structure for openness.* Fairfield, CT: Fairfield Public Schools.

DelGiorno, B. J., and Tissair, M. E. 1970, Dec. The research team approach to learning (ReTAL), part II. *School Science and Mathematics, 70:* 833.

Dewey, J. 1938. *Experience and education.* New York: Macmillan.

Fuller, R. G. 1980. *Piagetian problems in higher education.* Lincoln, NE: ADAPT, University of Nebraska.

Gabel, D. (Ed.). 1989. *What research says to the science teacher, volume 5: Problem solving.* Washington, DC: National Science Teachers Association.

Gagné, R. 1968. Varieties of learning and the concept of discovery. In L. S. Shulman and E. S. Keislar (Eds.), *Learning by discovery: A critical approach.* Chicago: Rand McNally.

Gagné, R. 1977. *The conditions of learning.* New York: Holt, Rinehart and Winston.

General Learning Corp. 1974. *Teacher's guide: Inquiry role approach.* Morristown, NJ: Silver Burdett.

Glaser, R. 1968. Variables in discovery learning. In L. S. Shulman and E. S. Keislar (Eds.), *Learning by discovery: A critical appraisal.* Chicago: Rand McNally.

Hassard, J. 1990a. Cooperating classroom. *Science Scope, 13*(6): 36–40.

Hassard, J. 1990b. *Science experiences: Cooperative learning and the teaching of science.* New York: Addison-Wesley.

Helgeson, S. L. 1989. Problem solving in middle school science. In D. Gabel (Ed.), *What research says to the science teacher, volume 5: Problem solving.* Washington, DC: National Science Teachers Association.

Helgeson, S. L., Blosser, P. E., and Howe, R. W. 1977. *The status of pre-college science, mathematics, and social science education, 1955–1975. Vol. 1, Science Education.* Columbus, OH: Ohio State University, Center for Science and Mathematics Education.

Herman, G. D., and Hincksman, N. G. 1978. Inductive versus deductive approaches in teaching a lesson in chemistry. *Journal of Research in Science Teaching, 15*(1): 37.

Hunkins, F. P. 1989. *Teaching thinking through effective questioning.* Boston: Christopher-Gordon Publishers.

Johnson, D. W., and Johnson, R. 1975. *Joining together: Group theory and group skills.* Englewood Cliffs, NJ: Prentice-Hall.

Johnson, D. W., Johnson, R., Holubec, E., and Roy, P. 1984. *Circles of learning.* Alexandria, VA: Association for Supervision and Curriculum Development.

Johnson, D. W., Maruyama, G., Johnson, R., Nelson, D., and Skon, L. 1981. Effects of cooperative, competitive and individualistic goal structures on achievement: A metaanalysis. *Psychological Bulletin, 89:* 47–62.

Kagan, S. 1989, Dec./1990, Jan. The structural approach to cooperative learning. *Educational Leadership, 47*(4): 12–16.

Koran, J. J., Koran, M. L., and Baker, S. 1980. Differential response cueing and feedback in the acquisition of an inductively presented biological concept. *Journal of Research in Science Teaching, 17:* 167.

Lawson, A. E., Abraham, M. R., and Renner, J. W. 1989. *A theory of instruction: Using the learning cycle to teach science concepts and thinking skills.*

NARST Monograph, Number One. National Association for Research in Science Teaching.

Lawson, A. E., and Wollman, W. W. 1975. Physics problems and process of self-regulation. *The Physics Teacher, 13:* 470.

Lampkin, R. H. 1951. Scientific inquiry for science teachers. *Science Education, 35:* 17.

Liem, T. L. 1987. *Invitations to inquiry.* Lexington, MA: Ginn Press.

Martin, M. 1985. *Concepts of science education.* New York: University Press of America.

Piaget, J. 1971. *Biology and knowledge.* Chicago: University of Chicago Press.

Rakow, S. J. 1986. *Teaching science as inquiry.* Bloomington, IN: Phi Delta Kappa, Fastback 246.

Renner, J. W., Abraham, M. R., and Birne, H. H. 1985. The importance of the form of student acquisition of data in physics learning cycles. *Journal of Research in Science Teaching, 22:* 303–325.

Renner, J. W., Cate, J. M., Grzybowski, E. B., Atkinson, L. J., Surber, C., and Marek, E. A. 1985. *Investigation in natural science: Biology teacher's guide.* Norman, OK: Science Education Center, College of Education, University of Oklahoma.

Schwab, J. J. 1960. Enquiry, the science teacher, and the educator. *The Science Teacher, 27*(6): 6–11.

Science Curriculum Improvement Study. 1974. *Science Curriculum Improvement Study (SCIS): Teacher's handbook.* Berkeley, CA: University of California at Berkeley, Lawrence Hall of Science.

Seymour, L. A., and Padberg, L. 1975. The relative effectiveness of small group and individual settings in a simulated problem solving game. *Science Education, 59:* 297.

Seymour, L. A., Padberg, L. F., Bingman, R. M., and Koutnik, P. G. 1974. A successful inquiry methodology. *The American Biology Teacher, 36*(9): 349.

Shulman, L. S., and Keislar, E. S. (Eds.). 1968. *Learning by discovery: A critical appraisal.* Chicago: Rand McNally & Co.

Slavin, R. E. 1989, Dec./1990, Jan. Research on cooperative learning: Consensus and controversy. *Educational Leadership, 47*(4): 52–54.

Stake, R. E., and Easley, J. A. 1978. *Case studies in science education.* Urbana, IL: University of Illinois Center for Instructional Research and Curriculum Evaluation.

Suchman, R. 1966. *Developing inquiry.* Chicago: Science Research Associates.

Suchman, R. J. 1975. The Illinois studies of inquiry training. *Journal of Research in Science Teaching, 2:* 230.

Tamir, P. 1975, Jan. Invitations to inquiry and teacher training. *American Biology Teacher, 37:* 50.

Watson, S. B. 1992. The essential elements of cooperative learning. *The American Biology Teacher, 54*(2): 84–86.

Watts, M. 1991. *The science of problem solving.* Portsmouth, NH: Heinemann Educational Books.

Welch, W. W. 1983. Experimental inquiry and naturalistic inquiry: An evaluation. *Journal of Research in Science Teaching, 20:* 95.

Welch, W. W., Klopfer, L. E., Aikenhead, G. S., and Robinson, J. T. 1981. The role of inquiry in science education: Analysis and recommendations. *Science Education, 65:* 33–50.

Wittrock, M. C. 1968. The learning by discovery hypothesis. In L. S. Shulman and E. S. Keislar (Eds.), *Learning by discovery: A critical appraisal.* Chicago: Rand McNally.

Wright, E. L. 1981. Fifteen simple discrepant events that teach science principles and concepts. *School Science and Mathematics, 81*(7): 575.

Demonstrations are commonly used in science instruction. They can be planned to illustrate concepts and principles, to gain students' attention, and to initiate inquiry.

CHAPTER 5

Demonstrations

A science demonstration is a powerful instructional strategy. This strategy can be planned to increase students' interest in a lesson, illustrate a concept or principle, make a point, answer a question, review an idea, initiate inquiry and problem solving, or introduce a unit or lesson. Whatever their function, demonstrations should be well planned in advance for use in instruction. Carefully planned demonstrations can greatly enhance the learning environment and make special contributions to the teaching of science.

Overview

- State and explain the functions of science demonstrations.
- Illustrate and discuss the use of demonstrations to correct misconceptions.
- Explain the advantages and disadvantages of demonstrations.
- Discuss the important factors to consider when planning a demonstration.
- Provide suggestions for presenting effective demonstrations.
- Propose ideas for special equipment for certain types of demonstrations.
- Indicate sources of tested demonstrations.
- Explain how videotapes and video cameras can be used in demonstration work and in science instruction.
- Present an evaluation procedure to assess the strengths and weaknesses of a demonstration.

▼ Mrs. Gross, an eighth-grade science teacher, asked her students to immediately take their seats when they entered the classroom and remain quiet. Without saying a word she proceeded to pour water from a bottle containing a little phenolphthalein into another bottle containing a small amount of sodium hydroxide. The students observed that the liquid changed color and became bright pink. She then poured the pink liquid into a bottle that contained hydrochloric acid. When the pink liquid became colorless the students became very curious and asked Mrs. Gross for an explanation but she continued to remain silent. Instead she passed out to each student samples of common cleaning agents normally found around the house, such as liquid soap, ammonia, and window-cleaning fluid. Mrs. Gross finally broke her silence by asking each student to add a small quantity of phenolphthalein to each liquid and observe and record their observations. Mrs. Gross did not discuss their results at this point even though they asked many questions. She then gave each student samples of acidic liquids, including vinegar, orange juice, and lemon juice. She asked the students to add a little phenolphthalein to each sample and record their results. The students became more curious and Mrs. Gross was easily able to introduce the topic of acids and bases.   ▲

This example shows how a demonstration can introduce students to a new topic. The type of demonstration presented was a discrepant event used to arouse student interest and curiosity. Thus motivated, the students conducted laboratory exercises in which they obtained further information about identifying acids and bases using chemical indicators.

The following example illustrates the use of the discrepant event to introduce a unit on the earth's atmosphere in an earth science course.

▼ Mr. Case always initiated the topic of the earth's atmosphere with this demonstration. He took two sheets of newspaper and a thin piece of wood about 4 inches wide and 2 feet long (figure 5–1). Mr. Case then laid the piece of wood on the

FIGURE 5–1 Demonstration to introduce the topic of the earth's atmosphere. When the wood strip is struck with a hammer, the strip breaks. The students, however, expect the newspaper to fly into the air.

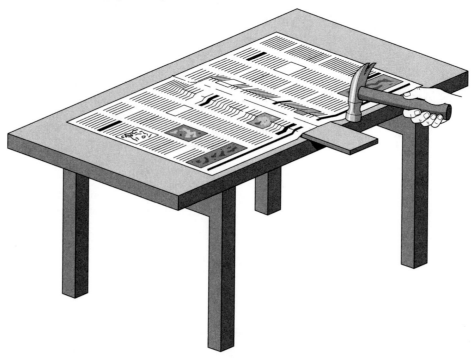

table so that 6 or 7 inches projected over the edge. He then placed the two sheets of newspaper over the wood strip about an inch back from the edge of the table. After carefully spreading and smoothing the newspaper over the strip and the table, Mr. Case told the students that he was going to strike the wood strip with a hammer on the section that extended over the table's edge. He asked the students to predict what would happen to the newspaper after he struck the wood. Most students expected the newspaper to fly into the air but were surprised to see that the wood strip broke instead. Mr. Case followed this event with a series of questions, asking the students to describe what had happened and to infer why the wood strip broke. Mr. Case directed their thinking and eventually led them to consider air pressure as an explanation for the event. Mr. Case explained that air exerts a pressure of approximately 15 lbs per square inch at sea level, and this pressure was exerted on the top surface of the two sheets of newspaper. The students were amazed to learn how much pressure was exerted on the surface of the newspaper after they calculated it. The students then proceeded to draw their own conclusions. ▲

Demonstrations can be very effective teaching strategies in science instruction and provide excellent ways to introduce science units and lessons. Demonstrations are concrete experiences that can be considered advance organizers for structuring subsequent information and activities into a meaningful, instructional framework for students. Effective demonstrations can focus students' attention, motivate and interest them in a lesson, illustrate key concepts and principles, and initiate inquiry and problem solving. When used discriminately, they enhance the learning environment. When used indiscriminately or as a substitute for laboratory work, their purposes can be defeated.

FUNCTIONS OF DEMONSTRATIONS

Demonstrations may be used in several ways in science instruction, each of which makes its own special contribution. They can be used to initiate student thinking and to formulate problems. They can be planned to illustrate a point, prin-

ciple, or concept or to answer a particular question. They are also useful for reviewing and reinforcing an idea. They are excellent to use to introduce a lesson or a unit of study. They can be employed to address misconceptions.

Initiating Thinking

Many science demonstrations, when presented for the first time to a group of students, will leave them searching for an explanation. This type of demonstration, often called the *discrepant event,* can stimulate student interest and curiosity. To achieve the atmosphere of inquiry, the teacher must present the event as a problem to be investigated. "The intuition-offending event must be presented to the learner in such a way that the science principle underlying the event is not immediately revealed" (Liem, 1987). Psychologists believe that discrepant events promote higher-level thinking and meaningful learning, and, when properly presented, students will have better recall and retention of information. Even a simple presentation can initiate student inquiry.

Illustrating a Concept, Principle, or Point

Demonstrations are concrete experiences that can be used to illustrate a science concept, principle, or point. To illustrate a concept, for example, some teachers first state the concept and then parallel the statement with a demonstration that illustrates the idea. This is a direct and explicit way to clarify and explain ideas. This approach is used *only* to illustrate the concept, law, or principle and not to develop the idea.

Consider a demonstration to illustrate what effects different concentrations of ions have on conductivity, as well as to reinforce the principle that conductivity is related to the ion concentration in a solution. The materials and apparatus are set out on the demonstration table as shown in figure 5–2. The teacher first tells the students that the conductivity is directly related to the ion concentration of the solution and that if the ion concentration is sufficient, an electrical current can flow, causing the light bulb to glow. The

FIGURE 5–2   Electrolyte demonstration to illustrate the effects of different ion concentrations on conductivity

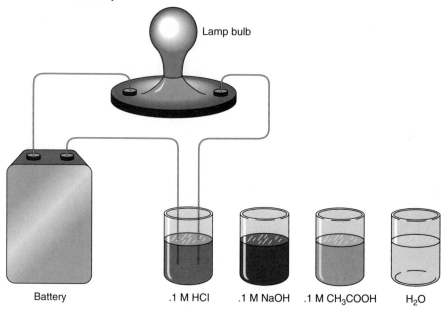

Lamp bulb

Battery   .1 M HCl   .1 M NaOH   .1 M CH₃COOH   H₂O

teacher describes the ionic properties of each of the solutions (.1 M HCl, .1 M NaOH, .1 M CH$_3$COOH [acetic acid], and H$_2$O) and then inserts the two electrodes to test conductivity. The teacher records the results of these tests on the chalkboard, and writes summary statements to reinforce the principle that conductivity is related to the ion concentration in a solution.

Probably the most common use of a demonstration is to illustrate a point. In such cases, the demonstration is usually not part of the lesson plan. Often, the apparatus has to be set up on the spot and in a matter of minutes. If the materials and/or apparatus cannot be gathered quickly, it is wise to postpone the demonstration until the next day, so that the continuity of the lesson is maintained and the students' attention is not lost. The disruption of a lesson for long periods of time can cause students' inattention and classroom management problems.

▼   The students had observed the eclipse of the moon the previous night and asked the teacher to explain the event. Mrs. Scanlon proceeded to set up a readily available model to demonstrate the relative positions of the earth, moon, and sun during an eclipse.   ▲

Answering a Question

Students sometimes ask questions that can be answered easily with a demonstration. These situations usually occur spontaneously and require teachers to have a large repertoire of demonstrations from which to draw. Such demonstrations should not be conducted spontaneously unless the teacher is assured that the demonstration will be presented successfully and that the materials are readily available. Collecting materials for an unplanned demonstration can waste class time. Perform the demonstration the next day if preparation is a problem.

▼   During a unit on oceanography, the topic of skin diving arose, and a student asked about the bends. Mrs. Rose tried to explain that a case of the bends was due to the release of nitrogen gas bubbles in the blood caused by a person ascending too quickly from an atmosphere of high pressure (greater depth) to an atmosphere of ordinary pressure (surface). Realizing that her statement had little meaning to the students, she recalled a demonstration but decided to perform it the next day because materials were not readily available.

The next day Mrs. Rose conducted the demonstration at the beginning of the period. She placed a beaker of water in a vacuum chamber. When she turned the vacuum pump on, the water appeared to boil. She indicated that the "boiling" was due to gases coming out of the solution because the air pressure in the vacuum chamber was lower than the pressure in the water. She explained that the bubbles released were analogous to the nitrogen bubbles that are formed in a diver's body as the diver moves up from an area of high pressure (greater depth) to a region of low pressure (surface). ▲

Reviewing Ideas

Demonstrations can be the center of excellent review sessions when combined with oral recitations. They can be performed to reinforce important ideas that were previously addressed during a lesson or laboratory exercise.

▼ A chemistry class attempted to copperplate by electrolysis during a laboratory period. Some students performed the procedure successfully; others did not. Two weeks later, during a recitation session, Mr. Wexler presented the same laboratory exercise as a demonstration using the same materials. During the demonstration, Mr. Wexler asked a number of questions relating to the reactions involved in electrolysis. Many of the students were able to clear up a number of points that had previously confused them about the process. ▲

Introducing and Concluding Units

As mentioned earlier, carefully planned demonstrations can be used to introduce new topics and provide stimulating introductory discussions. They can be presented to explain and clear up concepts that have already been taught or are being taught. They can promote creative thinking on the part of the students by presenting them with a related or an apparently unrelated event for explanation.

Interesting and exciting demonstrations are excellent ways to conclude units of study. They can serve as a climax as well as a summary of the high points of a unit. They can also motivate and interest students in further study.

▼ To end a unit on air, Mr. Kuhn usually gave a demonstration involving the preparation of oxygen. After he collected several bottles of pure oxygen, he demonstrated how chemical substances such as sulfur and magnesium burned in an atmosphere of pure oxygen. The students were amazed to see the different ways certain substances burned in oxygen. The teacher then explained that high school chemistry involves many laboratory exercises and experiments that are equally exciting. ▲

Addressing Student Misconceptions

Many teachers have the notion that students have no preconceived ideas about scientific concepts and phenomena. In reality, students are *not* devoid of such ideas and often possess misconceptions regarding certain scientific principles and phenomena. Demonstrations can be used effectively to address such misconceptions. In this type of instruction, teachers can elicit predictions from their students regarding a scientific principle or phenomenon, and request explanations for their predictions. After students' conceptions have been put forth and recorded on the chalkboard, a demonstration can be performed that will contradict the students' predictions. This contradiction not only motivates students to resolve the discrepancy but also allows them to formulate their own logical explanation for the phenomenon.

▼ To introduce the concept of burning to her eighth-grade science class, Mrs. Tripp gave the students a pretest to identify their preconceptions of burning. The pretest results showed that most students believed that all types of burning involve the production of gas and that this gas is released into the atmosphere and therefore the burned material loses weight. Following the pretest, Mrs. Tripp conducted a demonstration specifically designed to address this misconception about burning. She placed an equal-arm balance on the demonstration table and suspended a piece of steel wool from each arm so that the arms balanced. Mrs. Tripp then asked the students to make predictions regarding which direction the balance would lean if she burned one piece of the steel wool.

The students were asked to record and explain their predictions on a worksheet she had provided. She proceeded with the demonstration by directing the flame of a gas burner at one piece of steel wool. The students observed that the balance arm tipped down on the side that was burned, indicating that the mass of the steel wool increased when it combined with oxygen. The students recorded their observations and tried to come up with an explanation for the discrepancy between their prediction and the result of the demonstration. Finally, through discussion and appropriate questioning, Mrs. Tripp guided her students to the conclusion that burning does not always involve a decrease in mass. Her students observed that burning is a reaction of a substance with oxygen to form new substances. (See BouJaoude, 1991.)   ▲

This example illustrates the use of a discrepant event to modify a misconception. The demonstration was the strategy used along with probing questions and discussion to lead students to more valid conceptions. Often, teachers will demonstrate the discrepant event to arouse interest and curiosity, and then ask the students to conduct a laboratory exercise to change the misconception.

▼   Mr. Green's seventh-grade science class was studying a unit on earth and space that included a lesson on the phases of the moon. To introduce this lesson, Mr. Green asked the students if they had observed the moon over the past several evenings. Several students said that a half moon was now visible and that a week ago the moon was less than half. Mr. Green asked if they knew why the moon had different "shapes" or phases. Many students attempted to answer the question, asserting that the earth's shadow on the moon causes the various phases. To alter this misconception, he performed a simple demonstration. To represent the moon he used a grapefruit-sized styrofoam ball in which he had inserted a pencil to use as a handle.

A table lamp provided a light source to represent the sun. Using the students to represent the earth, he asked them to stand in front of the light source and then darkened the room. He then moved the ball in a counterclockwise direction around the source of light. He started in the new moon position where the ball was directly between the students and the light source. As the ball was moved around the light source the students were able to observe the complete cycle of the phases of the moon. Mr. Green completed the lesson by discussing the demonstration and by using questioning strategies to reinforce their knowledge.   ▲

ADVANTAGES OF DEMONSTRATIONS

Demonstrations have several advantages that make them highly useful in teaching science. They can be used to motivate students and gain their attention. Students understand and remember better what they see firsthand rather than what they read or what is described to them. Demonstrations allow the teacher to guide and channel student thinking to produce certain desired learning outcomes. They permit teachers to conduct activities that are unsafe for students to perform. Finally, they are economical of materials and apparatus as well as of class and teacher time.

Student Interest and Attention

A prerequisite for effective learning is to get students' attention and involvement. Demonstrations are probably the best strategy to employ to accomplish this. They can provide an exciting event that will capture most students' attention. And, if conducted properly, a demonstration can involve all the students in attempting to answer a question or in observing an event. Demonstrations can orchestrate questions, explanations, and concrete materials to produce a meaningful presentation that actively involves all or most of the students in a science class.

Guiding Thinking

Demonstrations are useful to guide and channel students' thinking. They allow teachers to guide thinking by arranging the external and internal conditions of learning in a classroom through the use of concrete materials and carefully planned questions. Students' thinking can be focused directly on the intended learning outcomes, such as illustrating a laboratory technique or science principle. Or students' thinking can be guided to discover a relationship or to answer a question through inquiry. A demonstration gives a teacher more control over the learning environment than does either a lecture or laboratory work.

Safety Considerations

Demonstrations enable the teacher to conduct activities that may be too dangerous for students to carry out themselves. The teacher may decide that laboratory conditions in a middle school science classroom are inadequate for allowing students to work with open flames or to heat liquids. A laboratory exercise that requires students to boil water in a beaker, for example, can be transformed into a demonstration that the teacher performs, thus precluding the possibility of students knocking over beakers of boiling water. Some chemical reactions produce strong and toxic fumes. A chemistry teacher can demonstrate these reactions under a well-ventilated hood rather than permit each student to conduct the exercise. It may be better to use high-voltage apparatus and radioactive materials in demonstrations rather than in laboratory exercises conducted by students. It is always advisable for the teacher to be safety conscious at all times and make students aware of the way in which safety precautions are being carried out.

Economizing Resources

Some items are too expensive or delicate for general use. Because of cost, many high school physics departments, for example, have only one oscilloscope, which is used to analyze electrical impulses in large group presentations. Similarly, biology laboratory apparatus used to measure and record physiological processes are expensive, and few schools have even one. Precision analytical balances are usually found in small numbers in chemistry laboratories and are used by chemistry teachers primarily for demonstration purposes. A human skeleton is useful in biology teaching but, because of its high cost, it is not often found in many laboratories. When supplies and equipment are limited, teachers must often use a demonstration instead of laboratory activity.

Economy of Teacher Time and Effort

A demonstration may save the teacher considerable time and effort. It is certainly easier to prepare materials for one laboratory activity than for twenty duplicate ones. It is also easier to perform one activity than to supervise twenty activities. And it is also more convenient to store materials needed for one activity than for twenty. However, this type of economy has its drawbacks. If substituting a demonstration for individual laboratory work results in lessened interests and understandings, the savings in time and effort may actually be costly.

Economy of Class Time

A demonstration can also save class time. Some laboratory activities can be given as demonstrations without any loss of student understanding. This eliminates time spent setting up apparatus, disseminating and collecting materials, and giving directions. The teacher can also perform the demonstration more smoothly, quickly, and efficiently than the students. Smoothness and swiftness of the pace of a demonstration might defeat the desired outcomes, however, because students may not necessarily understand what is taking place if they cannot carefully observe each step during the event.

LIMITATIONS OF DEMONSTRATIONS

Demonstrations also have limitations that make them useful for only certain types of learning situations. Visibility is always a problem when demonstrations are presented. The use of science materials and apparatus and complicated assemblages may cause students to become confused and prevent them from understanding relationships. The lack of student participation in a demonstration can reduce student interest, understanding, and feeling of involvement.

Visibility

All students must be able to see the details of the apparatus being used and all the details of the procedures and results if they are to benefit from the demonstration. This may not be possible when the demonstration is conducted at a fixed location in the classroom, or if students are seated at some distance from the demonstration table. In certain instances in which visibility is questionable, the teacher may wish to assemble the students around the demonstration site so that they can see all the details of the apparatus and observe the events while the demonstration is in progress. Passing materials around to the students beforehand may be an alternative to gathering around the demonstration site. But if there is a doubt about the scale of materials used, the teacher should consider conducting

the activity as a laboratory exercise if cost and safety permit.

▼   As part of a unit on digestion, Mrs. Coleman explained to her seventh-grade science class that the presence of starch in food substances can be determined by placing a small amount of tincture of iodine on the substance and observing the color change to blue-black. She indicated that she was going to place pure corn starch in a test tube half-filled with water and then add two drops of iodine to the solution so they could observe the color. She waved the test tube in front of the students and announced that the color had changed to blue-black. She then repeated the test using a slice of potato, but was stopped when the students informed her that they could not see the color change and wanted a closer look. She then passed the test tube around the class before she proceeded with the next phase.   ▲

Few science classrooms are equipped with a closed-circuit TV system, but the use of such a system has the advantage of presenting close-up views of demonstration apparatus and reactions that cannot be seen easily from the back of the room. Several TV monitors placed in strategic areas throughout the classroom will provide easy viewing.

▼   Mr. Johnson wanted to demonstrate standing waves in a glass cylinder (see figure 5–3). He used an audio oscillator, an amplifier, and a speaker, assembling them as follows. He placed cork dust in the glass cylinder and pressed the cone of the speaker against one end of the tube. He closed off the other end with a rubber stopper. When he operated the speaker, standing waves were set up and the cork dust accumulated at the nodes. He

FIGURE 5–3   Apparatus for demonstrating standing waves in a glass cylinder

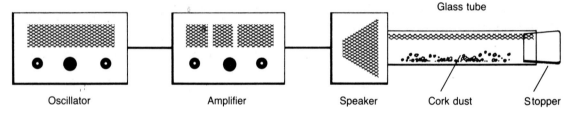

Oscillator   Amplifier   Speaker   Cork dust   Stopper

changed the frequency of the oscillator, which caused the cork dust to become rearranged in the glass tube. He used an overhead projector so that the whole class was able to observe what was taking place in the glass tube. The class watched the cork become rearranged in the tube when the speaker was operating.

Many students did not understand how or why the amplifier and oscillator were used. They asked many questions, which indicated to Mr. Johnson that they were confused and lacked an understanding of what had taken place during the demonstration. He decided to use some students to repeat the demonstration, while he carefully explained the function of the amplifier and oscillator during their presentation. Mr. Johnson encouraged the class to ask questions during the presentation to clear up any problems.   ▲

Student Understanding

Some demonstrations are presented so quickly that students miss the point. In such instances the teacher fails to involve the students in the demonstration and does not check for their understanding of what has occurred. Most students are reluctant to ask questions when they have been unable to follow the steps during the demonstration. They become confused and miss important points. The swiftness of pace and the smooth way in which demonstrations are presented may often be deceiving to students and, consequently, the desired outcomes are not produced.

Maintaining Student Interest

There are few opportunities for active student involvement during a demonstration. It is difficult for the teacher, for example, to control mental participation of students while they are inactive in their seats. As a consequence, the teacher often loses their attention during a demonstration unless he makes provision to involve them through questioning procedures and other techniques to maintain their interest and assure their understanding. Inattentiveness and lack of interest on the part of students can defeat the purposes of a demonstration.

Elaborate Equipment

The use of elaborate or professionally made apparatus adds a note of authority and often makes the results of a demonstration difficult to question. Elaborate demonstrations that require complex equipment and complex procedures are often very convincing to students. This situation may cause students to become uninvolved in the demonstration and misunderstand its purpose.

Controlling Discussions

It is not uncommon for a few students having special abilities or interests to dominate discussions resulting from a demonstration. In such cases, many students are left out and do not feel involved. These situations are to the detriment of many students and the teacher must recognize them to allow maximum participation of all students.

PLANNING A DEMONSTRATION

A demonstration is a performance and many factors must be considered before the event takes place. One of the first things a teacher has to decide is whether the strategy is the best way to handle a particular topic. After this has been decided, then factors such as gathering suitable equipment, visibility, audibility, student participation, contrasts, and climaxes have to be well planned by the teacher in order to have a successful demonstration.

Preliminary Considerations

The first thing that a teacher should consider when planning a demonstration is whether the strategy, with all its disadvantages, is the best way to address a certain topic. It certainly would be inadvisable, for instance, to use the demonstra-

Teachers with a great deal of ingenuity can construct models and special apparatus for demonstration purposes. Teachers can also suggest ways in which students can participate in this type of activity.

tion to show students the test for the presence of starch in carbohydrates using iodine when a laboratory exercise would be more effective. In this instance, visibility must be taken into consideration because the results of the test cannot be seen easily by students seated in the back of the room. Also, the exercise is not costly or dangerous for students to perform.

A second point to consider is whether the topic lends itself to the demonstration strategy. Are there elements of surprise or suspense? Will the demonstration take place at a reasonable pace? Is there enough variety to maintain interest while the procedure is taking place? Are there long periods of inactivity (e.g., boiling water or extensive filtrations)?

A third consideration is the size of the materials. Can the equipment be seen by all students? A demonstration is useless unless all students can see what is taking place step by step and observe the results so they can draw their own conclusions.

Remember that a demonstration is presented because it may result in benefits to students, not

because it can fill in a gap in a lesson. The demonstration has to be planned so that its primary objective is to produce permanent changes in behavior in students.

Gathering Materials and Equipment

Materials and equipment for a demonstration should be planned and collected well in advance of the actual presentation. Last-minute preparation may prove frustrating and cause either a delay of the presentation or no presentation at all. For example, the teacher may find that the apparatus has been lost or damaged in storage. Chemicals may be too old, exist in the wrong concentration, or be in short supply. When special equipment is required, the teacher may have to construct it or order it from a supply house. If materials have been ordered, enough time should be allowed for delays in shipping. In other words, potential problems should be recognized well in advance so that the demonstration can take place as planned. It is good policy to inventory supplies and equipment early each year so that materials can be ordered well before they are needed.

Ensuring Visibility

Small items should not be used for demonstration purposes unless the teacher makes special provisions to make them visible. Often overhead projectors are used for this purpose. Simple but large-scale apparatus are best for demonstrations, but even when large equipment is used, the teacher must take care to ensure that small and important details are made visible to students.

The apparatus and/or materials on the demonstration table must be arranged to avoid blind spots. Some blind spots are due to the relative positions of the students to the table and some are caused by the way materials are placed on the table. Before presenting a demonstration, the teacher should view the setup from various areas in the classroom to determine whether there are problems in viewing.

The background behind the apparatus is an important consideration. A partially erased chalkboard makes a very unsuitable background; on the other hand, white or colored cardboard or cloth backgrounds with suitable supports provide adequate contrast for viewing.

Along with background, proper lighting is also necessary. Backlighting to silhouette the apparatus can be very effective for viewing demonstration materials. A brilliantly lighted backdrop behind an apparatus will increase the contrast between opaque and transparent objects.

Spotlights properly arranged to focus on demonstration apparatus can produce unusual results. In addition, color filters used with spotlights can result in dramatic effects. In instances where spots are located directly above the demonstration table, side lighting may be needed to help students distinguish various elements of the apparatus setup.

A demonstration setup can also be arranged in areas other than the demonstration table. Set-ups on low tables or the floor permit students to view them from a different perspective. Materials suspended from the ceiling permit an upward view. Large-scale materials can also be assembled outdoors where students can move around them.

Focusing Student Attention

The demonstration should be set up in an uncluttered setting to allow students to focus their attention on the details of the procedure. Some teachers start with a bare demonstration table and then proceed to remove needed items from a container or other source. Items taken from a box, the contents of which are not visible, can introduce an element of surprise, for students will continue to wonder what will be removed next and how it will be used in the demonstration.

If a large-scale apparatus is required for the demonstration, some teachers assemble it first and then cover it with a cloth or large cardboard box before the students enter the room. The students will automatically focus attention on the "unveiled" setup.

Another technique used to focus attention on the demonstration is to set in operation, before class begins, a scientific novelty or device that is used for the demonstration. The students will immediately become interested in what is to take place during the lesson.

Planning Variety in Demonstrations

Variety can be accomplished by modifying the demonstration procedures. Each modification can have different effects on students. Variety can be provided as follows

1. The teacher may present a demonstration already viewed by students.
2. The teacher may set up a demonstration at the suggestion of students in order to solve a problem that has been generated through discussion.

3. A student may present a previously viewed demonstration. The student may have to practice the demonstration after school to present it during the class period.

4. A student or several students can carry out a demonstration under the teacher's direction. In this case the student or students have no idea of what is involved in the procedure until the teacher gives them oral directions. A demonstration conducted by students should be simple and clear so that the procedure or series of related procedures is easily carried out.

5. A student or teacher from another classroom or an outside expert may be used to present a demonstration. Guests should be selected carefully. Experts may use vocabulary unfamiliar to students, or they may pace a demonstration so quickly that students do not obtain the basic understandings while it is being presented, or they may not even recognize the purpose of the demonstration.

6. The teacher can use special equipment such as an opaque projector, video camera, or microprojector to add novelty and surprise during the demonstration procedure.

7. The teacher can present demonstrations outdoors or in another room or in the corridor, which can be effective in capturing students' attention.

8. Unusual noises, lights, or motions can motivate and interest students. The occasional use of such devices must be justified in terms of the amount of time and energy required during planning and the desirable outcomes that are attained by students.

Trying Out the Demonstration Beforehand

Last-minute preparations will normally result in confusion and difficulties. As mentioned earlier, a teacher who scurries about looking for a piece of equipment, or cuts a piece of glass, or looks for misplaced chemicals during a demonstration will lose students' attention and valuable class time. Materials should be ready to permit the demonstration to go smoothly.

Unexpected complexities can arise even during a simple demonstration. A damaged pulley or dead dry cell can sabotage a demonstration. The only way to be certain that a demonstration will proceed smoothly is to set it up beforehand, try it out, and then use the same materials during the actual presentation. The availability of backup materials such as additional dry cells, pulleys, and glassware will often allow the demonstration to proceed as planned without problems. One cannot overemphasize the importance of trying out the demonstration well in advance to be certain that it will proceed smoothly when it is actually delivered.

PRESENTING A DEMONSTRATION

A demonstration by its very nature will allow only a few students to actively participate during the procedure. Consequently, the teacher is faced with problems that are different from those encountered in laboratory work and other strategies.

Establishing the Purpose of a Demonstration

The recognition by the students of the purpose of the demonstration is essential to assure maximum participation and to obtain the desired outcomes. The purposes of a demonstration should be kept as simple as possible and given in short, direct statements. All statements of purpose, complex or not, should be written on the chalkboard or overhead projector. Verbal statements are inadequate unless the demonstration deals with a situation of special interest to the students. For example, a demonstration that has the promise of an explosion, odd sounds, or other unusual event will automatically permit students to accept the purpose of the demonstration. Students enjoy spectacular and unusual events during a demonstration. They also like the feeling of suspense.

Interesting problems often arise from the action of the demonstration materials themselves. The students will discover the purpose of the demonstration from these problems.

Some teachers tell their students in advance of the demonstration what the outcome will be. This practice is not usually recommended but is sometimes necessary to make the outcomes more meaningful. When it is employed it may deny students the chance to identify their own problems, to speculate, to plan methods of attack, to draw conclusions based on their *own* findings. Use this practice discriminately.

Acquainting Students with Materials and Procedures

Students require help in recognizing the materials that will be used during the demonstration as well as to learn the function each item plays during the procedure. Do not make the assumption that students know the materials and their functions. Each item must be identified and its purpose made clear either before the demonstration or during the procedure itself.

During the demonstration, it is good practice to stop occasionally and summarize what has taken place. In demonstrations requiring several steps or several associated activities, stop periodically to summarize the results. Ask students to make the summaries, if appropriate. The results can be given orally as they are recorded on the chalkboard. Use both techniques to reinforce comprehension.

Pacing a Demonstration

Students, particularly those in middle school, do not have long attention spans and consequently have difficulty sitting quietly for long periods of time. Therefore, it is desirable to plan short demonstrations that move along at a somewhat rapid pace. Avoid procedures during the course

of the presentation that may cause long pauses or delays. For example, if the demonstration requires boiling water, prepare it beforehand so that it is available at the appropriate time. Demonstrations that require long periods of time can be made interesting and meaningful if they are planned to be delivered in stages separated by brief intervals during which students engage in other activities.

Creating suspense is a very useful tactic to hold students' attention. Events leading to a dramatic climax such as an explosion will maintain student interest, give them a feeling of involvement, and motivate them to meet the objectives of the demonstration.

Humor can also play an important part in the success of the demonstration. Situations can be devised and planned to end with unusual, unexpected, or humorous outcomes. Students enjoy being entertained while learning in a relaxed situation. The teacher must be aware as to whether the students are benefiting from the presentation. Facial expressions, laughter, and student comments are helpful in judging the pace and effectiveness of a demonstration.

Students should be permitted to participate physically during the demonstration procedure. There are few situations in which the teacher is required to do all the manipulations, and, even then, students can assist the teacher. If a number of manipulations or phases are involved in the procedure, first, allow one student to carry out a phase and then allow others to conduct the subsequent manipulations or phases.

Ensuring Understanding of the Procedure

Events can happen so rapidly during a demonstration that a student may miss the whole point of the procedure. Momentary lapses of attention, blocked vision, or laughter may have occurred that made essential remarks or words inaudible, thus preventing students from grasping the significance of the demonstration.

The teacher must use all types of techniques to help students avoid missing important points. Ask questions such as, "Describe what I just did during this step," or "What observations have you made at this point?" When asking such questions, do not signify approval or disapproval but proceed to ask another student the same questions. If a disagreement occurs, then it may be necessary to repeat the demonstration or step.

Students often are not certain of the procedures used even when utmost care is taken to describe them. Brief reviews at various times can help students clarify their thinking. A review can be made in several parts with one student reiterating the purpose, another describing the materials, and another describing the procedure.

Written or oral summaries are useful in promoting student understanding. If a demonstration consists of a number of distinct phases, interim summaries are helpful. Data and charts placed on the chalkboard can help summarize results and discussions.

When materials or products of a demonstration must be viewed or examined by students, pass them around the class in duplicate and fully labeled. They can also be exhibited for later examination if time is limited or if materials or products are too cumbersome to pass around.

Achieving the Purpose of a Demonstration

Science teachers must coordinate several elements to achieve the purpose of a demonstration. They must present a concrete event, get students' attention, and ask questions. This requires a carefully planned and executed event.

With some demonstrations it is important that students recognize the purpose at the very beginning of the presentation for it to be maximally effective. The purpose of the demonstration should be communicated as simply as possible in short, direct statements. These can be written on the chalkboard, with the key terms and ideas emphasized. The students should have the purpose of the demonstration fixed in mind. Oral reminders or written statements on the chalkboard to which the teacher can refer during the demonstration processes are essential for successful outcomes. As long as the students understand the purpose of a demonstration, it is likely that the teacher will be able to maintain interest and guide students' thinking during the demonstration process.

The purpose of a demonstration in some instances can become more meaningful and obvious when applications in everyday life are presented as part of the presentation. Near the end of the demonstration and before the conclusion, reinforce students' understanding of the concept or principle under study by discussing its application in everyday life. Be prepared to describe several common situations that illustrate the idea. This will make the instruction relevant to students and make the purpose of the demonstration better understood. It will also help students realize the importance of science and technology to society. During such discussions, first ask students to describe applications with which they are familiar and then discuss those that have been prepared for discussion.

Oral statements of the purpose of a demonstration are generally not recommended unless the activity involves a simple situation of special interest to the students.

▼ "I want to show that friction depends on the forces pressing two surfaces together," said Mr. Davis to his ninth-grade science class. He then proceeded to drag a block of wood across the demonstration table with a spring scale to measure the force to overcome friction at a constant speed. He did this several times, using different loads on the block each time. Each time he asked a student to come to the table to read the force of friction as indicated on the spring scale. The students recorded their readings on the chalkboard. At their desks, the students plotted a graph of friction as a function of the normal force, using the data on the board. Mr. Davis ended the lesson by summarizing the relationship between friction and force and writing this relationship on the chalkboard. He then announced that the class would continue to study friction over the next several days. To focus attention on the purpose of the demonstration, this is what he wrote on the chalkboard:

Friction

Definition: Resistance to relative motion between two bodies in contact. Friction increases as the force pressing two surfaces together increases. ▲

Those who wish to use demonstrations to initiate student inquiry should say very little at the beginning of this instruction. They can solicit student involvement by asking a simple question.

▼ Mrs. Christas brought out a wooden disk and a brass ring. She placed the solid cylinder and the hollow cylinder at the top of an inclined board and at the same time asked the class, "Which one will roll down the slope faster?" Some students thought the solid cylinder would reach the bottom of the incline first; others thought the hollow cylinder would roll down faster. Most thought the two objects would reach the bottom at the same time. ▲

The students ask some interesting questions when they observe what happens during the demonstration. From the questions that arise, students discover the purposes of the demonstration.

▼ Mrs. Howe set up a demonstration before her ninth-grade class convened. She clamped one end of a copper tube to a laboratory support. She placed the other end of the tube in a position almost touching another support. Wires connected each support to a dry cell and a bell. (Note that when the tube touches both supports, a circuit is completed and the bell rings.) Mrs. Howe set up the demonstration so that the tube attached to one support was only a small distance away from the second support; consequently, the bell does not ring. Mrs. Howe directed a student to come to the demonstration table and heat the copper tube with a Bunsen burner. Within moments, the bell began to ring and continued to ring as heat was applied. When the heat was removed, the bell stopped ringing.

The students wondered why the bell rang when heat was applied to the copper tube. One suggested that the heat produced electricity, but this was questioned by some students because of the presence of dry cells. Another student suggested that the copper became a better conductor when heated. Mrs. Howe acknowledged the responses and then asked the students to trace the circuit. The students discovered the small gap between the tubing and the support, motivating them to formulate a new hypothesis, which they tested and verified.

Mrs. Howe ended her inquiry demonstration with a brief summary of important points that she intended to emphasize. She wrote the points on the chalkboard to lead into the day's lesson. ▲

Initiating inquiry through a demonstration is an effective technique for science teachers to use, but it requires some skill. Individuals using this technique can adopt a "hands-in-pocket" attitude, which is difficult to master and use. It demands that teachers give directions to students without predicting results. It requires that they ask questions that encourage careful observation and intelligent speculation without telling the students what they will see or what conclusions they should draw. It demands that teachers evaluate students' statements without being influenced by what they believe students should say or think.

▼ Mr. Able placed a container on the demonstration table and removed from it a box of baking soda, a bottle of vinegar, a spoon, and a drinking glass. After he showed the materials to the seventh graders, he asked, "Has anyone ever seen an experiment using these materials?" Several hands went up. He selected one boy to come to the demonstration table and show how the materials had been used in the previous activity. As the boy came forward, Mr. Able moved to one side out of the center of attention. After the boy had demonstrated the action of vinegar on baking soda, Mr. Able asked another student who had never tried the experiment to repeat what the boy had done.

Mr. Able then took other materials from a container—a lemon, a tomato, a can of grapefruit juice, a box of alum—and in each instance asked for suggestions for more illustrations of neutralization reactions. The students made a number of suggestions, carried them out, and recorded the results on the chalkboard.

During the demonstration, Mr. Able did not manipulate anything other than to take items from a container. He gave no information except to introduce a new word: *react*. Figuratively speaking, his hands were in his pockets at all times. The lesson ended with a limited but valid generalization, which was later used for more extensive work. ▲

This demonstration illustrates how a teacher maximized student involvement in a simple exercise. He got the students to manipulate the materials and suggest ways to do this. He got the students to answer and even ask some questions. This approach produced suspense when the solutions bubbled over the glass container. The students felt anticipation when materials were removed from the box.

Some teachers tell their students in advance the outcomes of a presentation. This practice is sometimes justified, but if it is used frequently it denies students the opportunity to formulate problems and hypotheses, to plan procedures, and to draw conclusions based on their findings.

At the end of a demonstration, however, the teacher should ensure that all of the students understand the purpose and important points of the activity. As mentioned, events often happen so quickly during a demonstration that a student may miss important points. As already discussed, a number of distractions in the classroom may also prevent students from observing what has taken place.

The teacher should plan various techniques to ensure that students understand the demonstration. One technique may require students to describe what they have seen without the teacher signifying approval or disapproval; the teacher then asks another student whether she has made the same observations. If there is disagreement among the students, it may be advisable to repeat all or part of the demonstration.

Students are not always certain of the procedures that have been followed during a demonstration. It may be necessary to review these procedures by giving several students the opportunity to describe what has taken place. Brief reviews will help students clarify points. A review can be planned so that various students can make contributions with one student stating the purpose, another describing the materials, and still another giving the procedures.

Oral and written summaries by students or the teacher are extremely important. When a demonstration consists of several distinct phases, it is desirable to ask students to summarize what has taken place at the end of a given phase. Oral summaries are sometimes suitable, but it is usually best to write the data and summaries on the chalkboard.

Acquainting students with the functions of the materials and procedures in a demonstration can further enhance understanding of its purpose. This can be done during the presentation, but it may detract from its pace and continuity. Consequently, the teacher may wish to familiarize students with the apparatus and mechanics at the end of this activity. At this point, it may be helpful to gather students around the demonstration site to acquaint them with the materials, give them an opportunity to handle the materials, ask them to manipulate the apparatus, and to explain how the apparatus works.

SPECIAL EQUIPMENT FOR DEMONSTRATIONS

Certain demonstrations require special apparatus and equipment. Often, a basic apparatus is assembled to meet certain requirements and later disassembled and reused for other demonstrations. Sometimes special apparatus must be designed and built for specific purposes.

Building Apparatus

Teachers with a great deal of ingenuity can construct unique types of apparatus for demonstrations. Professional journals such as *The Science Teacher, The American Biology Teacher, The Journal of Chemical Education, The Physics Teacher, Science Scope,* and others contain many ideas regarding the construction of unique devices that can be used in demonstration work.

There are a number of advantages in using teacher-constructed devices for demonstrations. Most important is that the devices are designed to illustrate and emphasize specific points. Because the teacher will be well acquainted with the device, he will use it with great ease and finesse. Students will respond with interest in the presentation when they observe the enthusiasm that the teacher displays while using the specially built equipment.

Students can also construct demonstration apparatus. They benefit from the experiences of planning and constructing the devices and the presentation they make. Their peers will also take special interest in the presentation. The teacher may suggest ways in which students can participate in constructing apparatus for demonstrations. Many ideas can be obtained from various sources or directly from the teacher. Sometimes students merely need a description of a device in order to construct it.

Demonstration apparatus can be used to illustrate how commonplace devices such as the gasoline engine or electric motor work. Those used to illustrate scientific principles can often be constructed by students as a class project or as a special assignment.

▼ Mrs. Fiorello constructed a device to demonstrate that the strength of an electromagnet depends on the number of turns in a coil and on the current flowing through the coil. She used two steel bolts wound with insulated wire and connected in series so that the same current flowed through each. She mounted these in a frame to ensure visibility. She asked several students, one at a time, to use the electromagnet to determine what it would pick up. Some determined its strength by observing how many paper clips it picked up when a single dry cell was used, when two dry cells were used, and when three were used. One student suggested that an ammeter be introduced to show the changes in current as the number of dry cells was increased. ▲

Teacher- or student-constructed demonstration apparatus should be designed to be as simple as possible. It is very easy to introduce confusing elements while attempting to emphasize certain points. For example, the device for illustrating the principles of an electromagnet described above could be confusing if the students do not realize that the current is the same in both electromagnets.

Students generally understand the demonstration if the apparatus used is closely related to things with which they are familiar. For instance, the effect of tension on vibrating strings can be illustrated by using an actual guitar or banjo instead of an elaborate piece of equipment purchased from a scientific supply house. Also, students can build a simple device to show this same effect.

Equipment for Specific Ideas and Concepts

Weight and Other Forces. Spring-type platform scales are excellent for weighing objects and for determining downward forces. Spring balances of different sizes and sensitivity can be used to determine weights and other types of forces. Dial scales, such as those used in the kitchen or nursery, are very useful for many types of demonstrations. Scales with large dials should be selected so they can be seen easily by students from all parts of the room.

Atmospheric Pressure. For demonstrations concerned with atmospheric pressure, a vacuum pump is needed, either hand- or electrically operated, as well as pump plates and bell jars of various sizes. Direct dial-type pressure gauges, which can be seen easily by students at their seats, should be used for demonstrations.

Submerged and Floating Objects.   You can use large, rectangular aquariums for demonstrations involving buoyancy, hydrostatics, and Archimedes' principle.

Chemical Reactions.   Demonstrations involving chemical reactions are often not easily seen by students at their seats unless large-scale items are used. Glassware for demonstrations should be large enough to be seen at the back of the room. Specific gravity cylinders and battery jars are excellent substitutes for test tubes when heat is not required. When heat is involved, oversized test tubes and large flasks ensure visibility.

Electric Currents.   Large-scale galvanometers, voltmeters, and ammeters with large dials are necessary for demonstrations in electricity. The galvanometer is a versatile instrument because it can be converted to a voltmeter or an ammeter by using suitable shunts and series resistances. Large-scale instruments permit readings to be made from most parts of the classroom.

Temperature Changes.   Suitable equipment for demonstrating temperature changes by direct reading is expensive because it involves elaborate thermocouples and galvanometer combinations. Normally, these are too complex to be used in Grades 6–9 and are probably impractical for work in high school chemistry or physics. Direct readings can be made by using various types of commercially made thermometers, but the scales are too small to be read at a distance. When students cannot make readings at their seats, ask one of them to take the readings and place the information on the chalkboard. If class size permits, students can also take readings at the demonstration table.

Projection Equipment.   Special projection equipment is sometimes needed to permit the entire class to view material that is normally seen under a microscope. Microprojectors can be used to view microscopic materials such as one-celled organisms, cross sections of leaves, root tips, thin tissues, and so on, but viewing is often less than desirable unless the projected image is of good quality. This necessitates a well-darkened room. Each student can view the material to be examined through the microscope under the microprojector beforehand so that the teacher can point out important structures. Pertinent points can be made regarding what students should look for when viewing the material through their microscopes.

▼   The students in Mrs. Shaw's biology class had already viewed paramecia through a microscope using material from a hay infusion they had prepared. To continue the study of protozoa, the teacher ordered an amoeba culture from a biological supply house. The students prepared several slides from the culture and asked the teacher if they could use the microprojector to view the material. The whole class was able to observe various structures and movements of the organisms. The students later prepared slides to view amoebas under their microscopes for further study.   ▲

An overhead projector is useful to project material such as leaves or petals to show their shapes. The silhouettes produced will not show details of the structures, but shapes of the materials can easily be seen. Details of structures of live material that are translucent can also be observed using an overhead projector. Before using the overhead projector for demonstration purposes, the teacher should determine beforehand whether the quality of the projection is good enough for class viewing.

▼   Mr. Jespersen wanted assurance that all students in his earth science class understood how crystals grow. He was concerned that the laboratory experience alone would not permit all students, particularly the slower ones, to understand the process. To make the process more understandable, he decided to conduct a demonstration that allowed students to observe the growth of thymol crystals with the aid of an overhead projector.

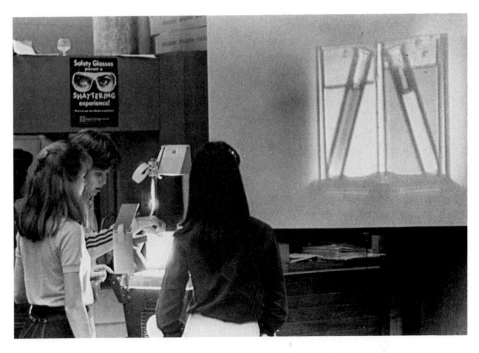

The overhead projector can be used for demonstration purposes. The teacher should determine beforehand whether the quality of the projection is good enough for class viewing.

He proceeded to melt solid thymol in a small petri dish over a hot plate. He then placed this dish on the overhead projector and permitted it to cool for a short period. He then added small crystals of solid thymol to the melt. The students observed the growth of the thymol crystals projected on the screen. They saw the rapid growth layers of thymol on the small crystals of thymol that had been added. Accretion was very visible on the crystal faces. Mr. Jespersen's discussion of the demonstration introduced the study of igneous rocks. ▲

Most observations made during a demonstration involve the sense of sight. In some cases the senses of hearing, smelling, and feeling are as important while making observations. Sounds, for example, can be amplified by using microphones connected to an amplifier and loudspeaker. The materials used for a demonstration or the products of a demonstration can often be passed around the classroom so that students can feel, touch, or smell them to make proper observations.

▼ Miss Shehadi amplified the sound of a student's heartbeat by placing a microphone connected to an amplifier on his chest. She asked the students in the sixth-grade class to listen to the heartbeat so that she could point out the nature of various sounds that were being amplified. Miss Shehadi then directed the student to jump up and down in place ten times, after which she amplified the sound of the student's heartbeat and again pointed out several sounds that were associated with this physiological process.

Following the demonstration, Miss Shehadi grouped the students for laboratory work. After demonstrating the use of a stethoscope, she gave each group of students one to use for a laboratory activity. The students then proceeded to examine their own heartbeats under different conditions specified by the teacher on a laboratory exercise form. ▲

VIDEOTAPES AND VIDEO CAMERAS

The educational value of using video cassette tapes depends on the teacher's skills in integrating the content of the tape into the lesson. There are a number of uses for videotapes in science instruction. As with films, filmstrips, slides, and film loops, videotapes can be used to present content, techniques, demonstrations, laboratory skills, and so on. Videotapes on various science topics are commercially available from publishing companies, scientific supply houses, and companies dealing with educational materials and equipment. Prepared videotapes can also be obtained free of charge from a variety of sources including industry, power companies, public libraries, teaching centers, research facilities, universities, and many others. For example, some medical schools have prepared videotapes that show surgical procedures in the areas of the human heart and brain, the dissection of a human cadaver, and laboratory procedures and techniques involved in medical research. These presentations take students into the operating room or research laboratory and permit them to view situations that make learning more concrete and relevant. This type of visual learning is appealing, interesting, and beneficial to most students. These tapes can usually be borrowed without difficulty.

Commercially prepared videotapes or tapes available from other sources should be carefully reviewed to see that they serve a purpose in instruction. The descriptions given in advertisements are often inadequate and cannot be used to determine the content, quality, or use for instruction. Do not purchase a tape before reviewing its contents or use a tape just because it is free of charge.

Teacher-made videotapes can also serve important uses in instruction. The teacher can film carefully prepared demonstrations, close-ups of microscopic material, laboratory techniques, and skills and use them immediately during the same lesson for review or discussion or for future instruction.

Of course, the preparation of tapes requires a video camera. New types of video cameras are equipped with close-up lenses that allow the taping and viewing of objects that are only a few centimeters in size so they appear on a TV screen to be two or three times their original size. Macrolenses permit the taping of super close-ups resulting in objects being magnified eight to ten times their original size. Close-ups allow the teacher to point out and discuss details not easily presented by other methods. The teacher can use the video camera to prerecord the event, demonstration, or other activity for use in a future lesson. The activity can also be viewed on the television screen while it is in progress without recording it.

Macrolenses can be used to record or display close-ups of plant and animal material and small objects such as resistors, rocks, quartz crystals, and other items that cannot easily be seen by students seated in the back or other parts of the classroom. Macrolenses are also useful during demonstration work to help students view chemical reactions that are taking place in a test tube or beaker. They can also magnify and project diagrams, photographs from books, catalogs, magazines, and other sources. Once recorded, they can be used for future instruction.

The stop action found on most videotape recorders allows the teacher to pause or freeze an action while the class is viewing the tape in order to discuss detail or allow measurements. This feature permits a great deal of flexibility when using prerecorded tapes. There are many instances when a teacher may want to pause at a particular point during a demonstration in order to further amplify on what is being said and shown.

Video cameras can be attached to a microscope so the whole class can view the material. This view can be recorded if the teacher desires or it can be a one-time presentation that is not recorded.

Videocassette recorders are useful for recording programs that are aired on television. There are programs such as *Mr. Wizard's World* that deal with science demonstrations and laboratory work. Copying the program requires permission from the network that broadcasts the program. This permission is essential before using the film for instruction. In general, the network will give permission to record and use the program if the request is made by the science teacher. A letter seeking permission should indicate that the tape will be used for instruction and should include the times it will be used and the disposition of the tape once it is used. The letter should be signed by the science teacher and a school administrator.

Science documentaries, nature programs, science demonstrations, and other science productions on television can be very useful in instruction. The advantage, of course, is that their worth for instruction can be determined while the program is being broadcast. If the teacher determines that it is not useful, the tape can be reused for another purpose.

The use of video presentations for classroom instruction requires the installation of several TV monitors strategically placed throughout the classroom to assure that all students can view what is taking place during a demonstration procedure or while a tape is being shown.

SOURCES OF TESTED DEMONSTRATIONS

There are a number of sources from which a teacher can obtain descriptions and procedures for demonstrations in physics, chemistry, biology, earth science, and general science. Science education journals, particularly *The Science Teacher, Journal of Chemical Education, The Physics Teacher, The American Biology Teacher, Geo-Times, The Journal of Geological Education,* and *Science Scope,* through the years have published demonstrations that have been conceived and tested by classroom science teachers. The national curriculum groups of the 1960s and

1970s such as the Biological Sciences Curriculum Study (BSCS), The Physical Sciences Study Committee (PSSC), Chemical Bond Approach (CBA), CHEM Study, The Earth Sciences Curriculum Project (ESCP), and Harvard Project Physics (HPP) published teachers' guides, and in some cases, source books that contain excellent examples of tested demonstrations. These guides and source books may be out of print but might be available in some schools where these curriculum projects were taught. Some recent source books and handbooks that contain suggested demonstrations for biology, chemistry, earth science, physics, general science, and for middle school grades are listed in the bibliography at the end of this chapter.

EVALUATION OF A DEMONSTRATION

The form in figure 5–4 can be used to assess the strengths and weaknesses of science classroom demonstrations. The form contains eight major categories that must be considered when evaluating the performance of a demonstration. Within each of these eight areas is a list of the desirable factors that contribute to the successful demonstration. To evaluate a demonstration, simply rate each of these factors within the category and sum all of the point values within the category. Inspection will reveal the degree of performance in that category. Only the totals of each major category should be considered. The results of such an evaluation can give the teacher feedback concerning a demonstration's strengths and weaknesses.

The evaluation form can be used to assess most classroom demonstrations. Special demonstrations involving discrepant events, presentations before small groups, or special apparatus should be evaluated by taking into account the special situations involved. The form, however, can be modified to accommodate these situations.

FIGURE 5–4   A form for evaluating the success or effectiveness of a demonstration

FACTORS FOR EVALUATING A DEMONSTRATION

Rating Scale

3	Excellent	1	Adequate
2	Good	0	Poor

	3	2	1	0

A. Preparation
  1. All materials and apparatus are readily available.
  2. The apparatus is in good working condition.
  3. The teacher has replacement materials as backup in case demonstration does not work because of faulty materials.
  4. The teacher clearly understands and carries out the procedures with confidence.
  5. Proper safety precautions have been followed.

Total points earned _____
Total possible points ___15___

B. Visibility
  1. The demonstration table is clear of extraneous and irrelevant materials.
  2. The apparatus and materials being used are of adequate size for class size.
  3. The room/demonstration is adequately lighted.
  4. The demonstration can be seen easily by all students.
  5. The teacher does not obstruct students' view during demonstration.

Total points earned _____
Total possible points ___15___

C. Communication and Audibility
  1. The teacher speaks at a moderate pace.
  2. The volume of the teacher's voice is projected to all areas of the room.
  3. The teacher enunciates clearly.
  4. The teacher has poise.
  5. The teacher speaks with confidence and authority.

Total points earned _____
Total possible points ___15___

D. Logic of Presentation
  1. The teacher clearly states the purpose of the demonstration at the beginning or end or at an appropriate time.
  2. The teacher clearly describes and simultaneously shows the steps of the demonstration.
  3. The steps involved in the demonstration are evident.

Total points earned _____
Total possible points ___9___

FIGURE 5–4   (*continued*)

	3	2	1	0
E. Questioning				
1. The teacher asks students many questions.	—	—	—	—
2. The teacher asks questions of students throughout the room.	—	—	—	—
3. The teacher waits at least 3 seconds before calling on students to answer a question.	—	—	—	—
4. The teacher asks questions at appropriate cognitive levels.	—	—	—	—
5. Questions are probing in nature and stimulate students to think.	—	—	—	—
6. Questions direct students through a logical thought pattern.	—	—	—	—
7. Questions permit students to draw their own conclusions.	—	—	—	—
8. Questions stimulate students to initiate further investigations.	—	—	—	—

Total points earned _____
Total possible points __24__

	3	2	1	0
F. Time Allotment				
1. The length of the demonstration is appropriate for the expected outcomes.	—	—	—	—
2. The demonstration is planned so that student interest is maintained throughout.	—	—	—	—

Total points earned _____
Total possible points __6__

	3	2	1	0
G. Context				
1. The teacher introduces the demonstration at the appropriate time during the unit or lesson.	—	—	—	—
2. The demonstration clearly fits into the context of the lesson or unit.	—	—	—	—

Total points earned _____
Total possible points __6__

	3	2	1	0
H. Conclusion				
1. Discussion at the end of the demonstration includes review and, if appropriate, applications in everyday life.	—	—	—	—
2. The principal concept or idea is clearly evident at the end of the demonstration.	—	—	—	—

Total points earned _____
Total possible points __6__

SUGGESTED ACTIVITIES

1. Make a card file of demonstrations. Use the bibliography at the end of this chapter to locate demonstrations. Include a variety of demonstrations for various purposes: to initiate inquiry; to illustrate a concept, principle, or law; to introduce a unit; and to motivate students.

2. Ask experienced science teachers for descriptions of successful demonstrations that they have used in their teaching.

3. Develop a checklist that you can use as a guide in presenting an effective demonstration to a small group. Compare your checklist with those developed by other members of your science methods class, making necessary changes. Points that you might consider in the checklist are purpose, visibility, participation, safety, pace, and understanding.

4. Use your demonstration effectiveness checklist (see figure 5–4) to analyze and critique a demonstration given by an experienced science teacher.

5. Prepare a demonstration and present it to your science methods class. Ask your classmates to use a checklist to critique the demonstration. Videotape the demonstration. This will permit you to critique your own demonstration with the checklist.

BIBLIOGRAPHY

Bell, W. L. 1990. Chemistry of air bags. *Journal of Chemical Education, 67*(1): 61.

BouJaoude, S. 1991. A study of the nature of students' misunderstandings about the concept of burning. *Journal of Research in Science Teaching, 28*(10).

Brown, J. 1984. *333 more science tricks and experiments.* Blue Ridge Summit, PA: Tab Books.

Chiappetta, E. L. (Ed.) 1987. *Ideas and activities for physical science.* Houston: University of Houston, College of Education.

Dantonio, M., and Beisenherz, P. C. 1990. Don't just demonstrate—motivate. *The Science Teacher, 57*(2): 27–29.

Erlich, R. 1990. *Turning the world inside out: And 174 other simple physics demonstrations.* Princeton, NJ: Princeton University Press.

Friedhoffer, B. 1990. *Magic tricks, science facts.* New York: Franklin Watts.

Frier, G. 1981, Sept. The use of demonstrations in physics teaching. *The Physics Teacher, 19*(6): 384–386.

Fuchsman, W. H., and Garg, S. 1990. Acid content of beverages. *Journal of Chemical Education, 67*(1): 64.

Gillium, D. E., and Herrmann, M. S. 1990. Determining the zinc coating weight on steel. *Journal of Chemical Education, 67*(1): 62.

Gil-Perez, D., and Carrascosa, J. K. 1990. What to do about science misconceptions? *Science Education, 74,* 531–540.

Herbert, D. 1981. *Mr. Wizard's supermarket science.* New York: Random House.

Herbert, D. 1983. *Mr. Wizard's 400 experiments in science.* North Bergen, NJ: Book Lab.

Hilton, W. A. 1981, Sept. Demonstrating as an aid in the teaching of physics. *The Physics Teacher, 19:* 389–390.

Howes, R., and Watson, J. K. 1982. Action packed demonstrations to wake up large classes. *The Physics Teacher, 20,* 40–41.

Joseph, A., Brandwein, P. F., Morholt, E., Pollock, H., and Castke, J. F. 1961. *Teaching high school science: A sourcebook for the physical sciences.* New York: Harcourt Brace Jovanovich.

Killgore, J. 1987. Slime, glop, putty, and goo: Amazing fluids that defy even Newton. *Science World, 43*(14): 23.

Liem, T. L. 1987. *Invitations to inquiry.* Lexington, MA: Ginn Press.

Lunetta, V. N., and Novick, S. 1982. *Inquiry and problem-solving in the physical sciences: A sourcebook.* Dubuque, IA: Kendall-Hunt.

Morholt, E. F., and Brandwein, P. F. 1986. *A sourcebook for biological sciences* (3rd ed.). New York: Harcourt Brace Jovanovich.

Philips, W. C. 1991, Feb. Earth science misconceptions. *The Science Teacher, 58*(2): 21–23.

Shakhashiri, B. 1983. *Chemical demonstrations Vol. I. A handbook for teachers of chemistry.* Madison: The University of Wisconsin Press.

Shakhashiri, B. 1985. *Chemical demonstrations Vol. II. A handbook for teachers of chemistry.* Madison: The University of Wisconsin Press.

Shakhashiri, B. 1989. *Chemical demonstrations Vol. III. A handbook for teachers of chemistry.* Madison: The University of Wisconsin Press.

Shalit, N. 1981. *Science magic tricks.* New York: Holt, Rinehart and Winston.

Shamp, H. W., Jr. 1990. Model misunderstandings. *The Science Teacher, 57*(9): 16.

Shimada, H., Yasuoka, T., and Mitsuzawa, S. 1990. Observation of paramagnetic property of oxygen by simple method—a simple experiment for college chemistry and physics courses. *Journal of Chemical Education, 67*(1): 63. '

Summerlin, L. R., and Ealy, J. 1985. *Chemical demonstrations: A sourcebook for teachers.* Washington, DC: American Chemical Society.

Thompson, C. L. 1989, Jan. Discrepant events: What happens to those who watch. *School Science and Mathematics, 89,* 26–29.

Van Cleave, J. 1990a. *Biology for every kid, 101 easy experiments.* New York: Wiley.

Van Cleave, J. 1990b. *Chemistry for every kid, 101 easy experiments.* New York: Wiley.

Van Cleave, J. 1990c. *Earth science for every kid, 101 easy experiments.* New York: Wiley.

Wood, R. W. 1990a. *Physics for kids: 49 easy experiments with optics.* Blue Ridge Summit, PA: TAB Books.

Wood, R. W. 1990b. *Physics for kids: 49 easy experiments with acoustics.* Blue Ridge Summit, PA: TAB Books.

Wood, R. W. 1990c. *Physics for kids: 49 easy experiments with electricity and magnetism.* Blue Ridge Summit, PA: TAB Books.

Wright, E. L. 1981. Fifteen simple discrepant events that teach science principles and concepts. *School Science and Mathematics, 81*(7): 575.

When lectures are well planned and delivered, they can be an efficient way to teach large groups of students. However, lectures that last an entire class period can cause students to be bored, inattentive, and mischievous.

CHAPTER 6

Lecture, Discussion, and Recitation Strategies

Although a lecture presentation can be an efficient, effective way to teach large groups of students, the method is abused when science teachers substitute it for a more appropriate form of instruction, such as laboratory work. The recitation strategy is useful in assessing students' background and learning in a particular area. It can also be used to develop concepts, principles, and ideas during a lesson. To be effective, a recitation needs to be well planned, well structured, and not boring or threatening to students. Well-organized discussion sessions are useful in facilitating cognitive and affective gains in students. Discussions can be conducted to deal with societal issues, promote inquiry, and develop problem solving skills. Discussion sessions, whether led by teachers or students, require very careful planning in order to realize certain desired outcomes. Audiovisual aids should be used to make science instruction more meaningful and interesting to students.

Overview

- Discuss the use of the lecture method in science instruction.
- Point out the difficulties in asking oral questions and suggest ways to overcome them.
- Present ways to conduct effective recitations (question-and-answer sessions).
- Suggest procedures that can be used to conduct meaningful discussions led by teachers or students.
- Discuss the use of audiovisual aids in science instruction.

A great amount of teacher talk and student talk takes place during the course of science instruction. Teachers provide directions for laboratory work, present demonstrations, conduct discussions and recitations, and present lectures. Students ask and answer questions, participate in discussions, explain demonstrations, conduct demonstrations, and present oral reports. Whatever the approach, speaking is, without a doubt, used extensively in science instruction in middle school and secondary school science courses.

To be successful in science teaching, teachers must be able to make their oral presentations interesting and meaningful to their students. Some teachers have the natural ability to deliver excellent oral presentations. Others lack this ability. These teachers must realize their limitations and try to improve their communication skills to be effective in science instruction.

THE LECTURE METHOD

The lecture method of teaching is commonly used in middle school and secondary school science teaching. It is a traditional method that deserves consideration as a teaching strategy in science instruction. When using this method, a teacher must be able to justify it over other methods available for science instruction. Any teacher who employs this strategy must cope with its limitations and use its strengths to best advantage.

Science teachers who use the lecture method must be convinced that the knowledge or information that is to be presented for background or other reasons is important to students and that the only person who can provide this information is the teacher. In other words, the teacher has the knowledge to give to the students, and the easiest and most efficient way for students to acquire it is through a lecture presentation—not by reading, discussion, or any other method. This assumes that the teacher can present this information in a stimulating and interesting way,

so that students have the desire to learn it. The teacher, therefore, is the stimulus for students to learn information that the teacher feels is important.

Science teachers must be sensitive about how students receive the information through the lecture method. Students can receive information either by rote or in a meaningful way (Ausubel, 1961). If teachers do not know the difference between rote and meaningful reception when using the lecture strategy in instruction, this can cause problems. Ausubel stated, "It is fashionable in many quarters to characterize verbal learning as parrot-like recitation and rote memorization of isolated facts and dismiss it disdainfully as an archaic remnant of discredited educational tradition" (1961, p. 15). It is hoped that teachers who know the distinction between rote and meaningful reception will not misuse the lecture method. They should be able to employ various techniques to make lectures interesting and meaningful to students by building on what they already know. Ausubel believed that weaknesses attributed to the lecture method are not due to the method itself but to the abuse of the method by the teachers who use it (1961, p. 16).

Some science teachers believe that the lecture method is an efficient way to teach a large group of students. But in many cases the teacher misuses the lecture method by using it as a substitute for more appropriate forms of instruction. Teachers who substitute the lecture for laboratory work are denying students the opportunity to learn the methods of science and prevent them from understanding the ways scientists go about their work. Students cannot acquire this understanding through a lecture alone.

There are instances in science teaching when large group instruction may be efficient. Certainly a lecture-demonstration presented to a large group of students can save both time and money. It saves more time, energy, and money to present one demonstration before 32 students than to prepare a laboratory activity to be individually conducted by 32 students. The lecture may be an efficient way to convey information to students who have difficulty reading textbooks or do not do their reading assignments. For these students the teacher can provide the necessary information, emphasizing the key points and allowing students to ask questions during the presentation.

Even though the lecture method has certain immediate advantages—a large amount of material can be covered in a short period of time, and it is an efficient use of time and money, there are a number of disadvantages. During lecture presentations, the learning process does not proceed as efficiently as one would assume. A great deal of information presented by the teacher is not internalized and understood by students. Unfortunately, the "pouring in" of information into the minds of adolescents and preadolescents is often incomplete. They lack the necessary cognitive structures and knowledge base to internalize many of the abstract concepts included in many science lectures. Teachers who do not develop the expertise to match the subject matter they deliver to the intellectual competence of their students will be ineffective with the lecture strategy.

Adolescents and preadolescents are restless by nature, preoccupied with immediate problems, and often handicapped by limitations of vocabulary and background of experience, and they have short attention spans. The teacher should consider that lectures to such an audience can be dull and meaningless unless these factors are taken into consideration during planning. Middle/junior high school science teachers in particular, who use the lecture method for science instruction must be very conscious of the disadvantages in using it as a substitute for more appropriate forms of instruction, such as laboratory work and other hands-on activities. Students need to have the feeling of involvement during a lecture in order to gain and maintain attention. To do this, the teacher needs to vary the pace of the presentation, occasionally interject humor, use visual aids when appropriate, and permit students to ask questions. Teachers

who use this method successfully are those who deliver short, effective, and interesting presentations. Lectures that last an entire period can cause students to be inattentive, bored, and mischievous, resulting in discipline problems.

Organizing Students

Generally, formal lectures are reserved for large groups of students. A small number of students should be handled in more informal ways. In such cases, it may be more appropriate for students to be involved in the delivery of the information rather than the teacher being the sole deliverer. For example, a teacher may assign a student or a group of students to present a lecture-demonstration before the class, or the teacher may assign a student to present a report to the group.

Lecture sessions involving 30 or more students probably will have to be organized on a formal basis since there is less teacher-to-student and student-to-teacher interaction than there would be in a small group situation. Lectures involving large groups of students discourage student questions as well as student discussion so that teachers may not get a good sense of whether students are assimilating the information being presented. Teachers can often determine whether students understand the material presented by observing their facial expressions as well as the amount of restlessness that is being displayed. If teachers observe such signs, they should periodically stop and ask appropriate questions to determine whether the students understand the material.

The teacher who delivers lectures to large groups should encourage questions from students during the presentations so that points can be cleared up immediately rather than later. The teacher also should make use of small group discussions following lecture presentations, at which time students can discuss various aspects of the lecture. It is important to mention that small group discussions led by students require proper planning. If students are to lead the discussions, the teacher must select and train students as leaders who have the respect of other students, have leadership capabilities, are articulate, and are willing to take their responsibilities seriously. Student leaders will need periodic guidance and advice by the teacher when discussion sessions are being planned as well as when discussion sessions are in progress in order to ensure productive results.

Suggestions for Preparing Lectures

A lecture that is well organized, well planned, and delivered well is an excellent way to review and summarize content, expand content, develop insights, and clarify points that are confusing to students. A lecture of limited duration can be used to introduce a lesson or a unit, develop a concept or a principle, or present science as a selectively and sequentially organized body of knowledge. When organizing a lecture, one must take into account the knowledge that students possess, their cognitive development, and the conceptual complexity of the subject matter under consideration. As was pointed out in Chapter 3, when planning instruction, one must consider the conceptual development of students. Accordingly, a student has a better chance to achieve certain learning outcomes when there is a good match between a student's developmental level and the cognitive complexity of the science content under study. When the content is too complex and abstract for the students involved, that is, the complexity of the subject matter and cognitive structure of the students do not match, no learning or very little learning will take place. To plan successful lectures, it is extremely important that science teachers know the cognitive requirements needed by the students to learn the material to be presented during the lecture.

A teacher's lecture notes can be planned in prose form or they can consist of a skeleton outline of the key points that serve as a reminder of what should be covered. Many teachers feel comfortable about delivering a lecture only when they have written the complete lecture in prose form. There are several reasons why a teacher may wish to use this form. Teachers who present new material gleaned from several sources may want to have a prose version of the complete lecture to make certain that they will cover all the material thoroughly and correctly. A prose version also provides them the security of knowing that all the information is at their finger tips should difficulties arise during the presentation. A detailed version of the lecture material can also incorporate reminders where audiovisual aids should be used during the course of the lecture. Teachers also can identify the questions that can be asked at various intervals during the lecture to determine if the students understand the material that is being presented. The prose version can be as detailed as required by the teacher. Lectures written in prose form can be used more than once and can be updated periodically.

Science teachers who are secure about using the lecture strategy often use an outline to serve as a reminder of the points to be covered during the presentation. Generally, these are experienced teachers who are comfortable with their knowledge of the content and need only a skeleton outline to guide them through the lecture. The outline can consist of a title and main headings with key terms and ideas organized under the headings. Some teachers prefer an outline that incorporates the content in more detail rather than using single words or phrases as reminders. This outline can consist of complete statements of ideas and terms and be organized under specific headings. Places where questions can be asked and where audiovisual aids are to be used can also be identified in the outline.

Organization of a Lecture

A lecture that is organized around a few major ideas or concepts presented in logical sequence has been found to be successful when presented to adolescent and preadolescent students. When too many ideas are presented during a given period, the lecture will be ineffective. The presentation should be short, 10 to 20 minutes at the most, and include activities that require the students to use the ideas or concepts presented. There should also be periodic summaries during the course of the lecture and a conclusion during which the teacher emphasizes the major points that help students identify the important understandings and relationships. Lectures can be divided using the following organization: The *introduction,* the *main body, summaries within the presentation,* and the *conclusion.*

The Introduction. The introduction of a lecture is very important to direct students' attention to what will be presented. This introduction may take the form of questions posed to students before the lecture to focus attention on the topic. The dialogue that takes place between the teacher and students and among the students will give students a sense of what will be presented and emphasized. Some teachers like to use instructional objectives as an introduction to a lecture. In such cases, the instructional objectives can guide students through the lecture and become a basis for learning new concepts and ideas.

The use of advance organizers is another way to introduce a lecture. Advance organizers help students incorporate new information into their cognitive structure. They help explain and interrelate the material they precede (Ausubel, 1963, p. 91). Two types of advance organizers can be used in the lecture method, the expository organizer and the comparative organizer. The expository organizer places the information to be learned into perspective with other information that is conceptually related. For example, for a lecture on the circulation of the blood, an expository organizer might include a brief description of other systems in the body such as the lymphatic system and the renal/urinary system. A comparative organizer for the same lecture might compare the circulatory system with a hot water heating system in a house. The simplicity of a hot water system provides a concrete anal-

ogy for students to begin the study of a similar concept involving the human body.

Successful lecturers stage their presentation by involving students immediately with some technique that gets their attention. They may use questions, advance organizers, demonstrations, short stories, a discussion involving the students on a topic that may be familiar to them and pertinent to the lecture which follows, and instructional objectives that focus on the ideas and concepts to be covered during the presentation.

The Body of the Lecture. One mistake that teachers make in organizing a lecture is trying to cover too much material in a short period of time. Teachers often overestimate students' abilities to grasp inordinate amounts of subject matter during a lecture session. When a teacher attempts to rush through a large amount of material involving difficult relationships that the students do not understand, the students may become confused and discouraged. A teacher may believe that the relationships are obvious to students, when in reality they are not. Usually relationships are not obvious to students and must be pointed out and made obvious. This means that the teacher must take the time during the presentation to ask questions to determine the students' understanding of the material presented and correct whatever is necessary before going on with the lecture.

The presentation of a few ideas at a time will give teachers the opportunity to develop the relationships so that they are meaningful and understood by the students. Remember that the processing capabilities of middle/junior high and high school students are limited. Their interest will wane once they get lost in the abundance of information that is being presented.

It is best to have a simple plan of organization for a lecture—the simpler the better. A complicated sequence can only cause confusion. The more examples that are used to illustrate points during the course of the presentation, the better. The most effective lectures are well organized, simple, enriched with visual aids and many examples, and short.

Summaries Within the Presentation.   Teachers should build questions into the lecture presentation that motivate students and establish the relevancy of the material being presented. Asking questions will also preclude the possibility of lecturing for long periods of time without a break. Students should also be encouraged to ask questions during the lecture and not necessarily at designated times. If the teacher has a preference as to when questions can be asked, this preference should be announced before the lecture begins. Some teachers prefer questions at designated stopping points or at the end of the lecture but it is best to allow questions at any time during the presentation.

The Conclusion.   The conclusion of the lecture is the place for the teacher to summarize major points and ask additional questions that can lead to future lectures or activities. The emphasis on the important points during the conclusion will help students identify relationships needed to undertake future assignments and to be involved in other activities used during future instruction.

Suggestions for Presenting Successful Lectures.   The success of a lecture depends largely upon the collaboration between the lecturer and the audience. The successful lecture is an interaction in which the lecturer offers information and receives attention, and the audience offers attention and receives information (Clarke, 1987).

Adolescents are easily bored unless certain stimuli are introduced during the course of the presentation. Such actions as the teacher moving about the classroom, using humor, periodically pausing, making gestures, and varying the pace of the lecture can help to prevent boredom. Lectures that are presented at a slow pace will bore students, while those delivered at a fast pace will prevent students from understanding the lesson.

A moderate pace is recommended to assure more learning. The use of audiovisual aids such as the chalkboard, overhead projector, videotape recordings, films, models, and charts during the lecture presentation can increase cognitive gains in students. Students will better understand and retain the information offered during the lecture when a variety of stimuli are used during the course of instruction.

To assure a successful lecture, the teacher should use various techniques to make certain that there is a continuous interaction between the teacher and the students in the audience. The following suggestions adapted from Clarke (1987) should be helpful in maintaining audience attention.

1. Emphasize important ideas by changing the rate, volume, and pitch of the voice.
2. Emphasize important statements by using pauses that allow time for the audience to respond.
3. Enunciate words clearly.
4. Avoid repetition of words.
5. Avoid using such words or phrases as *um, er, like well, ah, you know, uh-huh,* and *okay.*
6. Maintain eye contact with the audience.
7. Scan the audience to observe reactions.
8. Interject humor and signs of curiosity and interest, and other indications of your personality.
9. Keep the flow of visual aids smooth and free from distraction.

CONSTRUCTION AND USE OF ORAL QUESTIONS

Teachers must be able to formulate good oral questions for any given purpose. This means that questions have to be well conceived, concise, and clearly stated. Students can answer only the questions they understand; they cannot respond to confusing, awkwardly worded, or verbose questions. They have to understand the intent of the question in order to respond correctly.

Carefully prepared questions can facilitate an organized question-and-answer session.

Oral questions can be used at any time during the course of instruction. They can be used to determine the experience background of students at the beginning of a unit or when a topic is to be introduced. They can be used for review and drill during oral recitations or to guide discussions during and following various types of instructional activities such as laboratory exercises, field trips, demonstrations or lectures. Higher-level questions—those that require students to give broad responses—can facilitate cognitive and affective results in students. This type of questioning during discussions helps students to clarify their values and make decisions on societal issues and problems. Higher-level questions can also be used to develop problem-solving and inquiry skills.

Constructing Clear Oral Questions

An oral question should be simple and direct so that students can grasp the meaning and intent immediately. Long, complex statements are often awkwardly worded and, consequently, confusing to students. Such statements are also difficult for students to keep in mind during a discussion or recitation, and their intent can be forgotten unless they are repeated or written on the chalkboard.

Oral questions must be constructed using words that are familiar to the students who must answer them. A word that they have not encountered before can make the question meaningless. Science teachers often forget that students have a limited vocabulary and use scientific terms and other words that students do not understand.

There are many instances when a teacher can improve a question by simply replacing an unfamiliar word in the sentence with one familiar to the students.

| *Original* | *Improved* |
| What is the etiology of the disease? | What is the cause of the disease? |

In the example above, the word *etiology* can be replaced by the more common word *cause,* which has the same meaning. It is not necessary for biology or life science teachers to use the language of physicians who might use the word *etiology* to describe diseases.

| *Original* | *Improved* |
| Why does the mixture of the two chemical substances effervesce when the test tube is agitated? | Why does the mixture of the two chemical substances bubble when the test tube is shaken? |

In this example, the word *bubble* is substituted for *effervesce,* and the word *shaken* is substituted for *agitated.* The statement has the same meaning and is more simply stated.

A teacher must exercise great discretion in attempting to use simple and common words when asking questions, because it may minimize the vocabulary and concept development of the students. The technical vocabulary of science that is part of the curriculum should be taught. Just because some terms are multisyllabic or new does not mean they should be replaced by simpler terms or avoided. The vocabulary of science should be used in the context of clear and simple language.

| *Original* | *Improved* |
| What is your notion of an astronomical unit? | What is the definition of an astronomical unit? |

In the example given, the original question is cluttered. "What is your notion of ..." can be easily replaced by "What is the definition of ...," but the term *astronomical unit* should remain, however complicated it may appear. *Astronomical unit* is a basic term used in the study of astronomy, and it should be freely used.

Finally, teachers should use questions that are structurally correct and complete.

Original	*Improved*
Proteins are made where?	Where in the cell are proteins made?
The equilibrium constant tells you what?	What information does the equilibrium constant provide concerning the rate of a chemical reaction?

In each instance the original questions are vaguely stated, and the students can provide many plausible answers. Rewording the questions gives the students a clearer understanding of exactly what is expected.

Lower- and Higher-Order Questions

Many systems are used for classifying different types of questions. One of the most widely used classification systems, discussed in chapter 18, "Constructing and Administering Science Tests," is Bloom's Taxonomy of Cognitive Objectives (Bloom, 1956). The six levels of Bloom's taxonomy are given in table 6–1. Along with each level are the cognitive activity, key concepts, and sample questions. The knowledge and comprehension levels represent the lower levels in the taxonomy at which questions can be developed. The application, analysis, synthesis, and evaluation levels represent the higher levels of the taxonomy at which questions can be developed. The lower-order questions obviously require less thinking on the part of the learner than the higher-order questions.

Science teachers can improve their question-asking skills by learning to classify their questions and by using a mixture of lower-order and higher-order questions (Rowe, 1978). Each type

TABLE 6–1   A taxonomy of classroom questions

Level	Cognitive activity	Key concepts	Sample questions
1 Knowledge	Remember, recall, or recognize facts, ideas, information, or principles as they were taught.	Knowledge, recall, memory	1. Define photosynthesis. 2. Who discovered a cure for rabies? 3. What is the autumnal equinox?
2 Comprehension	Comprehend, interpret, or translate information or ideas.	Describe, explain, illustrate	1. How would you measure the distance between the earth and a planet in the center of a neighboring galaxy? 2. What does the graph tell you about the influence of x-rays on bacterial cell growth?
3 Application	Solve problems, find solutions, and determine answers through the application of rules, principles, or laws.	Application, solution, determination	1. Determine the resistance in the circuits from the data given. 2. Find the molarity of the solutions, given their normality.
4 Analysis	Distinguish the parts from the whole, identify causes, find support and evidence. Construct hypotheses and draw conclusions.	Reason, think logically, induce, deduce	1. What are the effects of the two drugs on the mobility of the goldfish? 2. Present evidence that demonstrates the harm that has been caused by nuclear power plants.
5 Synthesis	Produce, design, make, and construct products. Synthesize ideas, produce ways, and determine how to. . . .	Make, produce, create, write	1. Produce a scenario about life in your city if heart disease were eliminated. 2. Design an experiment to determine how much energy can be saved by using storm doors in a home in the winter.
6 Evaluation	Judge, appraise, assess, or criticize. Substantiate on the basis of a set of standards or criteria.	Evaluate, judge, critique, substantiate	1. Evaluate the government's research program on mind-altering drugs from a moral and ethical point of view. 2. Judge the merits of the research based on your criteria for conducting research.

Based on data from *Taxonomy of educational objectives. Handbook I. Cognitive domain,* by B. S. Bloom, 1956, New York: David McKay Co., Inc.

of question has its place in teaching science and one type should not be used exclusively in preference to other types. Teachers too often ask recall and factually oriented questions and avoid asking questions that are thought-provoking and probing. Students should also be asked oral questions that require them to analyze prob-lems, synthesize ideas and make assessments, criticize, and make value judgments.

Closed and Open Questions

Some educators have classified questions on the basis of whether they are "closed" or "open."

This classification is based on the type of responses the questions evoke.

Closed questions elicit short responses that are lower order in cognitive complexity. They focus on specific and correct answers. Closed questioning techniques are used to build vocabulary, for information background, to conduct inductive inquiry sessions, to review, and to summarize. They are convergent in that they are used to get certain responses from students—for example, "What formula do you use to determine the capacitance in a series circuit?" or "Describe the plate tectonics theory."

Open questions are the opposite of closed questions in that they elicit divergent thinking. They are suitable to initiate and sustain discussions. Open questions provide opportunities for students to give a broad range of responses. They stimulate creative and personal responses, deemphasizing the notion of correct and incorrect. Open questions can be used to create an atmosphere of sharing, clarification, and openness in the science classroom—for example, "Should the school institute a program to inform students about drugs?" or "What do you think about biological warfare?" The use of open-ended questions can give students an opportunity for greater freedom and participation in learning science.

Probing Questions

There is a tendency for teachers to solicit brief responses from their students when asking oral questions. Teachers will usually ask a question, receive an answer, and then move on to the next question or activity. A great deal can be lost using this approach. However, much can be gained by asking the students to provide a more detailed response to a question. This can be accomplished by using probing-type questions.

Probing questions encourage students to expand their response to a question and then the teacher follows the response with another question along the same line. A student may answer several questions in succession and the ques-

tioning will end when the teacher feels that enough probing has taken place with that student. The teacher may then continue the line of questioning with another student or several students in order to cover the topic under consideration. This procedure is used to elicit deeper thinking and greater clarity in student responses.

The following sequence of questions is based upon a reading assignment given by the teacher regarding the dangers of nuclear power plants and the effects of radiation.

Teacher: Can nuclear power plants be dangerous?

Student: I know that they can be dangerous.

Teacher: On what basis do you make that statement?

Student: Well, look at what happened at Chernobyl in the Ukraine in the Soviet Union several years ago.

Teacher: What happened?

Student: An explosion in one of the reactors there caused one of the worst accidents concerning nuclear reactors that affected the lives of thousands of people.

Teacher: How did the accident affect the lives of people?

Student: A number of people died but I don't remember how many people received large amounts of radiation. Also, thousands of people were forced to move from their homes.

Teacher: Why was it necessary to move people out of the area?

Student: Because the people's health was in danger. Also, the environment around the reactor was unsafe.

Teacher: Why was the area unsafe?

Student: Because of the amount of radiation released, healthy crops could not be grown and healthy animals could not be raised in such an environment.

Teacher: Can you comment further on your statement?

Student: I think the animals would probably produce abnormal offspring or no offspring at all because the dangerous radiation could cause chromosomal aberrations. Plants would be affected in the same way.

Teacher: What about people? How are they affected by dangerous types of radiation such as released at Chernobyl?

Student: Based on the Chernobyl incident, the dangerous radiation can have very harmful effects such as thyroid cancer, leukemia, and anemia.

Teacher: Is there anything more you want to add to this list?

Student: Yes, some people who were exposed had their bone marrow destroyed and had to have bone marrow transplants. They could not produce red blood cells and would have died without having the transplant so they could produce red blood cells.

Teacher: In the long run, what other effects will be observed in people in the area who survived the incident?

Student: I don't know.

The last question was more difficult than the others because the answer was not directly found in the reading assignment so at this point the teacher has probed to the point where other questions relating to the dangers of nuclear power plants and radiation had to be directed to other students.

During the question-and-answer session, the science teacher raised a number of open-ended questions about the dangers of radiation from nuclear power plants. The student had a great deal of knowledge about the Chernobyl incident based on assigned reading. After probing with a series of lower-order questions, the teacher had to stop questioning the student because the teacher believed that additional questions to the same student using the same line of questioning would be fruitless. The teacher then directed questions to other students to stimulate and motivate them to learn about the topic. This activity could be followed by an assignment involving further reading to obtain a background on the topic for further discussion.

Probing questions permit students to provide clear responses and to support what they say. They encourage analysis, synthesis, and the evaluation of a problem. Probing questions can also stimulate inquiry and the clarification of personal values.

Pausing and Waiting

Not only do teachers tend to ask too many questions, but they ask them too rapidly. Melnik (1968) reported in a study of question-asking in the classroom that some teachers asked as many as 150 questions per class hour. Although these instances represent the extremes, teachers tend to ask many lower-order rapid-fire questions. This is one of the reasons why question-and-answer sessions are ineffective.

Wait time has been studied extensively by science educators (Rowe, 1978) and is defined as the duration of time between speakers. The pause that follows a question by the teacher and a response by the student is referred to as Wait time 1. The pause that follows bursts of students' responses is referred to as Wait time 2. Data gathered from many classroom situations indicate that the average wait time between a teacher's question and a student's response and the pause that follows students' responses, is approximately 1 second (Rowe, 1974a). It is obvious from these data that question-asking sessions are rapid fire and that the quality of discourse is at the lower levels of cognitive complexity. This type of question-asking/student-responding behavior precludes effective use of discussion sessions that emphasize reflective thinking and inquiry.

Science teachers should make more use of open questions that elicit divergent thinking, but they must develop the proper skills to use them effectively. This means that they must learn to increase the silence and wait time in their question-and-answer sessions when using higher-order questions. A series of closed questions in general does not necessarily require silence and wait time between question and response. For open questions, wait time from 3 to 5 seconds should be employed to give students a chance to think and reflect. The increased wait time between teacher- and student-talk, student- and student-talk, and student- and teacher-talk will probably cause some uneasiness, but the payoffs of this change are worth the effort. Rowe (1974a, pp. 89–91) reported the following benefits when wait times were extended beyond 3 seconds:

1. The length of student responses increased.
2. The number of unsolicited, but appropriate, responses increased.
3. Failure to respond decreased.
4. Confidence increased.
5. The incidence of speculative responses increased.
6. The incidence of students comparing their findings increased.
7. The incidence of evidence-inference statements increased.
8. The number of questions asked by students increased.
9. The slow learners responded with greater frequency.
10. The variety and type of verbal behavior of students increased.
11. The teachers demonstrated greater response flexibility.
12. The teacher's expectations for the performance of students rated as relatively slow learners changed.
13. The number and type of questions asked by the teacher changed.

Tobin's (1984) study of wait time in middle school grades indicates several advantages of extending teacher wait time in whole-class situations. He showed that when a teacher extended wait time for an average of between 3 to 5 seconds:

1. the quality of teacher and student discussions was improved and student achievement was increased
2. longer pauses between teacher questions and student answers were used for stimulating higher cognitive levels of thinking
3. teachers had a tendency to use strategies to push students into further discussions rather

than "mimicking or evaluating student responses"
4. the length of student responses increased and the number of students involved and reacting also increased.

According to Tobin (1984), the results of the study indicate that teachers should use an average wait time of 3 to 5 seconds when classroom discussions are intended to stimulate higher cognitive processes. Based on this study he recommends that teachers use extended wait time in whole-class situations "to improve the quality of the learning environment and increase achievement in subjects such as mathematics and science where higher cognitive level outcomes are often a concern."

Whole-Class Interaction During Questioning

During the course of science instruction the teacher usually interacts with the whole class. Lectures, demonstrations, laboratory activities, field trips, and other forms of instruction will usually involve all the students in interacting with the teacher. Most forms of instruction require the teacher to ask questions and the students to respond to questions for one reason or another. When questions are posed to the whole class during question-and-answer sessions, some teachers will often accept chorus responses or cater to those who raise their hands. During the same class a teacher may favor a bright student to respond to a higher-level question and disregard others whom he feels will not be able to answer correctly. There are also instances during the lesson in which the teacher will entirely ignore some students. One may also observe that the teacher neglects students in the back of the room and concentrates on those seated in the front, or to the left or right of the room. One may find that the teacher recognizes only a few students to respond to certain types of questions because these students are generally able to give the correct responses, maintain the pace of the lesson, and permit content coverage.

Target Students. The classroom practices just discussed are not uncommon in science instruction; similar situations are described in a study on "target" students by Tobin and Gallagher (1987). This important study has direct application to science instruction and classroom practices in general. This study investigated whole-class interactions involving 15 teachers, 200 science lessons, and students from Grades 8 to 12 and revealed a number of interesting observations during question-answer sessions.

1. Most teachers tended to direct most questions at random and to the whole class. This type of questioning favored those who raised their hands to respond or those who called out the answer.
2. Teachers said they felt that the students who raised their hands needed to be recognized and that these were students who most likely knew the answer to the question.
3. Teachers attempted occasionally to call on students who did not raise their hands, but the students most involved were those who called out the answer, raised their hands, or signaled by nonverbal means a willingness to respond.
4. Teachers generally asked lower-cognitive-level questions at random and usually directed higher-level questions to more able students—target students.
5. Two types of target students were identified: those selected because the teacher felt that the students would give responses that would "facilitate learning and content coverage," and those who raised their hands to respond to questions or called out responses. These students also asked questions and evaluated the responses of other students.
6. A small number (three to seven) of target students, mostly male, was identified during whole-class interactions. These students asked more questions than others and were

asked to answer higher-level questions. They also received better feedback from the teachers.
7. Target students were either those who volunteered answers or those who were selected by the teacher to respond.

This study has certain implications in actual practice and in the training of future science teachers. The investigators state, "a belief that is popular among educators . . . is that the pace and level of instruction is directed towards the middle of the class in order to cater to the majority of students and to cater also to the needs of both more- and less-able students. . . . Our observations of secondary school classes in Western Australia and subsequent observations in the US schools (Gallagher, 1985) indicate that the steering group consists of the most able students" (Tobin & Gallagher, 1987, p. 74). They go on to say that, because only a relatively small number of students in the class monopolize whole-class interactions, the cognitive level of most classes tends to be low (Tobin & Gallagher, 1987, p. 74), and that, in general, the cognitive demand of the academic work was also low.

Based on the study by Tobin and Gallagher (1987), science teachers should be very conscious of what occurs during whole-class interactions, and particularly of the level of learning that is taking place during instruction. They must make certain that all students are involved during instruction and that no particular type of student is disregarded or favored for any reason. When high achievers dominate classroom discussions, maintain the pace of a lesson, or monopolize the teacher's attention and time, the lower achievers are short-changed. All students should be given the opportunity to interact and be involved at all times and at their cognitive level. They should be given the chance to solve problems, analyze data, and apply what they have learned, regardless of their abilities. This is a difficult and challenging task. Perhaps the practice of using whole-class interactions as the primary method of instruction must be reassessed to foster a higher cognitive demand of academic work for all students. Teachers must innovate or try different and new instructional strategies to involve all students of all ability levels.

Using Oral Questions to Assess Student Background

The teacher can use oral questions to assess the experience background of students in a class. The responses to the questions can suggest the extent of the foundation that students already have in an area and on which the teacher can build. If such information is critical, then it behooves the teacher to phrase each question in such a way as to determine which students have the background and which students do not.

Teachers often use oral questions to help students recall past experiences that can be used to introduce a new topic during the progress of a unit.

▼ Mr. Jasper's seventh-grade science class was studying electricity. They had just completed a series of activities in the science laboratory involving series and parallel circuits. He wanted to determine how many of them had had experiences in their homes with fuses and circuit breakers. Mr. Jasper passed out one fuse to each student in the class and asked, "Does anyone know what I just passed out?" Several students raised their hands. One of the boys yelled, "It's a fuse!" "That's right," said Mr. Jasper. "How many of you have seen or heard of fuses before?" Most of the students raised their hands. "How many have seen electric fuses in your homes?" About half of the students raised their hands. Several of the students described where electric fuse boxes were located in their homes. Some even remembered the number of fuses they saw in the fuse box. Other students said, "We don't have fuse boxes, but our electrical system has circuit breakers." Mr. Jasper said, "Good. It so happens I have a circuit breaker right here." He held the circuit breaker before the class and then passed a number of them around the class. "How many of you have seen a circuit breaker before?" A few students raised their hands. Some students said, "I know we have circuit breakers, but I have never seen one out of the circuit breaker

box." Mr. Jasper said, "Let's talk about fuses and circuit breakers." Then the students were anxious to tell about the experiences they had in their homes. One girl told of the time she plugged a cord from an electric toaster into an outlet, which caused a spark that frightened her and then she said that all of the lights went out in the kitchen. She went on to say that her father went to the fuse box and examined each fuse to see which one had burned out and that he replaced it with a good one. Another student told of the occasion when all the lights in the lower rooms of the house went out and that her mother did not know what to do, so she called an electrician. The electrician showed them both how to replace fuses in the box. Another student who was anxious to talk about circuit breakers said, "We don't have all this kind of trouble in our house when the lights go out. We just go to the circuit breaker box and determine which circuit breaker needs to be reset."   ▲

Through his questioning, Mr. Jasper set the stage for the study of resistance and electric fuses and circuit breakers. Some of the students who had experiences with the circuitry in their homes had seen fuses and were now thinking of them in terms of these experiences. Others, who were unfamiliar with fuses but had experience with overloading circuits in their homes, knew about circuit breakers. Other students who had no experiences with fuses or circuit breakers gained vicariously through the stories told by their classmates. They all wanted to learn more about fuses and circuit breakers, and Mr. Jasper had a good idea of how to proceed based on the students' apparent background on the topic.

THE RECITATION METHOD

The recitation method is characterized as a question-and-answer session during which the teacher asks a series of questions and the students provide the answers based on what they have read in their textbook or other assignments or what has been presented by the teacher during a lecture, demonstration, or other activity. Through this procedure the teacher assesses whether students have learned what was presented in the textbook, lecture, or other assigned material. Often recitation periods become sessions of review and drill in which few students actively participate and many sit quietly without making any contributions. The recitation is a respectable teaching strategy if the teacher does not abuse or overuse it or make it into an activity that could be threatening or boring to the students. It can be used to develop understanding of concepts and principles or for elaboration and explanation of ideas.

Techniques

The teacher should explain to students the purposes of a recitation when using it in instruction. The teacher should make clear that the intent of recitations is to provide feedback to the teacher and that this feedback can give the teacher the bases for elaboration and explanation of ideas. The students should be made to understand that there is no need to be threatened during a recitation session, even though the teacher is checking on the students' knowledge and understanding.

Once the students experience that they can trust the teacher through his actions, they will feel at ease and look forward to the sessions. This means that the teacher should praise students for giving correct answers during the recitation and avoid reprimanding them for incorrect responses. The teacher should show the students that he is using the method to help them rather than to evaluate them for report card purposes. This does not mean, however, that no evaluation is to take place, for when the teacher praises a student, he is evaluating the student. When students give wrong answers, the teacher is also evaluating, but this is not cause for reprimanding them or rejecting them. However, it *is* cause to determine why the wrong answers were given. Maybe a series of questions can be asked that will determine whether the correct response can be drawn out of students in another way.

Conducting the Recitation. The success of a recitation depends on several factors. Most important is the quality of questions asked. Questioning is more effective if the teacher formulates the questions beforehand, because he can give more attention to the type of question to be asked, its construction, and the sequence in which it will be asked during the recitation. The questions should be written out on 3-by-5-inch cards, one to a card, which may then be organized in a particular sequence and shuffled and reorganized to change the sequence as necessary. Prior planning avoids lags during the recitation, because the teacher does not have to think about the next question or its construction. This permits the recitation to proceed at a reasonable pace.

The teacher should set the stage to maintain order during a question-and-answer period. Teachers should require complete attention and respect during a recitation period. This does not mean that absolute silence must occur or that students sit inactively in their seats. It does mean that the teacher must keep the students alert, motivated, interested, and anxious to concentrate during the session. Teachers who do not recognize these factors will let question-and-answer sessions deteriorate into useless activity.

Teachers should tell students how they will be recognized to respond to a question. Students should know that they need permission to contribute a response. Teachers should be consistent in how they acknowledge those who request permission to answer a question. The usual signal, the raised hand, should always be given preference. If one student raises her hand and another gives a response without permission, the teacher should recognize the first student and ignore the second one, even though her answer may be correct. If several hands are raised at the same time, it is best to ask the student who raised her hand first to answer the question. The teacher should make it clear to the other students who raised their hands why he called on a particular student to give the response in preference to any of them. If two or more students raise their hands simultaneously, it is good practice to give preference to students who have not contributed much during a particular session. Sometimes, if two or three students raise their hands simultaneously, the teacher just arbitrarily selects one to respond and manages to acknowledge the others.

▼ "I see that Laura, Bill, and Mary are ready to answer the question. Let's let Mary answer this one because she has not had a chance before." ▲

There are instances when a few students—three or four—tend to dominate a question-and-answer session. It is difficult to ignore these students, because they are usually the better students and probably the most cooperative. And yet, a teacher wants to ask other students to participate. This is when the teacher must be very tactful.

▼ "Robert and John want to answer this question, but they have just answered two questions. Let's give June a chance to answer, and then both of you can tell the class whether you agree with her." ▲

Once the teacher knows the students in a class, he can determine why certain students have a tendency to be quiet and not participate during the recitation. The student may be a slow learner, uninterested, or merely shy. To bring this type of student into the picture, the teacher may formulate special questions that he knows the student can answer correctly because of her background. This technique will help this type of student to counteract feelings of isolation and become a participating, active member of the group.

Questions directed specifically to students who are obviously not paying attention to what is taking place will immediately cause them to pay attention. This technique may prevent students from becoming involved in activities that will create discipline problems.

The teacher should treat students who give incorrect responses with the same courtesy as those who give correct ones. It would be unfair to arbitrarily condemn a student who gives an incorrect answer just because it was not the expected response. Perhaps the question was poorly stated or the student misinterpreted it. The student may have understood the question but found it difficult to express the correct response orally. In such cases, the teacher should analyze why the response was incorrect or partially correct or even determine whether there are actually two possible answers to a question even though the teacher was expecting a particular response. If the student is completely wrong, there is no excuse for sarcasm, scolding, or contemptuous remarks.

If several students respond to a question simultaneously without raising their hands and by shouting the correct response in unison, the question is probably too simple and probably not worth asking. Chorus answers should not be acceptable, because this type of response can cause the question-and-answer session to degenerate into situations in which very little learning takes place. If this type of response happens frequently, the teacher should examine the questions to determine their worth.

Good question-and-answer sessions are lively and proceed at a reasonable pace. Teachers should limit active participation to one or two students at a time and should not permit the questioning to drag or the answers to become too involved and lengthy, which results in boredom and students thinking about other things.

The teacher should use every possible means to encourage maximum participation. He should ask questions clearly and in firm tones so that all students can follow the line of questioning. It may be necessary to repeat the question more than once. The questions should be short, concise, and simply worded so that the students can keep them clearly in mind. If the responses of students are not audible to the rest of the class, it is important that the teacher repeat the response so that all the students understand what has been said. It is also good practice to periodically repeat a question to make certain the students have it in mind.

The teacher should be aware of the mood of the class during the recitation, making sure that the students are not bored and inattentive and that they are anxious to contribute. The teacher should give the students an impression of enthusiasm. He should be alert to recognize students who wish to contribute, as well as those who are inattentive. Consideration should not be given only to those students who vigorously raise their hands; all students should be drawn into the picture. Teachers should also be careful not to allow themselves to be drawn into a discussion involving one or two students, thus losing the rest of the class. And, of course, the teacher should avoid having one, or two, or even three students dominate the recitation period.

Normally, it is best that the teacher deliver the question first and then designate the student who is to respond. Naming the student first and then posing the question is an invitation for the rest of the students to relax and become uninterested. There may be occasions when the teacher has a particular question in mind for a designated student. Perhaps the student has been working on a project that involves an experience related to the question. In such cases, it is imperative that the teacher make the student aware through his tone and action that he is not attacking the student by asking the question, but that there is a definite reason for asking the question. A teacher's tone when delivering the question can avoid giving the student the impression of attack or threat.

Teachers should not use the threat of asking a question as a way to keep students attentive. Such statements as "You'd better pay attention or else you're next in line for a question" or "Keep quiet or I'll ask you the next question" should be avoided. If the teacher calls on the inattentive students, he should not repeat the question for them. If they can't answer, the teacher should go on to another student without comment. The inattentive ones will soon get the message and become alert to the fact that they might expect the next question. If a student fails to give the correct response, the teacher should ask another student, who may give the correct answer. However, the teacher should not neglect the student who had given the incorrect response; she should be given a chance to another question as soon as possible.

There is great potential for boredom and monotony during a recitation, and it is advisable for a teacher to avoid having the students recite previously learned material and leaving it at that. The recitation is often boring for the bright students, who already know the material, and meaningless to students who perhaps did not read the textbook or complete the assignment on which the recitation is based. The recitation may be the opportune time for the teacher to expand, elaborate, and comment on student responses. It can be the time when synthesis can take place and when the teacher can relate the various points raised by the students. Using the recitation in this way can give students a chance to think about the questions and responses and take them from what they know to what they do not know.

A question-and-answer session should be terminated after a reasonable length of time. This type of activity does not interest the student for long periods. Students become restless and start shifting their attention to other things. By the end of 10 minutes, students will show these signs of inattention and the number of students anxious to participate will start to dwindle. At this time shift to another activity, preferably not one that makes the teacher the center of attraction. Laboratory work, a film, and possibly a demonstration involving students are excellent activities to use following a recitation session.

The teacher should be very careful about formulating the questions for the recitation. The questions should be posed so that the responses will have conceptual connections. By asking questions in a particular sequence, the teacher can lead the students to an understanding of these connections and eventually the concept.

▼ Mrs. Guevera gave a textbook reading assignment involving force and resistance. On the following day, she gave the class a short introduction based on the assignment and then proceeded to ask questions for 15 minutes. She prefaced her questions with the following comments and then started the questioning based on the assignment. "The reading in your text involves a discussion of force and resistance. I want to find out how well you understand the main ideas in the sections I assigned. Let me ask you these questions to start the lesson." The teacher proceeded by asking, "Look at the book on the desk. Is it moving? Mary, is this book moving or is it at rest?" "It is at rest." The teacher then asked, "Is this book moving now, Mary?" after she pushed it with her hand. "It's moving but only after you pushed it," commented Mary. "Would this book have moved if I hadn't pushed it?" "No, unless someone else pushes." "So here is an object at rest, but we caused it to move. What caused it to move? Gerald?" "The push caused it to move," said Gerald. "What do we call this push that caused this book to move? Tom?" "It's called a force," said Tom. "What is a force, then, Dominic?" "A force is a push." "Is a pull also a force, Jacob?" "Yes," said Jacob. "All right then, Jacob, give me a definition of a force." "A force is a push or a pull," said Jacob. "What else can you include in that definition? Let me put it on the blackboard as far as you have stated it. You said that a force is a push or a pull. What does it do, Jacob?" "It causes something to move," said Jacob. "All right now. Give me the complete definition, Susan." "A force is a push or pull which causes something to move," said Susan. "Fine. What other things in this room are moving?" The students gave Mrs. Guevera a number of objects in the room that were moving. She called on the students one by one and placed the name of the object on the chalkboard. The list included fingers, the second hand of the clock in the room, the heart, arms, air, and eyes. The rest of the class continued to make additional contributions to the list.

The teacher then asked what objects are at rest. The students contributed one by one. The list included chairs, desks, the clock box on the wall, the demonstration table, and the ring stand on the demonstration table. "Fine." Mrs. Guevera also asked others to make contributions to the list. She then asked, "Now, does this mean that those objects at rest cannot move? Timothy?" "No, they can all be made to move, but something has to cause

them to move." "What can cause them to move, John?" "A force can cause anything to move, even this building. This building could be dynamited and then it will move. The dynamite can cause the force," said John. "Now let me ask another question about forces. Does a force always have to produce motion? Gerald?" "No," said Gerald. "Explain that answer, Mark." "Well," said Mark, "if I push down on my desk top, I am exerting a force, but no motion occurs." "Then what has to happen in order for movement to occur, Jim?" "For a force to cause motion it has to be greater than a resistance force." "Excellent," said the teacher. "But you have now introduced a new word—resistance. What is resistance, Joe?" "Resistance is the force that prevents the desk from moving," said Joe. "What do you mean, Joe?" said the teacher. "Well, the floor must be preventing the desk from moving, since the desk is attached to the floor, so the floor must be exerting a resistance force on the desk." "Excellent, Joe. Then what is resistance, Bertha?" Bertha said, "I don't know. I don't understand it." "All right, let's see if we can help Bertha understand this. What is resistance, Hans?" "Resistance is a force that opposes motion. It can slow an object down or it can prevent it from moving." "Good, Hans. Bertha, do you understand that?" said the teacher. "Let me ask you another question, Bertha. Please stand up first. Try this. Can you exert a force—a push or a pull—great enough to move your desk? Try to move your desk, Bertha." Bertha attempts to move the desk but cannot because it is attached to the floor. Bertha said, "I can't, it's attached to the floor." "How could you move the desk then, Bertha?" "Only if I were strong enough to pull it off the floor." "Good. Then what is the resistance in this case?" asked the teacher. "It's the force on the desk from the floor, and I can't move it because I can't exert enough force to pull it up." "Now you understand, Bertha. Can you exert enough force to overcome the resistance force on your desk top? Mora?" "Yes, but it is not very heavy, and I'm strong enough to lift it."

"Now, let us summarize what we've learned. We've learned what a force is." The teacher asked for a restatement of the definition of a force. "And we've also learned what resistance is." The teacher asked one of the students to define resistance. "We've also learned that a force does not always produce motion and that for something to move, the force has to be great enough to overcome any resistance. Fine. Now let's perform a demonstration involving forces." Mrs. Guevera then moved toward a demonstration apparatus that she had set up before the class began, involving the use of pulleys to be used to lift objects to illustrate force and resistance. ▲

Mrs. Guevera's recitation revolved around a reading assignment from the textbook. She introduced the topic by making comments that were meaningful to the students. She asked very simple questions to start the lesson and then gradually developed the concept of resistance. In all instances, she never answered a question herself. She gave many students an opportunity to respond. Seldom did she ask the same student more than one question unless it involved the completion of a train of thought. The lesson proceeded at an excellent pace. She managed to use objects in the classroom to illustrate the points. She involved certain students by asking them to perform simple exercises to make certain they understood what was taking place. Mrs. Guevera stopped at an appropriate time, after the ideas were developed, and proceeded to another activity—the demonstration involving pulleys—to further illustrate the lesson on force and resistance.

In summary, some suggestions for conducting recitation sessions follow:

1. Prepare questions in advance in the sequence in which they will be presented.
2. Use questions that require sentence or phrase responses.
3. State the question to the class, and then call on a specific student for a response.
4. Give everyone a chance to answer questions. Avoid asking only the brighter students to answer questions.
5. Acknowledge incorrect responses by asking another student to provide the answer. Do not ridicule or chastise students who give incorrect responses.
6. Ask another student to add to incomplete or partially correct responses.

7. Repeat the responses to all questions so that all students can hear what has been said.

THE DISCUSSION METHOD

The classroom discussion is one of the most powerful strategies that a teacher can use to facilitate cognitive and affective gains in students. This approach can be used to provide students with opportunities to clarify their own values and make informed decisions with regard to everyday social problems. It can also be used to promote inquiry and develop problem-solving skills.

There is a strong agreement among science educators that contemporary social issues must be included within science instruction. In this context, the discussion method provides ways to facilitate the exchange of ideas on societal issues and topics that require the students to make judgments and clarify their own values. Through the discussion, the student will come to understand the issue with increased depth and comprehension. By listening to the positions of others and by presenting her position, the student will be able to clarify her own values as well as gain insight and understanding concerning the views of others. In short, discussion of social issues will facilitate students' understanding of each other and their ability to make informed decisions about science-related societal issues.

The science teacher can also use the discussion method to promote inquiry. Through discussion the teacher can challenge the students to discover things for themselves by setting up situations that can stimulate and prompt them to ask questions, assimilate and analyze information, and draw their own conclusions. The discussion method provides an opportunity to involve students in discovering new knowledge for themselves. The advocates of the discovery method draw support from cognitive psychologists Piaget and Bruner. Bruner in particular, who has been a great proponent of this method, states that "the emphasis on discovery indeed helps the child to learn the varieties of problem solving, of transforming information for better use, helps him to go about the very task of learning" (1962, p. 87).

The teacher who knows how to ask good questions that motivate students to learn, think, and discover will be able to conduct fruitful discussion sessions where students will be able to understand a topic in depth and with clarity. Teaching, according to Bruner (1962, p. 83), must provide the "intellectual potency to stimulate students to find things out for themselves and thus gain knowledge on their own." This can occur through discussions, but students also learn the procedures to discover this knowledge. And through discussions the teacher can show them ways to understand generalizations in order to gain further insights about the topic under consideration.

The experiences a student gains through classroom discussion are valuable at many levels. Open discussion among students tends to enhance mutual respect, self-esteem, cooperation, and communication skills. In addition, discussions that are directed toward the solution to a problem facilitate cognitive gains such as critical thinking, inquiry, process skills, and overall achievement. Once a student learns that she can produce knowledge on her own by being an independent thinker, she will be able to use these methods throughout her whole life—during adulthood when there will not be a teacher to lead and prompt her. In short, the discussion helps students become independent thinkers and consequently, good citizens in a democratic society, where freedom of thought and expression are highly regarded and cherished.

Teacher-Led Discussions

The teacher who leads a discussion must be able to handle many different situations with tact. It is the teacher's task to select the participant who wishes to speak when several students want the floor at the same time. The teacher must be patient as he listens to rambling and irrelevant comments. He must be able to end them as promptly as possible without being discourteous and then direct attention back to the subject under consideration.

The topic has much to do with the success of a discussion. Some topics are not appropriate for class discussion; others are very suitable. First, the topic must be one about which students have knowledge. Only then is it possible for them to do more than speculate idly. Controversial issues on which students take a stand also can make excellent subjects for discussion. These issues may deal with matters of current interest such as the abortion issue, the creationism-versus-evolution controversy, drugs, and others. They may deal with topics centered around questions that require problem-solving skills to answer—for example, statements such as: "Some feel that all complex behaviors are learned, while others believe that many behaviors are inherited. Which position do you support and why?" "Sedimentary rocks can be distinguished from other classes of rocks on the basis of bedding, color, fossils, ease of breakage, and porosity. Which do you feel are the poorest criteria to use as distinguishing characteristics and why?" Excellent discussions often center around conflicting laboratory data.

Topics that are concerned with indisputable facts do not lend themselves to discussion unless students are uncertain about the facts under consideration. It would not be possible, for instance, to discuss the properties of air if the students in a class already know these properties. The only discussion possible in such a case would involve a review of the properties of air.

A teacher is frequently faced with the problem of dealing with students who contribute irrelevant and facetious remarks. In such a case it is good practice to treat all remarks as serious ones. Sometimes students blurt out statements that seem irrelevant to the discussion. Courtesy demands that the teacher recognize the student and perhaps with a brief comment indicate to her why the point that was made does not have relevance. If the student is serious, she will un-

Discussions can be effective when they raise questions and problems that students feel are worth solving.

derstand why the comment did not fit the discussion. The student who continuously interjects facetious remarks can be dealt with in a private conference at the end of the class period, during which the teacher can explain that the student will not be given an opportunity to participate in future discussion if the practice continues.

Teachers should not try to answer all questions that arise during the discussion. Only a few teachers have the background necessary to respond to all questions with care and accuracy. More important, when a teacher answers a question, it tends to bring to an end the intellectual activity that initiated the question. If a student asks, "Will this piece of ice float in kerosene?" and the teacher responds by saying "Yes," there is nothing more to say or do. But if the student is told, "Why don't you try it and find out?" the teacher is initiating a real learning situation.

Relevant questions can be used to set the stage for problems to be solved by the class, by

groups, and by individuals. Sometimes interested students can be detached from the rest of the class and assigned to a group to work on a problem. More often, the teacher must explain why a particular question cannot be dealt with during the class but that it would be possible to make arrangements for the student to follow up her interests at a later date. Much depends on the nature of the question and the teacher's ability to see if the question has potential for future discussion.

A class discussion generally wanes after several minutes of interaction. Most students have difficulty giving their undivided attention to discussions that last an entire period, because, in reality, only one student can actively participate at a time; others sit trying to listen intently to what is being said. The larger the discussion group, the less often a student can participate actively. The topic would have to be unusually stimulating to maintain students' attention on one topic for a significant amount of time.

What happens in many situations involving discussion is that a few students dominate the discussion while others sit and contribute little or nothing. The teacher must take care to avoid boredom, which may result in discipline problems. Short discussion periods are generally more effective than long ones and avoid the possibility of students engaging in daydreaming, conversation with other students, mischievousness, and other activities that can degenerate into discipline problems.

Preliminary Considerations

Discussions have some very important functions in the science program. Unfortunately, science teachers use the term *discussion* when they are engaged in almost any type of verbal interaction with their students—questioning, lecturing, stating procedures to be followed, and so on. When a teacher says, "We will now discuss the topic of population growth," what he really means is, "I am going to lecture to you on the topic of population growth." A teacher sometimes says, "Let's discuss how we will attack this problem," meaning "I will now tell you precisely what to do to perform this experiment." When the teacher says, "Let's discuss the findings of the experiment," he means, "I will tell you what conclusions you should draw."

During a classroom discussion, all students must be free to express their viewpoints. This means that students must have the background and preparation to make contributions. The teacher, too, must have the interest and background of knowledge about the topic.

The degree of participation in a discussion will vary with a topic, the ability of the students in the discussion group, and the psychological readiness of students to discuss the topic. The participation will also depend on the teacher and his skill in conducting discussion sessions successfully.

Students cannot discuss a topic in depth unless they are interested in it and have the so-phistication to understand it. The maturity of students should be taken into consideration when the topic is selected. Topics that are of interest to ninth graders may not be of interest to twelfth graders, and vice versa. The teacher should select topics with care before attempting to organize discussion sessions based on them.

The personality and preparation of the teacher can either promote or stifle a discussion. A teacher who is highly opinionated, authoritarian, and an extreme disciplinarian cannot promote free discussions. Teachers who are not threatening to their students, present a relaxed atmosphere, and can withhold their opinions will be successful at using this method.

Teachers must have adequate preparation to be able to ask the right questions and be able to pose questions that will lead the discussion. The teacher must be able to construct questions that students can understand and that will motivate them.

The physical environment in which a discussion takes place has a great deal to do with the success of a discussion. A dingy, hot room with uncomfortable seats will produce poor results. On the other hand, a light, airy, attractive room produces an atmosphere that will encourage students to engage in discussion activities.

True classroom discussions imply that the teachers do not dictate or seek to influence the opinions of students as they lead discussions. If they do, they are actually lecturing. Free discussion is the most effective way to raise problems and deal with controversial issues. Free discussions give students a chance to exchange ideas and interact with their fellow students so that they can draw their own conclusions, make their own judgments, and develop and express their values. Free discussions are ideal to obtain verbal summaries from students, and they are the only means to stimulate and bring out teacher-student planning.

Good classroom discussions require special skill. The teacher must have the skill to maintain attention centered on one single point without discouraging students who make irrelevant remarks. It also takes skill to control situations so that a few students do not monopolize a discussion, preventing others from making contributions. Asking students who are constantly eager to speak to hold their comments until others have had a chance to make their contributions requires a certain amount of tact. It also takes a greater amount of skill on the part of the teacher not to influence the discussion by stating his own opinions.

Organizing Students

It is important for teachers to realize that they cannot conduct discussions under certain conditions. Because discussions require considerable interaction and interchange among students, it is necessary to limit the number in the group. A skilled teacher can probably conduct a good discussion session with somewhere between fifteen and twenty students. Too large a group prohibits maximum participation, so that the session becomes a lecture where students occasionally make comments and ask questions. The optimum number for a discussion group is between ten and twenty. Below ten the teacher will discover that the group will not be large enough nor diverse enough to construct a fruitful discussion. The pool of knowledge from which a small group of students can draw is limited, and therefore a discussion can be stifled because of the limited amounts of information that the students can contribute.

In order for students to exchange ideas and comments, it is advisable to seat them in a circle or other arrangement so that they are close enough to interact. The physical arrangements should permit students to speak so that others can hear without raising their voices and to sit face to face so that they can observe each other's facial expressions. Physical arrangements are important to create the proper environment for expressing viewpoints and raising questions.

If the teacher plans a discussion around inquiry sessions, it is recommended that the seating arrangement be around the demonstration table or other areas where activities involving the entire class will take place. The seats and apparatus should be placed to allow interaction among the teacher and students and maximum view of the demonstration table and other props. If the teacher wants the students to handle materials and conduct an experiment in small groups before discussion, he should have a number of sets of materials available to avoid confusion, disruption, and disorganization. In this way students can work on a problem in small groups and later reorganize in a class session for the actual discussion of the problem under consideration.

Discussion sessions should also take place in areas where students and teachers have easy access to reference materials. The availability of library and laboratory references will facilitate the collection of information at appropriate times as the discussion warrants. Students will then come to realize that it is important to have data readily available to continue the study and discussion and that without them the discussion would not be fruitful. The teacher should stop a discussion at the appropriate time and ask students to break up into small groups or independent study sessions to gather the data, after which the class can reconvene for further discussion. This ease in changing from large to small groups or independent study sessions and vice versa will facilitate meaningful discussions among students. Sometimes, when outside specialists are called in to participate in the discussion or to be involved in small group sessions, this need to regroup efficiently is essential.

Small groups or independent study sessions may be especially useful if there are specialists at hand to help conduct discussions. These specialists may be other teachers in the school or interested persons from the community who are specialists in the area under consideration. One teacher may find it difficult to answer all the possible questions that may arise during independent study sessions, and the help of specialists would certainly be in order to guide students and answer their questions. In this way students will be able to contribute more and with greater authority during large group sessions.

The success of the discussion method depends on the flexibility of the teacher, students, physical arrangements, and scheduling. A disruption of any of these may diminish the effectiveness of a discussion.

Initiating Inquiry Sessions

Discussion can be used to motivate students to discover things for themselves. They are effective when they raise questions and problems that the students feel are worth solving.

▼   "I know that many of you smoke cigarettes," Mrs. Ardito announced to her class. "And I also know that many of you who are not smoking now will smoke later in your lifetimes. I am certain that you have heard various arguments against smoking. As a matter of fact, those of you who smoke, and perhaps some of you who do not, realize that the federal government requires that manufacturers print on packages of cigarettes that cigarette smoking may be detrimental to your health. Let's try to find out what is behind some of these arguments against smoking."

For the next several minutes a number of students gave the rest of the class information and misinformation they had gathered incidentally from their reading and through conversations they had had with adults and fellow students. Most of the students finally concluded that none of them knew much about the subject but wanted to find out more. One of the girls in the class declared that she wanted to find out more about the dangers of smoking and asked Mrs. Ardito to suggest some reading material on the subject. Other students expressed the same interest, so Mrs. Ardito passed out a bibliography she had prepared on the topic. She assigned all students to read certain articles and asked certain students to read additional ones.

Mrs. Ardito then proceeded to acknowledge one of the boys, who told the rest of the class that his father had to give up smoking because of ulcers that his father had acquired. Another boy indicated that his uncle died of lung cancer, which the uncle's physician attributed to heavy smoking. Other students contributed by letting their fellow stu-

dents know about other instances where an individual had a medical problem that was attributed to smoking. The discussion finally led the students to ask various physicians to give them information about medical problems that they believe were attributed to smoking. The boy whose father had given up smoking because of ulcers said that he would ask his father's doctor what smoking had done to his father.

The discussion finally ended in a planning session. The problem was defined; several suggestions were made by the students regarding the procedure they should follow at this point. Some students indicated that they wanted the chance to read first before proceeding. Others said that they wanted to interview parents, doctors, and insurance company representatives to understand the problem more fully. Mrs. Ardito realized that she had motivated them to study the topic. She asked the students to read the material she had assigned, and based on this reading, she would continue the planning session to determine how they would further proceed.   ▲

Mrs. Ardito made an excellent choice of topic for discussion, because many adolescents are faced with the problem of smoking. Most of them resent being told by their parents and teachers that they should not smoke, but many of them welcome the opportunity to discuss their problem. The topic as introduced by Mrs. Ardito was one for which the students had adequate background to make a useful discussion possible. Some students had had some contact with the problem, directly or indirectly, which made it more vivid. Some even had well-formed opinions about the subject. On the other hand, most students realized they lacked information and data to discuss the problem in depth. Controversy even occurred, which motivated the students to want to investigate the problem further.

Controversial issues such as smoking, drugs, or abortion made good topics for problem-setting discussions. Matters of current interest can also be used to initiate this type of discussion.

Teachers can set up demonstrations or display unusual materials to initiate problem-setting discussions.

▼ Mrs. Garvey, the science teacher, brought to class a large wasps' nest she had found in her back yard during the winter. She held the nest up so that the whole class could see it, and then she passed it around to see if the students recognized what it was. Some of the students thought it was a hornets' nest, others thought it was a wasps' nest, and some of them even insisted that it was a bees' nest. Mrs. Garvey gave each student a piece of material that covered the nest and asked for a description of it. Several students attempted to describe it and finally concluded that the material was paperlike, brittle, and gray. The teacher then cut the nest in two to display its interior. Some dead insects were found, which were placed in a petri dish and passed around the class. Many of the students decided that the insects were wasps; others were convinced that they were hornets. Those who thought the nest originated from bees conceded that the insects were not bees. They wanted to determine definitely whether the insects were hornets or wasps. They talked about the procedure they should use to determine how to identify the insects. They then discussed the possible function of the comb, proposing that it was used for honey, even though no honey was found in the comb. The students listed a number of questions they wanted to have answered about the nest and its origin. They made several proposals as to how to proceed to answer the questions.   ▲

It is not always good practice to start a new topic with a discussion. Students do not always have sufficient background or interest to make the discussion profitable in the early stages of a study. Other approaches to generate interest and motivation can be used that are generally more suitable and more valuable. Initial discussions to start a topic should be used sparingly and only when the teacher feels the students are prepared through background and past experiences so that they can make significant contributions.

Conducting a Guided Discovery Discussion

In a guided discovery session the teacher identifies a problem and structures the lesson around activities that will direct students to discover new knowledge. Guided discovery lessons involve a great deal of planning and structure on the part of the teacher. The teacher must have considerable skill in constructing pertinent questions and the ability to ask the right questions at the right time to guide the students toward the solution of the problem without inhibiting their thinking. The questions should be thought-provoking and useful to resolve the problem under study.

The teacher must construct questions beforehand to ensure that the discussion runs smoothly. The questions should also be outlined to be given in a particular sequence. The teacher should plan more questions than he will probably be able to use during a session. It is better to overplan than to underplan the lesson.

Planning discovery lessons requires expenditure of time. The teacher who has not had practice in formulating and asking questions will need a great deal of practice in writing discovery lessons and developing his question-asking ability. With experience, the chore becomes less burdensome.

▼ Mrs. Zeidler performed a simple demonstration involving a discrepant event. She filled a collecting bottle full of water to the brim and placed a 3-by-5-inch piece of cardboard (an index card works as well) over the mouth of the bottle. She then inverted the bottle holding the cardboard firmly over the mouth and removed her hand, which had been supporting the card. The students observed that the water remained in the bottle even though the cardboard was not supported in any way. Mrs. Zeidler then asked, "What keeps the water from running out of the bottle?" John said, "Did you use glue to stick the cardboard to the bottle?" The teacher permitted John to inspect the cardboard and the bottle to determine for himself that no glue had been used. Then Mrs. Zeidler said, "You probably still don't believe me. I'm going to ask you to perform the experiment yourself." So she divided the class of twenty students into five groups of four. She had already placed a collecting bottle and a 3-by-5-inch piece of cardboard on each of five laboratory tables. The class then moved to the laboratory tables, where they proceeded to repeat the demonstration.

The students began by asking one of the members of the group to repeat the operations that Mrs. Zeidler had used during the demonstration. Three of the five groups permitted all the students to conduct the experiment, all of whom were successful. Two groups were unsuccessful in their attempts. Mrs. Zeidler did not provide any help or explanation. These students still believed that some trick was involved.

Mrs. Zeidler then asked all the students to reconvene as a class. She asked one of the students who successfully performed the experiment to repeat in front of the class what he had done. This attempt was also successful. She asked one of the students from an unsuccessful group to perform before the class. He again failed. She asked him to repeat it but this time required one of the students who had successfully performed the operation to observe what the unsuccessful student was doing. The student making the observation noted that the demonstrator had not placed the cardboard flush against the mouth of the bottle. He asked the student to correct this, and the demonstration was repeated with success.

Mrs. Zeidler then posed the question, "Why did the water remain in the bottle when the bottle was inverted and no support was holding the cardboard in place?" John answered, "The water was not strong enough to push the cardboard out of the way." Mrs. Zeidler asked, "Do you mean strong?" John said, "Maybe I mean heavy enough to push the cardboard down." "Who agrees with John?" Several hands went up. Several students had no opinion. Tom, one of the students who disagreed with John, asked, "Why doesn't gravity just pull the cardboard down?" Mary retorted, "The cardboard sticks to the lip of the bottle because the cardboard absorbed some water and the moisture acted like glue."

Mrs. Zeidler asked the class, "How can we test both John's statement and Mary's statement?" Jerome suggested that they attempt to show Mary was right by moistening a piece of cardboard and sticking it on the blackboard. Mrs. Zeidler asked Jerome to try this before the class. He did so and the cardboard fell to the floor. Mrs. Zeidler asked, "What does this prove?" John said, "This proves

nothing, because the blackboard is not glass and the cardboard is not in the same position on the blackboard as it is when the bottle is used." Mary said, "Let's try sticking a piece of wet cardboard on a piece of glass and hold it upside down." The teacher obtained a piece of glass about eight by ten inches and asked Mary to conduct the demonstration. The cardboard fell to the floor.

Mrs. Zeidler said, "All right, what is holding the cardboard on the bottle filled with water?" Robert said, "What about air?" Mrs. Zeidler replied, "Let's discuss some properties of air to possibly answer Robert's question. What is air?" Jonathan said, "Air is matter; it is colorless, odorless, and tasteless." Mrs. Zeidler said, "What are some characteristics of matter?" Dorianne said, "It has mass, weight, and takes up space." Norman said, "Well, when we talked about air last week, we performed an experiment where we weighed a deflated balloon and then weighed it again after it was inflated. The air must have mass because it took up space in the balloon." Jerome was acknowledged and said, "We know that it has weight because when we weighed the inflated balloon, it weighed more than it did before it was inflated." Mrs. Zeidler said, "Fine. What else do we know about air?" Christopher said, "We know that air exerts pressure." Mrs. Zeidler said, "Christopher, how do you know this?" Christopher responded, "Air must exert pressure because it stretched the inside walls of the balloon." Theresa said, "What about inflating tires? Isn't there a lot of pressure in the tire when it is inflated?" Norman said, "Sure, because when I inflate my bicycle tires I set the pressure gauge at the air pump at 55 lb. I make sure I don't put more air in the tire because I'm afraid I'll blow up the tire." Mrs. Zeidler asked, "If there is air pressure in an inflated tire, then is there air pressure outside the tire?" Kathy responded, "There is. Wherever air is present, it exerts pressure." Mrs. Zeidler said, "Is there air pressure around this bottle?" Mabel said, "There must be, because air is everywhere. There is air around our bodies and in this room and outside." "Is there air pressure being exerted on the cardboard, then?" Gary said, "Yes. If air pressure is exerted everywhere, including the bottle, then it must be exerted on the cardboard." Mrs. Zeidler said, "How is air pressure measured?" Christopher answered, "It must be in pounds because the pressure gauge at the gasoline station that I use says air pressure is in pounds."

Mrs. Zeidler said, "Has anyone else an idea of how air pressure is measured?" David said, "I don't know how air pressure is measured, but I do know that when I use my hand pump, I push air into my bicycle tire. I then use my father's pressure gauge, which tells me that I have 55 lb of pressure in the tire." Larry raised his hand, and Mrs. Zeidler said, "What do you want to add?" Larry stated, "I just looked up air pressure in the index of our book, and on page 328 it states that air pressure is measured in pounds per square inch." Mrs. Zeidler asked, "How many pounds per square inch?" Larry said, "Fourteen and seven tenths pounds per square inch." "That's correct," said Mrs. Zeidler. Then she asked, "How many pounds of pressure are being exerted on this 3-by-5-inch piece of cardboard?" Timothy said, "We first have to find out how many square inches there are in the piece of cardboard." "How many are there, Timothy?" asked Mrs. Zeidler. "If we multiply 3 times 5, the number of square inches is 15 square inches for that piece of cardboard." Mrs. Zeidler asked the class if Timothy was correct. Most of the students shouted yes in agreement.

Mrs. Zeidler asked, "If this cardboard measures 15 square inches, then how many pounds of pressure are being exerted on this cardboard? How do we determine this?" Norman said, "If we multiply 15 square inches by 14.7 lb, we will find out the total amount of pressure forcing the cardboard against the mouth of the bottle." "How much is this? Come up to the blackboard and calculate this for us, Patricia," remarked Mrs. Zeidler. Patricia calculated a total pressure exerted on the cardboard as 220.5 lb. "Is there enough force to hold the card in place? There must be, because the card stays in place every time we perform the experiment." Mrs. Zeidler asked Norman to summarize what had been learned up to this point. Other students helped Norman with the summary.

Mrs. Zeidler then asked the students to suggest other types of experiments involving the bottle of water and cardboard. Some students suggested trying the same demonstration using large bottles and heavier cardboard. Others suggested filling the bottle practically full of water, and some even went so far as to find out whether the cardboard would remain in place without any water in the bottle. ▲

Mrs. Zeidler's demonstration involved a discrepant event. She gave students an opportunity to develop and test hypotheses regarding the event. She asked open-ended questions, which were very well worded. She allowed the students to discover the concepts themselves and gave them very little information. Her lead questions were given in the right sequence and permitted the students to think about other situations involving air pressure. Her questions gave the students a chance to do some thinking, using information they had gained through a prior laboratory exercise and through the experiences in inflating bicycle tires.

She also asked them to think up variations of the original demonstration and to formulate new hypotheses, which they could test.

How would you categorize the teaching episode described above? Would the teaching be more aligned with an inductive approach (learning cycle) or a deductive approach to instruction? How effective was Mrs. Zeidler, the science teacher, in modifying the students' conception of air pressure? Refer to the discussion of the learning cycle found in Chapter 4 to help answer these questions. Also Chapter 3 provides further information to help categorize Mrs. Zeidler's lesson.

Discussions for Student Planning

Students often become interested in science topics that are associated with events about which they have read in the local newspaper or news magazine or that have been brought to their attention by television, their parents, or other individuals. It is not unusual for a student to ask the science teacher questions concerning the event or topic during the science period, particularly if it pertains to the area being studied. Teachers have the tendency to take students' interest lightly by saying, "We don't have time to discuss this now," or "I don't know anything about it." Or they may answer the specific questions as quickly as possible and pass up the opportunity to permit students to satisfy their interests and curiosity. Some of the topics brought up in this way may be of great concern to students, and teachers should capitalize on this. It may be appropriate to postpone the discussion of the topic if the event or topic has nothing to do with what is being studied, assuring the students that they will have an opportunity to delve into the topic at a later date. If, on the other hand, the topic is directly related to the unit under study, then it is the opportune time for students to plan and conduct discussions on the subject.

Students can plan their activities on a topic of interest with a minimum of guidance and direction by the teacher, but they need time during the class period and outside of class to plan their attack—talk over procedures, make decisions, and assign responsibilities to their fellow students.

▼ There was a serious water shortage in the city. The students in Mr. Charles's class had heard a great deal about the situation from their parents, other adults, and television. By coincidence, the class was studying a unit on water, and one of the students asked Mr. Charles during the class if he could spend a little time discussing the topic. Mr. Charles asked students if they had any ideas about the causes of the shortage. He learned they knew very little about the causes of the shortage and even less about the water supply system that was servicing the community.

Mr. Charles suggested that the students take the time to list the questions they would like to have answered concerning the situation. He also suggested a general reading assignment in the textbook; the reading was not only directed toward broadening their thinking, but also permitted them to think about other questions that they could add to their lists. The students brought in specific questions that they wanted to answer. They discussed these, and as a class they made suggestions of things to do. Some suggested making a map of the watershed, the storage reservoirs, and the aqueducts that deliver water into the city. Some thought

it advisable to contact the local newspaper to obtain local rainfall records. Other students planned a bulletin board display of the city's water system. Through discussion they suggested that they perform experiments and demonstrations described in their textbook and other sources. Mr. Charles also suggested some when he was asked to contribute. They planned to make studies of the amounts of water needed for running dishwashers, flushing toilets, taking showers and baths, and watering lawns. Mr. Charles also suggested that they find out how much water is lost through a dripping faucet or a continuously running toilet.

As suggestions accumulated, some of them gave rise to new questions to be answered. Finally, the students planned the whole discussion around important questions to be answered. They portioned out the tasks, each student volunteering to work alone or with one or two others on pertinent projects. ▲

The teacher who allows students to plan discussions using minimum guidance and direction must be willing to permit the students to plan their own approaches and procedures. The teacher may have his own ideas, but they should be used only to amplify or round out those of the students.

Discussing Results of Experiments

Class discussion revolving around the significance of data derived from experiments or demonstrations can be among the most worthwhile activities of a science class. Students who have comparable field or laboratory experience are able to communicate effectively with each other. They can make collective judgments, think through issues, study exceptions and discrepancies, and qualify the conclusions they make. Through these discussions, they may find discrepancies in their data, which stimulates them to repeat their experiments and make observations under different and more carefully controlled situations.

▼ During a laboratory activity, the ninth-grade science students had tested the force needed to draw some loaded toy trucks up an inclined plane.

They had concluded that less force was needed to draw the trucks up the inclined plane than was needed to lift them up vertically.

After the students had done some reading, they discovered that the mechanical advantages they had determined in the laboratory did not agree with those the textbook reported, even though they thought they had followed the directions carefully. One of the students discovered a discrepancy. He noted that they had neglected to use low-friction metal rollers described in the text, and he believed that friction was obviously the cause of the differences. The teacher provided them with rollers, and the students repeated the experiment, this time obtaining results closer to those reported in their text.

The discussion that followed led the students to consider "efficiency," and a new problem arose as a consequence. Efficiency seemed to increase as the slope increased. New experiments were planned to test this, using various devices such as metal rollers, toy trucks, blocks of wood, and roller skates. ▲

The discussion of the results the students obtained in the first series of experiments gave them an understanding of some of the factors that had an effect on their results. They then devised methods and procedures to find the cause of the discrepancies between their findings and those reported by the text. The reading in the text revealed that they had neglected to use a certain device, which could have had a bearing on their results. Fortunately, the teacher was able to provide the students with the device so they could follow the exact procedures described in the text. However, even if the device was not available to incorporate in their experiments, they would still have understood the reason for the discrepancy and the need to qualify their conclusions.

Discussing results and data permits students to see how the scientific method operates. Complex directions for highly refined experiments are apt to be followed in cookbook fashion with little understanding of what is actually happening. Simple experiments permit students to see discrepancies quickly, because they can under-

stand what is actually taking place. This understanding allows them to think of possible refinements and new plans of attack to obtain additional data. This is the scientific method in action.

Discussions for Review and Summary

Generally, teachers use question-and-answer techniques to conduct reviews and oral summaries. However, if the teacher is willing to let students conduct their own reviews and summaries on occasion, he will find that they are very capable of handling short periods of free discussion for this purpose.

Students are usually able to deal with limited amounts of information at any given time, especially if this information has recently been covered by class work. They may not have the maturity to conduct major summaries and reviews without help. For major and lengthy reviews, the teacher should provide students with an overall plan to guide the discussion but at the same time permit them to organize the discussion and assign the various tasks to the students in the class.

▼ Mrs. Lopez conducts "buzz sessions" for review and summary purposes. She gives the students a list of topics that make up the unit to be reviewed. The students organize themselves in groups of three and four, and each group is assigned a topic for which it is responsible. During their buzz sessions the students discuss and review what they have learned about the topic they have been assigned. They use the text and their notes as a basis for discussion. They prepare themselves to review the topic before the entire class and anticipate questions that the students might ask to help clarify various important points. ▲

Student-Led Discussions

What has been said about teacher-led discussions also applies to those led by students. The suitability of a topic, participation, and duration of discussion are important considerations when students lead discussions. Discussions conducted by students can degenerate into a useless activity unless the teacher gives them the guidance they need to conduct a successful discussion. The student selected to lead the discussion should be a natural leader, one who has the respect of the students in the class, who is articulate, and who will take responsibility seriously.

The most successful discussions led by students are those that center around their experiences. Student-led discussions are effective for planning attacks on problems as well as discussing data derived from previous laboratory work. Students cannot effectively lead lengthy discussions that require very extensive backgrounds of reading and experience. Student-led discussions centered around societal issues or controversial problems probably would not be successful.

▼   Mr. Druger had a group of twenty gifted students who were interested in physics. He gave them many responsibilities—more than what he would assign to students in a more typical class. After the completion of a unit on light, he gave each student in the class a copy of an instruction booklet concerning a well-known make of camera. Mr. Druger explained that he was going to give them an opportunity to use some of the physical principles they had learned while studying the unit on light. Then he announced that they were going to be involved in a unit on photography and that the students were going to organize it. Mr. Druger proceeded to appoint a student leader and asked her to organize and conduct a planning session to decide what kinds of activities they were going to be involved with.

Under the student leader's guidance, the students came up with a number of problems to attack—the meaning of depth of field, the focal lengths of lenses, significance of f-stops, and light intensity factors. The students divided into groups to work on a topic and devised experiments to help them understand the topic under study.   ▲

Small group discussions led by students are sometimes more successful than class discussions. There are more opportunities for students to interact. However, care must be taken to en-

Small group discussions led by students are often more successful than class discussions. However, teachers must assign leadership roles to students who will take their responsibilities seriously and have leadership capabilities.

sure that the discussion will run smoothly. Again, teachers must assign leadership to students who will take responsibilities seriously and have leadership capabilities. The teacher must also provide guidance, giving advice as it is needed and having individual conferences with the student leaders so that they know what to do and what the expected outcomes should be. Without guidance, the small group discussion will not be successful.

The teacher, during most classroom discussions, serves as a moderator. In summary, to lead a good discussion session the teacher should be able to do the following:

1. Seat students in a circle or horseshoe arrangement so that they interact easily and can observe each other's facial expressions. Facial expressions also indicate a type of communication among participants.
2. Keep the discussion moving at a reasonable pace.
3. Keep the discussion pertinent to the topic under consideration.
4. Encourage all students to participate. Do not allow two or three students to monopolize the conversation.
5. Acknowledge all contributions that the students make.
6. Reject irrelevant comments with tact.
7. Summarize frequently or permit students to summarize as often as is feasible.
8. End the discussion when the students begin to lose interest.

USING AUDIOVISUAL AIDS IN SCIENCE INSTRUCTION

Audiovisual aids can be used at any time during science instruction as long as they enhance instruction. They can be employed during a lecture presentation, a recitation, a demonstration, a discussion, laboratory work, project work, and other forms of instruction. Visual aids can be integrated during instruction for purposes of elaboration, explanation, clarification, illustration, or emphasis. Visual aids are excellent to use to summarize a lecture or a discussion. They are often employed by teachers to introduce a lesson, a unit, a lecture, a discussion, or a laboratory activity. They can be used to make points clearer and more meaningful to students, present information, introduce a problem, or initiate a discussion. They can bring nature into the classroom and expose students to new experiences. Whatever the use, the science teacher must first determine their effectiveness in instruction. The type of visual aid to be used will depend largely upon what it will do to improve science instruction and learning.

Many types of visual aids require special equipment. All audiovisual aids involve seeing or hearing or both, and some require touching or feeling along with seeing or hearing or both. Visual aids include 16 mm films, 2-by-2-inch slides, filmstrips, video-cassette recordings, and film loops. To project 16 mm films, a 16 mm film projector is required. A videotape recorder and a television set will project a videotaped film directly on a TV monitor. To project 2-by-2-inch slides requires a slide projector made for this purpose and to project film strips requires a film strip projector. Film loops are projected by using a film loop projector.

Audiotape recorders can be used to record sounds in nature such as bird calls and frog calls, as well as classroom activities while they are in progress and science programs that are being broadcast on television or radio. Television sets are employed to view programs of interest as they are being broadcast, or view videotape recordings of TV broadcasts or those that are commercially available.

Overhead projectors are useful to project written material that a teacher would normally place on a chalkboard. The projector is placed in front of the classroom where the teacher operates it while facing the class.

Other aids commonly used in science instruction include the chalkboard or blackboard, charts, and models. These aids do not require the use of specialized equipment.

Educational Films

Educational films or videotapes for VCRs are usually in the form of complete film productions. Film loops (single concept films) are short productions lasting somewhere between 1 and 5 minutes. Film loops are used to present a single concept or event that can be used in instruction to present a problem or initiate a discussion.

Films are useful in science instruction to help students gain new experiences some of which are unique and impossible to obtain in any other way except through a film presentation. They are also useful during science instruction to pose problems, provide an overview for a unit of study, introduce a unit of study or a lesson, summarize outcomes and learnings, and review material covered during a lecture, discussion, recitation, or laboratory exercise. No single film can be used for all functions mentioned; a film is generally suitable for one function at a time.

Using a Film for Instruction. Any activity that occupies 15 or more minutes of instruction time should be carefully planned. A film should be shown only if it serves a purpose for instruction and learning. It should not be shown simply to fill up time during a class period. It should be as important an activity as a demonstration, laboratory exercise, discussion, or recitation.

A teacher should undertake several preliminaries before showing a film. These include previewing the film, reading the manual that accompanies the film, preparing students to view the film, and planning follow-up activities.

Previewing the film is necessary to determine whether it will serve the purpose for which it is intended. By previewing the film the teacher can learn whether it can be used to introduce a lecture or a discussion, to set a problem, or to summarize. The teacher can also determine whether the students can benefit from viewing it. He can ascertain what preparation the students need for understanding the vocabulary used in the film and the types of experiences students should have before they view the film. During the preview, the teacher can plan instructional material that will guide students as they are viewing the film. The material can be in the form of a guidesheet that includes questions which the students should be able to answer by viewing the film. A guidesheet can also identify important vocabulary that students should learn as a result of seeing the film.

The teacher should carefully read the film manual before viewing the film if one is available. The manual usually provides a synopsis of the film, its objectives, preliminary activities, follow-up activities, and the vocabulary students should have to make the film meaningful.

Preparing Students. Students often need first-hand experiences to make a film more understandable. They may need to perform a laboratory exercise, view a demonstration, conduct an experiment, or go on a field trip in order to understand the film presentation. They may need help with technical words so that they will not encounter difficulties in understanding the subject matter of the film. Students often are not familiar with the materials that will be shown during the film presentation; therefore, they may have problems in interpreting what they see. The teacher may be able to show the actual material before the film is viewed to make the presentation more meaningful.

Showing a Film. The presentation of a film should run smoothly. No time should be wasted to bring the image into focus and to adjust sound when using a 16 mm projector. Videotape films sometime need track adjustments as well as sound adjustments and these should be made before the presentation. The equipment should be tested beforehand to ensure that it runs properly and set in advance of the presentation.

It is usually a good idea to show a film first, without interruption, so that the students can concentrate on the subject of the film, and then reshow it if students have questions that can be answered by viewing it again. Later showings of the film are useful to follow-up discussions or a lecture or for review.

Teachers often find that they can make better use of the film for instruction when they can eliminate the sound. This provides the teacher with opportunities to make important comments and call attention to important points and certain events during the film presentation. Some projectors and videotape equipment can be stopped at certain places while the film is in progress so that one frame at a time can be viewed for discussion purposes. Some equipment will allow sequences to be projected in slow motion so that students can view details that would not otherwise be evident.

Follow-up Activities. A discussion session after a film should be well planned. This is the time when students can ask questions about important points that were confusing to them during the presentation. The teacher can also initiate discussions to stimulate inquiry and initiate problem-solving situations after a film has been shown. A discussion the next day may require a reshowing of the film.

It is not uncommon for teachers to neglect follow-up activities after a film presentation. This often happens when they believe that the film is self-explanatory. A film that does not stimulate discussion or initiate inquiry or problem solving is probably not worth showing in the first place.

Filmstrips and Slides

Filmstrips and slides are important in science teaching because they can provide a great deal of versatility during instruction. They can be used effectively during lectures, discussions, recitations, project work, and laboratory activities. Because of their versatility, they can be used at any point during instruction. The teacher can spend several seconds or 10 minutes of instruction on one slide or one frame of a filmstrip, using it as a focal point of a lecture, a discussion, a recitation, or a laboratory activity. Recitations

and discussions for review can be made more interesting by using slides and filmstrips.

Filmstrips or a series of slides are useful to introduce a lecture, stimulate a discussion, illustrate a lecture or discussion, clarify a point, or illustrate an application after an idea or concept has been developed. They are excellent to use for making comparisons that would otherwise be difficult and for making an answer to an unexpected question more understandable. Slides and filmstrips that are used while giving oral directions can make an assignment more interesting and vivid.

Slides and filmstrips can also be the basis for a review session—to review previously presented material and to review material presented through firsthand experiences such as those obtained on a field trip or through a demonstration or laboratory exercise.

Slides of charts can be stored more conveniently than charts and can be viewed more easily than charts, particularly for large group instruction. Complicated diagrams, graphs, and other intricate illustrations can be photographed and made into slides to be stored and used as needed. Slides of charts and intricate diagrams are available from commercial supply houses.

Overhead Transparencies

The overhead projector is a useful piece of equipment for projecting transparencies that the teacher can easily prepare before instruction takes place or during the course of instruction. The equipment is placed in front of the classroom where the teacher operates it while facing the students. The overhead projector is equipped with transparent sheets or rolls of plastic on which the teacher writes with a wax pencil or a felt pen as conveniently as writing on a chalkboard. Drawing diagrams with colored wax pencils or felt pens can provide special effects for clarity and emphasis. The material placed on the plastic sheets can be erased easily so that the plastic can be used over again.

Permanent transparencies for the overhead projector can be prepared in various ways. Special types of transparent permanent inks for writing or drawing on plastic sheets are available in various colors to produce special effects. Diagrams, drawings, and written material on plastic sheets can be filed and available for future use. Photocopy machines are able to produce transparencies directly on plastic sheets. Material to be projected is typewritten, handwritten, or drawn on a sheet of paper and then copied on clear plastic instead of paper using a photocopy machine.

Another type of permanent transparency is called the overlay. These are made by drawing parts of a picture or diagram in different colors or the same color on different plastic sheets. As one section of the diagram or drawing is being viewed, another section on a different sheet is placed directly over the one being shown. In this way the teacher is able to gradually construct a complete diagram or picture by placing one section upon another as the material is being presented.

A series of articles in the *Science Teacher* on the use of the overhead projector in the teaching of chemistry entitled "The Tested Overhead Projection Series (TOPS)" describes a number of chemistry demonstrations designed to be performed and viewed before a class using an overhead projector. This requires that the demonstration material be prepared for projection and that the overhead projector be converted for vertical stage projection. (For more information concerning this technique, see Hubert N. Alyea, "Tested Overhead Projection Series," *The Science Teacher,* March, September, October, November, and December, 1962.)

The overhead projector can also be used for projecting live material like plant stems and leaves. It can project demonstrations such as those involving magnetism to show the magnetic fields around a bar magnet or a horseshoe magnet. In this case a magnet is placed on the glass table of a projector and iron filings are sprinkled on the magnet resulting in a very clear picture of the magnetic field around a magnet. In earth science, the growth of crystals can be viewed on the projection screen by an entire class using an overhead projector. There are other uses for the overhead projector in teaching biology, physics, and earth science. With a little imagination and ingenuity, science teachers will be able to develop demonstrations that can be viewed with the use of the overhead projector. Some uses can be found in the science education literature, such as *The Science Teacher, The American Biology Teacher,* and *Science Scope.*

Television

Television has a number of advantages for classroom use. Students can hear and view events as they are happening on the spot. Catastrophes such as those created by earthquakes, tornadoes, hurricanes, and volcano eruptions can be seen on television as they are occurring. Scientific events such as a space shuttle launching or men landing on the moon, or scenes on the moon, the planet Mars, and elsewhere in space are extremely interesting and timely and introduce students to issues involving science, society, and technology. News programs dealing with abortion, the AIDS crisis, and the TB epidemic expose students to important information regarding societal issues that can affect them in their daily life.

Prerecorded commercial videotapes and TV programs recorded on videotape by teachers and students can be played through a television monitor with the use of a videotape recorder/player. This equipment is inexpensive and gives the teacher a great deal of flexibility in instruction since programs can be recorded when they are broadcast and viewed by students at other times when they fit the curriculum. TV stations often require that one seek permission to use recordings of TV programs for instructional purposes. Normally, permission is easily obtained by writing the TV station, indicating the name of the program, when the program was broadcast, dates when the recording will be used for instruction, and in what class it will be used. The letter should be written on school stationary and signed by the teacher and the principal.

Closed circuit television has a tremendous range of uses, including the magnification of devices, specimens, and other materials. By using a video camera and a television set, a teacher can more or less create a closed circuit TV operation in one classroom. A videocamera connected to TV monitors will permit students to view a demonstration that is taking place at the front of the classroom. A videocamera equipped with special lenses can be used to enlarge small items so they can be easily viewed on a TV monitor. Monitors placed at strategic parts of the classroom for optimum viewing—front, back, sides, and at various levels will permit students seated to see what is taking place at the demonstration table. For example, a dissection of a frog performed at the demonstration table can be viewed in detail by all students as it is taking place. Microscope attachments are also available so that the equipment can be used as a microprojector. Closed circuit television has great potential in science instruction and teachers should determine its advantages and disadvantages in their own teaching situation.

Audiotape Recordings and Videocamera Recordings

There are a few commercial recordings of sounds in nature such as those produced by animals, insects, and birds. The most successful of these are the recordings of bird calls and songs. Teachers use the latter to help students learn to identify birds by their songs. Ordinarily, the teacher will play the recording and at the same time project pictures that correspond with the bird on the screen to enable students to associate the sound with the bird.

Recordings of musical instruments may be used in physical science or physics classes to study changes in pitch and the quality of sounds produced by different types of vibrating materials. For physics classes there are recordings to help test the fidelity of audio equipment that help students to develop a better understanding of the characteristics of sound.

Portable audiotape recorders are very versatile machines that enable one to pick up sounds almost anywhere, in nature, in obscure areas not serviced with electricity, and in the classroom. The recordings can be reproduced on site and as needed. Teachers can make good use of this equipment to record material useful in instruction.

Another versatile piece of equipment is the videocamera. The camera is portable and can be used much as one would use a 16 mm or 8 mm movie camera. One excellent feature of this equipment is that the film is contained in a cassette and can be placed directly in the camera. Once the material is photographed, it is immediately available for viewing using a videotape recorder/player and a TV monitor. The film does not require special processing to be viewed. The videocamera is equipped so that it can record sound as material is being photographed. Cameras are also available with a selection of telephoto lenses, some of which are useful to photograph close-ups of parts of equipment and machinery, animals, and plants. These electronic devices can be used to record action scenes on site accompanied with sound. They provide opportunities to photograph and record scientific events, field activities, demonstrations, lectures, laboratory work, student reports, and other classroom activities.

There are many makes and sizes of videocameras on the market and one should be careful when selecting the equipment. Some equipment is relatively inexpensive and will perform as well as expensive makes. Some cameras use film that is enclosed in a regular videocassette cartridge that can be placed directly in the videotape recorder/player for viewing. Others use a film enclosed in a small cassette. For viewing the film through a TV monitor, this format requires that the cassette be first placed in an adaptor that is the size of a regular videocassette cartridge and this package is placed in the videotape recorder/

player. Some smaller format films can be viewed on the TV monitor by using the videocamera as the recorder/player instead of the videotape recorder/player. The film is placed in the videocamera properly connected to the TV monitor and this combination permits viewing through the television set.

The Chalkboard

The chalkboard is the most commonly used instructional aid in teaching science. It is available in most science classrooms and laboratories, and permanently installed at the front of the classroom and other places where instruction takes place. Portable chalkboards are also available. These can be moved to classroom areas where one is not permanently installed but is needed for instruction.

Science teachers use the chalkboard because of its versatility. It is available at a teacher's discretion and used extensively to explain concepts using diagrams, symbols, words, and sketches. Concepts and ideas can be developed gradually using illustrations on the chalkboard accompanied with oral or written explanations. Illustrations, drawings, and sketches can be altered as needed while a presentation is being made.

To use the chalkboard effectively, note the following points:

1. Make certain that the chalkboard is free of extraneous material.
2. Prepare difficult or complicated diagrams in advance. It is often advisable to first lay out the diagram on paper to determine space limitations on the chalkboard. Complicated diagrams that need to be gradually developed can first be placed on the board and then lightly erased, leaving a vague image. As the teacher presents the lesson, the diagram is developed by tracing over the lines of the image.
3. Try to avoid glare on the chalkboard so that students can view it without difficulty.
4. Always check to see whether enough chalk is

available. Also make certain that erasers, rulers, and compasses are at hand as needed.
5. Write as large as possible so that all material written or drawn can be seen from all parts of the room where students are seated or standing. Rooms that measure 30 by 40 feet or 25 by 50 feet require letters or figures approximately 3 inches in height to be seen easily by all students in the classroom.
6. Do not stand in front of the material being written or drawn on the chalkboard. Stand to the side so that what has been written or drawn on the chalkboard can be viewed easily, and use a yardstick or pointer to draw attention to parts of the diagram, graphs, or pictures.
7. Do not talk while facing the chalkboard. Always try to speak in the direction of the class.
8. Use colored chalk when needed to provide contrasts. Yellow and green chalk are excellent to use to provide contrasting colors.

CHARTS

Charts are standard equipment in most science departments. Some teachers use them extensively during lecture presentations, discussions, and recitations. Charts have limitations, but they also have certain advantages. The diagram on a chart is generally superior to the one that a teacher may draw on the chalkboard and can be used as a substitute for the teacher-prepared diagram. Good charts possess visual appeal and clarity when they have sharp lines, bright colors, and bold printing. Charts should be simple and uncluttered and large enough so that the print can be read from all parts of the room. Crammed charts are confusing, unattractive, hard to read, and difficult to use. Charts should be hung on properly lighted walls. A chart should never be placed where students have to look toward a window, because the glare will prevent proper viewing. In using charts remember that the students must know what they are looking at; otherwise students may find the charts difficult to interpret. Oversimplified diagrams can introduce misconceptions; complex ones can cause confusion.

MODELS

Models are usually superior to teacher-made diagrams, charts, and flat pictures because they are three-dimensional. They do, however, have the same limitations as diagrams, charts, and pictures. If students do not know what a model is supposed to represent, they will find it difficult to interpret.

A model for classroom use should be large enough so that it can be viewed from all parts of the room, particularly if it is used to illustrate points during a lecture or serve as the center of class discussion. Small models should be used for individualized or small group work.

The degree to which a model resembles a real object is an important consideration. Avoid oversimplified models, because they can give students the wrong impressions of what the model is supposed to represent.

SUGGESTED ACTIVITIES

1. Prepare a short lecture with visual aids to present to your peers. Ask them to videotape the lecture. Critique the presentation with them while viewing the tape.
2. Write out directions that might be given orally for (1) a reading assignment in a textbook, (2) a simple laboratory exercise, and (3) getting ready for a test. Discuss these presentations with other students in your class. Ask suggestions for improvement.
3. Prepare a brief account of one of your own experiences that you think might help students understand a specific principle. Present this to other members of your science methods class. Ask for suggestions and criticisms.
4. Write out a set of questions you might use in guiding a discussion of the results of a field observation or an experiment. Conduct the discussion with the students in your science methods class to determine whether questions can effectively guide the discussion.
5. Write out a set of questions you might use to guide students through an oral inquiry session involving a discrepant event. Present this to your science methods class for discussion and criticism.
6. Observe a science teacher leading a class discussion. What percentage of the students in the class participate in the discussion? What percentage dominate the discussion? What percentage pay little or no attention to the discussion? Critique the entire session. Prepare a report and discuss it with your science methods class.
7. List five topics around societal issues that lend themselves to general class discussions led by the teacher. List five topics that can be led by student leaders. Plan one of the teacher-led discussions to include (1) the type of lead questions to be used to initiate the discussion, (2) questions to be used during discussion, (3) activities that students can conduct in small groups, and (4) activities for large group sessions. Discuss the plan with the members of your science methods class. Ask for suggestions and criticisms.
8. Plan a discussion on the topic such as AIDS or another controversial societal issue to be led by the teacher. Plan follow-up discussions in groups led by a student. Remember that student leaders require training before they can be used in this capacity. Plan a training program for student leaders on the topic under discussion.
9. The chalkboard is commonly used in sci-

ence instruction. Prepare a lesson that involves a concept requiring the use of a complicated diagram that you want to place on the chalkboard. Prepare the complicated diagram on paper beforehand and then place it on the chalkboard by developing it as you explain the concept before your class. (See the directions in this chapter for developing a difficult or complicated diagram on the chalkboard.)

10. Educational films and videotapes are often useful as advance organizers to introduce a lesson or a unit. Select a videotape or a 16 mm film that provides a good introduction to a lesson. Prepare students for the film or video presentation by giving them a series of questions based on the film so they know what to look for during the presentation. Discuss the questions beforehand and clear up any difficulties that may arise relative to vocabulary and subject matter of the film. Discuss the film after the presentation. Present this lesson before your science methods class.

11. Practice making transparencies for using with an overhead projector. Also make an overlay of a diagram or a drawing using different colors for emphasis and special effects.

BIBLIOGRAPHY

Ahl, A. S. 1983. Determinism or probability—or teaching students how to ask questions. *The American Biology Teacher, 45*(2): 102.

Ault, C. R. 1989. Problem-solving in earth science education. In D. Gabel (Ed.), *What research says to the science teacher, problem solving: Vol. 5* (pp. 46–47). Washington, DC: National Science Teachers Association.

Ausubel, D. P. 1961, Jan. In defense of verbal learning. *Educational Theory, 11:* 15.

Ausubel, D. P. 1963, June. Some psychological and educational limitations of learning by discovery. *New York State Mathematics Teachers Journal, 13*(3): 90.

Barman, C. 1982. Some ways to improve our overhead projection transparencies. *The American Biology Teacher, 44*(3): 191.

Barman, C. R. 1990. *An expanded view of the learning cycle: New ideas about an effective teaching strategy* (monograph and occasional paper, series No. 4). Council for Elementary Science International. Washington, DC: Office of Education.

Bloom, B. S. (Ed.). 1956. *Taxonomy of educational objectives. Handbook I. Cognitive domain.* New York: David McKay.

Blosser, P. 1975. *How to ask the right questions.* (How to Do It Series). Washington, DC: National Science Teachers Association.

Bruner, J. S. 1962. *On knowing.* Cambridge, MA: Harvard University Press.

Clark, R. E., and Salomon, G. 1986. Media in teaching. In M. D. Wittrock (Ed.), *Handbook of research on teaching* (3rd ed.). New York: Macmillan.

Clarke, J. H. 1987. Building a lecture that works. *College Teaching, 35*(2): 56.

Cooper, J. M. (Ed.). 1990. *Classroom teaching skills* (4th ed.). Lexington, MA: D. C. Heath.

Dick, W., and Carey, L. 1985. *Systematic design of instruction.* Glenview, IL: Scott Foresman.

Dillon, J. T. 1984. Research on questioning and discussion. *Educational Leadership, 42:* 50.

Gall, M. 1984. Synthesis of research on teacher questioning. *Educational Leadership, 42:* 40.

Gallagher, J. J. 1985. *Secondary school science* (Interim Report). East Lansing: Michigan State University, Institute for Research on Teaching.

Heinich, R., Molenda, M., and Russell, J. D. 1993. *Instructional media and the new technologies of instruction* (4th ed.). New York: Macmillan.

Henson, K. T. 1988. *Methods and strategies for teaching in secondary and middle schools.* New York: Longman.

Lawson, A. E., Abraham, M. B., and Renner, J. W. 1989. *A theory of instruction: Using the learning cycle to teach science concepts and thinking skills*

(NARST Monograph No. 1, p. 9). National Association for Research in Science Teaching.

Mac Donald, R. E. 1991. *A handbook of basic skills and strategies for beginning teachers.* White Plains, NY: Longman.

Melnik, A. 1968. Questions: An instructional-diagnostic tool. *Journal of Reading, 11:* 509.

Nair, I. 1987. Chernobyl: Asking the right questions: A technical discussion. *The Science Teacher, 54*(8): 24.

Osborne, R., and Freyberg, P. 1985. *Learning in Science* (pp. 101–111). Auckland: Heinemann.

Rowe, M. B. 1974a. Wait-time and rewards as instructional variables: Their influence on language, logic, and fate control. Part I. Wait-time. *Journal of Research in Science Teaching, 11*(2): 81–94.

Rowe, M. B. 1974b. Relation of wait-time and rewards to the development of language, logic, and fate control. Part II. Rewards. *Journal of Research in Science Teaching, 11*(4): 291–308.

Rowe, M. B. (Ed.). 1978. *What research says to the science teacher* (Vol. 1). Washington, DC: National Science Teachers Association.

Tobin, K. 1984. Effects of extended wait-time on discourse characteristics and achievement in middle school grades. *Journal of Research in Science Teaching, 21*(8): 779.

Tobin, K., and Gallagher, J. J. 1987. The role of target students in the science classroom. *Journal of Research in Science Teaching, 24*(1): 61.

A scientifically and technologically literate individual appreciates the interrelationship of science and technology and their relationship with society.

CHAPTER 7

Science, Technology, and Society

The development of technological literacy can result only from curricula that go beyond those that center upon traditional science course subject matter. This goal calls for new programs that link science and technology together, stressing the application of knowledge. These programs must focus on the work of engineers and technologists as well as scientists. Consequently, they must engage students in a variety of learning activities, including those that involve the design of products and systems. The new programs of study must also provide opportunities for students to analyze the impact of science and technology on society, and to determine their costs and benefits.

Overview

- Present a rationale for including technology in school science programs.
- Explain technology and how it relates to science and society.
- Illustrate several ways that designing products and systems can become an integral part of science course instruction.
- Provide examples of strategies and techniques that can help students become aware of scientific/technological/societal issues—how to investigate them and make decisions regarding their findings.
- Address important considerations for dealing with controversial issues and personal values in the classroom, especially evolution and creationism.

RATIONALE

Science and technology exert a profound influence on society. Scientific knowledge influences our thinking about human, political, and social affairs. It also affects how we think about ourselves. In general, scientific knowledge has had and continues to have both positive and negative impacts on individuals as well as society as a whole. Technology also influences individuals and society. It plays an important part in providing products that affect the quality of life. Social, political, and economic changes in society are often affected by new technologies.

Some people fear technology. They believe that the problems resulting from modernization through technology are growing faster than the solutions to these problems. In effect, the attitudes, needs, and values of these people, individually and collectively, can influence the use of existing technologies and the development of new technologies. New technologies often arise because there is a need or demand for them in society or because present technologies need to be improved upon. Society, therefore, exerts its control over technology by assessing its worth and calling for its products or rejecting them.

Science and technology must be understood better by everyone in our society. The development of accurate understandings of this relationship can be accomplished by first dealing with the unique attributes of each enterprise, then addressing their implications for society. It is therefore important to realize that to be scientifically and technologically literate one must understand and appreciate the interrelationships of science and technology as well as their relationships with society. The task of developing a scientifically and technologically literate society should become a major goal of science education.

A continued outcry from all sectors of society demands relevance in the classroom. This societal pressure mandates that science curricula introduce students to careers in the scientific, medical, and engineering fields, and provide instruction to assess their cultural influence. Therefore, science teachers must go beyond the content and methods of science and demonstrate how the application of scientific knowledge can affect the welfare of society. This opinion was voiced by Hurd (1961, p. 399) and other science educators more than 30 years ago when they proposed support for a society-centered biology curriculum:

> Our efforts as biologists ought to be toward the unity of biology with life as a whole; teaching what is required to illuminate biosocial problems and sustain a humane world under conditions of favorable culture and a supporting natural environment. Students want a biology course that says something about their own growth and development and shows their own well-being. Students seek experiences that will bring them into contact with the realities of man's existence and provide insight for future planning.

Students are constantly exposed to environmental and social problems either through direct involvement or indirectly through the communications media. These social issues and problems relate to such areas as drug and alcohol abuse, cancer, AIDS, obesity, overpopulation, abortion, and hazardous waste. In order for science education to be relevant, students must be able to examine these issues and to apply the scientific principles and processes to them. They must be given many opportunities to discuss their beliefs and values, and to propose solutions to "real world" problems. It's only through this type of exposure and instruction that students will be able to make informed decisions with regard to important societal issues and problems.

According to Gottlieb (1976), the traditional lecture/laboratory, science instructional format is antithetical to the educational goals of developing intellectual autonomy in students. Even the widely used Biological Sciences Curriculum Study (BSCS) biology texts were criticized as (1) containing too much biological knowledge with little attempt to help students establish relevancy to the endeavors of humankind, and (2) having no consideration for the interactions of biology with the social, economical, and political problems of contemporary society (Andrews, 1970). The factual knowledge of science is of no value if it is only stored and forgotten.

Science teachers must develop the content background of students so they can use this knowledge in their daily lives. The big question is: What type of content and how much do we teach our students? Some science educators believe that we are attempting to teach our students too much science information and trying to prepare them for careers they will never pursue. Yager (1985, p. 21) says that preparing students for courses or careers in science, medicine, and related professions is misguided because "for at least 98 percent of all high school graduates such justification is not only inappropriate, it borders on being a hoax." He suggests that we focus on social issues and what takes place in society. Yager urges science teachers to begin science lessons and units with real problems facing society and let the students define the problems, propose the solutions, gather the information, and derive the solutions.

Bybee (1986) asserts that the science-technology-society (STS) movement currently under way has the potential to educate young people for the world in which they live—now and in the future. An STS curriculum offers a broad set of educational goals that are appropriate for most students who enroll in science courses at the middle and secondary school levels. He believes that attention to these goals will help to restore the public's lost confidence in science education.

Rubba (1987) also believes that STS has the potential to provide students with a variety of essential skills that they can apply to their everyday life. However, this will occur only if science teachers go beyond the awareness and the discussion of issues. He urges that science teachers help their students investigate issues and involve them in inquiry, problem solving, and action learning. Rubba stated: "The issue is secondary;

it is the vehicle to the development of investigations and action skills which students can incorporate into a decision-making strategy applicable to the STS issues they will face throughout life" (Rubba, 1987, p. 183).

Central to the whole issue of incorporating STS into the science curriculum is the attempt to provide students with the knowledge and skills to make decisions that will affect their lives. Ultimately, we as science teachers want people to make good decisions, which will impact them directly. We want young people, for example, to consider the consequences of taking drugs when they are urged to do so by others, and then to act in favor of their own welfare. We want young adults to consider the importance of having a plentiful supply of good drinking water in their community, to study water-supply issues, and to vote responsibly when and if they are brought up in the community. We want teenagers to consider the consequences of not using seat belts when they drive so they will "buckle up" when they get into a motor vehicle.

How do we as science teachers provide students in our science classes with information and skills to act responsibly? Ajzen and Fishbein (1980) indicate that human beings are rational and systematic about their behavior and use the information that they have internalized to make decisions. They stress that the antecedents of behavior are intentions, attitudes, and beliefs—in that order. If we subscribe to their model of reasoned action, we can begin by helping students to acquire information that provides them with a belief system which permits them to learn about the consequences of many behaviors that they may choose in their lives. We can organize learning environments that help students learn about the psychological and physiological effects of certain drugs such as crack and cocaine, for example. We can help them to clarify their attitudes and those of their peers toward taking these drugs. We can even permit them to resolve their own intentions to take drugs. Although it is naive to believe that we can prevent everyone who enrolls in a science class from taking drugs, we can certainly help students acquire some knowledge and skills that can be used to make important decisions now and in the future.

Some science teachers have always introduced technology and social issues into their teaching. They have made their courses relevant, going beyond the facts of science and the assigned science textbook. Unfortunately however, many science teachers hesitate to incorporate STS into their courses of study because (Jarcho, 1985, p. 17):

- They are wedded to the text which stresses pure science.
- They feel unprepared to teach topics they know little about.
- They resist change.
- They realize they will get into values and beliefs.

These are reasons for concern. When science teachers resent change, feel unqualified to handle new topics and approaches, and are afraid to get involved in issues that focus on values and beliefs, then we are shortchanging our students. These concerns have to be taken seriously and become focal points for change so that initiatives can be taken to make science courses more meaningful and useful to students.

WHAT IS TECHNOLOGY?

Just as science is not easy to define, neither is technology; furthermore, the differences between science and technology are not clear-cut. Science and technology are inherently intertwined. As a consequence, they convey different meanings to the professional as well as to the layperson. In general, science can be regarded as the enterprise that seeks to understand natural phenomena and to arrange these ideas into ordered knowledge, whereas technology involves the design of products and systems that affect the quality of life, using the knowledge of science where necessary. Science is a basic enterprise that seeks knowledge and understanding. It is aligned with observation and theory.

Technology, on the other hand, is an applied enterprise concerned with developing, constructing, and applying ideas that result in apparatus, gadgets, tools, machines, and techniques. The products of science are called discoveries, while the products of technology are referred to as inventions.

Technology began when humankind invented tools and processes to make work easier and life better. Early technology was simple compared with today's high-tech products. Simple tools were made by primitive people to aid in hunting and fighting. Many of today's devices are complex, such as computers that present and manipulate information in amazing ways, making it possible for humans to explore the far regions of our solar system or to create new genes that perform special functions in living organisms. After centuries of interplay between technology and science, there exist a myriad of designs, goods, and services that serve humankind.

Today, technology and science are often so intimately related that they rely on each other. This interplay of science and technology was clearly evident during the American Industrial Revolution. Increased use of internal combustion engines and electrical devices before 1890 put into place the foundation for considerable technological advancement. During the years that followed, the internal combustion engine was developed and Westinghouse marketed the first electric generator. Such technological advances underscore the usefulness of scientific knowledge in the fields of thermodynamics and electricity.

Americans have only to look back to the beginning of this great republic to realize the important role that technology had in its development. Most people are familiar with the names of individuals who participated in promoting one or more technologies in America, such as Benjamin Franklin, Thomas Jefferson, Eli Whitney, Alexander Graham Bell, Thomas Edison, George Eastman, Ellen Swallow Richards, and Henry Ford. Individuals like these and others made possible mass production, which enabled mass consumption that changed our society (Purcell, 1981).

The practical and utilitarian nature of science and technology, and their implications for science education, are evidenced in the advice given by Benjamin Franklin to Dr. Thomas Cooper, who in 1812 was planning a chemistry course at Carlisle College:

> The chemists have not been attentive enough to this. I wish to see their science applied to domestic objects, to smelting, for instance, brewing, making cider, to fermentation and distillation, generally to the making of bread, butter, cheese, soap, to the incubation of eggs, etc. And I am happy to see some of these titles in the syllabus of your lecture. I hope you will make the chemistry of these objects intelligible to our housewives. (Meier, 1981, p. 23)

Recent technological achievements that can be addressed in school science programs are: computers, microchips, compact disc players, automatic cameras, digital watches, guided missile systems, superconducting materials, nuclear power, genetic screening, organ transplants, artificial organs, superhighways, and plastics.

Engineers design products and services that benefit society, often drawing on scientific information to assist in their work. They also engage in inquiry, use their imagination, and figure out solutions to problems. Engineers experiment, control variables, and make keen observations. These men and women possess a body of knowledge about their enterprise along with knowing many scientific disciplines. Most important, these individuals create their products and processes.

In addition to the benefits of technology, the costs must also be considered. Large-scale production of goods and services consumes valuable resources such as fossil fuels, forests, recreational areas, plant and animal species, and drinking and irrigation water. These resources are being depleted and are becoming more costly and difficult to obtain. Nations throughout the world must make decisions regarding the use of all resources so that this generation does not use or misuse valuable raw materials, leaving future generations without them. The problems and issues which reside with the use of natural resources that are essential to society must be part of a curriculum improvement movement to improve scientific literacy. The moral and ethical implications of how to apply science must be decided upon by society, hopefully a society that is well informed about critical science and societal issues.

INCORPORATING STS INTO THE CURRICULUM

Science teachers have several options for incorporating STS into their curriculum. They can include the study of products and systems to improve students' understanding of technology. They can also involve students in the investigation of issues that are related to science and technology, permitting them to analyze situations and to make decisions.

Designing Technological Products and Systems

The development of useful products and services illustrates the imagination of engineers, scientists, and technicians who attempt to affect the quality of life. Individuals who develop goods and services think, reason, and imagine in ways similar to those who theorize about natural phenomena. Often, these individuals learn to better understand basic science concepts and principles when they develop products that build upon established knowledge. Fortunately, many types of products and systems can be designed by students to help develop their technological literacy.

Products.   Some science teachers have always engaged their students in building gadgets to further their involvement in science courses. For example, consider Mrs. Fellabaum's approach to instruction in her eighth-grade general science course.

▼ More than half of my course centers on basic physics, which includes electricity, light, motion, and sound. I begin the study of each of these topics by asking students what they know about these areas and what they would like to learn about these topics. The discussions are followed with several laboratory exercises to develop fundamental concepts and principles. With some background, the students begin to participate in design projects.

When we study motion, for example, I always involve students in the construction of "mouse trap" cars. These vehicles are powered by the basic mouse trap sold at hardware stores. They are made from scratch or modified from toy cars and trucks. Since I attempt to associate this experience with speed, acceleration, and friction, one of the primary objectives is to design a vehicle that will accelerate the fastest from a stopped position to a 2-meter line. Another objective is to design a vehicle that will roll the farthest. Students have the option as to which type of vehicle they want to design. These activities are of high interest and they seem to reinforce the topic we are studying.

I do not stop at the design phase of this activity because I want to further the hands-on and minds-on aspects of science and engineering. Therefore, all students are asked to write a short paper that requires them to analyze a real vehicle and explain the features that contribute to its acceleration, speed, efficiency, and so on. Some students use their family car for the analysis, others go to automobile dealerships and get the brochures on new cars. One of my students videotaped drag races from a TV show and used these visuals to illustrate features that contribute to their acceleration. This activity also provides the opportunity for us to invite local guest speakers to discuss how people design, maintain, and repair automobiles. We have on occasion taken a field trip to a local garage where dragsters are built. Surprisingly, the girls as well as the boys are motivated by this approach to science and technology education. ▲

Library work is another form of inquiry that many science teachers use to study technology. Students can find a great deal of reading material to improve their understanding and stimulate their interest in this topic. The following list can help students organize their thinking and re-porting during the study of technology and its relationship to science and society:

■ Tell how the technology was discovered.
■ Explain how the technology works and include the scientific principles upon which it is based. Include diagrams.
■ Give some of its beneficial uses, such as entertainment, medical, defense, education, and so on.
■ Cite any potential dangers of the technology.

Table 7–1 lists technological devices that students can gather information about from books, magazines, journals, and encyclopedias.

Systems. A teacher can focus an activity upon systems made up of products and devices that already exist—systems which can be modified and improved to meet the needs of a given community. For example, a critique of a mass transit rail system can be incorporated in a physical science course that addresses energy, mechanics, and motion. Consider a mass transit system for an urban area that is congested with internal combustion engine vehicles that consume large quantities of fossil fuel and pollute the atmosphere. The location of the rail system, the type of power used to move the vehicles, and the economics of the system are but some aspects of the problem.

TABLE 7–1 Technological products that students can study and explain to further their understanding of science and technology

Antenna	Integrated circuit
Artificial intelligence	Laser gun
Automatic focusing lens	Magnetic tape
Cardiac pacemaker	Microwave oven
Chip	Nuclear reactor
Compact disc player	Quartz clock
Computer wafer	Radar detector
Digital watch	Solar panel
Fluorescent light	Solid state wafer
Fuel cell	Superconductor
Geiger counter	Touch-tone telephone
Hologram	Transistor

Including science/societal issues and problems in the curriculum gives students opportunities to discuss their own values and make informed decisions concerning matters that affect their lives.

Addressing Issues and Problems

The issue/problem focus is prominent in the National Science Teachers Association's (1991) position statement for STS programs. These guidelines emphasize that students identify issues and problems of local concern, study them, and make appropriate decisions based on their inquiry. Although the terms *issue* and *problem* are used interchangeably within STS instruction, they have a different meaning. An issue is an idea on which people hold different beliefs and values. A problem is a situation that is at risk for a given population, for example, the poisoning of fish in estuaries that are receiving toxic waste from industry. In any event, the study of issues and problems has taken place for many years, primarily in the area of social studies and environmental education.

Social studies educators have long addressed issues and values in their curricula. A useful model of approaches in values education was described by Superka et al. (1976, pp. 4–5) to include five strategies: "(a) inculcation, (b) clarification, (c) moral development, (d) analysis, and (e) action learning." More recently science educators, working in environmental education, have put forth a model that parallels the work in social studies, but that appears to

relate more directly to science instruction: "(a) science foundations, (b) issue awareness, (c) issue investigation, and (d) citizen action" (Ramsey, Hungerford, & Volk, 1989, p. iii). The five instructional strategies described below are taken primarily from these models.

Inculcation. One approach to values education is to instill certain values in students directly, which means that the teacher identifies a given attitude or belief and attempts to impart it to students. Or the teacher identifies a given behavior and attempts to develop this behavior through a variety of activities. The inculcation approach would most certainly incorporate lecture and persuasion. Although inculcation is *not* a recommended approach to use for values education, it is used inadvertently by many teachers. The following example illustrates some of the dangers inherent in this approach.

▼ Mr. Clemens teaches high school chemistry in a community where steel mills and petrochemical plants have existed for many years. He often criticizes these industries for polluting the land, rivers, and air in the community. On every possible occasion, Mr. Clemens cites how the waste materials of the steel and chemical industries are "poisoning" the environment. He often invites local environmental activists to speak to his classes about the problems of the smokestack industries and their effects on urban America. Mr. Clemens requires all students to write a report on the problems of environmental pollution in their community.   ▲

This vignette illustrates how a science teacher attempts to inculcate students with one point of view—the evils of industry and how industry is damaging the environment. Although few would deny that industry has contributed to environmental pollution, industry has also made great progress to clean up and to protect the environment. Many industries carefully monitor the wastewater they discharge into rivers and lakes, and control the concentration of toxic substances in their waste emissions. Mr. Clemens could do a better job of educating his students regarding pollution by providing an opportunity for industries in the community to explain what they do with their waste and how they attempt to protect their workers and the surrounding community from potential health hazards. He should provide students with data and opinions from a variety of sources—industry representatives as well as environmentalists—giving students the opportunity to modify their own beliefs and to make their own decisions.

One primary purpose for addressing societal issues is to encourage students to find out about science and technological problems so that they can evaluate them and draw their own conclusions. The inculcation approach to values education prevents this from occurring. It is obvious that the use of inculcation is contrary to the purposes of engaging students in discussions that will help them to enhance their reasoning skills and to improve their understanding of social problems. The purpose of such discussions is to get students to clarify their own beliefs and to become aware of the beliefs of their peers and others. They are supposed to permit students to alter their beliefs if necessary, after they have acquired new information and encountered the beliefs of their peers. These experiences can cause many students to modify their views and to open up their thinking. Inculcation does not

permit the type of dialog and openness that is required to produce these learning outcomes.

Issue Awareness.   One approach to helping students become more aware of issues and problems is through the clarification of these ideas. The clarification of personal values related to science and societal issues is one method that many science teachers might feel comfortable using when they address controversial issues. This approach provides students an opportunity to become aware of their own beliefs and those of other students. Issue clarification can be accomplished through discussion techniques, which might be considered safe and prudent, especially in communities where parents are very sensitive to what their children are being taught in school.

A relatively easy way to implement values clarification is to find a newspaper article on a social issue related to a science topic currently under study. Place the title of the article on a sheet of paper or on the chalkboard and list questions for students to answer. A good example of a social issue directly related to a science topic is "Banning smoking in public places." This can be discussed in a life science class during the study of respiration. The following questions can be listed for discussion on this topic:

1. Do you smoke?
2. Does anyone in your family smoke?
3. Do you like to be in a room where people are smoking?
4. Should people be allowed to smoke in places such as:
   ■ airplanes
   ■ buses
   ■ basketball games
   ■ restaurants
5. What health problems does smoking cause?
6. Should the government attempt to discourage smoking?
7. Should laws be passed to prevent smoking in public places?

Another values clarification technique is to use societal issue vignettes. Richard Brinkerhoff (1986) recommends that science teachers prepare vignettes about science- and technology-related problems and use these vignettes frequently throughout the school year. His ideas have been implemented by some science teachers as part of their daily science teaching.

▼ Ms. Radle has prepared overhead transparencies of numerous societal issue vignettes. At the beginning of each class period, she projects a vignette so that students can start reading it when they take their seats (see example in figure 7–1). Students begin immediately to respond to the vignette so that by the time the bell rings to start class, they are already thinking about the societal issue. After the roll is taken, Ms. Radle engages the students in a short discussion of the issue. This usually takes about 5 minutes, but on some occasions the discussion takes longer and may even last the full period. Each student keeps a journal on societal issues, which the teacher collects at the end of each week and examines for expression of feeling, clarity of writing, factual accuracy, and logical thought. ▲

Teachers who use this approach to develop awareness about science and societal issues indicate that this is one of the most positive aspects of their science course. Students develop knowledge and awareness of topics that they would not have become familiar with, except through this type of instruction. These written exercises and oral discussions stimulate interest in science. The beginning-of-class technique used by Ms. Radle and others also promotes effective use of class time because students begin to work as soon as they enter the classroom. Ordinarily, students would not begin to work on academic tasks until after roll is taken and everyone settles down and stops talking.

There is little doubt that values clarification activities can be used to initiate the discussion of science- and technology-related issues in science classrooms. They can be infused into existing curricula and implemented in a manner that takes little time away from other types of instruction. However, science teachers must not stop at the awareness level but must continue on so that students will inquire further into issues and extend their learning.

Moral Development.   The moral development approach is used to improve the moral reasoning ability of students through the examination of issues. This approach is similar to the analysis approach in that it emphasizes reasoning and thinking. However, the moral development approach focuses on personal moral values such as fairness, justice, equality, and human dignity; the analytical approach focuses on social moral value issues such as global warming, nuclear weapons, and world hunger (Superka et al., 1976).

The instructional approach that is used to develop moral reasoning centers around a moral dilemma, which is a factual or a hypothetical story. The moral dilemma vignette is discussed in small groups before it is discussed by the whole class. The dilemma can be introduced in the form of a reading or a film, and it involves characters who are confronted with a personal moral dilemma. A dilemma is a situation that requires one to choose between two rather equal

FIGURE 7–1   An overhead transparency of a vignette on science and societal issues. Adapted from *Extinction* by P. Erlich and A. Erlich, 1980, New York, Random House.

Today's Science and Societal Issue

Chimpanzees are used to test new vaccines against human viral diseases. It seems that only infant chimpanzees can be imported successfully into the U.S. The mothers are usually killed during capture. The demand for chimpanzees is great and, therefore, the species is in danger of becoming extinct. Unfortunately, no other species of animal has been found to serve medical research the way the chimpanzee does. Is the conquest of human disease worth the destruction of this species?

---

choices, each of which is accompanied by a difficult consequence.

Acquired Immune Deficiency Syndrome (AIDS), for example, is a disease that lends itself to discussion in science courses because of its moral implications and biochemical nature. AIDS could become the most serious epidemic since the plague. Understanding this disease requires study to understand its biological, clinical, and social aspects. Some students have already been confronted with the AIDS dilemma. In certain communities, students who have AIDS have been prohibited from attending school because of the possibility of their spreading the virus to other students. In other communities, although students with AIDS have been permitted to attend school, they have been shunned by classmates. This situation has been difficult for youngsters with AIDS because they are faced not only with a serious illness but also with rejection by their peers.

A moral dilemma or case study relevant to students in a middle or secondary school science class might read as follows:

▼ Billy Ray is a ninth-grade student who has AIDS. He got this disease as a result of a blood transfusion that he received during reconstructive hip surgery. Until he got AIDS, Billy Ray was popular among his classmates, but now many of the classmates in the small-town high school that he attends avoid him. In fact, some of the classmates do not want Billy to attend school because they believe he may somehow infect them with AIDS. They have gone to the school administration and made such a fuss that the principal has requested that Billy stay home and be tutored. Billy is very upset and despondent over this situation.

You are Billy's friend and want him to attend school. You are comfortable about carrying on your association with him. You, as his friend, want to do something about his situation but have some reservations. On the one hand, if you say nothing and permit the principal to prevent Billy from attending school, you are doing Billy an injustice and will also have a guilty conscience about this situation. But if you try to get some support for Billy to stay in school, many students may turn against you.

Since you live in a small town where prejudice and narrow-mindedness exist among many members of the community, what do you do? ▲

Situations like this are very real and exist in many communities. Science teachers can address the moral reasoning of their students by involving them in moral dilemmas in which students are asked to reason through these situations that may have personal relevance. They can give students information regarding the transmission of the AIDS virus and how people get this disease. Science teachers can dispel many misconceptions about other diseases as well. Many other topics, such as drug abuse, abortion, and obesity, can be used to develop moral dilemmas that will get students to focus on issues of personal concern.

The degree to which students can deal with moral dilemmas depends upon their intellectual and moral development. Much of the original work in this area, which will be discussed later in this chapter, has been done by Jean Piaget and Lawrence Kohlberg. In essence, however, they found that the moral reasoning level of the individual will greatly influence how he/she responds to moral issues and social problems.

Issue Investigation. The investigation and analysis approach goes beyond clarification in that it helps students learn a great deal about issues and technologies so they can understand them better. This approach stresses organization of factual information as well as the presentation of attitudes, beliefs, and values. The analytical approach requires students to participate in inquiry and to find out about ideas by doing library and field work, and by determining other people's beliefs. It also encourages students to separate fact from opinion and to become aware of the values of individuals who hold different views regarding an issue. This approach promotes scientific inquiry and higher level thinking—very important outcomes for science programs. The investigative/analytical approach culminates in students making decisions about important issues.

The analytical decision-making model developed by Oliver and Newman (1967) can be used by science teachers to provide in-depth experiences to students on environmental and technological issues. The following guideline can be considered for these purposes:

1. Identify and clarify the basic question.
2. Gather facts about the issue under study.
3. Evaluate the "factual" data.
4. Evaluate the relevance of the "factual" data.
5. Propose a tentative decision.
6. Determine the acceptability of the solution.

The analytical decision-making model can be used to involve students in examining important issues that will directly affect their lives. Some states, for example, are considering whether or not to use some of their land for long-term storage of radioactive waste from nuclear power plants. This is an important decision because of the impact it can have on the local economy, and because radioactive waste triggers such strong reactions from people regarding the environment and public health. The community near any proposed storage site must weigh the risks and benefits of the land use.

Students who undertake assessing whether or not the storage facility should be built near a given hypothetical community must investigate many aspects of the problem. They must learn about radiation and half-life, and study the safety record of similar facilities in other states and countries. They must compare the potential danger associated with nuclear power with that of other industries. This inquiry will lead students to study the hazards of using coal and petroleum to produce electricity, and the problems of disposing of chemicals such as polychlorinated biphenyls, lead, arsenic, and other industrial waste products that have been going into landfills.

Students must also research the beneficial effects of a storage site to the community. How will this project benefit the community economically? Local business will increase tremendously, at least during the construction of the storage site, to provide food and lodging for the con-

struction workers, technicians, and engineers. Perhaps roads and rails will have to be improved or built. Then there will be the maintenance of the facility once it has been completed and placed in use.

Students should investigate the attitudes of the people in the community regarding the nuclear waste storage problem. Although they cannot survey the citizens of the hypothetical community for which the proposed site is being considered, there are studies that have polled a variety of groups regarding their opinions about nuclear power and radiation. These results will surely be a significant factor in the decision to accept or reject a nuclear waste storage site in the community.

Students who undertake the analysis of issues and problems often need to find out the attitudes and beliefs of others regarding these ideas. This necessitates the development and administration of questionnaires. They can ask a set of questions that assess how knowledgeable individuals are about a given situation, which would address factual knowledge. They can ask a set of questions that assess what individuals believe to be true. And the students can ask a set of questions that assess how the individuals feel about a situation, which would address their attitudes. Beliefs are tied more closely to factual information, while attitudes are tied more closely to feelings. Attitudes represent one's evaluation of an idea or event. The beliefs and attitudes of individuals provide insights into their values regarding the environment and society.

Analyzing newspaper and magazine articles is an excellent way to enhance student inquiry (Hungerford et al., 1988). Figure 7–2 presents a format that students can use to guide their analyses and organize the information they derive from written material or oral presentations. First, it is useful to know who the individuals are who are communicating and presenting their views. Then it is easier to determine the position these individuals have taken on an issue. Listing the main points of each person's argument further clarifies each person's position and elucidates

FIGURE 7–2 Format for examining an issue from newspapers, magazines, or TV shows

---

The issue under analysis: _____

_____

The individuals involved:

The positions taken:

Main points of their arguments:

Beliefs and values:

---

what he or she knows. Finally, it is revealing to determine each person's beliefs and values.

An important learning outcome of issues investigation and analysis is decision making. Given all of the information that students have gathered, what will they decide to do regarding the situation they have studied? For example, if students have been studying the feasibility of placing a solid waste dump in their community, what do they feel should be done? Do they believe that more waste disposal areas are needed for their community, but that these facilities should not be constructed in their backyard? Decision making can result from a paper prepared by students or from a debate conducted before the class. Students should be asked to make these and many other types of decisions, and then to take some type of responsible action if this is their desire.

Action Learning. Action learning involves more than thinking, reasoning, clarifying, and decision making. It stresses taking social action in the community, which extends instruction beyond the classroom. Action learning requires student participation in activities that benefit society. This approach is emphasized in the mission statement of the North American Association for Environmental Education (Tanner, 1987, p. 5), which "stresses the need for active participation in solving environmental problems and preventing new ones."

Some science teachers have, for many years, engaged students in ecology projects in their community. They have taken interested students to streams and beaches to clean up these areas and collect samples of organisms to take back to the classroom for study.

▼ Mr. Bloom observed a great deal of water standing in a drainage ditch located several blocks from the middle school in which he teaches life science. The ditch was an unsightly area because of the dirty water and the accumulation of trash. Mr. Bloom had a hunch that some sewage water might be seeping into the ditch because of the age of the city's sewage system, and that if they tested the water, perhaps fecal material would be found in it. When he proposed to students in his classes the idea of cleaning up the drainage ditch and analyzing the water for contamination, he got many volunteers who wanted to participate in the project.

On a Saturday in October, Mr. Bloom took a large group of students, along with several parents, to the ditch to collect the cans, cups, and paper products that had accumulated there. He also directed some students to collect water samples to be given to the city water department to analyze for contamination and to be examined in the school's science laboratory.

Mr. Bloom had made prior arrangements with the city public works department for the use of their rakes, shovels, and trash bags to aid in the clean-up operation. The public works department also picked up the trash that the students collected. The city water department tested their water samples for fecal material, and concluded that sewage was seeping into the drainage ditch.

The students derived a great deal of satisfaction from this project, especially when they observed the public works department digging up the broken sewage pipes and replacing them. They also benefited from examining water samples for microorganisms using their microscopes. Mr. Bloom has taken many groups of students to various parts of the city to participate in clean-up projects and ecological studies because of the interest students have in this type of activity. ▲

Action learning is definitely a way to get students involved in issues that are important to their lives and to the community in which they live. This approach helps students to become active participants in society and to perform useful services. However, considerable thought and planning must go into action learning activities. Some activities may be potentially dangerous and science teachers must weigh their risks as well as their benefits.

Action learning, along with issue awareness, moral development, and issues investigation offer science teachers many ways to make science relevant. These approaches provide numerous options from which to select a method that best accommodates their teaching style, student maturity, and community needs.

CONSIDERATIONS FOR STS INSTRUCTION

The examination of technologies and the discussion of social issues in the science classroom require careful planning and good judgment. Science teachers must select topics that are directly related to the curriculum so that investigation into their relationships with science and society is legitimate. They must select topics that are relevant to students and that will impact their lives. The STS approach provides many opportunities to show teenagers who scientific and technological knowledge can make them better informed about themselves and the world they live in. However, in order for this approach to be successful, science teachers must implement instruction that considers the maturity level of the students. For example, middle school students will not be able to participate in discussions at the same level as senior high school and college students. In addition, science teachers must also be knowledgeable about the technologies and social issues that they wish to discuss in the classroom.

General Guidelines

Relevancy.   Science teachers should identify and study technologies and issues they wish to

include in their curriculum. These topics should be related to concepts and principles that are part of the syllabus. In this manner, the science subject matter under study will be reinforced by discussions that focus on these topics. For example, nuclear energy will have a different focus in a biology course than in a chemistry course. A biology teacher might stress radiation effects on biological tissue and how to determine the amount of radiation people receive from background radiation, dental X rays, etc., during a given year. A chemistry teacher might stress radioactive decay and the types of radiation that are given off during the decay of various radioactive isotopes. However, both teachers might address the pros and cons of the transportation of nuclear weapons by rail through their city.

Science teachers should make every attempt to organize science subject matter so that it is relevant to the lives of students as well as pertinent to the curriculum. Experienced and successful science teachers often tell how excited students get when instruction is directed toward content and topics that teach students about themselves. Students want to learn about their bodies and how they can maximize their potential to compete with peers. That is why drugs, sex, nutrition, sports, and cosmetics are topics of high interest to teenagers. Try to select topics of immediate concern to students and try to avoid the approach that says: "This information might not seem important now, but as you learn more you will understand."

Teacher's Abilities and Skills.   Science teachers must employ good interpersonal and group facilitator skills when conducting discussions on controversial issues. They must create a positive and open classroom climate that will encourage students to discuss their feelings and beliefs (Barman, Rusch, & Cooney, 1981). Science teachers must train themselves to be good listeners so that their students will engage in open and meaningful discussions. They must avoid judging the responses of students or putting them down in any way, which is similar to what is expected of science teachers when they conduct inquiry and recitation sessions. If students are embarrassed by what they believe their teacher or their peers will think of them when they share their feelings on a given topic, little or no dialog will take place. Science teachers must use great care when they attempt to distinguish fact from opinion during discussions. Students often confuse opinions for what they believe to be factual information, which obviously can distort their perceptions and beliefs.

Science teachers, especially those new to teaching, should ask administrators and colleagues for their opinions about the inclusion of issues that might be problematic to discuss in science class. They should discuss their plans with the principal to determine the possibility of parents' concern about what is taking place in their children's science classes. It is the principal who usually receives complaints from parents regarding the experiences their children receive in school. Principals have a great deal of experience dealing with problems of this nature and can be a good sounding board for instructional plans. Experienced science teachers, who have taught in a given school or community for many years, can also be very helpful in making recommendations on the inclusion of social issues in the science curriculum.

Students' Intellectual Development. The age and ability level of students must be considered when a societal issue is chosen for classroom discussion. Many social issues that are scientifically relevant cannot be presented to low ability students or to those who are simply not mature enough to discuss them. Adequate treatment, as used here, is meant to imply more than a superficial mention of the underlying complexities of the social issue. A disregard for student maturity very often creates more distortion and misunderstanding than educational benefit. For example, discussions of genetic screening may not be appropriate for a group of sixth-grade students because they do not have the sophistication to handle the issue. It is true that such students can discuss the issue at some level,

but the intricacies of the concept of personal freedom that underlies this issue would most likely be ignored or glossed over when dealing with such a group.

The teacher must be very careful that student age and ability do not preclude a relatively complete discussion of the issue at hand. It is important to remember that inclusion of a social issue in science instruction is for the purpose of (1) helping students learn and understand science content and (2) enabling students to make informed decisions with regard to the scientifically based issues that will confront them in life.

Jean Piaget's work on the development of intelligence provides a conceptual framework to help science teachers understand the potential and limitations of their students to discuss certain ideas. His work indicates that intellectual competence is tied to the development of general reasoning patterns. The degree to which these thinking skills have been developed will set limits on adolescents' thinking.

Science teachers should be aware that intellectual development has both cognitive and social dimensions. Although cognitive and affective reasoning develop together, there is not a one-to-one correspondence between these areas of competence. People in general, and young people in particular, may be more advanced in certain aspects of their physical world than in certain aspects of their social world. And, of course, the opposite is also true.

Lawrence Kohlberg furthered Piaget's work in the area of moral reasoning. He illustrates how moral reasoning is linked to values in the moral dilemmas he has developed. Each dilemma involves a character who finds himself in a difficult situation and must choose between two conflicting values. It is these values that constitute one's moral reasoning. The ten basic moral values that Kohlberg incorporates into his moral dilemmas are given below:

The Ten Universal Moral Issues

1. Laws and rules
2. Conscience
3. Personal roles of affection

4. Authority
5. Civil rights
6. Contract, trust, and justice in exchange
7. Punishment
8. The value of life
9. Property rights and values
10. Truth

(Reimer, Paolitto, & Hersh, 1983, p. 84)

Kohlberg uses these ideas to develop situations in which to study how people reason in the social dimension. He also believes that educators can enhance moral development by exposing students to moral dilemmas, asking them to present their reasons for how they would deal with a particular dilemma. (Refer back to an example of this in the moral dilemma case study on AIDS, discussed in this chapter.) This would require students to express and clarify their values.

Perhaps Kohlberg's biggest contributions to this field of study are the six "stages of moral judgment" that he has formed from his investigations in this field of study. His six stages as presented in figure 7–3 have been adapted from Barman, Rusch, and Cooney (1981, pp. 11–15), and Reimer, Paolitto, and Hersh (1983, pp. 58–61).

Kohlberg's stages of moral development provide a useful framework to select and incorporate into the curriculum the four recommended approaches to values education discussed earlier in this chapter. Analysis of these stages will permit one to maximize the match between any one of the values education approaches—awareness, moral development, investigation, and action learning—and the students in a given science class. Although one may find that a large proportion of the students demonstrate decisions based on the Conventional Level, often students will display behavior either above or below their principal level of moral development. Zeidler and Schafer (1984) suggest that moral reasoning is influenced by the context or setting of a moral dilemma. For example, a group of students may demonstrate a higher level of decision making on an environmental issue that they are familiar with and interested in than an issue that they

FIGURE 7–3 Kohlberg's six stages of moral development

Preconventional Level: Self-Centered Needs

Stage I: Avoidance of punishment. In this stage, people follow rules to avoid punishment. They are obedient and do what they are told to avoid getting into trouble.

Stage II: Self-benefit. In this stage, people follow the rules to serve their own best interests. They are aware of the motives and needs of other people, and treat others in accordance with how that treatment will benefit themselves.

Conventional Level: Conformity and Maintenance of Law and Order

Stage III: Concerns for acceptance. In this stage, people behave to meet the expectations of others. They are greatly influenced by peers and peer pressure, and behave in order to meet the expectations of those close to them. Being good is important and demonstrating loyalty, trust, and respect.

Stage IV: Authority orientation. In this stage, people act according to what appears to be "right or best" for society. They are influenced by authority and rules, and have the best interests of society as a whole in mind, as opposed to their own personal interests.

Postconventional Level: Consideration for the Values of Others

Stage V: Greatest good for the greatest number. In this stage, people believe that the rights of others are of uppermost importance. They will act to change laws if they feel that the laws are injurious to the rights and well-being of society. They believe that laws should be useful and serve everyone—"the greatest good for the greatest number of people."

Stage VI: Conscience is the guide. In this stage, people believe that individuality is supreme and the rights and freedom of each individual should be preserved. The equality of human rights must be maintained. They will do everything in their power, including breaking the law, to promote these universal principles.

have little knowledge of and interest in. Therefore, it is possible to get students to become more sophisticated in their beliefs and actions by involving them in values education instruction that centers on their interests and considers their maturity.

Evolution versus Creationism— Dealing with Nonscience Tenets

The evolution/creationism controversy illustrates the influence society can have on science education. For more than half a century biology textbooks have de-emphasized the teaching of evolution because a very small percentage of people believed that the study of evolution threatened their religious beliefs regarding the origin of humankind. In addition, many science teachers have been so intimidated by the problems surrounding this societal issue that they have either de-emphasized teaching evolution or have excluded it from their course. Consequently, this section has been written to give science teachers a strong rationale for including evolution in their instruction and for not addressing nonscience tenets in their curriculum.

The biological theory of evolution was discussed and written about before 1800, but the first formidably structured statement concerning variation among organisms was made by Charles Darwin with the publication of *On the Origin of Species* in 1859. Darwin's theory was presented as a grand and unifying theme of the biological sciences. Darwin was immediately criticized for his, apparently, atheistic views, although he clearly reflected his theistic beliefs in the document. Thus, the current debate pitting evolution and creationism against one another was born. The controversial issue concerned with evolution and creationism centers around whether creationism should be presented as a valid alternative to evolutionary theory within science instruction. However, the issue did not always take this form. In the 1920s, the debate was whether to present evolutionary theory at all in the public school science classrooms. The textbooks of that

period (1910 to early 1920s) covered evolution, but did so forthrightly.

The original legal battle over the teaching of evolution in public schools took place in Dayton, Tennessee, in 1925 where John Scopes was accused of teaching evolution, which constituted a violation of a Tennessee law prohibiting such instruction. Much notoriety was given to the trial, as Clarence Darrow defended Scopes against the Tennessee prosecution led by William Jennings Bryan. Much has been written about this landmark trial through the years. What has surfaced is a rather humorous account of a circus-type atmosphere. Indeed, the trial became known as the "Monkey Trial" and was the subject of a play and a movie. In reality, the legal aspects of the trial were rather undistinguished. There was clearly a law against the teaching of evolution, and Scopes had clearly violated it. (Actually, he only assigned the pages for students to read.) Thus, he was found guilty. Furthermore, Scopes had agreed to participate in the "incident" to test the merits of such a law. However, the legend of events past very often supersedes reality. Since Scopes was fined only $100, the trial has been considered a victory for evolutionists. However, even though many antievolution laws were repealed subsequent to the trial, the controversial nature of the issue prevented comprehensive coverage of evolution in biology textbooks before the 1960s. Then, as now, teachers often consciously avoid the teaching of evolution in an effort to avoid community "backlash." This has been the legacy of the Scopes trial.

One of the most significant court decisions of recent times was Epperson *v.* Arkansas in 1968.

Government in our democracy, state, and nation must be neutral in matters of religious theory, doctrine, and practice. It may not be hostile to any religion or to the advocacy of no-religion; and it may not aid, foster, or promote one religion or religious theory against another or even against the militant opposite. The First Amendment mandates governmental neutrality between religion and nonreligion. (Epperson *v.* Arkansas, 393, U.S. 97, 1968.)

The current debate concerning the teaching of evolution is not with the banning of such instruction but rather the provision of "equal time" for the creation story as a valid alternative. Thus, the approach has changed, but the controversy continues. Mandates forcing science teachers to provide instruction regarding creationism strikes at the heart of academic freedom. Science teachers are exercising this academic freedom and not choosing to teach creationism. Furthermore, as Skoog points out:

> Equal time policies are not necessarily fair or educationally sound. . . . Science teachers cannot treat all knowledge equally. We must select content based on its power to explain the natural world scientifically and its ability to unify, illuminate, and integrate other facets. We should not include ideas that cannot serve these functions. . . . The courts have ruled that laws prohibiting the teaching of evolution are unconstitutional and that teachers have no obligation to shield students from ideas that may offend their religious beliefs. Furthermore, it is possible to excuse students from classes where topics, offensive to their religious beliefs, are being presented (Skoog, 1985, p. 8).

This issue brings us to the nature of scientific theory. Creationists refer to evolution as only a theory. They appear to use the term as it is used in the American vernacular, imperfect fact. Such use of the term implies a hierarchy of confidence that runs downward from fact to theory to hypothesis to guess. Actually, scientific theory is not placed in such a hierarchy. In science, theories are structures of ideas that explain and interpret scientific facts (data). Thus, using the term theory to mean an imperfect fact is not the sense in which scientists use it when they call evolution a theory. Scientists do not believe that evolution is an "imperfect fact."

What constitutes a scientific theory actually touches the core of the current creationism/evolution controversy. Creationists feel that creationism is just as much a theory as evolution and that it is as scientific. Therefore, they feel that the creation story should be presented in

science instruction as a valid scientific alternative to evolutionary theory. On the other hand, evolutionists feel that creationism is not science at all, but rather religion, and should not be included in the science classroom. In 1982, Judge William Overton interpreted the constitution and in doing so, sided with a variety of individuals in the Arkansas case that tried the constitutionality of a law requiring equal treatment of creationism and evolution. The verdict was simple; it was decided that creationism is religiously based and inclusion of it within public education would violate the separation of church and state.

Since this issue is extremely controversial, it would appear to serve as an efficient vehicle for classroom discussion. Nevertheless it is *not* a good issue to debate because the creation aspect of the issues is not part of the science curriculum and according to court decisions and science educational guidelines it should not be part of the curriculum.

Perhaps the National Science Teachers Association position statement concerning "Inclusion of Nonscience Tenets in Science Education" is a worthwhile guide for science teachers to follow when dealing with a topic such as creationism versus evolution.

> People have always been curious about the universe and their place in it. They have questioned, explored, probed, and conjectured. In an effort to organize their understanding, people have developed various systems that help them explain their origin, e.g., philosophy, religion, folklore, theaters, and science.
>
> Science is the system of exploring the universe through data collected and controlled by experimentation. As data are collected, theories are advanced to explain and account for what has been observed. Before a theory can be included in the system of science, it must meet all of the following criteria: (1) its ability to explain what has been observed, (2) its ability to predict what has not yet been observed, and (3) its ability to be tested by further experimentation and to be modified as required by the acquisition of new data.
>
> NSTA recognizes that only certain tenets are appropriate to science education. Specific guide-

lines must be followed to determine what does belong in science education: NSTA endorses the following tenets:

I. Respect the right of any person to learn the history and content of all systems and to decide what can contribute to an individual's understanding of our universe and our place in it.

II. In explaining natural phenomena, science instruction should only include those theories that can properly be called science.

III. To ascertain whether a particular theory is properly in the realm of science education, apply the criteria stated above.

IV. Oppose any action that attempts to legislate, mandate or coerce the inclusion in the body of science education, including textbooks, any tenets which cannot meet the above-stated criteria.

(Adopted by the NSTA Board of Directors in July, 1985. National Science Teachers Association, 1992–93, p. 205.)

Since this problem never seems to go away, the National Academy of Science (1984) reaffirmed its position of this issue in their publication *Science and Creationism: A View from the National Academy of Sciences.* The Academy stresses that

... teaching the scientific theory of evolution alongside creationism is inappropriate. Teaching creationism is like asking our children to believe on faith, without recourse to time-tested evidence, that the dimensions of the world are the same as those depicted in maps drawn before Columbus set sail with his three small ships, when we know from factual observations that they are really quite different.

It is false, however to think that the theory of evolution represents an irreconcilable conflict between religion and science. A great many religious leaders and scientists accept evolution on scientific grounds without relinquishing their beliefs in religious principles. As stated in a resolution by the Council of the National Academy of Sciences in 1981, however, "Religion and science are separate and mutually exclusive realms of human thought whose presentation in the same context leads to misunderstanding of both scientific theory and religious belief." (NAS, 1984, pp. 5–6)

In closing this section we must stress the importance of evolution in science and the understanding of the universe. Although the topic is controversial, science teachers should not fall into the trap of attempting either to deal with evolution and creationism as a science and societal issue, or to demonstrate the inadequacies of creationism to explain the origin of life. This approach may be offensive and inflammatory to some students and their religious views. Stick to the discipline of science and teach it well.

SUGGESTED ACTIVITIES

1. Collect articles from books, magazines, newspapers, and professional journals that discuss new technological products and systems, and social issues. Keep a file of these ideas that can be studied in a given science course.

2. Identify several technologies that interest you. Explain their historical development and the scientific principle(s) behind them.

3. Analyze a middle or senior high school science textbook for the degree to which it incorporates technology and science-technology-societal issues in the text and student activities.

4. Develop a science unit that incorporates the following approaches to STS instruction:
   ■ Values awareness and clarification
   ■ Moral development
   ■ Issues investigation with analysis and decision making
   ■ Action learning

5. Select an earth science or biology textbook and analyze its treatment of evolution. Does the textbook present evolution as a fact or theory? How well does it explain evolution and use it to explain major scientific principles?

BIBLIOGRAPHY

Andrews, M. S. 1970. Issue-centered science. *The Science Teacher, 37:* 29–30.

Ajzen, I., and Fishbein, M. 1980. *Understanding attitudes and predicting social behavior.* Englewood Cliffs, NJ: Prentice-Hall.

Barman, C. R., Rusch, J. J., and Cooney, T. M. 1981. *Science and societal issues: A guide for science teachers.* Ames, IA: Iowa State University Press.

Brinkerhoff, R. F. 1986. *Values in school science: Some practical materials and suggestions.* Exeter, NH: Philips Exeter Academy.

Bybee, R. W. 1986. The Sisyphean question in science education: What should the scientifically and technologically literate person know, value, and do— as a citizen? In R. W. Bybee (Ed.), *1985 Yearbook of the National Science Teachers Association.* Washington, DC: The National Science Teachers Association.

Erlich, P., and Erlich, A. 1980. *Extinction.* New York: Random House.

Gottlieb, S. 1976. Teaching ethical issues in biology. *The American Biology Teacher, 38*(3): 148.

Hungerford, H. R., Litherland, R. A., Payton, R. B., Ramsey, J. M., and Volk, T. L. 1988. *Investigating and evaluating environmental issues and actions skill development modules.* Champaign, IL: Stipes Publishing.

Hurd, P. D. 1961. *Biological education in American secondary schools, 1890–1960.* Washington, DC: AIBS.

Jarcho, I. S. 1985. S/T/S in practice: Five ways to make it work. *Curriculum Review, 24*(3): 17–20.

Lux, D. G. (Ed.). 1991. Science, technology, society: Challenges. *Theory into Practice, 30*(4).

Lux, D. G. (Ed.). 1992. Science, technology, society: Opportunities. *Theory into Practice, 31*(1).

Meier, H. A. 1981. Thomas Jefferson and a democratic technology. In C. W. Persell, Jr. (Ed.), *Technology in America.* Cambridge, MA: The MIT Press.

National Academy of Sciences. 1984. Science and creationism: A view from the National Academy of Sciences. Washington, DC: National Academy Press.

National Science Teachers Association. 1991. Science/technology/society: A new effort for providing appropriate science for all. *NSTA Reports!* Washington, DC.

National Science Teachers Association. 1992–93. Inclusion of nonscience tenets in science. *NSTA Handbook,* p. 205. Washington, DC.

Oliver, D. W., and Newman, F. M. 1967. *Taking a stand.* Middletown, CT: Xerox Corp.

Purcell, C. W., Jr. 1981. *Technology in America.* Cambridge, MA: The MIT Press.

Ramsey, J. M. 1989. A curriculum framework for community-based STS issue instruction. In J. Penick (Ed.), *Education and Urban Society: Issues-Based Education, 1:* 40–53.

Ramsey, J., and Hungerford, H. 1989. So . . . you want to teach issues? *Contemporary Education, 60:* 137–142.

Ramsey, J. M., Hungerford, H. R., and Volk, T. L. 1989. *A science-technology-society case study: Municipal solid waste.* Champaign, IL: Stipes Publishing Co.

Reimer, J., Paolitto, D. P., and Hersh, R. H. 1983. *Promoting moral growth.* New York: Longman.

Rubba, R. A. 1987. Perspectives on science-technology-society instruction. *School Science and Mathematics, 3:* 181–185.

Rubba, R. A., and Wiesenmayer, R. 1988. Goals and competencies for precollege STS education: Recommendations based upon recent literature in environmental education. *Journal of Environmental Education, 19*(4): 38–44.

Skehan, J. W. 1986. *Modern science and the Book of Genesis.* Washington, DC: The National Science Teachers Association.

Skoog, G. 1985. Editor's corner. *The Science Teacher, 52*(1): 8.

Superka, D. P., Ahrens, C., Hedstrom, J., Ford, L. J., and Johnson, P. L. 1976. *Values education sourcebook.* Boulder, CO: Social Science Education Consortium.

Tanner, D., and Tanner, L. 1980. *Curriculum development.* New York: Macmillan.

Tanner, T., 1987, Oct. Environmental education for citizen action. In *S-STS Reporter,* University Park, PA: The Pennsylvania State University.

Yager, R. E. 1985. Preparing students for a technological world. *Curriculum Review, 24*(3): 21–22.

Zeidler, L., and Schafer, L. E. 1984. Identifying mediating factors of moral reasoning in science education. *Journal of Research in Science Teaching, 21*(1): 1–15.

Laboratory work engages students in finding out and learning how through first-hand experiences.

CHAPTER 8

Laboratory Work

Laboratory work is a unique type of instruction that must be an integral part of science teaching. This type of activity involves students in firsthand experiences that permit them to participate in science as a way of thinking and as a way of investigating. Laboratory work provides students with concrete exemplars of science concepts and principles, which can serve to reinforce course content. Although the laboratory has the potential to produce a variety of learning outcomes valued by science educators, it takes considerable expertise on the part of science teachers to produce the desired outcomes.

Overview

- Define laboratory work and explain its purpose.
- Describe five types of laboratory approaches that should be used in a science course.
- Discuss teaching tips that can result in successful laboratory experiences.
- Provide suggestions on how to inventory, order, and maintain laboratory materials and supplies.

Where would life end up if the conservation of mass did not exist?   ▲

Technical Skill Laboratory

Good laboratory techniques are essential to conduct successful laboratory activities and to collect accurate data. They require manipulative skills that involve the development of hand-eye coordination such as focusing a microscope, sketching specimens, measuring angles, and cutting glass. Good laboratory work also includes experimental technique and orderliness. Although laboratory work often relies on students' abilities to manipulate equipment, some is highly dependent on the use of special equipment and techniques; therefore, the emphasis of some laboratories should involve the development and use of these skills and techniques.

Science educators have placed too little emphasis upon developing proficiency in laboratory skills and techniques (Hegarty-Hazel, 1990). All students and science teachers should master the basic laboratory techniques and manipulative skills associated with the science area in which they are involved, some of which are presented in figure 8–2. Science teachers who plan and organize laboratory experiences ahead of time can identify techniques that require special attention. For instance, the microscope is used a great deal in biology laboratory work. Most adolescents have difficulty focusing the microscope and centering the specimen in the microscope field. Many experienced biology teachers provide their students with laboratory experiences on the care and use of the microscope. This work permits students to view such objects as newsprint and human hair under the microscope to learn how to focus the instrument and how to move objects under view. Because of the lens system, everything viewed is reversed, which confuses students; therefore, they need practice focusing the microscope. Beginners, for example, have a tendency to crush the glass cover slips under the objective lens when attempting to bring a specimen into clear view.

Psychomotor and mental practice are beneficial in improving the accuracy and precision of a student's laboratory measurements (Beasley, 1985). Physical practice with laboratory equipment provides concrete experience with the apparatus and procedure. It gives the student a set of experiences upon which to build images that represent the skill under development. Because time is always critical to laboratory work, first-hand exposure to the equipment is essential. Then, during class time and discussion sessions, mental practice of the skill and procedures under study can ensue. Consider one high school science teacher's experience.

▼ It is essential to precede the laboratory on measuring electricity with a laboratory on using the voltmeter and ammeter. If the students misread their meters when trying to determine voltage, current, or resistance, the lab is lost. This experience requires practice in reading meters first, then concentrating on collecting data and making calculations. It is too much to ask the teenagers that I have in my classes to determine the amps in a circuit before they are comfortable with the use of the meters. Remember, these youngsters do not use meters such as these in their daily lives. What makes matters worse is that the meters we use in our labs have three scales, which can be very confusing to students. Therefore, when the laboratory work in the area of electricity is carried out in stages that develop specific competencies, my students are successful and enjoy their science experiences.   ▲

The same science teacher goes on to say that she uses the following steps when teaching Ohm's law: (1) use the ammeter to learn how to connect it in series in the circuit and how to read the scale accurately; (2) use the voltmeter to learn how to connect this device across the circuit and how to read the scale accurately; and (3) use the meters to find amperes and volts, and then use the data for Ohm's law calculations.

Making a profile map can be a challenge for the concrete operational teenager participating in a middle school earth science class. Translating topographic information from contour lines

FIGURE 8–2   Examples of techniques and manipulative skills for science laboratory programs

General Science

Constructing and calibrating an equal-arm balance
Determining mass with an equal-arm balance
Constructing and measuring with a metric rule
Measuring temperature with a thermometer

Measuring volume with a graduated cylinder
Determining significant figures to indicate precision in measurement

Life Science

Dissecting a frog or a worm
Sketching an organism
Slicing a piece of tissue for microscopic examination
Preparing a wet mount slide
Focusing a microscope
Measuring dimensions under a microscope
Transferring a microscopic organism from one medium to another

Taking a pulse
Sterilizing instruments
Making a serial dilution
Germinating seeds
Using paper chromatography to separate chemicals
Counting the growth of microorganisms

Earth Science

Growing crystals
Orienting a map with a compass
Making and reading topographic maps
Making a profile map
Using a stereoscope
Using a Brunton compass
Testing the physical properties of minerals
Reading a classification chart for rocks

Estimating the percentage of mineral in rocks
Classifying fossils
Plotting data on a time chart
Analyzing soil
Reading an aneroid and mercury barometer
Reading an anemometer
Plotting the sun's path

Chemistry

Cutting, bending, and polishing glass
Boiling liquids in a beaker
Folding filter paper and filtering solutions
Heating liquids in a test tube
Pouring liquids from a reagent bottle and a beaker
Pouring hot liquids from a beaker
Dispensing caustic solutions

Transferring powders and crystals
Smelling a chemical
Preparing solutions of a given concentration
Titrating with a burette
Calibrating a test tube
Using an analytic balance

Physics

Soldering electrical connections
Connecting electrical devices in parallel and series circuits
Using electric meters
Measuring time intervals with electrical and water clocks
Measuring weight with a spring balance

Determining the focal length of mirrors and lenses
Locating images in mirrors
Constructing a simple telescope
Constructing a spectroscope
Setting up a siphon

to a vertical/horizontal profile is a task that requires considerable mental as well as manipulative ability. Most students need practice before they gain competence in this task and are able to realize the three-dimensional nature of a contour map.

Making sketches and drawings is an essential part of laboratory work. This type of activity not only provides a record of observations, but it reinforces visual images that pertain to essential concepts and learning outcomes in a science course. Here are some guidelines that can be

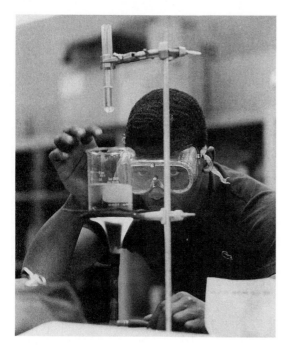

Good laboratory techniques are essential to conduct successful laboratory activities.

used when instructing students to draw what they see in a biology class (or any other science course):

- Draw, using a pencil and unlined paper.
- Make the drawing large enough so that important details are easily seen.
- Place the drawing near the left side of the sheet so that the labels can be placed on the right side.
- Print the labels one under the other.
- Use a ruler to draw lines from the labels to the drawing, and do not cross lines.
- Use light stippling or dotting to shade part of a drawing.
- Construct a title for the drawing at the top of the page.

The preceding discussion makes clear that many technical skills are necessary for improving students' ability to participate in and learn from laboratory work. Students need these skills to gain the unique learning outcomes associated with the science laboratory. If students have to struggle with basic manipulative procedures, they may lose sense of what they are trying to accomplish in the laboratory. Also, if students are so unfamiliar with basic lab procedures that they have to follow them in a rote manner, they will lose much of the cognitive and affective benefits of laboratory work because they are preoccupied with following directions.

Problem Solving Laboratory

In some instances, science teachers may want to engage their students in a problem solving laboratory where students identify a problem, design the procedures, collect information, and report the findings. In addition, this approach is recommended on psychological grounds because students are involved in organizing their own learning, thus inclined to think and understand what they are doing. In addition, most students take greater interest in their learning when they take part in organizing it.

Consider for instance, a set of problem solving laboratories that could be used at the middle or even at the senior high school level in the life sciences. The teacher might recommend that students identify an insect to study over a 1- or 2-week period. The students can be permitted to select an insect that they wish to study and determine the questions they wish to answer about the insect. Many students will be inclined to investigate what their insect eats. Most students will attempt to determine what reaction their insect will have to various stimuli. For example, students who select mealworms to study might determine how these insects react to water, vinegar, salt, soap, bran flakes, cola, sandpaper, glass, electricity, sound, heat, and light. The students can be encouraged to bring from home additional materials needed to study their insect. Some students may return to the laboratory after school hours to continue their inquiry beyond the regular time set aside for laboratory work.

Some science teachers use the problem solving approach with students who wish to satisfy their curiosity about certain situations or with students whom the teacher wishes to motivate to study science and technology. Science teachers can accomplish these ends outside of the regular laboratory time, usually after school when students can spend many hours engaging in firsthand experiences. For more information and examples on problem solving instruction, study the section on problem solving in Chapter 3 and especially the section on problem solving in Chapter 4.

PREPARING STUDENTS FOR LABORATORY EXPERIENCES

Students must be prepared for laboratory experiences in order to benefit from them. They need to know why they are expected to participate in an activity and what they will derive from it. Science teachers who report difficulties with laboratory activities either do not prepare the students for the laboratory work or perform only a very superficial job of preparing them. Teachers often permit the laboratory activity itself to dominate the situation and neglect other forms of instruction, such as prelaboratory discussion, giving directions, and postlaboratory discussions.

Prelaboratory Discussion

The prelaboratory discussion prepares students for a laboratory activity. This phase of instruction informs students on why, how, and what they will be doing. The prelaboratory discussion is an introduction to the laboratory because it gives the students a mind-set for the laboratory, but it does not give them any of the answers. This step in laboratory preparation explains how the laboratory activity relates to the topic being studied in the classroom. If, for example, a principle or a concept is being discussed in the classroom, it should be made clear to the students that the purpose of the laboratory is to verify the concept

or principle under consideration (provided it is a verification laboratory).

Consider the following discussion that might take place before a laboratory on qualitative chemical analysis to identify ions by the flame test.

Teacher: You are going to find out how chemists and laboratory technicians determine the presence of ions. We have been discussing ions in class, and now you have an opportunity to detect them in solution. You might think of yourselves as chemists working in a crime lab. I will write the ions on the chalkboard that you will test for in today's lab: sodium, strontium, potassium, copper, and calcium. Thomas, can you tell me what color will appear in the flame test for each of these ions?

Thomas: I'm sorry, but I can't remember them.

Teacher: It is not so important that you memorize the flame tests for all of the ions, but that you know where to find these tests in your textbook or on a chart, because these resources will be available to you when you are given a test on this subject. Will all of you look up the flame test in your textbooks at this time? [Pause.] Thomas, will you give me the flame tests for the ions listed on the chalkboard? I will write them down.

Thomas: Sodium is orange, strontium is violet, potassium is purple, copper is blue, and calcium is green.

Teacher: Thank you. Now let me show you how to test for these ions with this platinum wire. Pour a small amount of dilute hydrochloric acid into a clean watch glass. Dip the platinum wire into the hydrochloric acid and heat it in the burner flame. Why do I heat the wire first until it glows?

Student: To clean the wire so that it won't be contaminated.

Teacher: Good! Now I dip the wire into the solution with an unknown ion and then put the tip of the wire into the flame. Maria, what ion did I have in this solution?

Maria: Calcium!

Teacher: That's correct. Does anyone have a question about today's lab and how to proceed?

The prelaboratory discussion must give students the clearest possible picture and understanding of what they are to do in the laboratory. This will help the students concentrate to make the experience more meaningful. It will also prevent the experience from becoming a cookbook exercise in which the students must constantly refer to printed directions for guidance and thus become immersed in the mechanics of the laboratory instead of the excitement of finding out something for themselves. If special equipment or difficult procedures are involved, the teacher should show the students how to use the equipment and procedures and then call on some students to see how they perform these tasks. Prelaboratory discussions should be as short as possible, yet long enough to thoroughly orient the students to the laboratory.

Giving Directions

The directions for laboratory exercises must be explicit. They can be given orally, distributed in written form, or discussed during the prelaboratory session. Any combination of these can also be used. Oral directions may be adequate when one-step activities are involved and when the directions are simple enough to be remembered.

▼ Miss Stasson advised her earth science class that diluted hydrochloric acid reacts with substances containing carbonates. She gave each pair of students a dispensing bottle full of dilute acid and a tray of assorted rocks, minerals, bones, and shells. She directed the students to test the items for the presence of carbonates.   ▲

Sometimes summarizing directions on the chalkboard that have already been given orally is helpful.

▼ Mr. Bruhn set out test tubes, medicine droppers, a soap solution, and a liquid detergent. He showed the students how to test the effects of these

The directions for conducting laboratory exercises must be explicit and clear to the students. They can be given orally or in written form or discussed during a prelaboratory session.

solutions on hard water. He then summarized the directions on the chalkboard as follows:

1. Fill two test tubes nearly full of water.
2. Add 4 drops of soap solution to one test tube.
3. Add 4 drops of detergent to the other test tube.
4. Shake each test tube well.
5. Hold the test tubes up to the light and observe.   ▲

Written directions can be duplicated on paper and given to students or they may be found in the laboratory manuals used in the course. Regardless of the form, the activities should be broken down into several steps. Each step should consist of a brief set of directions followed by some questions, as shown in the following example:

▼ Strike one of the tuning forks against the palm of your hand. Observe the fork carefully.

1. What observations can you make?
2. With which of your senses can you make observations of the vibrating fork?
3. Can you count the number of times the fork vibrates in 1 minute? ▲

Postlaboratory Discussion

Students present and analyze their data during the postlaboratory discussion. Here the information can be analyzed and related to the objectives of the unit, the course, or both. The postlaboratory discussion is an excellent time to broaden students' understanding of the content and processes of science.

▼ At the completion of an exploratory laboratory on the topic of work, Miss Blaskiewicz assembled her class into groups. She asked a representative from each group to go to the chalkboard and report their findings on work (see figure 8–3). Miss Blaskiewicz also asked each group to identify the set of data that represented the greatest amount of work accomplished. The groups' activities were partially contrived by Miss Blaskiewicz to help the students understand work and to explore this principle by inspecting various data. Some students expressed some difficulty in believing that the girl who lifted the 4½ lb dictionary 90 times from the floor to 6 ft above the floor actually did more work than the boy who similarly lifted the 90 lb barbell a distance of 6 ft four times. Some students were also surprised that no work was done when the class "strong man" tugged at the student desk bolted to the floor but did not move it. By the end of the postlaboratory discussion, all of the students in the class realized that work is the result of a force moving an object over a distance and that if either the force or the distance is equal to zero, then no work is accomplished. Miss Blaskiewicz concluded the postlaboratory discussion by calling on several students to verify the work formula by using data illustrating positive and negative examples of work. ▲

The postlaboratory discussion presents an excellent opportunity to focus on important learning outcomes associated with laboratory work. For example, if the lab is designed to address conceptual knowledge, the teachers can check on misconceptions and the extent to which these alternate views are being affected. Students can be called upon to state what they believed regarding a given idea before the lab, and what they believe now as a result of their experiences and the data collection. Students can be requested to construct concept maps or concept circles to help them show relationships among key concepts. The teacher can call upon students to identify the science process skills that they used to conduct their investigation, and to spec-

FIGURE 8–3 **An example of postlaboratory data examined by Miss Blaskiewicz's class**

Work Formula: Work = Force × Distance

Group A   90 lb × 24 ft = 2160 foot-pounds
A boy lifting a 90 lb barbell a distance of 6 ft from the floor a total of 4 times.

Group B   4½ lb × 540 ft = 2430 foot-pounds
A girl lifting a 4½ lb dictionary 90 times from the floor to an overhead position 6 ft above the floor.

Group C   280 lb × 0 ft = 0 foot-pounds
A boy capable of lifting a 280 lb barbell, pulling up on a desk that is bolted to the floor.

ulate about how scientists and engineers might have conducted this laboratory exercise. A science teacher can make the decision to have students perform parts of the lab again as the result of the postlaboratory discussion. Performing a given laboratory experience only once may do very little to help some students. These individuals may need considerable practice and exposure to many basic procedures and ideas, especially those with limited science backgrounds or who speak English as a second language. Many students do not have time to complete their work thoroughly and to develop a reasonable level of competence in skills related to science and engineering.

ENSURING SUCCESSFUL LABORATORY OUTCOMES

Science teachers must carefully plan and organize their laboratory activities if students are to achieve important learning outcomes. They must give serious attention to the relevance of laboratory work, the degree of structure involved in activities, the methods by which students record and report data, classroom management, and evaluation of student work. Failure to give proper attention to these critical factors can undermine the value of laboratory activity in a science course.

Relevance of Laboratory Work

The association between course and laboratory work may not be evident to students during the course of daily instruction. Laboratory work often becomes a fragmented entity that seems to have little or no relation to the real world. This aspect of science teaching can become merely an activity to work through. However, laboratory activities that incorporate commonplace devices, which have immediate applications in the real world, are worthwhile to use. In laboratories where siphons, candles, electric bells, xylophones, household cleaners, mechanics' tools, over-the-counter medicines, and garden soils are

studied, students rarely question the value of the work or its association with scientific principles.

▼ When Mr. Teichelman's physical science class studies Newton's laws of motion, students bring to class a variety of model cars and trucks. Momentum and inertia are illustrated by using these common objects. For example, clay models representing people are made and placed on the vehicles, which are then sent down inclines. Upon impact with barriers at the end of the inclines, the clay models are thrown out, and students measure the distance the "people" are thrown. ▲

▼ Over the years Mr. Jordan has acquired a number of damaged thermos bottles that still hold water, along with a number of good thermos bottles. Each year his physics classes perform controlled experiments with the bottles to determine the effects of a vacuum in reducing heat transfer. ▲

The use of everyday materials demonstrates the applicability of science concepts and principles in daily life. These materials are usually inexpensive and easy to obtain. Many times, students who are unmotivated in science classes are those who readily volunteer to bring in objects that are useful in laboratory work. Chapter 3 stresses the importance of providing a relevant context for studying science concepts. If instruction begins with things familiar to students, the instruction is most likely to be related to the knowledge that students possess. Consequently, the instruction will begin with what students know, facilitating conceptual development. In addition, familiar objects provide a context that may be more interesting to students and serve to motivate their learning.

Degree of Structure in Laboratory Activities

Structure refers to the amount of guidance and direction teachers give to students. It usually takes the form of written directions or questions that are prepared on duplicated sheets or in laboratory manuals. Experienced science teachers often employ highly structured laboratory exercises, especially during the first part of a science course. Highly structured exercises provide students with a great deal of guidance to assist the teacher to manage the instructional environment. Science teachers emphasize that when they must instruct large numbers of students, short exercises that provide students with plenty of direction seem to work best.

▼ Mrs. Hall has earned the reputation of being a good earth science teacher because of her activity-based approach to teaching earth science. Her students know that they can expect a 15-minute laboratory almost every day. The laboratory exercises are duplicated on 8½-by-11-inch paper, and each exercise usually includes two or three short activities, each with a set of directions, accompanied by a diagram and a few questions. ▲

The following comments are from science teachers regarding the structure of their laboratories:

■ "Our labs are highly structured and the answers seem so obvious, but not to these sixth graders."
■ "In physical science we use highly structured labs that focus on specific objectives and reinforce specific concepts. This is important to our ninth graders who need help adapting to the high school environment."
■ "The labs for many regular students are highly structured with specific instructions and questions that are close-ended because most of my students are not very science oriented. They get easily frustrated if they do not understand exactly what they are to do."

Figure 8–4 presents an example of a highly structured science process laboratory handout. The handout contains the elements commonly used to construct laboratory activities, such as the purpose, materials, procedures, observations, and conclusions. The activity sections of the laboratory handout contain the procedures and spaces to record observations. This laboratory activity format is highly structured in that it tells the students exactly what to do—it gives the

problem, procedures, and data collection format. However, the students must carry out the procedures, analyze the data, and form a hypothesis.

A note of caution is in order regarding highly structured laboratory work. There is a problem with using too much structure over the entire course. If the teacher uses highly structured activities throughout a course, problem solving, conceptual change, modifying misconceptions, and motivation may be limited. Structure can stifle self-directed learning and decision-making behavior. Consequently, toward the middle and end of the course, science teachers should vary the structure of their laboratory work.

After students have acquired basic inquiry skills and techniques, the teacher should give them the opportunity to identify their own problems and devise their own procedures. Over time, students will learn to conduct complete laboratory experiments. Some science teachers have suggested that students can be given more autonomy during the eighth and ninth grades than during the sixth and seventh grades, and certainly many high school students are capable of conducting their own inquiries.

Some science educators and cognitive psychologists advise science teachers to permit students to be actively involved in designing and carrying out their lab work. As discussed in the section on conceptual change in Chapter 3, it is important to determine what students know about an idea before instruction begins; this also holds for laboratory procedures. It is recommended that science teachers involve the students in thinking about many ways to answer questions and find answers through hands-on activity in the laboratory. Students can be very creative in their investigations when they are permitted to suggest ways to test ideas in the laboratory. One effective teacher said this regarding laboratory work: "I'm changing my lab approach. I let my students suggest ways to do the labs and figure out how to set up the apparatus. I am tired of telling them everything and they get tired of my telling them everything."

FIGURE 8–4   An example of a highly structured science process-oriented laboratory activity form

---

The Effect of Temperature on Dissolving Time

Purpose   The purpose of this activity is to form a hypothesis regarding the relationship between temperature and dissolving time.

Materials

| sugar cubes | beakers | hot and cold running water |
| seltzer tablets | timers or clocks | |

Activity 1   Dissolve one sugar cube in water at two different temperatures and record the time it takes to completely dissolve each time. At least two trials should be performed at each temperature, and the mean dissolving time should be computed. Use the chart below to record your data.

Substance _____   *Dissolving time in seconds*

Temp in degrees Celsius	*Trial 1*	*Trial 2*	*Mean*
_____	_____	_____	_____
_____	_____	_____	_____

After analyzing these data, what hypothesis can you state regarding the relationship between temperature and dissolving time?

_____

_____

Activity 2   Repeat Activity 1 but use another substance, such as seltzer tablets. Record your data on the chart below.

Substance _____   *Dissolving time in seconds*

Temp in degrees Celsius	*Trial 1*	*Trial 2*	*Mean*
_____	_____	_____	_____
_____	_____	_____	_____

After analyzing these data, what hypothesis can you state regarding the relationship between temperature and dissolving time?

_____

_____

Activity 3   The teacher will assemble the class to discuss the data and the hypotheses. The data will be pooled and displayed on the chalkboard. What patterns do you observe in the data? How can you explain your hypotheses?

_____

_____

Conclusion   Construct a final hypothesis regarding the relationship between temperature and dissolving time.

Leonard (1991) suggests that science teachers attempt to "uncookbook laboratory investigations." He claims that too many procedures and written directions for students to follow can reduce advance organization by students, thus causing them to follow directions in a rote manner and to lose the meaning of the lab. Students can remember only so many directions and internalize only so much of someone else's procedures.

Student Recording and Reporting of Data

Students need assistance in recording and reporting their laboratory observations. When laboratory manuals are provided, the problem is somewhat reduced because laboratory manuals usually have a place for students to respond. When laboratory manuals are not provided, duplicated sheets or notebooks can be used to record data. Regardless of the form, recording must be kept simple. If students must devote too much time and effort in recording and reporting, they may develop an unfavorable attitude toward the laboratory.

Laboratory exercises vary in terms of their content and involvement. Consequently, recording and reporting these activities should vary. Some exercises focus on techniques and motor skills, requiring very little written activity. For example, exercises developing competence in using the balance, microscope, graduated cylinders, burettes, voltmeters, ammeters, force measures, dissecting instruments, and others require careful manipulative skills, but little is required in reporting the outcomes in writing.

Exercises that involve students in a great deal of inquiry and open-ended investigations may require a more extensive type of report. In these situations, students may need to identify strategies that they will use to answer research questions. They should be expected to explain the procedures used to collect data when reporting their investigations. In addition, students must report data in a form that best communicates the data's value to the investigation. This usually requires the use of many communication devices used in science, such as graphs, tables, formulas, and figures.

▼ Mr. Lawrence supplements the exercises in the commercially prepared laboratory manual used in his biology course with short investigations that students design and conduct. For these investigations, he requests students to work in pairs and present a 5-minute report to the class. The students must write very brief reports, indicating the question(s) under investigation, the procedure, the data, the analyses, and the conclusions. At the beginning, Mr. Lawrence keeps the investigations and reports as short as possible. As the year progresses, he requires each student to conduct his or her own investigation and to write up a report that is more comprehensive than a one- or two-page summary. ▲

Some science teachers require very little from students regarding the procedures used to carry out an investigation, because this information is usually written in the laboratory manual or on a handout. However, these teachers require their students to prepare a thorough explanation of: (a) the results, (b) the significance of the results, (c) how the laboratory relates to the subject matter content that is under study in the classroom, (d) how the laboratory reflects the nature of science, and (e) how the concepts and principles apply to other situations. Teachers indicate that students benefit greatly from these exercises when they are required to determine the importance and derive something meaningful from their laboratory work. They also mention that this assignment is very time consuming to grade.

A typical format for reporting science laboratory work is as follows:

1. Problem
2. Materials
3. Procedure
4. Results
5. Conclusions
6. Applications

This format can become a highly stereotyped form for reporting laboratory work. Such an unvarying format may result in boredom and resentment from many students. Furthermore, students often mistake these steps as being synonymous with the scientific method. Science teachers are advised to vary their requirements for reporting laboratory work.. Simple experiments require only simple records and reports.

Gardner and Gauld (1990) advise that teachers who emphasize correctness of data and conclusions might produce negative effects on students' laboratory performance and attitudes. They point out that students want to get the "right results" because their teachers use their results to grade them. Consequently, if students do not get good data or what is perceived to be the correct results, they may be penalized. This situation causes some students to copy data from other students so they will receive a good grade. Teachers should realize that getting the right answer can discourage curiosity and original thought. Science teachers must be acutely aware of the learning outcomes they are shaping from the type of laboratories they promote and the type of laboratory reports they require from students.

Management and Discipline During Laboratory Activities

Management is a critical factor for successful laboratory activities. This is especially true in the middle and junior high schools, where students are very active and perhaps cannot concentrate for extended periods. Laboratory room management may pose a special problem to the beginning science teacher, who may be a little lax in developing and maintaining rules for this type of activity. Some essential elements that need attention in the science laboratory include seating arrangements, grouping, discipline, and monitoring student activities. Desks and laboratory tables should be arranged so that they are not crowded, to allow for free flow of traffic. Keep students away from laboratory materials until they are ready to use them, especially during the time that the teacher is giving directions. Avoid placing work tables against walls.

Students can work individually, in pairs, or in small groups. The amount of equipment and materials usually dictates the working arrangement. Obviously, it would be best to have students work independently the majority of the time, but in most situations they must work in pairs or in small groups of approximately four students. Problems can arise when students within groups participate in very little laboratory work or when they interact between groups. Talking and fraternizing between groups usually results in a high noise level and disruptive behavior. It is best to require students to work and to talk only with those within their own group.

Noise level is a problem in open-space areas during laboratory activities. Noise creates distractions for classes in adjacent areas, causing fellow teachers to complain, and consequently resulting in negative reactions toward laboratory work from administrators. Science teachers instructing in open-space areas have had to work very hard to keep the noise level down. Those most successful at this task have instructed their students to speak quietly. These teachers help their students build a group esprit de corps, in which each group works quietly, guarding their findings, while remaining orderly.

Many laboratory activities can best be handled in groups. This is especially true for middle school students. Small-group laboratory activities will be most successful if every member is assigned a role. The following roles can be assigned to students within each group:

- *Coordinator:* Keeps the group on task and working productively.
- *Manager:* Gathers and returns equipment and materials.
- *Investigator:* Helps conduct the investigation.
- *Recorder:* Records data and keeps notes on the investigation.
- *Reporter:* Organizes and reports the findings.

The teacher should give students the opportunity to select roles that they wish to play in the investigation, giving them an opportunity to be actively involved in laboratory work. However, students should rotate their roles so that each is provided with a variety of experiences and responsibility for their learning. See Chapter 4 for more information on grouping and cooperative learning.

Some science teachers are more successful than others at getting students to participate in orderly and productive laboratory experiences. Although successful teachers might begin laboratory activities with a great deal of control and structure, they soon begin to encourage their students to take more responsibility for their work and conduct in the laboratory. The most successful teachers are those who spend less time controlling their students and more time structuring them and giving them more opportunities to learn on their own (George & Lawrence, 1982). These teachers maintain a classroom atmosphere in which students develop a sense of ownership and control over their work.

The prelaboratory discussion is an opportune time to begin to turn some of the control and responsibility over to the students. In this phase of the laboratory, students should be called on to explore the laboratory activity by sharing their ideas about it and even considering alternative ways to accomplish their tasks. It is also a good time to permit the students to determine how the laboratory activity fits into the course.

Teachers should avoid direct commands during prelaboratory discussions and during the actual laboratory activity. "Commands don't invite participation or action, they order it" (George & Lawrence, 1982, p. 230). Commands do little to set the tone for an enjoyable and productive laboratory experience. Whenever possible, commands should be replaced by other forms of control. Some examples are as follows:

Commands	Alternative statements
"Work fast today. We do not have much time."	"We have only 30 minutes to complete our lab today."
"Keep the racket down! You are making too much noise."	"The noise level is getting higher."
"Everybody stop what you are doing and clean up your lab tables."	"It is now 9:40. Time to clean up."

There are many instances during laboratory work when controlling and directing behaviors are desirable. When this is the case, the teacher should use controlling statements but should avoid sarcasm or judgmental remarks. Controlling statements tell students what to do, but they do not give a reaction to what students have already done.

Judgmental remarks	Controlling statements
"You fooled around yesterday; don't do that today."	"Let's work carefully today."
"Sarah, you talk too much, so you work with Kim today."	"Kim and Sarah, will you please work together today?"
"Can't you students ever pay attention to what I am doing?"	"Please observe the oscilloscope."

The teacher plays a major role in developing and maintaining a well-disciplined laboratory environment. This is essential in promoting student productivity and safety and avoiding complaints from other teachers and administrators. Consequently, the science teacher must keep students on task and maintain a reasonable noise level. Continuous interaction between teacher and students can facilitate this process. Walking from student to student or from group to group is also helpful. Such contact urges students to work and gives the teacher the opportunity to help students with problems. It is important to avoid spending too much time with any single group of students. It is also important to cover the whole room so that all groups of students receive the necessary attention.

Rules and policies regarding safety and behavior are essential to the success of the laboratory. They must be stated verbally early in the course, preferably during the first laboratory period. Once stated, they should be posted in clearly visible locations in the laboratory areas. Students should be aware that they will be expected to follow the rules consistently and without exception and that the teacher will be firm but fair about this expectation. The rules should include statements regarding conduct, safety, laboratory reports, use of equipment and materials, and grading, and they should be stated as positively as possible. Student input may be desirable when teachers are establishing rules of conduct; this will increase the probability that students will know the rules and, consequently, adhere to them. It may also be a good policy to provide a set of rules to parents so they know what behaviors the students are expected to exhibit in the laboratory. See figure 8–5 for a set of rules that can be used for developing a set of guidelines for science laboratory conduct.

Science educational researchers who have studied teacher and student behavior during laboratory work provide us with useful insights regarding the management of this special learning environment. DeCarlo and Rubba (1991) reported that the science teachers they carefully studied rarely asked stimulating questions during lab. This may be one reason why very little critical thinking results from laboratory work (Tobin & Gallagher, 1987). Shymansky and Penick (1981) found that nondirective teachers produced more on-task, lesson-related behaviors from students than directive teachers. Perhaps science teachers ought to consider a more hands-in-pocket approach to laboratory management, giving fewer directives and asking more questions that cause students to figure certain things out for themselves.

Evaluation

Evaluation of laboratory work as a part of the total science course grade is an essential part of

FIGURE 8–5   A list of rules that can be used to develop a set of guidelines for student conduct in the laboratory

Guidelines for Conduct in the Laboratory

1. Do your job well and assume your share of responsibility.
2. Keep the noise level to a minimum and speak softly.
3. Work primarily with members of your group, and avoid interacting with other groups' members.
4. Raise your hand if you need help from the teacher. The teacher will come to you; do not go to the teacher.
5. Horseplay is not allowed in the laboratory at any time.
6. Eating or drinking is not permitted in the laboratory.
7. Follow all safety procedures that are posted.
8. Copy the rules concerning "Safety in the Laboratory" in your notebook for reference.
9. Carefully handle all equipment and return it to its proper place.
10. Report faulty or broken equipment immediately to the teacher.
11. Do not waste materials.
12. All dangerous organisms, chemicals, and materials must be handled as directed by the teacher. If you have any questions, ask the teacher.
13. Make certain that all glassware is washed and dried before being returned to storage areas.
14. Keep your work area clean and organized.
15. Clean your tabletop before leaving the laboratory.
16. Remove litter from floor, particularly around the areas in which you work.
17. Strive for accuracy in making observations and measurements.
18. Be honest in reporting data; present what you actually find.

science instruction. There are several techniques to employ in this situation. Paper-and-pencil tests, laboratory reports, notebooks, practical examinations, laboratory behavior, and effort can all be used to determine the laboratory component of the course grade. At least nine areas regarding laboratory work can be used to evaluate students:

1. Inclination to inquire and find out.
2. Ability to ask questions that can be answered in the laboratory.
3. Desire to design procedures to test ideas.
4. Competence and mastery of technical skills.
5. Competence and mastery of science process skills.
6. Ability to collect accurate and precise data.
7. Willingness to report data honestly.
8. Ability to report patterns and relationships and to explain their significance.
9. Inclination to behave properly in the laboratory.

Short paper-and-pencil tests are usually used to evaluate laboratory work. Five to ten items are often sufficient to assess information learned or reinforced in the laboratory. These assessments can also determine how well students have attained process skills and science concepts. Laboratory reports and laboratory notebooks are used to assess students' ability to record data and report findings.

Laboratory practicals are an excellent way to assess students' knowledge of laboratory work. Laboratory stations can be set up where information or techniques can be assessed. The teacher must allow time to prepare the laboratory stations and must take care to ensure that students do not receive answers from their classmates. (See Chapter 18 for a more in-depth discussion of the types of tests that can be used to assess laboratory work.)

Science teachers use direct observation to assess student behavior in the laboratory. Some middle school science teachers give a grade for each lab period for conduct, for example satisfactory or unsatisfactory conduct. The effort demonstrated by students in laboratory work should be rewarded by science teachers, particularly at the middle and junior high school levels. Giving credit for demonstrated effort can develop and maintain positive student behavior in the laboratory, as well as reinforce laboratory work. The teacher, of course, determines what part of the total grade should reflect effort in the laboratory. In general, laboratory work accounts for 20 to 40 percent of the report card or course grade, and effort should be a part of this percentage.

MATERIALS AND SUPPLIES FOR THE LABORATORY

Several important recommendations should be considered in organizing and conducting laboratory activities. One essential aspect of a successful laboratory program is an accurate inventory of all equipment and supplies before classes begin. In most situations the science teacher is responsible for ordering and maintaining science equipment and supplies. This means that the teacher must know what is on hand, as well as the condition of materials and equipment. With this knowledge, the teacher can determine what needs to be purchased—to replace either supplies that have been depleted or equipment that has been damaged. Additional supplies and materials can be ordered for laboratory or hands-on activities as needed.

Determining the Inventory

A teacher in a new teaching situation is confronted with the task of determining an inventory of equipment, materials, and supplies. If the new teacher is fortunate enough to have had a predecessor who was conscientious about keeping a current record of materials, the inventory creates no problem. However, most teachers who arrive in new situations are not this fortunate.

A preliminary inventory accomplishes several purposes; not only does it inform the teacher about what is on hand, but it gives the teacher an opportunity to become acquainted with unfamiliar apparatus. This also allows the new teacher to review the materials needed for the laboratory activities that will be used during the course.

A card system to keep an up-to-date inventory of what is available should be implemented to serve a number of purposes:

1. It is a ready reference file for supplies on hand.
2. It enables the teacher to locate equipment easily, because each card indicates the area of storage space (shelf, file, etc.).
3. It makes taking an annual inventory easier.

Figure 8–6 shows a suggested form for an inventory card. Three-by-five-inch index cards are more than adequate for this purpose; the back of the card can be used for additional comments.

Points to Consider in Ordering

Knowing what is on hand is only a starting point in determining what is to be purchased. Before ordering, consider the following:

1. the budget for materials and equipment
2. the number of students who will be in a particular course

3. the amount of individual student laboratory work involved in the course or courses
4. the number of demonstrations that will be performed by the teacher
5. the equipment and supplies suggested in the textbook, workbook, and/or laboratory manual that will be used, or those required by the teacher's own course of study

Budget. In most schools, funds for the purchase of materials and equipment are limited. For this reason, the teacher will have to weigh the importance of various items before ordering them. That is, essential items should be first on the list of purchases. Items that are not so important can be purchased later or when money is made available.

In considering the purchase of certain items, the teacher should remember the cost involved in conducting individual laboratory work. If a particular item costs $200 and 10 items are needed for individual laboratory work, but the total overall budget is only $2,000, the teacher may wish to think twice before ordering 10 pieces. The teacher may instead choose to order one item and use it in a demonstration. If, on the other hand, an item costs $10 and 10 are needed for students to do individualized laboratory work, the teacher should consider the relative

FIGURE 8–6 Suggested form for an inventory card

INVENTORY CARD					
Item of equipment: _____ Catalogue no.: _____ Ordered from: _____ Cost: _____ Place of storage: _____					
Date of inventory	Number on hand	Condition	Date ordered	Number ordered	Cost

advantages of laboratory work over those of the teacher demonstration.

The materials and equipment needed to teach a new curriculum, for example, may require a considerable outlay of funds. Before embarking on a new program, investigate how much it will cost to become involved. The cost might be prohibitive or minimal, depending on the already existing inventory. Some of the commercially prepared kits containing materials, supplies, and apparatus for some of the national curriculum courses are rather expensive. However, many of the materials in the kits are readily available from supply houses, and a teacher can save a great deal of money by making her own kits for some of the study units. The teacher can also improvise by using materials purchased from hardware and department stores. There will be instances, however, when very specialized apparatus cannot be readily duplicated, requiring the purchase of the kit.

The use of expensive equipment for individual laboratory work should not be overlooked because of cost. The teacher should consider the purchase of a microscope each year, for example, until there are enough on hand to permit each student to handle and use a microscope. The amount of student laboratory work, the number of teacher demonstrations, and the number of student experiments can be determined only from the course of study, textbook, workbook, or laboratory manual that will be used.

If the teacher decides to use a particular laboratory manual, the equipment and materials needed to conduct each laboratory experiment are clearly indicated in the manual. One procedure that some teachers follow is described here:

■ Determine which experiments in the manual or textbook are to be conducted by the students.

■ For each experiment, make a list of items needed per student.

■ After making this list, determine the size of the class by obtaining an estimate of the num-

ber of students registered for the course from the school administration. If you are going to teach in a new situation, the principal will have a reasonably good idea of what the registration will be for a particular course.

■ By simple multiplication (number of students in the class times the items per experiment), determine how much to order.

Where to Obtain Equipment.   Equipment should be purchased from reliable companies as close to the school as possible. In most instances, the closer the company is to the school, the cheaper the cost for transportation, especially if the school has to pay the cost of mailing or express charges. An equipment company in California may sell an item for fifty cents less than one in New Jersey, but the cost of shipping such an item from California to a school in New York could result in a total cost of five dollars more than its cost from New Jersey.

It is also wiser to order, insofar as possible, from a single reliable equipment company. Many schools find that they get better service if the company knows the school as a regular customer. Being a regular customer may qualify a school for a discount. Large school systems may even have their own science supply department from which teachers can obtain materials and equipment.

When to Order Materials and Equipment. The science teacher should maintain a long range plan for ordering materials and equipment. Live material, for example, must be ordered at a specific time to ensure its arrival when it is needed. Live material ordered and delivered during the summer months for use in the month of January is impractical unless the specimens can be maintained easily during the intervening period. The teacher should record the ordering data on the inventory cards as a reminder to order such material far enough in advance as described in the previous section of this chapter.

Expendable material must be ordered yearly or each term. During the summer or before school closes, the teacher should indicate needs for the coming year to the principal or purchasing agent so that the material will be available for use in the fall. At this time, replacements for broken equipment should also be ordered, together with any new pieces of equipment not previously in the science inventory.

Maintaining Science Equipment

To make certain that equipment is in good condition, the teacher should inspect it after each use. Many pieces of equipment get out of adjustment during student use. They should be checked thoroughly when the students have finished using them. Many simple repairs can be done on the premises if the science department is equipped with proper tools. It is not important to have elaborate tools, but certain basic ones are essential. For example, the bulb of a microscope light source often needs to be replaced. This simple repair can be made with a screwdriver. Tools that permit a wider variety of uses can be added if the budget is adequate for their purchase. All tools should be of good quality. Cheap tools that break easily, become dull rapidly, and prevent good workmanship are a poor investment. Because good tools can represent a substantial investment, they should be stored carefully in a locked cabinet. If they are left lying around, they often disappear, either stolen or borrowed and never returned. The financial loss can be serious, and the inconvenience of not having them on hand when they are needed can cause problems. Some individuals have successfully stopped "borrowing" by spray painting all tool handles baby pink, and widely announcing it; such easily identifiable tools stick out like a sore thumb, and are less likely to "move" to other areas.

Certain power tools are very useful for laboratory work. An electric hand drill with appropriate accessories can have many uses. Larger, more elaborate power tools, such as saws, drill presses, and lathes, are usually available in school industrial arts facilities.

SUGGESTED ACTIVITIES

1. Interview a few new and experienced teachers to obtain their views on the purpose and importance of laboratory work in the courses they teach. Also, determine the frequency of the laboratory work in their courses. Discuss this information with other members of your science methods class.

2. Analyze the laboratory activities for a science course and classify the laboratories into one of the following categories: process skill, deductive, inductive, technical skill, or problem solving. Evaluate the laboratory activities based on their variety and appropriateness for the students who are using them. Discuss your evaluation with other class members.

3. Develop an instrument to evaluate how a science teacher conducts a laboratory. Include ideas discussed in this chapter, such as the prelaboratory discussion, applicability, structure, recording/reporting, management, and evaluation. Establish a set of criteria from which you can make a judgment regarding the degree of inquiry that takes place during a laboratory exercise.

4. Survey how science teachers and science coordinators in a school district conduct inventories, order supplies and equipment, and maintain and store laboratory equipment, chemicals, and supplies.

5. Discuss with the other members of your science methods class the science teaching facilities that you believe are good or exemplify effective science teaching.

6. Develop a laboratory exercise illustrating one of the five types of laboratories described in this chapter. Conduct the laboratory with a group of peers, or middle or senior high school students enrolled in a science class.

BIBLIOGRAPHY

Beasley, W. 1985. Improving student laboratory performance: How much practice makes perfect? *Science Education, 69:* 567–576.

Blosser, P. E. 1981. A critical review of the role of the laboratory in science teaching. Columbus, OH: ERIC/SMEAC Clearinghouse.

DeCarlo, L., and Rubba, R. A. 1991. *What happens during high school chemistry laboratory sessions? A descriptive case-study of the behaviors exhibited by three teachers and their students.* Paper presented at the Annual Meeting of the National Association for Research in Science Teaching, Lake Geneva, WI.

Driver, R., Guesne, E., and Tiberghien, A. (Eds.). 1985. *Children's ideas in science.* Buckinghamshire, UK: Open University Press.

Gardner, P., and Gauld, C. 1990. Labwork and students' attitudes. In E. Hegarty-Hazel (Ed.), *The student laboratory and the science curriculum.* NY: Routledge.

George, P., and Lawrence, G. 1982. *Handbook for middle school teaching.* Dallas: Scott, Foresman.

Hegarty-Hazel, E. 1990. *The student laboratory and the science curriculum.* NY: Routledge.

Henry, N. B. (Ed.). 1960. *Rethinking science education, 59th yearbook of the National Society for the Study of Education, Part 1.* Chicago: University of Chicago Press.

Hodson, D. 1985. Philosophy of science, science and science education. *Studies in Science Education, 12:* 25–57.

Hunt, P. N. 1984. Making the rounds with Dr. Semmelweis. *The Science Teacher, 51*(1): 33–37.

Hurd, P. D. 1964. *Theory into action.* Washington, DC: ERIC/SMEAC Clearinghouse.

Karplus, R. 1974. *Science curriculum improvement study: Teacher's handbook.* Berkeley, CA: Lawrence Hall of Science.

Kanis, I. 1991. Ninth-grade lab skills: And assessment. *The Science Teacher, 58*(1): 29–33.

Leonard, W. H. 1991. A recipe for uncookbooking laboratory investigations. *Journal of College Science Teaching, 21*(2): 84–87.

Lumpe, A. T., and Oliver, J. S. 1991. Dimensions of hands-on science. *The American Biology Teacher, 53:* 345–348.

Lunetta, V. N., and Tamir, P. 1979. Matching lab activities with teaching goals. *The Science Teacher, 46*(5): 22.

Norris, S. P. 1984. Defining observational competence. *Science Education, 68:* 129–142.

Padilla, M. J., Okey, J. R., and Garrard, K. 1984. The effect of instruction on integrated science process skill achievement. *Journal of Research in Science Teaching, 21:* 277–287.

Pella, M. O. 1961. The laboratory and science teaching. *The Science Teacher, 28*(5): 29.

Renner, J. 1966. A case for inquiry. *Science and Children, 4:* 30–34.

Rowell, J. A., and Dawson, C. J. 1983. Laboratory counter-examples and the growth of understanding in science. *European Journal of Science Education, 5*(2): 203–215.

Schrock, J. R. 1991. Dissection. *The Kansas School Naturalist, 36*(3): 3–15.

Schwab, J. J. 1964. *The teaching of science.* Cambridge, MA: Harvard University Press.

Shymansky, T. A., and Penick, T. E. 1981. Teacher behavior does make a difference in hands-on science classrooms. *School Science and Mathematics, 81:* 412–423.

Tobin, K., and Gallagher, J. D. 1987. What happens in high school science classrooms? *Journal of Curriculum Studies, 19:* 549–560.

Yeany, R. H., Yap, K. C., and Padilla, M. J. 1986. Analyzing hierarchical relations among integrated science process skills. *Journal of Research in Science Teaching, 3:* 277–291.

Science teachers are responsible for safety in the science teaching environment. They should anticipate all situations in the laboratory that require safety precautions.

CHAPTER 9

Safety in the Laboratory and Classroom

The science laboratory and classroom are places where accidents can occur because of the mishandling and use of apparatus, equipment, chemicals, and certain live materials that are often maintained in the laboratory. A safety program should include training teachers, students, and any other personnel involved in science instruction. Certain standard safety procedures should be implemented as a matter of course before, during, and after instruction and given high priority to ensure a continuous and effective safety program.

Overview

- Discuss a teacher's legal responsibilities with respect to safety in the science classroom.
- Explain the science teacher's responsibilities for safety in the classroom and laboratory.
- Indicate preparations the science teacher should make before the school year begins.
- Discuss the science teacher's responsibilities for promoting safety awareness.
- State the specific safety guidelines for teaching biology and life science courses.
- State the specific guidelines for safety in the chemistry laboratory and classroom.
- Explain the safety precautions that should be taken regarding the teaching of earth science.
- State the important safety guidelines for teaching physics and physical science courses.
- Suggest important safety measures regarding the use of radioactive materials.
- Discuss the importance of teaching a unit on laboratory safety to develop student awareness and responsibility.
- State the importance of offering workshops and seminars on safety, conducted by experts, to keep science teachers up-to-date.

Science teachers have recently devoted much time and attention to safety in the science laboratory and classroom. Their concerns about safety have been generated by the increase in the number of liability cases against teachers in recent years. Most of the grounds for the legal suits have centered around teacher negligence; that is, some teachers have failed to instruct students in the proper use of equipment, chemicals, and other materials, or to provide proper supervision.

Science teachers have a moral obligation to promote safety awareness in the science classroom and laboratory. It is important that they recognize the legal implications when students are not properly supervised and when students are not trained to be responsible for their own safety and the safety of others. The teaching of safety and the use of appropriate laboratory procedures for safety should be an integral part of all science instruction.

SAFETY AND THE LAW

The science teacher must be aware of the legal responsibilities regarding safety in the classroom. The very nature of the science classroom and laboratory increases the probability of student accidents, and teachers must take every precaution to ensure the safety of the students and themselves. If an injury occurs in the classroom or in the laboratory what do the courts commonly look for? First, the courts ask "Is there a duty (responsibility) owed?" And the answer to that is yes. Next, the courts want to know the standard of care that was provided. Three areas determine the standard of care in education: A teacher owes his or her students (1) active instruction regarding their conduct and safety in the classroom, (2) adequate supervision, and (3) consideration of potential hazards. Merely posting rules in a classroom is not enough. "The common practice of laboratory science teachers instructing students on safety rules and requir-

ing that they pass tests on this material makes good instructional and therefore legal sense [and] is recommended for every grade and subject" (Vos & Pell, 1990, p. 34).

The teacher must be well acquainted with state regulations regarding liability. Each state has specific statutes and requirements. If an accident occurs as a result of a teacher's noncompliance with such regulations, then the teacher is vulnerable to a legal suit, which injured students may choose to initiate. Such procedures can result in heavy fines or even the loss of a teacher's job. In addition to state regulations, the local board of education may have explicit policies and rules regarding the teacher's responsibilities. Very often, courts choose to uphold such policies, even though they may not appear in the teacher's contract, and the science teacher is held liable for breaches pertaining to the health and safety of students.

The teacher should also carefully read his contract and take special note of any responsibilities concerning student health and safety. Thus, the first and most obvious step that the diligent teacher must take is to carefully investigate safety regulations that have been specified by state statute, school board regulations, and the teacher's contract. Omitting this step is foolish, because any accidents that result from noncompliance may leave the teacher legally responsible.

Another source of teacher responsibility with regard to safety is common law, which refers to law that has been established by judges in actual courtroom cases. The law of negligence (which is the primary governing standard) derives mainly from common law. Because most states have waived common law immunity, the school board is usually named as the defendant in a suit, although it is not uncommon for a teacher to be sued based on common law (Vos & Pell, 1990, p. 38).

One such case in California involved an explosion in a chemistry laboratory. A student had inadvertently substituted potassium chlorate for potassium nitrate, even though the textbook directions called for potassium nitrate. An explosion resulted, injuring the student, who sued the teacher for not providing specific instructions about the dangers of chemical substitutions. The court held the teacher liable and stated, "A teacher's duty goes beyond merely providing students with general instructions." A number of other interesting and informative legal cases are reported elsewhere (Joye, 1978).

If an injured student sues the teacher on the grounds of negligence, the teacher can raise a legal defense called *contributory negligence,* which holds that the injured student behaved in a manner that contributed to the injury. The teacher must, however, offer evidence to show that the injured student's behavior constituted gross disregard for his or her own safety. If such can be established, traditional common law would prevent the student from recovering damages from the teacher. However, in some states this type of defense is being compromised by the doctrine of *comparative negligence,* which states that when contributory negligence is involved, both the plaintiff and the defendant will be held liable for damages consistent with their respective shares of the negligence. For example, if the plaintiff (student) committed 80 percent of the negligence, she could recover only 20 percent of the total damages from the defendant (teacher). Whatever the case, the teacher must be in a position to prove that he took every possible precaution to ensure the safety of the students.

Science teachers must always be prudent and demonstrate ordinary care in their teaching duties. During instructional activities in which certain safety hazards are known to exist, as in the use of corrosive or explosive chemicals in a laboratory exercise, for example, teachers must use *extraordinary care* to avoid any mishap that could injure a student. The teacher must prepare students for the activity, pointing out the dangers and stressing the importance of behaving properly in the laboratory. They must also supervise the students during the activity so that students carry out their procedures safely.

The best way for a teacher to avoid legal suits is, of course, prevention. Teachers must make a calculated effort to anticipate accidents and potentially risk-laden situations. In addition, the teacher should keep complete records of maintenance work, safety lectures, and specific safety instructions given during particular activities. This time and effort can reduce the teacher's liability. More important, the teacher can reduce the occurrence of classroom and laboratory accidents.

GENERAL SAFETY RESPONSIBILITIES

Science teachers are responsible for safety in the science teaching environment; however, many take the charge lightly. It is not unusual to find many potentially hazardous violations. Inoperable fire extinguishers, safety showers, overhead sprinklers, or eye wash fountains; inadequate lighting; unlabeled chemicals; and faulty equipment are but a few common hazards found in the laboratory. Teachers must be aware of such problems in order to maintain a safe environment. They must be alert about what constitutes a safety hazard.

Accidents that occur in the laboratory are usually minor. The most common types are cuts and burns from glass and chemicals (Gerlovich & Downs, 1981). Students taste chemicals, touch broken glass, handle hot vessels, and fail to wear safety glasses—all unsafe practices that can be avoided if teachers give students proper instruction and supervision.

The teaching of safety must be done systematically and constructively. Many teachers have introduced safety units into the curriculum to accomplish this. A study by Dombrowski and Hagelberg (1985) indicates that an instructional unit on laboratory safety increases student safety knowledge and reduces the number of unsafe behaviors in the laboratory.

Teachers are generally unknowledgeable about safety, which makes teaching the topic

FIGURE 9–1   Checklist for safety inspection of science facilities

	No Attention Needed	Attention Needed	Comments
Gas valves and shutoffs			
Water valves and shutoffs			
Electrical lines and shutoffs			
Electrical outlets			
Exhaust fans			
Chemical storage 1. Temperature of storage area			
2. Aisles cleared			
3. Chemicals properly stored			
4. Age of chemicals within expiration date			
Fire extinguishers 1. Placed properly			
2. Label and seal show recent inspection			
3. Special type available for chemical fires			
4. Special type for electrical fires			
5. Type for ordinary fires not due to electrical or chemical causes			
Fire blankets 1. Recent inspection			
2. Strategically placed			
Safety showers			
Eye wash fountains			
Sand buckets			
Student laboratory positions 1. Gas outlets for leaks			
2. Water outlets for leaks			
3. Sinks have proper drainage			
4. Electrical outlets are functional and properly grounded			

problematic. Consequently, it is recommended that teachers take every opportunity to learn about proper safety procedures in the laboratory (Gerlovich & Downs, 1981). They should attend special courses and seminars on safety, enroll in first aid and CPR courses, talk with other science teachers about safety problems, and read articles on safety in publications such as *The Science Teacher* and *Journal of Chemical Education*. The Manufacturing Chemists Association, the American Chemical Society, and the U.S. Department of Health and Human Services publish excellent material on safety.

Preparation Before the School Year Begins

Before the school year begins, the teacher should inspect the classroom and laboratory to determine sites of potential danger. A checklist (inventory) such as the one illustrated in figure 9–1 can guide the teacher throughout the assessment. A checklist ensures that the inspection will be organized and systematic and that items will not be forgotten. At the same time, the teacher has a record of direct responsibility and concern for safety in case of future litigation.

Conducting an inspection requires the cooperation of the school principal and must be coordinated to include the services of maintenance personnel, electricians, plumbers, and other skilled individuals. The teacher, in concert with the appropriate personnel, should inspect the laboratory and classroom areas for malfunctions involving gas lines, water valves, electrical lines and outlets, exhaust fans and hoods, and temperature controls in storage and classroom areas. Eye washes, fire blankets, fire extinguishers, and safety showers also require inspection. It is best to engage qualified individuals to detect and rectify problems in laboratory facilities. It is not the teacher's responsibility to correct electrical or plumbing problems; such problems must be handled by skilled personnel.

Before the school year begins, have fire extinguishers inspected and placed in areas where

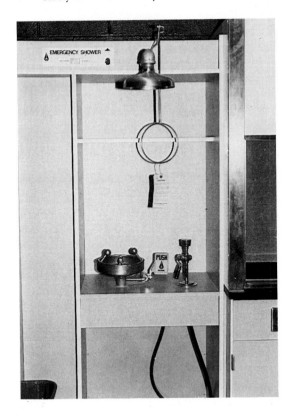

Clean and properly working safety showers and eyewash fountains should be available in the laboratory in case of emergency.

fires are most likely to occur. Examine labels and seals on fire extinguishers to determine whether they have been recently charged and are operable. Special types of fire extinguishers should be available in the laboratory—those that extinguish ordinary fires not due to electrical or chemical causes, and those that control electrical and chemical fires. Fire blankets are important in certain laboratory situations—particularly the chemistry laboratory. Place these in strategic areas in the laboratory and check them often to see that they are in good order.

Safety showers and eye wash fountains that have not been used for a period of time tend to malfunction because of accumulated sludge. The teacher is responsible for seeing that they are clean, functional, and available in various areas in the laboratory in case of emergency.

Sand buckets can be useful in case of fire. Periodically check them during the school year to see that they are clean, full of sand, and readily available in areas where fires are likely to occur. In general, they are standard equipment in chemistry and physics laboratories but not in the biology facility. Sand buckets should be standard equipment in all laboratories, regardless of the science discipline.

Inspect students' laboratory stations for gas leaks and electrical problems. All electrical outlets must be properly grounded and in good condition at all times. The teacher must correct electrical problems, preferably by engaging a qualified electrician. Gas leaks are detected by placing a soap solution around outlet joints and watching for the appearance of bubbles. Remember that merely "sniffing" the area is never adequate for detecting small gas leaks. Do not try to rectify gas problems without the help of qualified personnel.

Teachers must have easy access to master controls for gas and electrical sources in the laboratory, which may prevent serious mishaps when it is essential to cut off the power or gas supply quickly. The design of older facilities may make it difficult to satisfy this requirement, but if controls are accessible, they should be available to the teacher.

Arrange movable laboratory tables so that there is enough area around them to avoid congestion and cramping. The placement of movable tables should allow students to work freely and without interference from other nearby students. Too much congestion may constitute a safety hazard.

Before the school year begins, make a plan for an orderly evacuation of a classroom or laboratory in case of emergency. Seek the assistance of a fire marshal and other competent individuals. There are a number of points to consider when developing the plan. For example, many science areas usually have only one exit in a corridor and only one door that leads into a storage

Various types of fire extinguishers should be available in the laboratory, as well as fire blankets and sand buckets.

area. It is important to see that exit doors and aisles in the classroom and laboratory are clear at all times to permit orderly exits. Discuss evacuation procedures and distribute a written plan early in the school year. Post diagrams and narratives throughout the science area. Successful evacuations can be accomplished only if students, teachers, teacher assistants, and other concerned individuals know the plans and practice their use in one or two mock situations during each school term.

The storage area deserves attention before school begins. It is often a catchall for anything that is not needed in the laboratory or classroom. Many storage areas are so cluttered that passage is virtually impossible. Aisles and exits should be uncongested and the storage area itself in good order before the school year begins. This sets a good example for the students. Rules on the use and maintenance of the storage areas are necessary to maintain a safe environment.

Safety Responsibilities During the School Year

The science teacher is responsible for promoting safety awareness once the school year begins. This responsibility has to be taken very seriously, particularly if students will be involved in activity-centered science. More accidents are likely to occur when students are exposed to hands-on activities during laboratory work. Consequently, more care and supervision is required when students are handling equipment, chemicals, apparatus, live animals, poisonous plants, and other hazardous items. Special preventive measures are required to operate a laboratory in which students are directly involved with science materials.

In addition, class size must be considered when planning laboratory activities. Classes that exceed the recommended number of students for the available space pose a safety hazard (Rakow, 1989). Unfortunately, few states have legislation which mandates class size in the science laboratory. Guidelines for lab space and class size are available from various scientific supply companies.

Because the trend is toward the implementation of activity-centered science courses, it is imperative for the teacher to establish a safety training program for students involved in such courses. Safety is the teacher's direct responsibility; the teacher must consider it when planning and preparing courses of study and particularly when preparing and delivering science lessons. Audiovisual and reference materials regarding safety training and instruction in the secondary schools and middle schools are often available to teachers free of charge. *The Science Teacher, Science Scope,* and other journals include many articles on various aspects of safety (see bibliography at the end of this chapter).

Pitrone (1989, S22) suggests the implementation of a safety learning center to teach safety issues to middle school students. The safety learning center consists of four stations; the students participating in the center learn rules of laboratory safety and common hazard symbols, learn to identify some basic laboratory equipment, practice some simple laboratory procedures, take a quiz on safety, and sign a safety contract. The safety contract is meant to be proof that safety procedures were stressed in case such evidence is needed.

The teacher must always exhibit a sincere concern for safety to set a good example for students. Everything the teacher does that concerns safety should be visible and deliberate. The more safety is emphasized, the better students will understand the necessity of safety measures in the science teaching environment. Hampton and Thacker (1989) suggest performing a mock demonstration using unsafe laboratory procedures for the students. The teacher could then have the students list each unsafe practice, predict what accidents could happen as a result, and describe preventive measures that could be taken.

The teacher's attitude and behavior toward safety will influence the students; if the teacher takes the laboratory experience seriously, so will the students. Do not expect students to follow rules that the teacher does not follow. The teacher must be a good role model in maintaining a safe environment.

There are certain inexpensive general precautions that a teacher can take to assure a safer science teaching environment during the school year. Some suggestions follow.

1. Develop a set of general safety rules to be in effect at all times in the laboratory that includes a code for laboratory dress and requires safety goggles and gloves at appropriate times and forbids careless activity, horseplay, and the wearing of contact lenses during certain activities. Post the rules in prominent areas in the laboratory and discuss the rules regularly, especially before laboratory activities (even the simplest activities can be hazardous). Encourage the students to add their own rules as they become

more conscious of their own safety and the safety of others.

2. Safety posters are necessary in the science teaching environment. Posters that stress particular safety precautions are available commercially, but homemade posters may be more pertinent to the situation. The teacher can ask students to make safety posters as a project. Posters that include graphics are particularly effective when they illustrate the use of safety equipment such as a fire extinguisher, fire blanket, or safety shower. Post large signs that show the positions of safety equipment.

3. Schedule periodic safety inspections in the laboratory during the school year. Examine fire extinguishers, safety showers, fume hoods, and eyewash fountains. Check student laboratory positions to see that gas and water outlets and electrical sources are functioning properly. Report problems that require correction to the principal immediately. Do not allow students to be subjected to any unsafe situations.

4. Inspect storage areas regularly to see that aisles are clean, and chemicals are properly stored and labeled. Remember to store chemicals by class and then alphabetically.

5. Arrange laboratory furniture to allow enough working space for students to conduct activities. Inspect furniture to see that it is functional and not defective. Ask students to report any problems.

6. Inspect apparatus, equipment, and electrical devices before allowing students to use them. Electrical equipment and appliances must meet certain specifications and be approved by Underwriters Laboratories or another known organization. Homemade electrical devices can be hazardous; have them inspected by a licensed electrician before using them.

7. Report problems in lighting immediately to appropriate personnel. Good lighting is necessary for a safe working environment.

8. Monitor ventilation in the laboratory regularly. Again, report malfunctions as soon as possible. Do not conduct laboratory procedures requiring good ventilation while the system is functioning improperly.

9. Post emergency procedures that are to be followed in case of fire, explosions, chemical spills, violent chemical reactions, and chemical toxicity. Review specific emergency procedures with students before conducting an activity and periodically rehearse evacuation procedures. Stress the precautions that must be taken to prevent problems, especially the fact that corridors, exit doors, and aisles should be clear at all times to allow for orderly, efficient evacuation.

10. Instruct students in the use of safety equipment. Point out the location of such equip-

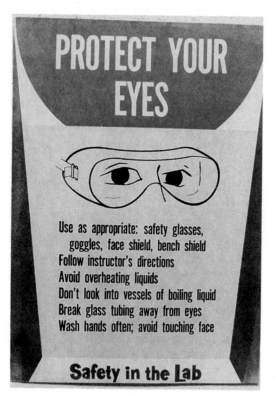

Safety posters are effective when they illustrate the use of goggles and other equipment.

ment and periodically demonstrate the use of fire extinguishers, safety showers, and eye fountains. Allow students to practice using the equipment.

11. Require students to behave properly in the laboratory. Ask them not to engage in horseplay or other careless activity and constantly supervise students to see that their behavior is consistent with your expectations. React quickly and bring unsafe practices to the attention of students. The teacher's reaction to an unsafe activity reinforces the seriousness and importance of the issue.

12. Demonstrate and stress the correct procedures before permitting students to use certain types of equipment, materials, and supplies. Describe the procedures while students are seated at their desks, but do not place an array of laboratory instruments in front of each student and then attempt to explain the procedures. Students tend to devote more attention to the equipment than to the instructions for use. Prevent students from handling certain objects until they become familiar with their use.

13. Prohibit students from performing unauthorized experiments or activities. Do not allow them to use materials, equipment, or supplies unless instructed to do so. Do not authorize students to open chemical storage cabinets, refrigerators, and so on. Only the teacher has this privilege.

14. Perform potentially dangerous student activities as a teacher demonstration before allowing students to engage in the activity. Point out possible hazards and show students how to avoid certain pitfalls while they are conducting the exercise.

15. Report and describe all accidents, even minor ones, to the administration. Keep details on file for reference in case of future inquiries or litigation. Request witnesses to supply their signed versions of the accident, including the circumstances that caused the accident, if known. Ask them to include the names of the individuals involved and a description of supervision the teacher provided during the time the accident occurred.

16. Do not allow students to transport chemicals outside of the laboratory without permission. Chemicals, equipment, and apparatus should be used only in areas that are constantly supervised by the teacher.

17. Control the use of chemicals by storing them in safe places and allowing only required amounts to be available in the laboratory as needed. Students are not to touch or use dangerous chemicals such as potassium, sodium, and mercury unless they have permission. Toxic chemicals such as benzene and carbon tetrachloride are no longer permitted for use in science instruction.

18. Plan and discuss cleanup operations prior to laboratory performance and reinforce them at the conclusion of the exercise to avoid problems such as crowding and rushing. Accidents occur when jobs are done in haste. Prepare and discuss logistical procedures such as how many people will work together at the sink, where to place glassware to dry, and how to safely dispose of chemicals. Monitor clean-up operations closely to elicit the appropriate student behavior during this potentially dangerous function.

19. Include first-aid kits as standard equipment in all science classrooms. Kits should be large enough to hold the proper materials and arranged to permit quick access to items without unpacking the entire contents. Carefully wrap unused materials to avoid contamination. A first-aid kit should contain the following items:

antiseptics	alcohol
sterile gauze	tweezers
tape	scissors
bandages	burn ointment
cotton	ampules of ammonia
gauze roller bandage	boric acid
	bicarbonate of soda

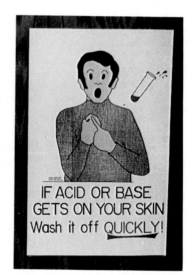

Safety posters are effective when they illustrate the use of safety equipment such as fire extinguishers and safety showers. They are also useful when they address important emergency procedures.

20. After reviewing first-aid procedures with students, post a list of the procedures near the storage rooms and other strategic areas. Do not allow students to use first-aid kits without permission. In the event of a serious injury, avoid the use of any medication or manipulation of the victim until the school nurse or a physician is consulted. The materials in first-aid kits should be used only for minor injuries, burns, cuts, or other conditions that are not serious. In all cases, carefully evaluate the situation before using any first-aid procedures or medication to avoid future litigation. Report all accidents immediately to the principal.

21. Do not trust investigations that are published in old books and science kits because the safety precautions are not emphasized (Hampton & Thacker, 1989, p. S4). Hampton and Thacker are middle school science teachers who stress the importance of safety procedures in middle school science.

22. Be careful in using demonstrations or hands-on activities in middle school science that require the use of chemicals or equipment that may pose a possible danger. Consult a high school or secondary school science teacher who majored in science and has the appropriate expertise if there is any question about danger in the use of such activities in the middle grades.

23. For emergencies in particular, but also to help avoid emergencies, maintain a file of names, phone numbers, and addresses of individuals, agencies, and companies that would be useful to contact in connection with maintaining a safety program, including such resources as school administrators, school science coordinators, industry experts, nearby university experts, suppliers, fire marshals, disposal firms, appropriate state/county/city agencies, and the like.

SPECIFIC SAFETY GUIDELINES FOR BIOLOGY

Precautions for Using Animals

Animals such as rats, mice, guinea pigs, hamsters, and rabbits must be handled gently and with thick rubber or leather gloves. There is always a danger that animals will become excited as they are being handled, particularly if they are injured or pregnant, or if foreign materials are being introduced into the cage. Animals should not be provoked or teased. If animals feel threatened they will defend themselves, sometimes to the point of biting or scratching. Animals will exhibit violent behaviors if poked with fingers, pens, and other objects through the wire mesh of their cages. Discourage such actions because they can result in the injury of an individual or an animal.

Only those who know how to handle animals should be responsible for them. This means that before giving students the responsibility for handling animals, the teacher must instruct students in their proper care and maintenance.

It is illegal to use animals for instruction that are poisonous or known carriers of disease. Avoid using poisonous snakes, scorpions, and Gila monsters. If snakes are used, be certain that they are not dangerous. In some cases, poisonous snakes can be treated to prevent the production of venom, so that they can be used without danger. In general, do not use wild animals that are known carriers of rabies and parasites unless reasonable cause can be shown. In such cases the animals must be inoculated before they are permitted in the classroom or laboratory. Obtain mice, rabbits, rats, guinea pigs, and gerbils from reputable supply houses that will guarantee inoculation against rabies.

A teacher should have good reason for keeping animals and insects in the classroom or laboratory. Safety, feeding requirements, replication of natural environments, and unjustified confinement must be considered before using animals in classroom work, all of which make the proper maintenance of animals in cages a problem. Maintain all animals in clean cages and feed and water them on a daily basis or as required. Make provisions for feeding animals on weekends and vacation periods. Often, teachers or responsible students take animals home over extended periods of time to ensure their proper care.

All animals used in teaching must be acquired and maintained in accordance with federal, state, and local laws. Most states require permits to acquire and maintain wild animals in captivity. If permission is granted to maintain a wild animal in the laboratory, it is important that the animal be returned to its natural environment as soon as its use is not required.

Most states have regulations regarding the use of animals in classroom work, which range from general to very strict. Texas law does not permit the dissection of higher vertebrates in the laboratory. California law states that schools or school-sponsored activities cannot involve live animals that have been experimentally medicated or drugged, forbids the induction of any pathological conditions, and prohibits injury to the animal through any treatment. In Maine, vivisection is prohibited for mammals, birds, tortoises, and turtles. Birds' eggs are not included.

Statements regarding animal use and treatment in the classroom have been published by the National Science Teachers Association (1978), the National Association of Biology Teachers (1980), and the American Humane Association (1980). Figure 9–2 lists guidelines adopted by the National Association of Biology Teachers (1980). In addition, numerous publications have addressed the controversy of using animals in the laboratory (Richmond, Engelmann, & Krupka, 1990 and Orlans, 1991). Options that may be considered as alternatives to dissection are also available (Mayer & Hinton, 1990).

Precautions for Specific Biology Procedures and Activities

Certain procedures and activities carried out in the biology laboratory require special mention. Activities involving the use of dissection instruments, sterilizing equipment and instruments, decayed and decaying plants and animal material, pathogenic organisms, and hypodermic syringes must be carefully monitored by the teacher. Activities involving blood-typing and

FIGURE 9–2   Guidelines for animal use (From "NABT Guidelines for the use of Laboratory Animals at the Pre-University Level" by National Association of Biology Teachers, 1980, *American Biology Teacher, 42*(7), pp. 426–427.)

I. Biological experimentation should lead to and be consistent with a respect for life and all living things. "Humane treatment and care of animals should be an integral part of any lesson which includes living animals."

II. All aspects of exercises and/or experiments dealing with living things must be within the comprehension and capabilities of the students involved. It is recognized that these parameters are necessarily vague, but it is expected that competent teachers of biology can recognize these limitations.

III. Lower orders of life such as bacteria, fungi, protozoans, and insects can reveal much basic biological information and are preferable as subjects for invasive studies whenever and wherever possible.

IV. Vertebrate animals can be used as experimental organisms in the following situations:
   A. Observations of normal living patterns of wild animals in the free living state or in zoological parks, gardens, or aquariums.
   B. Observations of normal living patterns of pets, fish, or domestic animals.
   C. Observations of biological phenomena (e.g., inducing ovulation in frogs through hormone injections) that do not cause discomfort or adverse effects to the animals.

V. Animals should be properly cared for as described in the following guidelines:
   A. Appropriate quarters for the animals being used should be provided in a place free from undue stresses. If housed in the classroom itself, animals should not be constantly subjected to disturbances that might be caused by students in the classroom or other upsetting activities.
   B. All animals used in teaching or research programs must receive proper care. Quarters should provide for sanitation and protection from the elements and have sufficient space for normal behavioral and postural requirements of the species. Quarters shall be easily cleaned, ventilated, and lighted. Proper temperature regulation shall be provided.
   C. Proper food and clean drinking water for those animals requiring water shall be available at all times in suitable containers.
   D. Animals' care shall be supervised by a science teacher experienced in proper animal care.
   E. If euthanasia is necessary, animals shall be sacrificed in an approved, humane manner by an adult experienced in the use of such procedures. Laboratory animals should not be released in the environment if they were not originally part of the native fauna. The introduction of nonnative species that may become feral must be avoided.
   F. The procurement and use of wild or domestic animals must comply with existing local, state, or federal rules regarding same.

VI. Animal studies should be carried out under the provisions of the following guidelines:
   A. All animal studies should be carried out under the direct supervision of a competent science teacher. It is the responsibility of that teacher to ensure that the student has the necessary comprehension for the study being done.
   B. Students should not be allowed to take animals to carry out experimental studies. These studies should be done in a suitable area in the school.
   C. Students doing projects with vertebrate animals should adhere to the following:
      1. No experimental procedures should be attempted that would subject animals to pain or distinct discomfort.
      2. Students should not perform surgery on living vertebrate animals.
   D. Experimental procedures should not involve the use of microorganisms pathogenic to humans or other animals, ionizing radiation, carcinogens, drugs or chemicals at toxic levels, drugs known to produce adverse or teratogenic effects, pain-causing drugs, alcohol in any form, electric shock, exercise until exhaustion, or other distressing stimuli.
   E. Behavioral studies should use only positive reinforcement in training studies.
   F. Egg embryos subjected to experimental manipulation must be destroyed humanely at least 2 days prior to hatching. Normal egg embryos allowed to hatch must be treated humanely within these guidelines.

FIGURE 9–2   *continued*

G. The administration of anesthetics should be carried out by a qualified science teacher competent in such procedures. (The legal ramifications of student use of anesthetics are complex, and such use should be avoided.)

VII. The use of living animals for science fair projects and displays shall be in accordance with these guidelines. In addition, no living vertebrate animals shall be used in displays for science fair exhibits.

VIII. In those cases where the research value of a specific project is obvious by its potential contribution to science but its execution would be otherwise prohibited by the guidelines governing the selection of an appropriate experimental animal or procedure, exceptions can be obtained if all three of the rules below are adhered to:

A. The project is approved by and carried out under the direct supervision of a qualified research scientist in this field.

B. The project is carried out in an appropriate research facility designed for such projects.

C. The project is carried out with the utmost regard for the humane care and treatment of the animals involved in the project.

---

field trips also have unique problems that can cause them to be potentially dangerous.

Care During Animal Dissections.   Certain cautions and procedures must be observed when conducting animal dissections.

Use only rust-free instruments that have been thoroughly cleaned and sterilized. Dirty instruments are not safe and may cause infections. Instruct students thoroughly in the proper use of the instruments. Scalpels are generally used for dissection; if none are available, single-edged razor blades can be substituted. Some teachers recommend using scissors instead of scalpels. The teacher should demonstrate the proper techniques for using scalpels and dissecting probes before permitting students to use them. It should be stressed when students use a cutting instrument such as a scalpel that the incision be directed away from the student's body.

Cuts can occur during the cleaning of scalpels and needles, so use care during the cleaning process. The use of rubber gloves while cleaning will protect the student against cuts and infection.

Biology teachers have recently been concerned about using specimens that have been preserved in formaldehyde because the chemical is thought to be a carcinogen. A considerable quantity of formaldehyde can be removed from specimens by soaking them in water for several days. If this procedure is used, change the water frequently. Use biological supply houses that preserve specimens using substitutes for formaldehyde to avoid exposing students to this hazard.

Using Live Material.   The following list describes procedures and precautions to follow when using live material.

1. Do not use decayed or decaying plant or animal material unless every precaution is taken to ensure its proper handling. Improper handling may expose students to infection or accidental ingestion.

2. Warn students not to touch their mouths, eyes, or any exposed part of their body while using decayed or decaying material. Disposable gloves and forceps should be used to prevent physical contact.

3. Store decayed or decaying material in the refrigerator if it is to be used over a period of several days.

4. Use fungi with care to avoid the release of spores into the classroom environment. Spores may cause some students to have allergic reactions upon exposure. Excessive numbers of spores can bring on asthma, sneezing, and other respiratory problems.

5. Avoid weeds and plants that may induce hay fever and other pollen allergies. Large amounts of plant pollen can produce adverse effects.

6. Do not maintain cultures of pathogenic bacteria, viruses, and fungi in the laboratory. To do so may expose students to infection, particularly if they are not trained in proper laboratory techniques for handling microorganisms.

7. Avoid the use of blood agar, which can induce the growth of pathogenic bacteria.

8. Avoid the use of viruses in the laboratory, because they may infect other living organisms in the school building.

9. Do not allow students to inoculate bacterial plates with human oral material. This simple exercise involves certain risks because it could lead to the production of pathogenic organisms.

10. Warn students to be extremely careful when using pipettes to transfer microorganisms. It is best to use safety bulbs instead of pipettes.

11. Use precautions when transferring by inoculating needles or loops. When a heated needle or inoculating loop is placed in a culture medium, the material tends to splatter and produce aerosols, causing the release of microorganisms into the air. Instruct students to remove as much of the culture material as possible before sterilizing the loop or needle.

Sterilizing.   A pressure cooker is excellent for sterilization purposes if an autoclave is not available. An improperly operated pressure cooker may pose certain dangers; one should thoroughly understand the directions for operating a cooker before attempting to use it. Clean the safety valve on a pressure cooker and make sure it is operable before using it. Strictly follow the safety limits indicated in the directions or on the pressure gauge. Turn off the heat so that air pressure in the cooker will gradually be reduced to normal level before removing the cover. Do not remove the cover until pressure is at a safe level. Do not allow students to operate a pressure cooker without proper supervision.

In the absence of an autoclave, sterilize glass petri dishes and test tubes that have been used for culture growth by placing them in a strong solution of Lysol™, creosol, or other chemical disinfectant for a period of time before washing. Wear rubber gloves when washing glassware or instruments used for inoculation.

The disposal of cultures in glass petri dishes can be a problem if an autoclave is not available. Use disposable petri plates instead of glass plates for laboratory exercises in which students carry out their own procedures.

Using Hypodermic Syringes.   The use of hypodermic syringes by students requires close monitoring. The syringe is a potentially dangerous instrument to use in science teaching and, because of its drug-related implications, the teacher must weigh the benefits of using it in instruction. If syringes are to be used by students, the teacher should warn them beforehand about such harmful effects as hepatitis or embolism resulting from an accidental skin puncture with an unsterilized syringe. The best advice is either to avoid their use altogether or use them in a controlled situation such as a teacher demonstration.

Securely lock syringes in storage areas when they are not being used. If students will be using them, the teacher must account for all syringes that have been distributed for a laboratory exercise. In general, the syringe is an instrument that must be selectively used and carefully controlled. All rules that apply to the use and handling of sharp instruments also apply to hypodermic syringes.

Blood-typing. Blood-typing has often been performed in the biology laboratory using student subjects. The question often arises as to whether this type of activity should take place in secondary schools. Before performing blood-typing on students, the teacher should ask the principal or other administrator whether there are any restrictions on using students as a source of blood. Some schools have imposed such restrictions following legal suits brought because of science laboratory mishaps.

If students are used as subjects, the teacher must obtain a written statement from parents indicating their willingness to allow the procedure as well as stating that their child is not subject to fainting, epileptic seizures, or other conditions. Only sterile lancets or needles to prick the finger from which blood is to be drawn should be used. Alcohol is applied to the entire finger using a saturated cotton wad or piece of cloth. Blood should be drawn from the fleshy portion of the fingertip. Alcohol is again applied, and the wound is covered with a bandage. Lancets or needles should not be resterilized by dipping them into alcohol. Each needle or lancet should be used only once.

It must be emphasized that conducting blood-typing using student subjects can be risky. This practice is not recommended, particularly with the focus on AIDS and hepatitis. If blood-typing is carried out in the school laboratory, teachers should obtain blood samples from other sources, including themselves. It is best to take the safest route and avoid the use of student subjects.

Precautions During Field Trips. Field trips have inherent problems requiring special attention to safety. Biology field trips take place in different types of environments, each of which has a unique set of problems. Consequently, it is essential that the teacher know beforehand what precautions are needed to conduct a safe field trip. This requires that the teacher first visit the area to evaluate it for potential hazards.

Students need a list of specific rules of conduct to follow during the course of a field trip. Each trip will probably require the development of a specific set of rules. The following are some general safety guidelines for a biology field trip.

1. Brief students about the area they will visit. Instruct them about the areas that they are prohibited from visiting without supervision, such as ravines, cliffs, and bodies of water.
2. Tell students the type of clothing they are permitted to wear or take with them. Appropriate footwear is essential to avoid accidents.
3. Instruct students about the plants they should not touch, such as poison sumac, poison ivy, and certain mushrooms. Permit them to become familiar with poisonous plants that they may encounter on the trip. If possible, bring specimens of such plants into the classroom before the trip so that students can learn how to identify them. Warn students not to touch specimens in the classroom.
4. Warn students not to touch or pick up reptiles or other animals or touch dead carcasses of animals or birds.
5. Caution students about eating any plant material in the field unless it is identified by an expert as safe. Poisonous plants are often very similar to edible ones, and even experts can make serious mistakes in identification. Alert students not to touch fungi and decaying material unless they are informed to do so.
6. A set of rules of behavior should be provided before taking a field trip. If there are special rules to follow when visiting a public park, make certain that students are provided with such rules before the trip.

Additional Considerations for Safety in Biology Teaching

Some additional safety considerations that apply to biology teaching are listed here.

1. Use indirect sunlight or a lamp to view a microscope. Avoid direct sunlight on the mirror,

which could cause damage to the eyes while viewing.

2. Avoid the use of alcohol burners in the laboratory; they are hazardous. Use Bunsen burners or hot plates instead.

3. Require appropriate clothing to be worn during the laboratory period and especially avoid long-sleeved or loose garments. Lab coats or aprons are advisable when students are working with caustic materials.

4. Require students to wash their hands at the end of every laboratory session to avoid ingestion of chemicals or poisonous plant materials.

Other suggestions for safety that are applicable to the teaching of biology can be found in various sections of this chapter. For example, the section on chemical safety describes a number of protective measures that can be implemented in the biology laboratory. The location and use of safety facilities such as showers, eyewashes, and fire extinguishers are discussed in detail in other sections. The discussion of earth science safety may be useful in planning aspects of biology field trips and laboratories.

SPECIFIC SAFETY GUIDELINES FOR CHEMISTRY

A number of activities commonly conducted by students in the chemistry laboratory are potentially hazardous. The teacher must weigh the risks involved before allowing the students to perform certain activities. If the risks are too high, there are alternatives; if alternatives are not possible, then abandon the activity or perform it as a teacher demonstration. In some cases, even as a teacher demonstration, the risks might still be too great.

The following are some examples of routine laboratory activities in which precautions or alternatives are necessary to assure a safe environment.

1. *Preparation of oxygen.* Oxygen is commonly prepared by using a mixture of potassium chlorate and manganese dioxide. The risk associated with this activity is that manganese dioxide is sometimes mistaken for charcoal because both are black powders. A mixture of charcoal and potassium chlorate is explosive when heated. To avoid any mixup, do not store manganese dioxide and charcoal in the same area and clearly label both.

Another activity used to produce oxygen involves the heating of mercuric oxide. The problem with this procedure is that mercury vapor, which is toxic, is released into the atmosphere during the process of heating. When many students conduct the experiment, a significant amount of vapor is released, which may cause a potential hazard. The teacher should conduct this activity as a demonstration or substitute an alternative activity.

2. *Preparation of halogens.* Inform students of the hazards of inhaling and preparing chlorine, bromine, and hydrogen chloride.

Warn students to heat the mixtures gently so that the amount of gas produced at any given time is not generated in large quantities, which makes it difficult to control the collection of the gas. Halogens should be collected under water and in well-ventilated areas to avoid subjecting students to appreciable quantities of the gases. Bromine, hydrogen chloride, and chlorine are harmful in large quantities and can cause damage to the mucous membranes and the skin. Some teachers prefer to perform the production of halogens as a demonstration instead of a student-centered activity to avoid risks.

3. *Ammonium dichromate volcano.* When ammonium dichromate is ignited, chromium trioxide is produced, which is toxic as well as irritating. This activity should be performed as a teacher demonstration under a hood to control the emission of the gas and the trioxide dust.

4. *Preparation of hydrogen.* The laboratory preparation of hydrogen requires certain

precautions; a mixture of hydrogen and air in certain proportions can be explosive. Collect hydrogen under water and in a pure state to avoid an explosion. To determine the presence of air in a collecting vessel, first collect small quantities of the gas in small test tubes. Place an ignited splint at the mouth of the test tube to determine whether the hydrogen pops when the flame is applied. Collect in small quantities until the popping stops, which indicates that air is no longer present in the generator. Once pure hydrogen is available, it can be collected in larger vessels.

5. *Determining the properties of potassium or sodium.* The properties of potassium or sodium are observed by allowing the substances to come in contact with water. Use very little of either of these elements at any given time. Sodium is less active than potassium and is probably better to use to demonstrate the activity of these types of elements in water. Place water in a porcelain dish instead of a glass container and then cover the container with a glass plate. Some teachers do not allow students to handle these elements and choose to conduct the activity as a teacher demonstration.

6. *Preparation of carbon.* The laboratory preparation of carbon using sulfuric acid and sugar can be spectacular. Warn students not to touch or handle the carbon produced because it may contain concentrated sulfuric acid. Nor should they handle the beaker in which the reaction occurred because it may be extremely hot from the reaction.

7. *Activities with carbon disulfide.* Carbon disulfide is affected by static electricity and under certain atmospheric conditions may ignite when exposed to the air. The use of carbon disulfide requires a well-ventilated area or a hood. Do not allow flames in the area of the activity and keep static electricity at a minimum. If students are using carbon disulfide, the teacher should dispense it in small quantities.

Even using carbon disulfide to dissolve sulfur requires precautions. Determine the amount of

static electricity in the atmosphere before conducting the activity, and conduct it only when atmospheric conditions are suitable.

8. *Iodine clock reactions.* The "Old Nassau" clock reaction is probably the most popular clock reaction performed in chemistry teaching. There are problems with this exercise, particularly if students perform it. Mercury salt is produced as a result of the reaction and often it is disposed of improperly. The question that arises is whether students should be exposed to mercury compounds because of the inherent dangers associated with them. A teacher demonstration eliminates some of the risk, provided the mercury compound is disposed of properly. Clock reactions that do not involve the production of mercury compounds are safer and should be substituted to avoid risks.

Other Chemistry Safety Precautions

The majority of accidents in science instruction involve activities in the chemistry laboratory or the use of chemicals in other areas of science. Many accidents occur because teachers are careless about requiring students to wear safety goggles. Students should also be required to remove contact lenses, even if they will be wearing safety goggles, during laboratory activities. Chemicals can get behind the lens and cause severe burns and eye damage. Proper dress and use of laboratory coats or aprons are often disregarded. The teacher should check students' attire to see that they are not wearing loose jackets, long neckties, or sandals.

The precautions listed below are further suggestions for a safe environment when using chemicals in the laboratory.

1. When inserting glass tubing into a rubber stopper, always lubricate the glass with glycerin beforehand. The glass tube is likely to break, possibly sending tiny glass fragments into the eyes or skin.
2. When bending glass tubing, keep the burner

flame low and heat the material gently. Do not force the glass to bend; it may suddenly break or shatter, possibly causing injury to an individual.

3. When using a match to light a Bunsen burner, always light the match *first,* then turn the gas on *slowly.* Turning on the gas before lighting the match can result in accumulated gas, which might be explosive. Keep arms, hair, face, and other body parts as far away from the burner as possible while it is being ignited.

4. When heating any substance in a test tube, point the mouth of the tube away from the body. Boiling often occurs quickly and without warning, causing the substance to spew the hot vapor on individuals in the vicinity.

5. Clean used test tubes meticulously. Residue left in test tubes could sabotage future experiments, either by altering results or, worse, by causing unexpected dangerous reactions.

6. Clean chemical spills immediately. There are potential dangers if laboratory tables are not cleaned. For instance, if spilled hydrochloric acid is not cleaned from the surface, an individual could lean on the table, causing burns to clothing or skin, or a spilled substance could accidentally be mixed with another spilled substance, which may result in a violent reaction.

7. Make absolutely certain that students do not attempt to remove a beaker, test tube, porcelain dish, or other glassware from a flame without using proper utensils. Tongs, test tube holders, and heat-resistant gloves are designed just for this purpose. No student should use bare hands during these tasks.

8. When diluting acids, always add the acid to the water, and not the reverse. The reaction of acid with water is exothermic; that is, large quantities of heat are released. If water is added to the acid, the water will tend to remain on the surface of the acid because it is less dense and, consequently, will not mix.

It may produce a violent reaction, which may cause the acid to splatter into the eyes or on the skin of individuals nearby. To add the acid to the water safely, place a stirring rod in the water and hold it at an angle. Pour the acid slowly down the length of the rod above the water. This procedure prevents splashing while pouring. Be sure to wear safety glasses.

9. When heating materials, always use open vessels during the heating process. Do not heat vessels that have been stoppered. When heating liquids, it is good practice to use boiling chips to prevent bumping.

10. When evaporating a toxic or dangerous solvent, use a well-ventilated fume hood.

11. Avoid subjecting flammable materials to any open flame. Open flames are dangerous in the presence of flammable substances.

In summary, all laboratory activities or demonstrations, whether performed by students or the teacher, must be considered potentially dangerous. Before using an activity, the teacher must first weigh whether the educational benefits are worth the risks that may be involved. The teacher must exercise extreme care no matter how many times he has performed the demonstration or conducted the laboratory; repeated performance of an exercise does not reduce the risks. Common laboratory exercises or demonstrations can be just as dangerous as those requiring unusual apparatus or infrequently used chemicals. It is essential that the teacher (and the student) always be on the defensive and practice preventive measures when working with chemicals.

Storing and Using Chemicals Safely

Many accidents occur in the classroom or laboratory because of improper storage and use of chemicals. Those who use chemicals in their teaching must be knowledgeable about what substances are potentially dangerous. To avoid unforeseen problems, know what facilities are required for safe storage of chemicals, the safety precautions needed when using them, and the safety procedures to use to properly dispose of them.

Certain chemicals should not even be used or stored in a science laboratory. With growing recognition of the hazards of some commonly used chemicals, the decision to store, use, or not use a specific substance requires a careful risk-benefit analysis (Coble & Hounshell, 1980).

Combustible substances, poisonous materials, acids and bases, and other dangerous chemicals have special storage requirements and, in most cases, must be securely stored to avoid potential accidents. Always securely lock combustible substances, such as methanol and ethanol, in metal cabinets. The same security is necessary for carbon disulfide, sodium metal, and potassium metal. Acids or bases must never be stored on wooden shelving cabinets or in a closet; store them on metal shelving. Never store chemicals that may react with each other in the same area. Glycerin and nitric acid, acids and cyanides, potassium chlorate and organic substances should not be stored in close proximity, even temporarily.

There is a list of poisonous substances that by law cannot be maintained or used in the lab, including substances such as benzidine, benzene, arsenic, vinyl chloride, and asbestos. Other poisonous substances that are considered to be carcinogenic cannot legally be part of the chemical inventory, including formaldehyde, carbon tetrachloride, phenol, xylene, and lead compounds. The Occupational Safety and Health Administration (OSHA) has published a list of substances that it recommends not be stored in a school science laboratory (OSHA, 1981).

Do not use chemicals that are suspected of causing physical or functional defects or are known carcinogens. Warn students of the possible harmful effects of such substances. See table 9–1 for a partial list of such chemicals.

TABLE 9–1   Some hazardous chemicals

Toxic substances
Ammonium dichromate
Ammonium thiocyanate
Arsenic and arsenic
 compounds
Barium salts
Benzene and benzene
 compounds
Beryllium
Bromine
Cadmium and cadmium
 salts
Carbon disulfide
Carbon tetrachloride (also
 possible carcinogen)
Chloroform
Chromic acid
Chromium trioxide
Cyanides (water-soluble
 cyanides)
Dimethyl sulfate
Hydrogen bromide
Hydrogen chloride
Hydrogen flouride
Hydrogen iodide
Hydrogen sulfide
Lead compounds
Manganese compounds
Mercury and mercury
 compounds
Molybdenum compounds
Naphthalene
Nickel and nickel
 compounds
Nitrogen dioxide
Styrene

Corrosive substances
Bromine
Hydro-halogens
p-dichlorobenzene
Sodium

Teratogens
Aniline
Benzene
Carbon
 tetrachloride
Lead compounds
Phenol
Toluene
Xylene

Irritants
Ammonium
 dichromate
Borane
Ether
Hydrogen
 peroxide
Methylene
 chloride
Nitrogen dioxide
Toluene
Xylene
Zinc chloride

**Carcinogenic
 substances**
Asbestos
Benzene
Carbon
 tetrachloride
Formaldehyde
Lead compounds
Nickel and nickel
 compounds
Phenol
Xylene

Sources: Adapted from *Aldrich: Catalogue of Fine Chemicals,* 1992, Milwaukee: Aldrich Chemical Co., Inc., and from "Unnecessary Risks," by D. Sievers, 1984, *The Science Teacher, 51*(6), p. 32.

Teratogenic chemicals, those that may cause physical or functional birth defects, should not be used. Again, teachers should warn students of possible harmful effects of these substances. These substances include aniline, carbon tetrachloride, phenol, and xylene (see table 9–1).

Other Suggestions for Safety

All chemicals must be treated as if they are potentially dangerous. Do not downplay the problems that can result if the skin is exposed to chemicals, or caustic fumes are inhaled. Nor should the teacher allow situations that may result in explosions or fires. The precautions needed to avoid hazards require knowledge about the nature of the chemicals stored in the laboratory as well as common sense in their use. The following suggestions can help prevent potential problems.

1. Keep laboratory and other areas where chemicals are stored or used well ventilated and maintain a relatively cool temperature.
2. Use approved safety cans and metal cabinets to store flammable liquids.
3. Store cylinders of compressed gases by type, and mark them as highly toxic, corrosive, or flammable. Store in cool and well-ventilated areas. Limit the amount of flammable liquids and gases maintained in the laboratory.
4. Store large bottles of acids and bases on shelves that are no more than two feet above the floor, and store tham away from each other to prevent corrosion and other chemical reactions.
5. Inspect chemicals annually to see whether they are properly identified or outdated. Properly dispose of contaminated, unlabeled, and deteriorated chemicals.
6. Do not leave chemicals in areas where students are working, unless they are going to use the chemicals.
7. Keep chemicals in storage until they are ready to be used.

8. Store only small quantities of flammable substances in areas where they will be used. Large quantities are difficult to handle in case of a fire or accident.
9. Do not concentrate large quantities of flammable substances in any area in the laboratory.
10. Do not store chemicals in hallways and other heavy traffic areas. Students are often curious about what is stored in boxes and cabinets. Mishandling can cause accidents.
11. Do not store chemicals on shelving above eye level unless there is easy access with ladders or stools. Individuals trying to remove chemicals from out-of-reach shelves run the risk of dropping chemicals and causing mishaps.

Using Reagents and Liquids.   Handle all corrosive substances with caution. Concentrated sulfuric acid, nitric acid, glacial acetic acid, and caustic alkalies are dangerous. When pouring liquid chemicals of this type, one must know the nature of the chemical in order to avoid potential problems. The following rules apply when using liquid corrosive chemicals.

1. Never place vessels anywhere near the face when pouring corrosive liquids into them. This will prevent splashing of the liquids into the eyes or onto the skin.
2. Use properly sized funnels during pouring operations—small funnels for small vessels, large funnels for large vessels.
3. Hold bottles containing the liquid very securely during the pouring operation.
4. Always pour into containers that are placed on protective surfaces. Never pour concentrated acid into a beaker that is held in one's hand or placed on a surface that can be damaged, such as marble or wood.
5. Always pour small amounts of acids or bases when making dilute concentrations of these substances. Stir continuously while pouring.

Always pour slowly and wipe up spills immediately.

6. Place reagents and liquids in small reagent bottles for students use. Large bottles are difficult to handle and may contribute to spillage and breakage.

7. After the liquid has been poured, return the chemical to the storage area to avoid improper handling by others.

Heating Substances in Test Tubes. Use the following rules when heating substances in test tubes.

1. Always use test tube holders to hold test tubes that are being heated.

2. Direct the mouth of the test tube away from others as well as oneself when heating substances in test tubes.

3. Agitate liquids gently during the heating process. Apply the heat just below the surface of the liquid in the test tube.

4. Heat toxic and noxious substances under a ventilated hood. Do not allow fumes to enter the classroom area.

5. Do not use open flames to heat flammable liquids.

6. Never heat a liquid in a stoppered tube.

Storing and Handling Sodium, Potassium, and Phosphorus. Chemicals such as sodium, potassium, and phosphorus require special attention because of their unique properties. The following suggestions can prevent potential dangers.

1. Store metallic sodium and potassium in kerosene or heavy mineral oil and avoid exposure to water solutions, containers of water, or sources of water. Even high humidity in the atmosphere can induce a reaction when the metals are exposed to the air.

2. Store white phosphorus under water and in a double container.

3. Cut white phosphorus under water because it may ignite in open air, causing serious consequences.

4. Use red phosphorus instead of white whenever substitution is permissible.

5. Burn residues of phosphorus under a ventilated hood before placing them in a waste container.

6. Do not expose sodium, potassium, or phosphorus to the skin; they are very corrosive.

Disposing of Chemical Wastes

To avoid unnecessary risks, remove waste materials from the laboratory that accumulate as a byproduct of scientific investigations and become useless because they are improperly labeled or have aged. In addition, chemicals that are no longer used or needed must be properly removed from the inventory.

Disposing of chemicals requires certain important procedures. Before disposing of chemicals, consider the federal, state, and local rules and regulations, the effect of the chemicals on the environment, their level of toxicity, and the degree to which they are hazardous.

Several publications contain information concerning chemical hazards and disposal. Publications of value to secondary schools include Dux and Stalzer (1989), Young (1987), and Armour (1989). These publications contain information regarding handling of hazardous substances. The following are some options for disposing of chemical wastes.

1. Do not pour strong acids and flammable liquids down a drain without first diluting them. Do not dilute acids by pouring water into them; pour acids into the water for dilution. Neutralize or dilute all hazardous wastes before disposing of them.

2. Do not pour volatile substances and chemicals that produce obnoxious odors down the drain. They may lodge in interconnected drains located in other areas of the building, causing odors in these areas.

3. Always label solid wastes as such and place them in suitable containers, making certain that the containers will not react with the wastes.

4. When there is doubt about the proper handling of solid wastes, seek advice from scientists at universities or nearby industries, or call commercial disposal firms. Chemical supply houses are also good sources of information.

5. Use care when disposing of carcinogens, radioactive materials, and other hazardous substances. If you do not know the procedures for handling these materials, seek help from the science coordinator, principal, or local fire department. They may provide the proper disposal method or suggest someone who knows.

6. Return unused chemicals to the supplier if possible.

7. Use an incinerator to safely dispose of certain substances after checking with an expert to see that the substance can be disposed of in this manner.

8. Dispose of some wastes by placing them in a landfill, but only after asking an expert whether the substance can be disposed of in this way.

It must be emphasized that the disposal of chemicals requires very careful planning. If there is any doubt about proper disposal, do not hesitate to ask experts for the safest way to remove wastes from school laboratories.

In certain geographic areas, a number of schools have formed disposal pools that are administered by science teachers to properly dispose of hazardous and potentially dangerous chemical wastes (Yarroch, 1980). These arrangements have proven to be very effective, particularly in areas with a high concentration of schools. Through these pools teachers can exchange information and, in some cases, provide expert handling and disposal of materials.

SAFETY IN THE EARTH SCIENCE LABORATORY

In addition to safety practices normally involved in all secondary science laboratories, the possible hazards associated with the teaching of earth science require special attention. Teachers are under the misconception that nothing dangerous can possibly be associated with the activities that occur in the earth science laboratory, but rocks and minerals commonly used in the laboratory can and do present certain health hazards.

The tasting of minerals and rocks, chemical procedures and analysis, crushing procedures, and the mere handling of rocks and minerals in the laboratory are potentially dangerous activities.

Warn students not to taste unidentified materials, rocks, or chemical substances. Before substances are tasted, they must be positively identified as not dangerous if ingested in small quantities. Students can make taste tests using minerals that contain halite (salt) or other common substances known not to be dangerous. Students should be cautioned not to taste minerals that are known to contain antimony, lead, copper, mercury, zinc, nickel, or aluminum. Many minerals containing these elements resemble and can be mistaken for halite, and are very dangerous when tasted (Puffer, 1979).

Minerals containing arsenic, and several that contain calcium, copper, lead, or zinc arsenites and arsenates, are poisonous. Acute arsenic poisoning can produce gastrointestinal disturbances, muscle spasms, dizziness, delirium, and coma.

Minerals containing lead also present health hazards, because lead is poisonous in all forms. The ingestion of lead minerals in large quantities can produce cramps, muscle weakness, depression, coma, and even death. Lead poisoning can be cumulative, and effects can range from moderate to very severe.

Ingesting common substances such as calcite has been known to cause death in human beings. On the other hand, the amounts of halite taken orally during taste tests will not normally cause any disturbance unless individuals are sensitive to salt. However, take care even in instances that are normally regarded as nonhazardous (Puffer, 1979).

Inhalation of certain mineral and rock dusts can also cause certain health problems. Minerals containing manganese, asbestos, and quartz are hazardous in dust form, so avoid inhalation. Manganese dust can induce headaches, weakness in the legs, and general irritability when inhaled. Silica dust can cause silicosis (a lung disease), which has symptoms similar to those of tuberculosis. Asbestos dust fibers are known to cause cancer and asbestosis in human beings. In such cases, tumors are induced to form in the lungs and peritoneum (Puffer, 1979).

Appropriate warning signs that indicate the presence of asbestos or dangerous mineral dust should be posted in the laboratory. The following are examples of such warning signs.

```
!WARNING!

ASBESTOS IS HAZARDOUS
TO YOUR HEALTH
```

```
AVOID BREATHING DUST

ADEQUATE VENTILATION REQUIRED
```

A charcoal block test that is often performed in the earth science laboratory requires extreme caution. Substances containing mercury, such as cinnabar, produce mercury vapors when subjected to charcoal block tests. The vapors can cause excitability, anxiety, irritability, delirium, and other symptoms. The mercury produced during such tests also creates handling problems; it is difficult to dispose of because it is insoluble in water, dilute acids, and common solvents. Do not allow students to handle substances containing mercury or subject them to any mineral containing the chemical. Minerals containing copper, lead, nickel, and uranium, when subjected to charcoal block tests, can produce hazardous substances. Use care when handling them.

Other Safety Precautions

The following safety guidelines pertain to earth science laboratory practices. The teacher should use them whenever appropriate.

1. The teacher and students should wear goggles when crushing rock with a hammer or other instruments.
2. Do not crush rocks or other minerals unless they are wrapped in a cloth. This precaution prevents rock fragments and dust from being dispersed in the laboratory area and may prevent injuries to the eyes and other parts of the body.
3. Require students to wear gloves when handling large rock samples, particularly when moving or crushing the samples. Jagged rocks can produce both surface and deep wounds.
4. Do not allow students to lift heavy rock samples alone. Instruct students to help each other when lifting large or cumbersome samples. Use dollies or other equipment to move large and awkward objects.
5. Warn students not to wear open-ended shoes or sandals during field trips. Require them to wear long pants and long sleeves for protection. Gloves are essential when collecting materials on field trips.
6. Before taking a field trip, provide a special set of rules regarding the conduct of students during the trip. Oral directions are not substitutes for written directions. Written rules

are constant reminders to students and ensure proper behavior in the field.

SAFETY GUIDELINES FOR PHYSICS AND PHYSICAL SCIENCE

Accidents from electrical sources in physics and physical science laboratories are not uncommon. The mishaps can range from minor burns to death. Burns caused by electrical sources are usually slow to heal and often require several months of treatment for recovery. Thermal burns caused by high temperatures near the body, such as those produced by an electric arc, are similar to sunburn and are usually not severe unless the body has been exposed for long periods of time.

Impulse-type electric shocks are not only unpleasant, but, in some cases, shock intensities produced by higher currents passing through the chest or nerve centers may produce paralysis of the breathing muscles. Excessively high currents will cause death, but it is also possible for amounts of current as low as 0.1 amps to stop the heart from beating. Currents that blow fuses or trip circuit breakers can destroy tissue and produce shock and damage to the nervous system. The following safety procedures are guidelines for working with electricity.

1. Know the total voltage and current of the electrical circuit before using a piece of electrical equipment.
2. Use extension cords that are as short as possible, properly insulated, and of a wire size suitable for the voltage and current involved.
3. Service electrical apparatus and devices only when the power is turned off. Make certain that power is not accidentally turned on during servicing.
4. Do not permit students to service electrical equipment or apparatus.
5. Do not permit students to be in the vicinity

of electrical apparatus or equipment being serviced.
6. Do not turn power on after servicing until all students are moved to a safe area. Notify students when it is safe to return to their positions.
7. Use properly insulated, nonconducting tools that are in good condition when working with electrical equipment. Use only appropriate tools—those that have specifications indicating that they can be employed for servicing electrical devices.
8. Properly mark all electrical equipment, using letters 2 or 3 inches high to indicate the voltage.
9. Make sure electrical contacts and conductors are enclosed at all times to avoid accidental contact and check them periodically for compliance.
10. Periodically inspect electrical outlets to see that they are in good order. Constant use may cause wear and loosening of outlets.
11. Avoid using metallic prongs, pencils, and rulers when working on an electrical device.
12. Do not wear rings, metal watchbands, or metal necklaces in the vicinity of an activated electrical device.
13. Never handle activated electrical equipment with wet hands or while the body is wet or perspiring.
14. Do not use highly volatile or flammable liquids to clean electrical equipment. There are cleaning solvents that can be used safely, but investigate whether they are suitable for electrical devices.
15. Allow only qualified electricians to perform electrical wiring and maintenance of electrical outlets and devices. Do not allow students or unqualified teachers to perform these functions.
16. Do not store volatile and flammable liquids in the vicinity of activated electrical equipment. The heat generated by equipment may cause a fire or explosion.
17. Do not handle electrical equipment that has

been in use for a long period of time. It may be very hot, which could cause serious burns, or the hot equipment may be dropped, causing damage to some of its parts.

18. Use electrical appliances that are approved by Underwriters Laboratories or another known laboratory.
19. Have homemade equipment inspected by a licensed electrician before using it. Always regard homemade equipment as potentially dangerous until it is checked by the electrician.
20. Use indoor equipment inside, not outside. The same is true for indoor and outdoor outlets, and indoor and outdoor electrical wires. Do not use outdoor equipment or wire when the ground is wet or when it is raining.
21. Make sure power tools used in the laboratory are in excellent working order. Check the cord or plug periodically for fraying. Check tool switches only when standing on a dry, insulated floor away from grounded objects.
22. Service tools that have shocked anyone or that have emitted sparks. Do not use such tools until they are in good working order.
23. Before using an electric motor, check to see that it conforms to the specifications of the National Electric Code. Three-phase motors require proper wiring for actual voltage levels in a particular facility.
24. Employ licensed electricians periodically to inspect and service all motors to detect possible defects that could cause problems.
25. Make sure all permanently installed motors have an accessible, quick disconnect switch.
26. Have all electrical devices properly grounded. Grounding can be complicated and must be done by a licensed electrician.
27. If possible, have ground fault interrupters installed to prevent possible electrocution. Many state electrical codes now require outlets to have a GFI (Harper & Bartlett, 1989).

Other Precautions for Physical Science Laboratories

Some important precautions and potential problems associated with electrical wall and floor outlets require special attention. Teachers often disregard the importance of keeping outlets clean and in good order. Teachers should also be aware of problems associated with electrical equipment such as high voltage step-up transformers, refrigerators, and vacuum pumps. Such equipment requires proper maintenance and use to avoid safety problems.

1. Have floor outlets inspected routinely to see that there is no accumulation of foreign material in the underfloor conduit. Place safety caps in unused outlets and, if necessary, have new gaskets installed to prevent damage to floor boxes. Avoid flooding areas around floor outlet boxes when mopping and advise the janitor to use only a damp mop in these areas.
2. Periodically check wall and table outlets and appropriately mark them as 110V, 220V, 440V, and other voltages. There should be a sufficient number of outlets in the laboratory to minimize the use of extension cords.
3. Teach most electrical concepts in the laboratory using low voltage in the 0–15 volt range rather than the 110–120 volt range from regular outlets. This range is recommended to reduce risks of shock.
4. Avoid conduction of 120-volt current through solutions. Use lower ranges.
5. Do not expose students to situations that require high voltage. High voltage from step-up transformers, induction coils, and static charges is potentially dangerous, and only the teacher should perform the activities requiring their use.
6. Recognize the potential danger of glass tubes such as TV picture tubes and cathode ray tubes that are to be evacuated. When conducting evacuation procedures, make

certain that students wear safety glasses and are properly garbed.

7. Provide belt guards for motor-driven devices such as vacuum pumps and rotors.

8. Warn students of the potential hazards in working with materials under low pressure, such as experiments and demonstrations using bell jars and thermos bottles. Require students to wear safety goggles when performing such activities. Also warn them to follow directions carefully and not exceed the specifications indicated for the activities.

9. Use refrigerators to store flammable liquids but make sure they have been approved for this purpose. Do not use any home refrigerator for this purpose.

10. Make sure ovens have proper, functional thermostatic controls. Do not dry out materials saturated with volatile substances in ovens that have open flames or exposed electrical heating coils. Dry such materials only in ovens suitable for this purpose.

RADIATION SAFETY

Secondary school biology, physics, and chemistry courses include topics that involve radiation. Physics and chemistry courses involve experiments that deal with radiation emission of radioactive isotopes, X-ray diffraction apparatus, Crookes tubes, laser beams, ultraviolet rays, infrared rays, and microwaves. Biology courses sometimes involve experiments exposing biological materials to radiation sources. Apparatus and materials that generate radiation can be hazardous unless precautions are taken to avoid radiation exposure.

To avoid excessive radiation exposure during activities involving radiation sources, limit the time of exposure to short periods. Lead sheets or bricks provide appropriate shielding between the students and the radiation source. To avoid great intensities of primary radiation, the distance between the radiation source and students should be as great as possible but still allow the activity to be carried out effectively. To avoid risks, the teacher should not attempt any activity without the advice of an expert.

The inclusion of laboratory activities that involve radioactive materials in secondary school science is controversial; some feel that such activities are inappropriate at this level. This is not the place to discuss the pros and cons of such activities. However, the teacher should carefully weigh the benefits of these types of activities against the potential dangers of subjecting students to radioactive sources. In any event, it is imperative that teachers know how to handle apparatus and materials so that the laboratory is a safe place for teachers and students to work. Teachers should be properly trained in the use of materials and apparatus before attempting to use them in the laboratory. Self-taught teachers must be certain that their techniques are safe and appropriate. Radiation experts and scientists, university professors, and other qualified individuals can provide invaluable assistance for formally trained and self-taught teachers. Furthermore, permission from the school administration to use radiation sources in teaching may be required. Some school districts have regulations that prohibit the use of such materials.

An excellent source for teachers who want to incorporate activities involving radiation is *Radiation Protection in Educational Institutions,* published in 1966 by the National Council on Radiation Protection in Washington, D.C. In addition, specific guidelines for the proper use of lasers in the classroom are available (Penn, 1989).

Using Live Animals in Radiation Work

Vertebrate animals should not be used in radiation work in secondary school science. Subjecting animals to small doses of radiation can induce cancer, tumors, and other conditions that cause discomfort or pain. Animals should be treated humanely, as specified in the guidelines of the National Association of Biology Teachers and the American Humane Association.

Storing and Using Radioactive Materials

Because of the potential dangers radioactive materials pose, the teacher must take special care when using, storing, or disposing of them. The following is a list of precautions that apply to radioactive materials.

1. Store radioactive materials in cupboards or refrigerators that are labeled *radioactive*. To ensure safety, have the cabinets properly shielded and locked at all times.
2. Refer to the United States Energy and Development Administration's list of radioisotopes approved for secondary school science activities before engaging in any activity.
3. Label radioactive material with the date received, type, and quantity of material.
4. Handle radioactive materials and equipment only while using proper shielding devices. Use thin plastic gloves when radiation to the hands is not a significant factor. The level of radiation used in secondary school work usually does not necessitate the use of remote handling procedures (i.e., specially shielded radiation chambers).
5. Thoroughly wash glassware subjected to radioactive substances in detergent solutions especially developed for this purpose.
6. Dispose of used materials promptly and appropriately.
7. Do not store radioactive material for any length of time beyond what is required. Place such materials in labeled containers and store in a locked room reserved for this purpose.
8. Consult an expert—a university or industrial scientist familiar with such materials or a radiation expert associated with a state or federal agency—if there are any questions regarding the disposal of radioactive wastes.

Personal Work Habits During Radiation Exercises

The use of radioactive materials requires special work habits, which are contained in the following list.

1. *Never* allow eating or drinking in a laboratory where radioactive substances are being used.
2. Do not lick labels or pipette solutions.
3. Require protective clothing.
4. Do not expose cuts or abrasions to radiation sources.
5. Thoroughly wash hands after using radioactive materials and equipment. Keep fingernails trimmed closely; radioactive materials can accumulate between the nails and supporting skin.

Suggestions for Use of Nonionizing Radiation

Using nonionizing radiation requires special precautionary measures. A list of these measures follows, including some special recommendations for using laser beams.

1. Laser beams are very dangerous. If the intensity is high enough, severe burns can result. Preventive measures are extremely important: laser beams can cause blindness in less than one second. The teacher must become skillful in handling the equipment and know the safety measures they require. The following are safety recommendations for the use of laser beams in the classroom.
   - Keep students away from all sides of the path of laser beams.
   - Warn students and other individuals not to look into the laser beam.

- Do not aim laser beams directly into the eyes.
- Do not allow laser beams to hit the exposed skin of an individual.

2. Ultraviolet radiation is harmful below 310 nm. Mercury arcs and other sources can produce radiation below 310 nm, and the teacher should take care to use proper shielding and adequate filtering materials.

3. Radiation from microwave ovens can cause severe damage. Although high frequencies cause heat sensation on the skin, low frequencies do not, and thus one is not aware that tissue damage is taking place. Microwave ovens should be equipped with adequate interlock mechanisms.

4. Ultrasonic beams of high intensity can also be extremely harmful; use them with caution.

SAFETY UNITS FOR STUDENTS

A study by Dombrowski and Hagelberg (1985) supports the fact that a unit on laboratory safety increases students' safety knowledge and reduces the number of unsafe behaviors. Safety units that can develop student awareness and responsibility toward safety are best presented during the early stages of a laboratory course. Throughout this period, students can learn how to use safety equipment such as safety showers, fire extinguishers, eyewash fountains, and fire blankets. They can be indoctrinated to use laboratory coats, eye goggles, and gloves at appropriate times. They can be shown how to handle broken glass, chemicals, and electrical equipment and how to light a Bunsen burner. Stress the importance of housekeeping as well as the necessity of maintaining a clutter-free work environment. During the course of the unit, the students can develop a set of safety rules, which the teacher can supplement.

The unit can vary in length, but three or four class periods would probably suffice. The unit should be general, dealing with aspects of safety that apply to the laboratory course that will be offered. The use of visual aids, demonstrations, and hands-on activities will meet the unit objectives. Active student involvement is necessary to make this unit an effective experience. Other safety considerations specific to a laboratory exercise can be dealt with as the course progresses.

The teacher can administer pretests and posttests using questions similar to those in figure 9–3 before and after the unit to determine whether the unit improves students' safety awareness, knowledge, and sensitivity. The same questions can be used for both tests, or questions can vary, depending on what has been stressed.

After teaching the unit, require students to sign a safety contract such as the one in figure 9–4. The students should not take the signing of this contract lightly. It is an agreement that the student will behave as required to maintain a safe environment. To make the document more meaningful, it is suggested that the parents read and sign the contract as well, so they, too, understand the implications of its contents. The safety contract is a valuable record for the teacher to have in case of future litigation. It will show that the teacher has been responsible in attempting to instruct, promote, and maintain a safe working environment for all concerned.

The teacher can effectively teach safety only with the proper background. Units cannot be presented in a haphazard fashion; they must be well organized and taught by knowledgeable individuals. A teacher's background knowledge should be extensive before embarking on a safety unit. Safety knowledge can be strengthened by taking courses, workshops, and lectures as they are offered by safety experts. Background can also be acquired by talking with science teachers, scientists, fire marshalls, and others who have the expertise on particular aspects of safety.

FIGURE 9–3 Sample laboratory safety quiz

LABORATORY SAFETY QUIZ

Directions: The following questions are either true or false. In the blank to the left of each statement, write the letter *T* if the statement is true, or *F* if the statement is false.

_____ 1. It is required by law to wear safety goggles in the area where chemicals are stored.

_____ 2. When diluting acids, water is always poured into the acid.

_____ 3. A chemical is considered dangerous only if it is toxic or flammable.

_____ 4. The teacher is the only individual responsible for safety in the laboratory.

_____ 5. A fire that involves a solvent should be extinguished with a carbon dioxide fire extinguisher.

_____ 6. The disposal of chemical wastes produced from a laboratory exercise should be done by flushing the material down the drain.

_____ 7. Prescription glasses can be used instead of safety glasses when working in the laboratory.

_____ 8. It is permissible for students to use beakers for drinking purposes after they have been sterilized.

_____ 9. It is permissible for responsible students to remove chemicals from the storage areas.

_____ 10. The student should be able to operate various types of fire extinguishers.

_____ 11. In general, it would be permissible for students to substitute one chemical for another in the case of a shortage of a particular substance.

_____ 12. It is permissible to store reagents and chemicals in student lockers.

SAFETY WORKSHOPS FOR TEACHERS

In an effort to conduct good safety practices, many principals are requiring teachers to attend special seminars and workshops conducted by safety experts.

These workshops are often conducted as an inservice component and also require the participation of principals, department chairpersons, and maintenance personnel. Some have dealt with the procedures to follow during a periodic laboratory inspection; others have different orientations and emphases.

In one such workshop involving an inspection of storage facilities and materials, participants were asked to discover specific instances of safety violations in the chemistry laboratory. Many of the violations were fabricated and included violations such as labels without dates, toxic substances stored in rusted storage cans, improper types of goggles, storage of acids on unstable shelving, and storage of chemicals under fume hoods. After extensive training, the

FIGURE 9–4 Sample student safety contract

STUDENT SAFETY CONTRACT

I agree to follow all instructions by my teacher regarding safety procedures during laboratory work. I will conduct myself in responsible ways at all times while working in the laboratory and not engage in any horseplay.

While in the laboratory I will do the following:

1. Carry out the required housekeeping practices.
2. Know the location of fire extinguishers, safety showers, eyewash stations, and safety blankets and learn how to operate and use them.
3. Wear appropriate safety goggles when performing potentially hazardous activities.
4. Wear appropriate dress, as required by the teacher, during laboratory activities.
5. Know where to get appropriate help in case of emergency.
6. Agree to follow all printed and verbal safety instructions as given by the teacher and/or principal.
7. Agree to take and pass the safety test administered by the teacher.

I have carefully read this contract and agree to its conditions. My parents have read and signed this contract, and fully understand the implications of its contents.

_____   _____
Signature of student   Signature of parent

_____   _____
Date   Date

---

teachers returned to their schools and conducted inspections to determine the extent of safety violations in their school laboratories. The violations reported were serious enough to require immediate action to correct the situations. The most important outcome of the workshop was the development of a safety guide, which was developed by a team of teachers selected from the original group of inservice participants. The guide, which was made available to teachers, helped significantly to improve safety conditions in the school laboratories (Dunkelberger & Snyder, 1985).

Safety workshops can vary in length from one day to several days. Some conduct a series of lectures and seminars delivered by various experts—science teachers, fire marshalls, scientists, and those skilled in electricity and plumbing. Others have a practical orientation that allows teachers an opportunity to operate safety equipment, conduct safety inspections, and listen to safety consultants and experts. Regardless of the approach, training for safety will make science teachers conscious of their responsibilities and provide them with the expertise they need to maintain a safe environment in their teaching situations. Teachers should take advantage of any opportunity to make themselves more knowledgeable about safety in their own teaching environments.

SUGGESTED ACTIVITIES

1. Prepare an inventory of the possible storage hazards associated with any one or more of the following courses: physics, chemistry, biology, earth science, physical science. After preparing the inventory, visit a local school and seek permission to examine the storage areas associated with the course you chose.

2. Outline a safety program that you would institute if you were a chairperson of a science department in a middle or secondary school. Ask a science chairperson in a local school to critique the program. Discuss the results with your science methods class.

3. Prepare the safety rules you would post in a prominent place in one or more of the following areas: (a) a chemistry laboratory, (b) a physics laboratory, (c) an earth science laboratory, (d) a biology laboratory. Prepare an example of the poster you would use for one of the areas. Ask a member of the class to make suggestions for improving the poster.

4. Observe a chemistry, physics, biology, or science laboratory while students are performing hands-on activities. What safety practices are obviously in effect? What safety hazards are evident while students are working? What precautions do students take to avoid accidents? Discuss the observations with members of your science methods class.

5. Ask permission to inventory the electrical equipment that is used for a physics course offered in a secondary school. Determine if the equipment meets specifications determined by law. If there is homemade equipment, has the teacher asked an electrician to determine its safety? Discuss the results with members of your methods class.

6. Design a safety contract that would be suitable for students taking middle school science. Compare the contract you have designed with the one found in Figure 9–4. What are the differences? Discuss the differences between the two contracts with the students in your science methods class. What did you stress in the contract you prepared that was not stressed in the other one?

BIBLIOGRAPHY

Aldrich Chemical Co., Inc. 1992. *Aldrich: Catalogue of fine chemicals.* Milwaukee: Author.

American Humane Association. 1980. *Guiding principles for the use of animals by secondary school students and science club members.* Denver: Author.

Armour, M. A. 1989, Mar. Chemical waste management and disposal. *Journal of Chemical Education, 65*(3): 64–68.

Armour, M. A., Browne, L. M., and Weir, G. L. 1985, March. Tested disposal methods for chemical wastes from academic laboratories. *Journal of Chemical Education, 62*(3): A93–A95.

Bealer, J. M. 1985, Dec. Either it's safe or it's not. *The Science Teacher, 52*(9).

Belt, W., and Dunkleberger, D. E. 1985, Sept. Safety considerations for the biology teacher. *The American Biology Teacher, 47*(6): 340–346.

Bretherick, L. 1979. Safety in the chemical laboratory: Reactive chemical hazards: Their causes and prevention. *Journal of Chemical Education, 56*(2): A57.

Carpenter, S. R., Kolodny, R. A., and Harris, H. E. 1991. A novel approach to chemical safety instruction. *Journal of Chemical Education, 68*(6): 498.

Chamberlin, R. 1989. An "A" for safety. *Science Scope, 13*(3): S16.

Coble, C. R., and Hounshell, P. B. 1980. A framework for evaluating chemical hazards. *The Science Teacher, 47*(5): 36.

Committee on Chemical Safety. 1976. *Safety in Academic Chemistry Laboratories* (3rd ed.). Washington, DC: American Chemical Society.

Couillard, D. 1982. Dancing mothballs—safe partners? *The Science Teacher, 49*(5): 26.

Cronin-Jones, L. L. 1990. A lab for all reasons. *The Science Teacher, 57*(3): 36.

Crouch, R. D., Jr. 1989, Nov.-Dec. A safe chemistry laboratory. *Science Scope, 13*(3): S6.

Dombrowski, J. M., and Hagelberg, R. R. 1985. The effects of a safety unit on student safety, knowledge, and behavior. *Science Education, 69*(4): 527.

Dunkelberger, G. E., and Snyder, S. 1985. Safety in the classroom: A model for training science teachers. *Journal of Chemical Education, 62*(1): 73.

Dux, J. P., and Stalzer, R. F. 1988. *Managing safety in the chemical laboratory.* New York: Van Nostrand Reinhold.

ERIC/SMEAC Clearinghouse. 1980. *Safety in the science classroom* (Information Bulletin No. 3). Columbus, OH: Author.

Garcia, T. 1989, Nov.-Dec. Earth science safety. *Science Scope, 13*(3): S20.

Gass, K. 1990a. Chemistry: Courtroom and common sense, Part I: Negligence and duty. *Journal of Chemical Education, 67*(1): 51.

Gass, K. 1990b. Chemistry: Courtroom and common sense, Part II: Negligence and duty. *Journal of Chemical Education, 67*(2).

Gentry, L. R., and Richardson, M. D. 1991, April. Guidelines for ensuring a safe science laboratory. *NASSP Bulletin, 75*(534): 90–94.

George, B., and Perkins, R. 1987. 10 steps to a safer new year. *The Science Teacher, 54*(1): 25.

Gerlovich, J., and Downs, G. E. 1981. *Better science through safety.* Ames, IA: Iowa State University Press.

Hagelberg, R. R., and Dombrowski, J. M. 1987, April. What research says: A survey of safety in high school science laboratories in Arizona. *School Science and Mathematics, 87*(4): 328.

Hampton, E., and Thacker, D. 1989, Nov.-Dec. A gram of prevention is worth a kilo of cure. *Science Scope, 13*(3): S4.

Hanssman, C. O. 1980. Safety is everyone's responsibility in the schools. *Journal of Chemical Education, 57*(3): 203.

Harper, C., and Bartlett, B. 1989. Putting in a plug for classroom safety. *The Science Teacher, 56*(9): 39.

Joye, E. M. 1978. Law and the laboratory. *The Science Teacher, 45*(6): 23–25.

Manor, C. R. 1982. Please don't eat the daisies. *The American Biology Teacher, 44*(1): 33.

Marsick, D., and Thornaton, S. F. 1988. Science teacher safety survey. *Journal of Chemical Education, 65*(5): 448.

Mayer, U. J., and Hinton, N. K. 1990. Animals in the classroom: Considering the options. *The Science Teacher, 57*(3): 26–30.

McFall, R. H., and Wilson, R. C. 1987. The blood type lab test: Can it spread AIDS? *The American Biology Teacher, 49*(6): 370.

National Association of Biology Teachers. 1980. NABT guidelines for the use of laboratory animals at the pre-university level. *American Biology Teacher, 42*(7): 426–427.

National Council on Radiation Protection. 1966. *Radiation protection in educational institutions* (Pub. No. 32). Washington, DC: Author.

National Occupational Safety and Health Administration. 1981. *Occupational health guidelines for chemical hazards* (GPO Stock No. 017-033-0037-8). Washington, DC: U.S. Government Printing Office.

National Science Teachers Association. 1978. *Safety in the science classroom.* Washington, DC: Author.

Orlans, B. 1980. Selecting an animal for the classroom. *The Science Teacher, 47*(2): 29.

Orlans, B. 1991. Forum: Dissection. *The Science Teacher, 56*(1): d12.

Patnoe, R. L. 1976. Chemistry safety check. *The Science Teacher, 43*(7): 28.

Penn, A. 1989. Shedding light on lasers. *Science Scope, 13*(3): S18.

Pitrone, K. 1989. Safety: A learning center. *Science Scope, 13*(3): S22.

Puffer, J. H. 1979. Classroom dangers of toxic minerals. *Journal of Geological Education, 27:* 150.

Purvis, J., Leonard, R., and Boulter, W. 1986, April. Liability in the laboratory. *The Science Teacher, 53*(4): 38–41.

Rakow, S. 1989. No safety in numbers. *Science Scope, 13*(3): S5.

Richmond, G. E., Engelmann, M., and Krupka, L. R. 1990. The animal research controversy: Exploring student attitudes. *The American Biology Teacher, 52*(8): 467–471.

Rosenlund, S. J. 1987. *The chemical laboratory: Its design and operation.* Park Ridge, NJ: Noyes Publications.

Sievers, D. 1984. Unnecessary risks. *The Science Teacher, 51*(6): 32.

Stroink, G. 1980. Radiation safety in the lab. *The Physics Teacher, 18*(3): 207.

Vos, R., and Pell, S. W. 1990. Limiting lab liability. *The Science Teacher, 57*(9): 34.

Yarroch, W. L. 1980. Hazardous waste disposal for schools. *The Science Teacher, 47*(1): 27.

Young, J. A. 1987. *Improving safety in the chemical laboratory: A practical guide.* New York: John Wiley and Sons.

Study outdoors the things that are best studied outdoors; study indoors the things that are best studied indoors.

CHAPTER 10

Science Projects, Science Fairs, and Field Experiences

Science instruction should not be limited to routine activities that are conducted in the classroom. Science instruction can also take place outside of the classroom and even after school hours. Projects, science fairs, and field work have a definite place in the science curriculum, since they can provide students with experiences and understandings that they would not normally obtain through routine classroom instruction. Planned and unplanned projects, science fair programs, and certain types of field work activities will not only enrich the science background of students but at the same time reinforce the goals of science teaching.

Overview

- Discuss the role and importance of science project work in middle and high school science courses.
- Present how science fair programs encourage students to pursue activities they would not normally carry out during regular class time.
- Describe the functions and benefits of field experiences in the science curriculum.

The following scenario is not uncommon in science instruction. The approach presents a view of science that is void of enrichment activities which can make science more meaningful and interesting to students.

▼   Mrs. Hampton teaches tenth-grade biology in a small, rural high school. She planned the following instructional activities for one school week in early September.

Monday

Mrs. Hampton presented a lecture on the annelids. During her lecture, she compared and contrasted the structure and physiology of annelids with those of previously discussed platyhelminths and nematodes.

Tuesday

Mrs. Hampton completed her lecture on annelids and then gave specific instructions to prepare students for a laboratory exercise on the dissection of the earthworm to be conducted the next day.

Wednesday

Students dissected the earthworm. Mrs. Hampton handed out a prepared laboratory exercise to students that specified the directions to be followed. Students were required to label diagrams of earthworm anatomy and provide information to answer a series of questions listed on the laboratory exercise form. Some of the students expressed their dislike for having to learn about worms.

Thursday

Mrs. Hampton reviewed various points regarding earthworm anatomy that the students should have learned during the dissection. Then she conducted a short review for the test on worms to be given the next day.

Friday

Test on worms.   ▲

The instructional activities just described have their place in science teaching because they can provide factual information upon which students can build their knowledge for subsequent learning situations. Unfortunately, many science teachers consistently use this instructional ap-

proach, resulting in a distorted view of the investigative nature of science.

Science should be taught so that students can view it as a dynamic activity rather than a static, uninteresting enterprise. Science instruction should not be limited to common, routine instructional activities—lecture, cookbook laboratory exercises, and testing. It should involve inquiry activities that go beyond classroom walls to give students a broader view of science and make science more meaningful and exciting.

Mrs. Hampton's plans for the week could have included activities that present science as a dynamic enterprise that involves discovery and inquiry. The following plan for the week illustrates an alternative instructional approach that gives students some firsthand experiences.

▼   **Monday**

Mrs. Hampton presented a lecture that compared and contrasted the structure and physiology of platyhelminths and nematodes.

Tuesday

Mrs. Hampton took her class on a field trip to a nearby stream where students collected samples of nematodes and platyhelminths. In addition, students gathered annelid specimens from soil samples taken from the grassy field adjacent to the school building.

Wednesday

The students compared the behavior and external anatomy of earthworms with nematodes and platyhelminths, using the live specimens they collected the previous day. The students also dissected preserved specimens of earthworms to investigate the internal structure. Students were required to label diagrams and answer questions that compared the internal structure and physiology of earthworms with those of round and flatworms.

Thursday

Mrs. Hampton reviewed the laboratory exercise and discussed the anatomy and physiology of annelids. She conducted a short review for the next day's examination on the three main worm phyla. During the discussion of annelids, one student wondered why earthworms often come to the surface of the soil when it rains. Mrs. Hampton helped this student design a project to investigate the reactions of earthworms to varying levels of moisture and light. The student was asked to prepare a report to present to the class after the project was completed.

Friday

Test on worms.   ▲

This scenario of Mrs. Hampton's instructional approach to teaching differs from the one presented in the previous scenario in that she enriched her biology course by including a number of stimulating experiences. The field trip to the nearby stream gave students a first-hand experience in collecting specimens to study in the laboratory. This activity no doubt generated a great deal of student excitement and enthusiasm. The students studied the behavior of worm specimens that they collected during the field trip and they compared the external anatomy of live specimens with preserved specimens. The interest of one student concerning the behavior of earthworms during heavy rain resulted in a student project on the reactions of earthworms to moisture and light. This activity helped to develop the student's inquiry skills and led to a project that was entered in the school's annual science fair. This approach to teaching science takes advantage of the beneficial outcomes of field experiences, which in turn may pave the way to involve students in science fair projects.

SCIENCE PROJECTS

Science projects are larger in scope than laboratory exercises, which are usually one or two periods in length. Projects usually require many hours of student involvement, spanning several days to several months. Projects may be conducted during or outside of class time, be unplanned or planned, and involve one student or the whole class. The examples that follow illustrate the nature of science projects and how they can be initiated.

▼   Mr. Gibson teaches in a rural school district in the Midwest. He was summarizing an ecology unit by discussing air pollution. When the subject of acid content in the atmosphere was brought up, Doug indicated that he heard that the new factory built just outside town was polluting the air. He asked Mr. Gibson if it was true that the factory was responsible for acid emission in the air.   ▲

How should Mr. Gibson respond to the student's questions? Should the teacher tell the student that he will find out the answer to the question or, if he knows the answer, should he give it to the student? If Mr. Gibson selected either of these options, he might stifle inquiry into the problem. However, if he responds by saying, "How can we find out if the factory is contributing to acid emission in the air?" he can

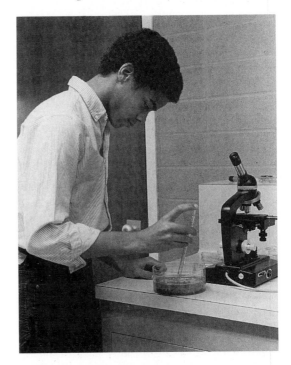

Projects can be conducted during or outside of class time, be unplanned or planned, and involve one student, or two or more students or even the whole class.

invite the student to answer his own question. This is a ready-made opportunity to allow the student to apply knowledge and skills gained in the classroom to a real-world situation. Mr. Gibson could design a class laboratory exercise in which the whole class pursues the answer to Doug's question. However, Mr. Gibson had not budgeted time for such an exercise, which would take at least a week to do properly. Also, Doug's question is quite specific and may be of interest to him alone. Surely Mr. Gibson does not want to plan his class activities around the specific interest of just one student. On the other hand, Mr. Gibson would like to provide Doug with the opportunity to investigate the answer to his question. The obvious solution to this situation is the science project. With inexpensive materials (sodium bicarbonate, methyl orange, glycerin, hydrochloric acid, paper towels, rubber band, thistle tube, bicycle pump), Doug is shown how to perform a fairly accurate atmospheric test. Mr. Gibson assists Doug in the design and use of the apparatus for this investigation at convenient times outside of the regular period (e.g., before and after class, before and after school, during free periods). Mr. Gibson does, however, plan a specific class time during which Doug can present his findings to the class.

Let us now summarize what has transpired:

1. Mr. Gibson's class is allowed to proceed uninterrupted to the next unit.
2. Mr. Gibson has been able to provide a learning experience directly related to a student's individual interest.
3. The student is given an opportunity to apply what he has learned in class to a relevant, real-world situation.
4. The student's quest for knowledge has not been cut short by classroom or time restrictions.
5. The student becomes highly motivated, and gains in the cognitive and affective domains are facilitated. Doug is pursuing a question of individual interest, and his feeling of achievement and self-worth can be enhanced by his

responsibility for imparting his findings to the class.

Science teachers like Mr. Gibson may be reluctant to interrupt their classes for unplanned learning experiences. However, these experiences are important and can be addressed through projects. It is not necessary that teachers forge through the planned curriculum and that they pass up ideal opportunities in class to involve students in inquiry activities. Science teachers should constantly take advantage of unplanned learning experiences. Very often, "side trips" or tangential inquiries produce the greatest cognitive gains because they tend to be more relevant to students' daily lives. This is why science projects are so valuable; they allow for pursuit of those highly motivating areas of interest but do not reduce the teacher's highly valuable class time.

In the previous example, the science project was assigned to an individual student. However, keep in mind that such projects can be performed by groups of students as well. Indeed, a group project may be preferred because it adds to the comprehensiveness and diversity of an investigation through assignment of tasks, cooperation, and interaction among individuals.

Planned Science Projects

There are alternative ways to reap the benefits of science projects than those just described. The science project is a valuable teaching tool, and the teacher should not limit or discard its use in response to the problems that may arise from unplanned assignments. Very often, students simply need more structure and direction. In short, the curriculum can be structured so that students' participation in a science project is ensured because they do not have the option of simply refraining from using this approach. Let us return to Mr. Gibson's ecology unit to illustrate this alternative classroom approach.

At the beginning of the unit, or even halfway into it, Mr. Gibson divides the class into groups. Each group is required to plan and complete an

Students should be given opportunities to work on projects of interest to them. Science projects that are conducted by students in the school science laboratory permit teachers to answer questions and provide guidance when needed.

ecology-related project. Mr. Gibson may do any of the following:

1. Assign a specific problem to each group.
2. Allow the groups to pick a problem from a list of suggested problems.
3. Allow the groups to develop their own problem to investigate.

Through one of these approaches, all students will gain exposure to a science project and its inherent educational rewards. How much flexibility the teacher allows with regard to the project topic depends on the teacher's preference and the particular class. For example, certain classes might be quite capable of developing relevant areas of inquiry. In these situations, the teacher would do well to allow groups of students to develop their own projects. This free-

dom would further ensure that the students would investigate a topic of interest. Other classes may need more direction and assigned problems to investigate; this may be the case in the middle and junior high school.

The science projects that follow can be performed entirely during school time in the science laboratory.

▼ Mr. Zappa has just completed a health unit about the use and abuse of barbiturates, amphetamines, marijuana, and alcohol. During Mr. Zappa's summary of the unit content, Mike asks about the truth of a newspaper article concerning individuals' addiction to "everyday" drugs such as caffeine, aspirin, and nicotine. Mr. Zappa could comment on Mike's query, but he decides that this question would be an appropriate topic for a science project. Furthermore, he feels that such a project would further enhance Mike's understanding of foreign substances and their effects on organisms. Mr. Zappa proceeds to help Mike organize the project. With little effort, Mr. Zappa helps Mike design a project that will evaluate the relative effects of caffeine, aspirin, nicotine, and alcohol on the heart rate of daphnia. This project is conducted during the regular laboratory period.   ▲

▼ Mrs. Volman teaches an advanced biology class on human anatomy and physiology. She likes to give her students various research projects to perform when they discuss the physiology of the digestive and urinary systems. She feels such assignments facilitate the students' understanding of the interrelationship of the two systems. Mrs. Volman divides her class into small groups of four to five students each. She gives each group a particular diet to which they are to adhere for a period of one week. Each day all the students of each group are to perform a urinalysis in the morning and afternoon. The results are then recorded and analyzed. Each group is then given time to present their findings and conclusions to the rest of the class.   ▲

Mr. Zappa's and Mrs. Volman's classes are illustrative examples of unplanned and planned in-school projects, respectively.

Many teachers prefer in-school science projects. Although such activities require more of the teacher's time and effort, the increased teacher supervision helps minimize "wasted" and undirected student behavior. The in-school project gives the teacher more supervisory flexibility than do those that are conducted outside of school. The teacher is able to decide freely on the level of independence to give to students; in the out-of-school projects the students are completely autonomous during the major portion of the project. Experienced teachers fully recognize that students vary in their ability and the amount of autonomy they can efficiently handle. Thus, the in-school science project is the most convenient way to allow students' pursuit of individual interests without sacrificing the teacher's opportunity to monitor students' work.

The student or group of students who pursue in-school science projects should be given the opportunity to present the results to the whole class. The motivational aspects of such an opportunity have already been discussed. In addition, criteria for the grading of such projects should be clearly stated. The classroom presentation, as well as a final written product, should be included in the total evaluation.

Class Projects

Class projects that employ large groups of students such as those described below are not common in science instruction. This type of cooperative learning requires that all students in the class work together toward a common objective. The participation of all students in a class toward a common goal may prescribe that students work in pairs, work in groups of three or more students, groups composed of boys and girls or solely of boys or girls. There may also be assignments where a student may work alone for a period of time. Grouping during project work, when properly conducted, can produce a number of beneficial outcomes, such as the following:

1. Permits close matching of assignments to abilities of students

2. Takes advantage of adolescents' and pre-adolescents' desire to work together
3. Encourages students to explore their leadership potential and qualities
4. Helps teachers become better acquainted with student personalities, strengths, and abilities
5. Provides tasks for shy and retiring students
6. Provides for a wide range of interests and talents
7. Gives direct experience with materials
8. Provides tasks to take care of special interests and skills of male or female students
9. Allows students to work cooperatively on a single task or on several tasks
10. Encourages verbal communication among students within a group or team
11. Exposes students to the democratic process
12. Encourages all students to actively participate in scientific inquiry

The next two examples show how students can work cooperatively on a class project for an extended period of time. (The teacher has the option of assigning the students to work in pairs, groups of three or four or more, all boys, all girls, or a mix.) In the first example, the science teacher suggested a class project in which the students might have an interest because they were studying a unit on the weather and the atmosphere. The teacher had given considerable thought to the project before suggesting it. It was not a spontaneous decision. The teacher was prepared to offer suggestions during the planning session, but because the students were interested in the topic, they came up with the idea of using an outside expert to help them design a weather station. The students also took the initiative to contact the weather forecaster and invited her to give them guidance over an extended period of time. During the project they also assumed many responsibilities that were assigned by the student project director or mandated by the group assignments.

▼   Mr. Geral teaches physical science to a group of ninth graders in a small suburban high school.

During the study of a unit on weather and the atmosphere Mr. Geral suggested that the students set up a weather station on the school grounds as a class project. The students became excited about the project and agreed that it would be a good learning experience for the whole class. To start the project, one student indicated that they should invite a local weather forecaster to help them plan the weather station. Another student said that the weather forecaster seen on local television often announced her willingness to visit schools to instruct classes on the weather. The students thought that it would be an excellent idea to invite the weather forecaster to their class to discuss the project. Several students volunteered to make an appointment with her at the TV station in order to explain the project and invite her to visit their class. Ms. Mahar, the weather forecaster, accepted the invitation and visited the class. She indicated that she would help them with the project provided that all the students in the class participated in the activities and that they took their assignments seriously.

After an introduction to weather observing and forecasting, Ms. Mahar told the students that some weather equipment was needed to establish the weather station and that some pieces could be constructed using cheap and simple materials. She recommended that for a class project it would be a good idea for them to make some of the equipment rather than to purchase it from a supply house. Ms. Mahar said that she would break the class up into small groups and give each group instructions for making one piece of equipment. She provided the instructions for making a rain gauge, a hygrometer, a psychrometer, and a wind velocity instrument. She said that other items needed, such as barometers and thermometers, were probably available in the science equipment inventory, but if they were not, she could provide them. She also said that she would provide the class with other specialized equipment needed for the station for an extended period of time. Ms. Mahar also gave a group of students the specifications for building a weather shelter in a specified area on the school grounds. With the help of the industrial arts teacher, they built the shelter and placed it in an area designated by Ms. Mahar.

After two weeks the weather station was set up and operating. Two students were assigned each day to collect and record data. They recorded the temperature and dew point, relative humidity, barometer readings, type of clouds, height of clouds, rain gauge measurements, wind direction and velocity, and other pertinent data.

The data were given to Ms. Mahar during her weekly visits. She showed them how the data were used in producing a weather map for forecasting the weather. She visited the class for four consecutive weeks and exposed the students to some of the principles of weather observing and weather forecasting. The students prepared their own weather maps from data Ms. Mahar collected from other weather stations all over the country. With Ms. Mahar's help, the students attempted to forecast the weather using the weather maps they had plotted. The students collected and recorded the weather data on a daily basis for the remainder of the term and found that their data compared well with data collected by professional weather observers.   ▲

The second example illustrates a situation in which the students came up with the idea of a class project on contour maps. The science teacher had not suggested the activity but was alert enough to realize the importance of capitalizing on the interests and motivation of his students. The students were academically above average and in the past had demonstrated that they were capable of working cooperatively on assigned tasks and handling problems in depth. The students carried out the project with some direction from the science teacher and a civil engineer, but many of the tasks were designed and assigned by the students themselves.

▼   Mr. Elm is an earth science teacher in a rural central high school. He allows students with good academic records to substitute earth science for the usual general science course available to ninth-grade students.

Early in the course, Mr. Elm provided the class with contour maps of various areas in the United States and spent a great amount of time teaching the students how to read and use them. When the class became proficient in reading and using the maps, one student suggested that the class actually make a contour map and a model of the area surrounding the community as a class project. Mr. Elm thought the project was feasible and asked the class to plan the strategy to carry out the project. The students elected a very capable student as project director. As project director, she appointed a committee of students to discuss the project and propose a plan of attack. Since a contour map of the local area was not available, initial discussion centered around the need for a careful survey of the area in order to produce the map. One committee member indicated that they needed the advice of a civil engineer in order to proceed with the project. Another student on the committee said that his father was a civil engineer who might be willing to visit the class to explain the nature of a survey. The engineer accepted the invitation and then proceeded to conduct a demonstration on the school grounds to show how he generated the data to make a survey using special instruments he brought to class. He allowed a number of students to use the instruments during the demonstration and they became very excited when they learned that the readings they took were accurate as confirmed by the engineer. At the end of the demonstration they asked the engineer if he would be willing to help them throughout the project. He accepted, and since the students exhibited such enthusiasm and a sense of responsibility, he loaned them a set of instruments to use during the project.

The survey required several class periods and many of the students also provided additional time after school to collect data. After the survey was completed, the students made a contour map of the area with the help of the engineer and the science teacher. The map was then used to construct a model of the surrounding area.

The students had to determine the material that could be used to make a model. For this phase of the project, they sought advice from the art teacher. She suggested papier mâché and plaster of paris since the materials were readily available and easy to work with. The girls in the class were very interested in this phase and worked diligently experimenting with the media.

After they determined procedures for building up the contours, they started to make the model. Each class member was busy performing one or more tasks. Some made molds, others mixed plaster, while others prepared trees from rubber

sponges. They constructed small houses from balsa wood and other materials.

As work progressed, the accuracy of the model was constantly checked against the data they had collected. The model measured about 6 by 9 ft and was made in three sections so that it could be moved easily.

The model was well executed and attractive with its miniature trees, woods, buildings, streams, and bridges. It was also very informative because of its wealth of detail.

Many of the students in the school were aware of the model that had been produced and asked if they could see it when they had a free period. The demand was so great that the principal decided to place it in the school lobby for about one week so it could be viewed by all interested students in the school. The project received much publicity in the local newspaper when the students were invited to place the model in a public area in the city hall. Several students who participated in the project were selected by the class to deliver a series of talks about the project before town officials and other community leaders. ▲

In summary, a class project should be considered in science instruction since it may add to the comprehensiveness and diversity of an investigation through student cooperation and interaction over an extended period of time.

SCIENCE FAIRS

During the spring semester of each school year thousands of science fairs are conducted in the elementary, middle, and secondary schools throughout the nation. These are big events for science teachers, students, parents, and other members of the community. Science fairs stimulate enormous interest in science. They provide students with the incentive to study problems in depth and to communicate their findings. Science fair programs give students a chance to pursue investigations that they would not ordinarily be able to carry out during regular science class periods because of limitations on equipment, space, and time. In addition to identifying

Science fairs stimulate enormous interest in science. They provide students with the incentive to study problems in depth and to communicate their findings.

the gifted science students, these events encourage all students to get involved in "sciencing." Science fairs not only display the talents and interests of students but also reveal the orientation of a school's science program and the type of science teaching that is taking place in the classrooms.

How does a science fair reflect a school's science program and what the teachers actually emphasize in their teaching? This question can be answered by categorizing the type of projects that students enter in the science fair. One scientist (Smith, 1980, pp. 22–23) who has been a judge at many fairs has observed that science fair projects can be placed in the following categories:

1. Model building (e.g., the solar system, volcanoes, clay models, and frog organs)

2. Hobby or pet show-and-tell (e.g., arrowheads, slot cars, dogs, and baby chicks)
3. Laboratory demonstrations right out of the textbook or laboratory manual (e.g., distillation, electrolysis, and seed germination)
4. Report-and-poster projects from literature research (e.g., fossils, birds, bees, astronauts, and the car)
5. Investigative projects that involve the students in critical thinking and science processes, such as measuring, consolidating data, and drawing conclusions (e.g., tests of reaction time, data on the effectiveness of various detergents, and comparisons of performances of vacuum bottles and insulated jugs)

With such variety in project types, judges are often bewildered as to how to compare and evaluate them. How can one objectively compare a model with an investigative project when each has different amounts of student involvement? Each of these categories is so different that it presents a serious problem for those who judge science fairs. It is difficult for individuals to judge students fairly when they are not instructed to evaluate projects on a comparative basis by category. Projects in a science fair are often evaluated haphazardly when judges compare investigative projects with models, models with charts, laboratory demonstrations with investigative projects, and so on. A fair should be organized in a way that permits judges to make awards within particular categories. However, this does not solve one of the major problems concerning science fairs: that is, what types of projects should be included in a science fair to make it an outstanding event in the community and one that emphasizes the scientific enterprise?

The emphasis of science teaching should be one of inquiry and investigation. This suggests that fair entries should be the types of projects mentioned in category 5 of Smith's list of science fair categories. Investigative projects require students to use many of the other activities (categories 1 to 4), but these projects require much more. They involve students in planning and thinking scientifically and not stopping short by reproducing a model from a textbook or simply copying information from an encyclopedia. Investigative projects require students to ask questions, plan procedures, collect data, and make conclusions based on information. These types of activities require students to think critically and to solve problems and minimize the conveying of science as a fixed body of knowledge, which should be deemphasized as a primary goal of science teaching.

It is recommended that teachers discourage students from preparing projects in the first four categories (models, posters, show-and-tell, and laboratory demonstrations) for a science fair program. Although they may stimulate student interest in science, these projects make science fairs uninteresting and monotonous and present a false image of what science really involves.

Students who enter projects in the first four categories described in the list are often placed at a disadvantage when judges attempt to evaluate them. In many instances, judges who have good credentials—engineers, scientists from universities and industries, and science teachers—will immediately discount the worth of these projects and spend little time evaluating them or talking to the students who prepared them.

Projects that students enter into a fair need not involve elaborate equipment or complex problems to be science fair winners. At first glance, projects involving expensive and elaborate equipment such as computers are impressive. The terminology students use to present demonstrations involving the computer is impressive. Nevertheless, the problems the students investigate and the conclusions they present often lack clarity and depth because the students become enamored and preoccupied with a complex piece of equipment and are less concerned with the investigation. They often become confused and muddle the investigation. This also happens when students are involved

with sophisticated equipment such as oscilloscopes, scintillators, and amplifiers. Investigations that ask simple questions and use simple and clearly stated procedures often make the best fair projects.

Judging Projects

The procedures and criteria for judging science fairs vary, but there are common elements that can be considered. Generally, a common set of categories is used to judge science fair projects. The following categories and their respective point breakdowns are often used:

1. Creativity (20)
2. Investigative procedures (30)
3. Understanding of the topic (20)
4. Quality of the display (15)
5. Oral presentation (15)

The first category, *creativity,* asks the question, How unique is the project? Is the project significant and somewhat unusual for the age of the student? To answer these questions, a judge must take into consideration the student's grade level and science background, as well as the type of science course the student is currently taking. A judge must also determine how the student arrived at the project—was it conceived by the student or suggested by a parent, a teacher, another adult, or another student? The judge will take the responses to these questions into consideration when assessing the creativity of the project.

The *investigative procedure* is the second category that is used when judging a project. The aspects of an investigation that a judge considers include the conciseness of the problem statement, the appropriateness and thoroughness of the procedures, the completeness of the information collected, and the accuracy of the conclusions. Often, students go through an elaborate set of procedures during their investigations and then either fail to answer the original question or provide an answer that is not substantiated by the data presented.

Understanding of the topic is another category that should be considered when judging a project. Did the student bother to learn about the topic? Did he go to the literature to investigate it? Can the student answer questions about the topic? Did he provide a list of references and bibliography used in connection with the investigation?

Quality of the display is a category that is generally used in judging projects. How well does the student present the project? Are the problem, procedures, data, and conclusions presented clearly? Is the display organized and understandable? Is it visually appealing? A judge must also determine how much help a student received in presenting the project. Students generally get help from others in making their displays, but they should acknowledge the work of others in writing and give credit to those who helped them. Teachers should make certain that students are given awards for what they have done rather than for what someone else has done for them.

Another category judges use to evaluate projects is the *clarity of the presentation,* in which they determine how well students can orally present their investigation. Can the student clearly communicate the nature of the problem, how he solved the problem, and how he arrived at the conclusion? Judges should give students the opportunity to answer questions and articulate what they have accomplished.

To direct the judges' attention to the various points discussed in these categories, an evaluation form that contains the following list of elements may be useful during the judging procedure.

- The problem was stated clearly
- Background reading was appropriate
- The hypothesis was clearly stated and reflected the background readings
- The experimental design demonstrated understanding of the scientific method
- Apparatus and equipment were appropriately designed and used
- Observations were clearly summarized

- Interpretations of data conformed with the observations
- Tables, graphs, and illustrations were used effectively in interpreting data
- Conclusions and summary remarks were justified on the basis of experimental data
- The log book was used to record experimental data, ideas, interpretations, and conclusions
- The bibliography contained a significant number of relevant and timely references
- Limits of accuracy of measurements were stated
- Work on the project suggested new problems
- Oral presentation was made in the time allocated with all phases of the project discussed
- The research covered all questions effectively and accurately
- The oral presentation made good use of visual aids
- The display board was effective in presenting the project (Goodman, 1985)

The elements in the preceding list can be rated by judges using a suitable scale, for example:

0–cannot make judgment

1–poor

2–fair

3–satisfactory

4–good

5–excellent

Goodman (1985) suggests a rating scale with a maximum of 100 points and that the awards be given based on the following scores:

60–69 Honorable mention

70–79 Third prize

80–89 Second prize

90–100 First prize

Select several judges to evaluate a given number of projects. Have the judges interview and assess the projects individually, rather than as a group, to prevent them from influencing each other. Ask the judges to provide supportive comments to students on the projects and also to recommend to students what they might do to improve their investigations. Students can use this information to better understand what they are attempting to accomplish through their procedures and their investigations.

Science fair judges should have good credentials. Selection of judges should be based on their scientific proficiency, their understanding of the scientific enterprise, and their familiarity with the cognitive abilities of the students being judged. Engineers, scientists from universities and industry, and qualified science teachers make excellent judges. However, the criteria by which projects are to be judged should be made clear to the judges before they begin to make their evaluations. It is good practice to provide them with a list of the criteria they are to use as well as some examples of questions they might ask to make judgments regarding each criterion. Encourage judges to talk to students and write comments and suggestions about each project on an evaluation form. Allow students to review the comments and suggestions so that they can further profit from the experience.

Taking Care of Inequities When Judging

In order to take care of inequities when judging science fairs, Levin and Levin (1991) suggest that entries be divided into categories or subject areas such as astronomy, bacteriology, ecology, physics, and so forth. This permits the number of projects to be judged per subject area to be small and the number of judges assigned to each area will also be small. Small groups of judges are assigned to each category, headed by one person who is so designated by the fair officials. At least two judges are assigned to interview each contestant in the category. The judges use a checklist provided by the fair officials to evaluate the projects that they are assigned to judge. The checklist is a guide that contains the criteria that should be used to evaluate the entries. Numerical scores are sometimes used in the checklist to help judges rank and compare projects that they judge. When all the projects in a category are evaluated, the judges in the area meet

to discuss their findings. Their rankings are presented to the group's head judge, indicating the projects which have been ranked the highest. The highest ranking projects are then visited by those judges who had not evaluated them in the first place. If time does not allow these judges to have a full judging session with the students involved in the projects, they must at least have a short discussion with them in order to become familiar with the projects that have been ranked the highest. The group of judges then discusses all of the entries that they have ranked the highest in order to share any details that may be of interest to the whole group. During this session the judges reach a consensus and present a list of winners, including honorable mentions, to the head judge of the category.

According to Levin and Levin (1991), the consensus process eliminates a number of problems that often arise when judges merely average numerical scores to make decisions. The consensus method allows the judges to become immersed in the judging process and thus allows them to judge each project carefully. Group discussions permit the judges to compare the strengths and weaknesses of each project. In this way, judges with little or no experience will realize their shortcomings when they overestimate or underestimate the worth of a project. This procedure and exposure will also permit inexperienced judges to be more competent when they judge projects in future fairs.

According to Levin and Levin (1991), the consensus method produces results that are more reliable and fairer as indicated by individuals who have had experience in judging other fairs using different systems. The procedures involved in other systems make it difficult to compare projects that are very similar or identical. The consensus method takes care of this problem.

The Competitive Aspect

A word of caution should be extended here to both experienced and inexperienced teachers. There is often too much emphasis on the competitive aspect of science fairs. Often, some of the less able students have done their best work and have invested a great amount of effort in their projects. These students take real pride in their projects and receive many benefits through their involvement. The science fair should offer less able students the chance to obtain recognition for what they accomplish outside of the regular science class—recognition that they cannot receive through normal class activity.

In some instances the competition gets so fierce that parents get overly involved in the science fair projects of the students. In such cases the projects actually reflect the thinking of the parents rather than the students. Children whose parents are affluent or professionals in medicine, engineering, and scientific fields have a distinct advantage and usually win science fair competitions. This is certainly unfair to less affluent students. Policies should be stated and enforced so that the students plan their own projects and carry out their own investigations with little or no outside help.

Fierce competition can be reduced when fairs are restricted to classrooms and schools as opposed to district-wide competitions. Grades or awards can be granted to a student based on a contract that the student has made with the teacher. This approach reduces competition among students and makes the science fair a more academically oriented activity.

Committees

The success of a science fair is determined by the types and strengths of the committees that are involved in its organization. The planning of a fair involves many facets—judging, publicity, advertising, awards, soliciting funds, program, and so on. The success of a fair is also contingent on obtaining responsible individuals to head as well as participate in the committee work. Selected parents, teachers, students, and members of the community can be placed on committees, depending on their interests and qualifications. Table 10–1 is a sample timetable for organizing a science fair. The following are some suggested committees:

1. *Planning.* Concerned with general organization of the fair; coordinates the work of other committees.
2. *Judging.* Responsible for selecting judges and orienting them regarding the evaluation of entries. A running list of judges should be kept, and evaluation of the individuals should be available for future use. Letters of appreciation for their help should be sent after the fair is over.

TABLE 10–1   Sample timetable for organizing a science fair

Month	Activity
September	Set a date. Identify a place. Determine a budget. Place on school calendar. Inform administrators, teachers, and students when it will take place.
October	Organize fair into divisions and categories. Prepare procedures for entry, guidelines, and rules. Identify criteria for judging. Communicate this information to teachers and students.
November	Verify the reservation for the facilities. Order medals, ribbons, and trophies. Design certificates of participation.
December	Identify judges and prepare letters of request.
January	Solicit judges. Distribute entry blanks. Contact news media.
February	Receive entry forms. Map out the facility on paper noting the following: space, table, and electrical needs. Assign numbers to projects and assign them a space and location. Construct a map of the facilities, spaces, and locations. Arrange for tables, electricity, etc. Notify students of their project number and location. Confirm judges' participation. Get the awards and certificates.
March	Conduct fair. Provide refreshments. Present awards. Send thank-you notes to judges and others who helped organize the fair and activities during the fair.

3. *Publicity.* Notifies news media regarding dates and activities of the fair. Prepares news releases.

4. *Awards.* Responsible for designating and ordering types of awards by category. Identifies appropriate individuals to make award presentations.

5. *Printed matter.* Prepares printed matter for the science fair including programs, floor plans, checklists for judges, and rules and regulations regarding students' conduct during the fair.

6. *Financing.* Solicits funds, contacting individuals and industry for contributions. Orients possible contributors by giving slide presentations and/or talks indicating categories of possible costs. Prepares a list of contributors for the publicity committee so they can be recognized in the news media.

7. *Set-up and take-down.* Prepares layout for the fair, obtains furniture, and arranges electrical wiring. Monitors safety during the fair.

8. *Refreshments.* Organizes groups to provide refreshments while judging is taking place. Organizes luncheons for judges and participants.

9. *Special programs for participants.* Plans activities for student participants while judges are conducting their evaluations. Activities include special tours, films, games, music, and athletic events.

10. *Applications and entry.* Sends out applications, receives entries, and contacts all possible participants about irregularities concerning their applications. Assigns numbers to projects. Checks projects in and out. Takes care of all correspondence and paperwork regarding entry into the fair.

THE SCIENCE OLYMPIAD

In recent years, the Science Olympiad has become popular in many areas of the United States. The Science Olympiad is a nonprofit organization that is dedicated to "improving the quality of science education, increasing student interest in science, and providing recognition for outstanding achievement in science education by both students and teachers" (Science Olympiad, 1992a, 1992b). Science Olympiads are organized around interscholastic academic competitions called tournaments that consist of a rigorous series of individual and team events for which students prepare during the academic year. The preparation for the National Olympiad is accomplished through classroom activities, research, training workshops, and regional and state tournaments having the olympiad format. Television game shows, olympic games, and popular board games are used as the format for competitions. The Olympiad programs also include open house activities such as demonstrations in science and mathematics and career counseling sessions that are provided by professors and scientists at the host institution (Science Olympiad, 1992a, 1992b).

State, regional, and national olympiads are organized into two divisions: Division B (grades 6–9) and Division C (grades 9–12). There are no state or national tournaments for the elementary division, Division A (grades 3–6), but schools have organized local, school district, or regional competitions for this division. Some schools conduct their own mini-olympiads for Divisions B and C, during which tournaments are held to select students to compete in regional and state tournaments. Students in the entire school are invited to compete for a place on the team going to the regional or state competitions.

National, state, and regional competitions involve approximately thirty-two events in various science areas—biology, chemistry, earth science, physics, and computers and technology. Each year the Science Olympiad Executive Board selects the events to be scheduled for the tournaments. These events are evenly distributed among three broad goal areas of science education: science concepts and knowledge, science processes and thinking skills, and science applications and technology. (See table 10–2.) The events are designed to use a variety of intellec-

TABLE 10–2 Titles of selected events included in Division B (Grades 6–9) and Division C (Grades 9–12), Science Olympiad 1992.

Science concepts and knowledge
Rocks and fossils
Anatomy
Balancing equations
Chemistry lab
Designer genes

Science processes and thinking skills
Metric estimation
Water quality
Measurements
Qualitative analysis
Map reading
Physics lab

Science applications and technology
Heat transfer
Computer programming
Aerodynamics
Egg drop contest
Meteorology
Bridge building

tual and practical skills. Some events require the recall of facts, others require concept development, a process skill, or the application of a concept. Some require the use of a specific skill, and others require a student to design and build a piece of apparatus (Science Olympiad, 1992a and 1992b).

Awards for first, second, and third place for each event at state, regional, and national competitions consist of olympic-style medals. Championship trophies are awarded to school teams in Division B and C that have the greatest number of total points during the national Olympiad. Scholarships are also awarded for selected gold medal winners at the national tournament in Division B and C.

State and regional tournaments are held mainly at colleges and universities throughout the country. The funding for the competitions has come from many sources. National tournaments have been subsidized by Army Research Office, U.S. Army R.O.T.C., IBM, Ford Motor Co., Dow Chemical, Du Pont, and other organiza-

tions. For further information concerning the Science Olympiad write to: Science Olympiad, 5955 Little Pine Lane, Rochester, Michigan 48306.

FIELD TRIPS AND FIELD EXPERIENCES

I remember vividly the first high school field trip that I conducted. . . . How well I recall surrendering to my biology students' insistence that "the weather is so nice, we should go on a field trip." I certainly agreed with their opinion of the weather for it was ideal and our biology program had been indoors for eight months. My hesitancy to enter the outdoor classroom with my students stemmed from my lack of acquaintance with local biota and my inexperience in designing and conducting meaningful field trip experiences for high school students. However, I have found over the years, that my preparation in this important aspect of teaching was no more inadequate than that of most natural science teachers who have recently graduated from our colleges and universities. This lack of preparation keeps many teachers of natural science indoors with their classes, aliens to the real world of natural science in the outdoors. (Keown, 1984, p. 43)

Field trips require a great deal of preparation and planning to be effective. Consequently, many teachers avoid using this strategy because they lack the know-how for conducting successful field trips. Science teachers should learn how to conduct effective field trips so as not to deprive their students of the benefits of such experiences. Teachers who are hesitant about conducting field trips should consult with teachers who have had successful results with the approach. Keown (1984) indicates that it is important for science teachers to share their knowledge about successful field experiences, especially the "how to" and the "where to."

Because field trips and field experiences are generally much more closely related to the experiences of teenagers than most science classroom activities, they tend to be much more meaningful. They usually take place in areas that are familiar to the students, such as hospitals, fields, streams, and parks. Instructional activities in these settings make students aware of organisms, phenomena, and activities that they did not realize existed and link science with the "real world." For example, a trip to a nearby field with a teacher experienced in botany and field plants can awaken students to a whole new world of flowers that they never noticed before, resulting in a new interest on the part of many students to undertake excursions on their own to locate plants and wildflowers.

The natural setting is appropriate for reasons other than stimulating student interest. Most living things react differently when removed from their natural habitat. Native plants are adversely affected by artificial light and indoor humidity. Animals are usually deprived of the proper diet, exercise, and seclusion that they require.

Field studies permit the firsthand study of many things, both natural and man-made, that cannot be brought into the classroom because of size or inconvenience. It is only in the field that students can study trees, power shovels, or rock formations. It is only on field trips that students can observe the behavior of birds in their natural habitat. And it is only outside of the classroom that they can investigate the operation of a factory assembly line.

Using familiar settings outside of the school allows students to engage in activities that are too noisy or violent to be used in the classroom. A soda acid fire extinguisher must be operated outdoors. Running a model airplane gasoline engine creates a loud sound that would be very disturbing to many classrooms if operated indoors.

Short Field Experiences

Some field experiences need only a few minutes for completion and can be accomplished within the limits of a class period. They can be conducted on the school grounds or in the areas adjacent to the school grounds. Let us now consider a few situations in which a field study is appropriate and preferable to the traditional in-class laboratory.

▼ Returning to Mr. Gibson's ecology unit, we find that he is now discussing the various organisms that inhabit the different levels of soil in a typical grassy field. Mr. Gibson shows his class a film on the topic and then further reinforces the content of the film by using a cross-sectional chart of soil that indicates the various organisms and their habitats. Mr. Gibson realizes the importance of multiple presentations of important content, and thus he has his students view slides and specimens of the various organisms that inhabit the soil. ▲

The preceding lesson is typical of most biology classes. However, there is an alternative approach that may be more motivating to students. Why not study soil in its natural environment? There is nothing wrong with films, models, and slides, but why not use these materials to reinforce and supplement what the students experience in the real world? Time is not an issue; the students could quickly collect a 1-by-1 foot soil sample within a class period. Transportation is not a problem; most schools have large grassy areas within short walking distance if not actually on the school grounds. Furthermore, the expense of specimens and supplies purchased through biological supply companies can be avoided or reduced when a class gathers its own samples. Finally, students gain expertise in soil sampling and other data-gathering techniques when they perform these activities themselves. If Mr. Gibson were to use this type of field study, his students would benefit more from actually working with the soil and organisms. The objects are concrete and more relevant than pictures and models. Furthermore, if real soil samples are used, the class could perform further analyses of the samples, such as determining population densities.

▼ Mr. Gibson is finishing up a unit on genetics and is explaining the applications of genetic principles to agriculture (e.g., corn hybridization). He has just purchased some charts illustrating corn hybridization. The charts illustrate the various reproductive organs of the corn plant in great detail. In addition, the life cycle of a corn plant is presented from seed to sexual maturity. Mr. Gibson uses these charts to conclude the instruction on the unit about genetics. ▲

Here the opportunity for a field experience has been ignored. Note that Mr. Gibson teaches in a rural midwestern school district. The farms of the surrounding area have many corn fields, and there are hybrid seed corn farms as well. Many fields are within walking distance of the school. Certainly Mr. Gibson could have taken his class on a short trip to such a field. Even the most insensitive observer would quickly note the unusual format of a hybrid seed corn field. Such an unusual arrangement would evoke numerous inquiries from the students, such as: Why are there alternating groups of short and tall stalks? Why is there a gap every four rows? Why are the stalks along the field edges all the same size? Why are only stalks with silk left standing? The answers to these questions and other factors would give Mr. Gibson's class a concrete illustration of how such a field applies the genetic principles that they have studied in class. The advantages of using materials in the environment, as opposed to classroom models, charts, and diagrams, are obvious. Consider, for example, the following: Will students learn more by inspecting actual tassels and silks or by studying diagrams? Will the format of the seed corn field illustrate forced cross-pollination more efficiently than a theoretical explanation? Will a theoretical explanation of the rationale for planting rows of corn at systematic time intervals serve as well as simply observing a detasseling machine in operation? Such questions seem foolish because no doubt exists as to the pedagogical superiority of the use of real objects as opposed to mere representations.

Planning Field Experiences

One must plan ahead in order to be successful with field trips and experiences. First, determine how and when these activities will fit into the curriculum. Second, survey the resources that are available for field experiences and, finally, carefully review the policy on field work.

Curriculum Adaptation. Conventional planning procedures usually begin with the identification of instructional objectives, followed by selection of the most suitable activities for attaining these objectives. With conventional planning, however, many opportunities for field work are neglected because their contributions seem insignificant. One teacher, Miss Rogers, describes an urban field study as follows:

▼ Many students have never seen a clover plant growing in a natural environment, although the plants are commonly found in vacant lots and fields. I take my students to a nearby lot and direct them to measure off one square foot and then count the number of plants in that area to determine how many four-leafed clovers they can find. They are also asked to compare the sizes and shapes of the leaves and to dig up several plants to see the root nodules. ▲

If Miss Rogers were teaching a conventional unit on reproduction, she might feel little enthusiasm for walking her students down three flights of stairs merely to see clover plants spreading by means of trailing stems. She might prefer to dig up clover plants herself and take them into the classroom to explain how the plants reproduce. However, Miss Rogers can get more instructional mileage by taking her students on a field strip to study clover plants in their natural environment. Here she can bring in many aspects of biology other than reproduction, such as nutrition and symbiosis, which will be taken up in the course at a later time. Thus, the field work will offer experiences that can be used to introduce new topics, providing advance organizers for future instruction.

Surveying Resources. Study indoors the things that are best studied indoors; study outdoors the things that are best studied outdoors.

This is a rule that one might use to govern field work.

Within and around each school, whether urban or rural, are hundreds of things worthy of study—resources far more valuable than those available in the most expensively equipped laboratories. Common things are often best to study; one does not need a volcano, a blast furnace, or a botanical garden for effective field work.

To make a survey of resources, it is best to begin within the school building and on the school grounds because there are comparatively few problems when trips are taken within school boundaries. One may then explore the immediate neighborhood. Table 10–3 lists some possible field study situations.

Administrative Policy.   Field experiences within the confines of the school property present few administrative problems. Most school systems permit teachers to take their students anywhere within these limits without special permission. Sometimes, however, a principal may want a written notification of intent to leave the classroom.

Excursions off school property, on the other hand, involve the problem of liability for accidents. Policies governing such trips have usually been established by the school officials. Most principals justifiably insist on a written notice, which includes the names of the students and the destination of the trip. Some systems require written permission from parents before students can be taken from the school grounds. This last requirement is a serious handicap in the case of short trips, and teachers should try to obtain blanket permission from parents for trips within the immediate vicinity of the school.

Conducting Field Experiences

An observer of a well-planned field study would be surprised at some teachers' apparent calmness during this activity. The observer might wonder how the teacher can be so composed with students moving about in all directions. If questioned about this, the teacher may respond with an oversimplified answer such as "I expect them to behave properly and they do." In reality, much practice and preparation precede the successful field experience.

It is true that the teacher expects students to control themselves during field work, but students are given a great deal of practice in self-control before such work. In addition, the planning before field work must be extensive to ensure that each student is sufficiently occupied and has little opportunity to get into mischief. The following example of a field study should serve to illustrate the logistics of such an endeavor:

▼   Miss Hunter took her seventh-grade class to study the plants and animals that live in and around a small pond a short distance from the school grounds. She made preparations well in advance. She organized the students into several groups: one group looked for submerged aquatic plants; another looked for aquatic plants on the surface of the water; a third group looked for aquatic birds; a fourth group looked for amphibians and reptiles; a fifth group looked for aquatic insects; and a sixth group looked for aquatic mammals or signs of their presence.

Within each group there was a division of responsibility. One student carried the special equipment, such as dip nets and field glasses; another student carried containers for specimens that might be collected; a third carried a field guide. One student took notes and another drew sketches for the display on aquatic life that had been planned.

Before making this trip, the students examined maps of the area to be studied, and they established limits within which they were to stay during the trip. Furthermore, all procedures to be followed were outlined before the field experience.

▲

In short, all conceivable situations and problems should be taken into account before any field experience. Initially, the planning might appear to be time consuming. Indeed, many teachers will forgo field work because they feel the effort is not worth the outcomes. This view, how-

TABLE 10–3 Examples of field study situations

School building

Heating plant: transfer of heat
Wood shop: electric motor ratings
Metal shop: gear ratios in machinery
Art department: temperatures in pottery kiln
Cafeteria: diet choices of pupils
Kitchen: bacterial counts from dishes and food
Corridors: emergency exit patterns
Music department: sound
Medical office: blood pressure measurement
Auditorium: acoustics

School grounds

Lawns: habitats of plants and animals
Shrubs: effects of light and shade
Trees: seasonal changes
Flagpole: shadow studies
Concrete walk: friction studies
Macadam pavement: absorption of heat
Soil around building: spatter erosion
Teeter-totters: lever action
Swings: centers of gravity
Snow bank: insulation effect of snow
Bicycles: gear ratios

Streets

Automobiles: analysis of exhaust
Bicycles: Doppler effect
Pedestrians: pedestrian safety habits
Driver-training car: stopping distances
Intersections: traffic patterns
Traffic lights: control mechanisms
Fire box: method of signaling
Electric lines: pole transformers

Residential dwellings

Gardens: topsoil and subsoil
Flower beds: phototropisms of flowers
Porch boxes: slips for regeneration
Lawns: earthworm castings
Lawn sprinklers: rainbows
Outdoor fireplaces: convection around fire

Community services

Water system: chlorination of water
Fire station: fire extinguishers
Police station: radar transmission equipment
Hospital: sterilization procedures
Church: pipe organs

Rural areas

Woodlands: analysis of soil
Ponds: temperature distribution
Streams: transportation of sediments
Fields: collection of insects
Roadsides: seed distribution
Ploughed soil: animals in the soil
Hillsides: erosion effects
Cliffs: rock structures
Beaches: wave action

Small businesses

Service stations: hydraulic lift
Garages: differential pulleys
Music store: electronic organs
Radio shop: oscilloscopes
Plumbing store: water softeners
Building supplies: insulation materials
Appliance store: refrigerators
Lumberyard: kinds of lumber
Transformer station: circuit breakers
Coal yard: types of fuel
Junkyard: electromagnetic cranes
Gravel pit: sedimentary deposits
Quarry: rock specimens

Hobbies

Astronomers: telescopes
Fish fanciers: aeration of water
Radio hams: antenna characteristics
Hi-fi enthusiasts: speaker characteristics
Beekeepers: life history of bees
Gardeners: fertilizers
Photographers: darkroom techniques
Rabbit raisers: inheritance

Construction projects

Excavations: soil-moving equipment
Foundations: concrete making
Bridges: derricks
Small homes: electric wiring
Office buildings: air conditioning systems
Schools: plumbing

Small manufacturing plants

Bakeries: action of yeast
Sawmills: structure of tree trunks
Dairies: pasteurization
Ice plants: use of ammonia gas
Greenhouses: photoperiodism, transpiration, plant growth
Foundries: metal casting

ever, is more a reflection of the teacher's apathy than anything else. In truth, the planning and conducting of field experiences is tedious, but the experienced teacher and students find it well worth the effort. As must be expected, the necessary planning becomes easier and more efficient with time as teachers and students become experienced in field work. But a word of caution is necessary. Teachers who superficially plan field work are inviting trouble, because a disorganized field experience is self-defeating. The purpose of the field experience is to achieve certain learning outcomes, and it should be treated as seriously as any classroom teaching strategy— it is not merely a day in the sun.

Psychology of Field Work.   The boys and girls whom teachers take on supervised field studies are the same boys and girls they have in the classroom; only the surroundings are different. It is the impact of these new surroundings that causes the students to react differently.

The four walls and the patterns of behavior in restricted environments that were established during six or more school years no longer exist during field work. The reflective surfaces that throw back the students' own noise and make them conscious of their actions are also gone, as are the reverberations that give strength and authority to the voice of the teacher.

Lectures in the field are difficult unless the teacher has a megaphone, a public address system, or an unusually loud voice. Discussions generally fail because students cannot hear each other and because there are so many distractions. Reading is difficult in bright sunlight. Writing is usually awkward. And students are too exhilarated to sit passively.

However, many constructive things can be done in the field. Students can explore, collect, take measurements, experiment, or do anything that demands physical activity. Therefore, when considering problems for field work, it is generally wise to select those that permit students to work with their hands in some way. Assume that if no provision for this type of activity is made,

some of the students will probably find things to do with their hands that teachers would rather they not do. During a five-minute walk on the way to a park, students in a seventh-grade class were observed doing the following, which were unrelated to the purpose of the trip:

1. Two girls walked along, their arms about each other.
2. Four girls picked up colored leaves.
3. Five boys picked up black walnuts and threw them.
4. Two boys tripped the girls ahead of them.
5. One girl chased the boy who tripped her.
6. One boy punched the boy ahead of him, was punched in return, and grappled with the boy momentarily.
7. Two boys continually pushed each other as they walked along side by side.
8. One boy broke twigs from bushes and a low branch and threw them at others.
9. One boy jumped up and pulled some leaves from a low hanging branch, then wadded up the leaves and threw them at students ahead of him.
10. One boy pretended to put a caterpillar on one of the girls ahead of him.
11. Three girls ducked and squealed at the caterpillar. A fourth girl slapped the boy's face.
12. One boy snatched the cap from another boy's head but was forced to return it by the teacher.

These are the normal reactions of preadolescents and early adolescents. The experienced teacher is familiar with such behaviors and is able to control them by providing rules and actions that will reduce their occurrence.

Teachers should remember that the attention span of young people is no greater in the field than in the classroom. In fact, it is considerably shortened by the many more distracting influences confronted outside the classroom. Therefore, students should not be held to one type of activity for long intervals; a variety of activities is desirable.

Finally, consider the overall time element of a field study. In general, several short experiences are usually more effective than one long one. Preparation and follow-up for short field studies should center on small problems rather than extensive ones. The factor of fatigue is also eliminated during short field experiences.

The Teacher's Role. Teachers should serve as consultants for students during field work. They should avoid lecturing or interrupting the work of students with comments of last-minute directions. Good planning easily prevents such common pitfalls.

Personal enthusiasm for field work is a great asset to science teachers. If they react to field activities with interest and excitement, the prospects are good that their students will do the same. Teachers must work along with the students, always displaying their own inquisitiveness and interest.

Generally, a teacher should not interrupt the work of students or become the focus of attention. Occasionally, however, something unusual arises and justifies an interruption. Teachers should remember at all times why they are conducting the field experience. They should try not to turn the outdoors into a confining space by adding undue structure and constraints. The students should be allowed to interact with their environment.

Extended Field Experiences

Thus far two types of approaches that use field experiences have been discussed—projects and field studies. There is a third instructional approach as well. Extended field trips have aspects in common with field studies and projects, but they also may present a number of problems. Properly conducted, however, they can be among the most rewarding educational experiences for those involved.

Time Constraints. The classification of a field experience as an extended field trip is largely based on the criterion of time. That is, the time required to conduct such a field trip exceeds the normally allotted time for a class period. There are two reasons for this extended time requirement. First, many field experiences involve locations that are not within short walking or driving distance from the school. Thus, much of the class time is used up just getting to and from the site. Naturally, it is assumed that the destination is worth reaching. Second, the field work to be conducted (nature walk, museum, tour, etc.) must require a few hours to complete, which would also preclude this endeavor from taking place during a single class period.

The extended field trip does, however, differ from a project in that the teacher wants all students to have the same experiences in the same learning environment. For example, the teacher would want all students to view certain exhibits in the museum. Therefore, such a trip must be directly supervised. Conversely, projects are normally individual or group pursuits. Finally, such extended field trips always involve locations off school grounds, usually long distances away, so that the issue of liability makes class supervision necessary. Projects performed out of school, on the other hand, are not legally considered formal school-sanctioned activities.

The extended field trip can be as short as a few hours or sometimes extend overnight. Overnight field trips such as camping are less common but are occurring with greater frequency in some school districts. The following discussion is limited to one-day excursions.

Transportation. Extended field trips are most commonly taken to such places as nature trails, zoos, museums, planetariums, botanical gardens, hospitals, and factories. Most schools do not have such resources within short distances, and if they do (e.g., a hospital) the guided tours usually take longer than an hour. The choice of transportation will depend on the distance to be traveled. If the destination is close enough for the students to leave after the school day begins and return to school before the end of the school day, there is the possibility of using school buses. Some schools commonly provide buses for field trips without question of limitation. Others allow a certain number of miles to each teacher, giving the teacher the option of using the mileage for a few longer trips or several short ones. Still other school systems permit the use of buses on an hourly basis. Some systems restrict the use of buses if the destination is so far away that departure is scheduled before school begins or return occurs after school ends.

There are many charter bus companies that can transport students on field trips. Occasionally, a school system budgets money for hiring buses, but in many cases students are asked to share the cost of transportation. The individual cost imposed on the student is usually nominal and normally does not create a financial burden on the student or family. However, the teacher should be prepared to make appropriations for students who have special financial problems. One way to offset costs is to find individuals, perhaps parents, who are willing to transport students in their personal automobiles. The use of private cars for field trips must be cleared by the administration for reasons of liability. Some school districts will not normally allow the use of private automobiles. Another point to consider is efficiency: if large groups of students are involved, the use of private automobiles becomes rather cumbersome and inefficient.

Administrative Policies.   The school policy for extended field trips is usually much more definitive and restrictive than it is for short field studies. Quite often, the teacher will have to submit a proposal to the school board or principal for approval. Overnight trips are either very rarely approved or are strictly forbidden. The teacher should be prepared to defend, in detail, the purpose of the trip and the outcomes she expects. This may be perceived as extra tedium for the teacher, but if the trip actually has instructional value the teacher should have already outlined expected outcomes.

Most school systems require written permission from each parent before students can be

taken from the school grounds. It is advisable for the teacher to require such permission regardless of whether the school requires it.

When a teacher is planning an extended field trip, she finds that specific rules for supervision are often stipulated. Such rules usually require an adult chaperone for every group of ten students. If more than one bus is being used for the trip, it is a good idea to have a teacher on each bus. Obviously, if other teachers are to be used as chaperones, they will need permission from the principal, because their absence from school will necessitate having substitute teachers for their classes.

A final point is necessary concerning student permission to participate in such a trip. Usually the school simply requires parental and administrative approval, but keep in mind that the students will be missing other classes while attending the field trip. Whether or not it is required, approval should be obtained from each student's teachers for the student to miss class that day. Some teacher might feel that a particular student cannot afford to miss class. Such a concern is usually valid and should not be challenged.

Legal Aspects.   In most states, if not all, a signed permission form from a parent does not release the teacher from liability for accidents on field trips. The standard of reasonable and prudent care applies on field trips just as it does in the classroom. The teacher must show that necessary precautions have been taken and adequate and appropriate supervision has been provided. There must also be evidence of sufficient planning and warning of possible dangers.

A permission slip is merely a way of informing parents that a field trip is to take place—not a release of responsibility. It does provide evidence that the parent had knowledge of the trip and that the student had the parents' permission to attend.

A field trip planning form such as the one in figure 10–1 is evidence that the trip has been well planned and that possible problems have

FIGURE 10–1   Extended field trip planning checklist

	Planned	Confirmed
1. Objectives of field trip	————	————
2. Date of field trip	————	————
3. Time of departure	————	————
4. Anticipated date of return	————	————
5. Anticipated time of return	————	————
6. Destination	————	————
7. Type of transportation: bus, car	————	————
8. Anticipated costs	————	————
9. Teacher supervisors	————	————
10. Parent chaperones	————	————
11. Teacher plans	————	————
a. Permission from school administrators	————	————
b. Permission letters to parents sent	————	————
c. Reservations made for tour	————	————
d. Reservations made with appropriate individuals to use the site	————	————
e. Rules of conduct for students	————	————
f. Permission letters from parents returned	————	————
g. List of special supplies needed for the trip	————	————
h. Dress requirements for students for trip	————	————
i. First aid kit	————	————
j. Lunches	————	————
k. Follow-up activities	————	————
l. References	————	————
m. Materials students need to bring for the trip	————	————
n. Evaluation of field trip	————	————

been identified. This also shows that proper supervision was also provided on the trip.

Preparation and Management.   The preparation of students for an extended field trip does not differ significantly from the preparation provided for any field study. Students should be told that they are expected to behave properly. If the trip is to a museum, zoo, botanical gardens, or similar institution, the facility will gladly send maps and information before the visit. This information should be given to students and reviewed in class in advance of the trip. For example, if the trip is to a museum, students can plan in advance what they wish to see. Although most extended field trips do not involve experimenting and manipulating, students can be given instructional materials to work on. Experienced teachers find that worksheets with questions referring to various exhibits throughout a museum direct student attention and ensure achievement of the intended outcomes. Such a technique can work equally well for planetariums, aquariums, nature walks, and so on. The key element here is that the teacher must be very familiar with the place being visited. If the teacher is not, it is hard to conceive of how she decided that such a trip was of educational value. Remember that long trips are advisable only when the learning opportunities justify the additional time and effort needed.

The best way to illustrate the "nuts and bolts" of the extended field trip is to describe two different types of such trips: a half-day trip to a local hospital and a full-day trip to the Museum of Science and Industry in Chicago. The planning and procedural steps are presented in sequential form, which can be used as a checklist of responsibilities.

▼   Mr. Dansert teaches biology for students in Grades 10 through 12 in rural Illinois. He has decided to take all his classes on a full-day field trip to the Museum of Science and Industry in Chicago. Because the trip takes three hours each way, his classes must leave before the school day begins and return after it ends. The sequence of events essential to the successful completion of such a trip follows:

1. Mr. Dansert takes an informal poll in all his classes to assess student interest.
2. A proposal is submitted to the school board and/or principal.
3. After approval has been granted, Mr. Dansert takes another poll among his students to assess how many buses are required, the cost per student, and how many chaperones are needed.
4. Three buses are needed for the 125 students who will be attending the trip; those not attending will have a study hall during the biology time period that day.
5. Three buses are reserved at least one month before departure.
6. Students are asked to get permission slips signed by teachers and parents.
7. Money is collected for the trip.
8. Eight chaperones are selected from parent volunteers to go along with three teachers (one per bus), making a total of eleven chaperons.
9. Mr. Dansert meets with the chaperones to discuss the format of the trip.

- Each of the eleven chaperones will be assigned approximately ten students.
- A schedule is worked out so that chaperones know the sequence of locations within the museum to take their students.
- It is suggested that each chaperone allow his or her students one hour in each section. Students should have freedom to move around individually in each section.
- Parent chaperones should be advised of their jurisdiction regarding what responsibilities they should assume. They should be informed of the types of behaviors they should expect the students to exhibit and what the students are expected to accomplish. They should be made aware of the procedures involved in case of emergency or unexpected problems and informed of liability involved. The teacher should not assume that parents know how to serve as adult exemplars or that they know how to manage groups of students.

10. Students sign up for specific chaperones. Depending on the students, they may be allowed to choose a chaperone or chaperones are assigned.
11. Necessary class periods are spent preparing for field trip procedures.
12. Museum worksheets are given to students to structure their experience and ensure that the proper learning outcomes are considered.
13. Buses leave at 7:00 AM and arrive at the museum at 10:00 AM. The museum has already opened at 9:30 AM.
14. 10:00 AM to 1:00 PM: Students view museum with their chaperones.
15. 1:00 to 2:00 PM: Students and chaperones eat lunch in designated lunch area.
16. 2:00 to 4:30 PM: Students continue to view exhibits with chaperones.
17. 5:00 PM: Buses leave Chicago and arrive at school at approximately 8:00 PM.
18. On the next school day Mr. Dansert begins review of the museum experience.   ▲

▼   Mr. Donahue wishes to take his twelfth grade advanced biology students on a tour of a local hospital. Only twenty students are involved but, because of time constraints, they will be gone from school for four class periods. Since students can leave the school grounds after the school day has started and will return before the end of the school day, school buses can be used.

Because of the number of students involved, and because the hospital will conduct a guided tour, Mr. Donahue does not require chaperones. Once again, permission is obtained from the school board and principal. The students obtain permission from their teachers and parents. Mr. Donahue reserves the school bus two weeks before the trip. Since Mr. Donahue is not taking all his students along, he designs an assignment for those classes that he will miss that day. The principal will either hire a substitute or place Mr. Donahue's students in study hall.

The following events occur on the field trip day:

1. Bus leaves school grounds at 10:00 AM and arrives at hospital at 10:45 AM.
2. Hospital tour starts at 11:00 AM and continues until 1:30 PM with a twenty minute break for lunch in the cafeteria.
3. Bus returns to school grounds at 2:15 PM.   ▲

Selection of Students.   The selection of students for a field trip may cause difficulties for some teachers. Such decisions often depend on the science teacher's philosophy and attitude and the policies of the school administration. If the trip is viewed as a reward of some sort, then teachers may bar students from participating who are known troublemakers and potential discipline problems. They may also bar students from participating who are uninterested, lazy, lack initiative, and do not perform to the teacher's expectations. These may include students who do not do their classroom and homework assignments or do a mediocre job even though they are capable of performing at a higher level. On the other hand, if the field trip is viewed as a unique and innovative activity and an important part of the course, then no students should be omitted. Indeed, slow learners and unmotivated students probably need the trip more than those of high ability, and some students, particularly slow learners, are not necessarily stimulated through traditional classroom procedures.

Teachers who instruct mainstreamed classes will be confronted with special problems when planning a field trip. First of all, they cannot and should not prevent students with disabilities from participating unless the students impose a danger to themselves or others. Depending on the field experience, teachers may have to make special arrangements for students with certain types of disabilities in order for them to participate. Certainly the school administration should be consulted regarding policy that has been established for students with disabilities.

Purposes.   Whether the field experience is included in a project, field study, or extended field trip, the teacher must not forget that it is merely an alternative to the classroom approach. The goals are the same; the field experience is not just a vacation from classroom tedium.

The teacher must carefully plan all field experiences. Students should know where they are going, why they are going, what they are going to do, and what skills and materials they are to master. Field experiences can and should be structured to ensure achievement of stated goals. This is most often accomplished through a particular task assignment, as is done with field studies and projects, or a worksheet, which is common to extended field trips.

The sample planning checklist such as the one in figure 10–1 may be helpful to make certain that all the necessary details of the trip have been attended to. The items are checked off as they are handled. The checklist helps the teacher remember what has been done and what needs to be done during the course of planning.

It is extremely important that the field experience "fit" properly into the instructional sequence. Students should be quite clear on how the field experience relates to what has been discussed in the classroom. Additionally, classroom activities after the field trip should refer back to the lessons and perceptions gained during the field experience. Field experiences have as wide a range of functions as any other instructional method. They can easily be used to set problems, perform experiments, reinforce classroom instruction, and review content already covered. The list is endless.

If field experiences are not planned properly or treated as serious instructional approaches, they will result in failure. However, a well-planned field experience can be the most powerful educational tool a teacher has. High-ability students are stimulated to explore areas that they had not even imagined, whereas low-ability or uninterested students can suddenly "see the light" that many years of classroom teaching have failed to impart.

SUGGESTED ACTIVITIES

1. Plan a project-centered unit in which the students work individually or in small groups on an investigation. The students may want to suggest their own investigation, or you can suggest one that you feel the students would enjoy and be able to conduct.
2. Prepare a project unit for a seventh-grade science class on a topic of your choice that requires that students work alone on individual projects. Make a list of the possible projects in connection with the unit. Discuss the topic and the list of possible projects with other members of your science methods class.
3. Plan a field trip to a gravel pit or another geological site and have the members of your science methods class perform certain assigned activities at the site you select. Plan and discuss the activities before the trip so that the students understand what is to take place at the site.
4. Plan a short field trip for members of your science methods class to identify ten trees on your college or university campus. Prepare an identification key for the trees you want to have identified and instruct the members of your science methods class on how to use the key before taking them on the trip. On the trip, assign each student at least five trees to identify using the key. Ask the students to compare their results. Some students may have difficulty using the key while they are on the trip. Students who are having problems can be instructed by those who have mastered the use of the key. When the students return to the classroom, ask them to suggest ways to conduct the trip more effectively.
5. With other members of your science methods class, visit two local science fairs, one devoted to projects prepared by high school students and the other devoted to projects prepared by middle school students. Independently evaluate a selected group of projects in each situation using evaluation forms you and your classmates prepare beforehand. Compare your evaluations with those of other members of your class.
6. Volunteer to be a judge at a local or regional science fair where the participants are middle school students. Prepare a checklist you would use to evaluate the projects and evaluate them using your checklist. Compare your evaluations with those of other judges who participated.
7. Visit a regional or state Science Olympiad. Compare a Science Olympiad with a local or regional science fair. What are the benefits that students derive from participating in a Science Olympiad? Which activity better meets the goals of science education, the Science Olympiad or the science fair? Why?

BIBLIOGRAPHY

American Association for the Advancement of Science. 1989. *Science for all Americans: Project 2061*. Washington, DC: Author.

Baca, B. J. 1982. Teaching biology field courses in the wake of environmental disasters. *The American Biology Teacher, 44*(1): 21.

Benson, B. W., and Kirby, J. A. 1982. Science fairs: Do your students measure up? *The Science Teacher, 49*(1): 49.

Bicak, L. J. 1982. Schoolyard science. *The American Biology Teacher, 44*(3): 153.

Biggs, A. L. 1982. An interdisciplinary course in Big Bend National Park, Texas. *The American Biology Teacher, 44*(4): 27.

Calinger, B., Champagne, A., and Lovitts, B. 1990. *Assessment in the service of instruction*. Washington, DC: American Association for the Advancement of Science.

Cramer, N. 1981. Preparing for the fair: Fifteen suggestions. *Science and Children, 19*(3): 18.

Fay, G. M., Jr. 1991. The project plan. *The Science Teacher, 58*(2): 40.

Fitzsimmons, C. P. 1983. Field trips within easy reach. *The Science Teacher, 50*(1): 18.

Giese, R. N. 1989. An open letter to science fair judges—focus on projects and presenters. *Science Scope, 12*(2): 38.

Goodman, H. 1985. At the science fair. In *Science fairs and projects*. Washington, DC: National Science Teachers Association.

Hamrick, L., and Hart, H. 1983. Science fairs: A primer for parents. *Science and Children, 20*(5): 23.

Harley, David W., Jr. 1990. Track and field work. *The Science Teacher, 57*(6): 58.

Johnson, J. R. 1989. *Technology report of Project 2061 Phase 1 Technology Panel*. Washington, DC: American Association for the Advancement of Science.

Keown, D. 1984. Let's justify the field trip. *The American Biology Teacher, 46*(1): 43.

King, D. T., and Abbott-King, J. P. 1985. Field lab on the rocks. *The Science Teacher, 52*(4): 53.

Lagueux, B. J., and Amols, H. I. 1986. Make your science fair fairer. *The Science Teacher, 53*(2): 24.

Levin, K. N., and Levin, R. E. 1991. How to judge a science fair: Use the consensus method. *The Science Teacher, 58*(2): 43.

National Science Teachers Association. 1985. *Science fairs and projects*. Washington, DC: Author.

Powell, R. 1987. Research projects in high school biology. *The American Biology Teacher, 49*(4): 218.

Quint, W. C. 1980. Science activities for school trips. *The Physics Teacher, 18*(8): 584.

Rivard, L. 1989. A teacher's guide to science fairing. *School Science and Mathematics, 89*(3): 201.

Schellenberger, B. 1981. Take your class outdoors. *Science and Children, 19*(2): 28.

Science Olympiad. 1992a. *Coaches' manual and rules, Division B (Gr. 6–9)*. Rochester, MI: Science Olympiad, Inc.

Science Olympiad. 1992b. *Coaches' manual and rules, Division C (Gr. 9–12)*. Rochester, MI: Science Olympiad, Inc.

Smith, N. F. 1980. Why science fairs don't exhibit the goals of science teaching. *The Science Teacher, 47*(1): 22.

Stronck, D. R. 1983. The comparative effects of different museum tours on children's attitudes and learning. *Journal of Research in Science Teaching, 20*(4): 283.

Teachworth, M. D. 1987. Surviving a science project. *The Science Teacher, 54*(1): 34.

Troy, T., and Schwaab, H. E. 1981. Field trips and the law. *School Science and Mathematics, 81*(8): 689.

Weisgerber, R. A. 1990. Encouraging scientific talent. *The Science Teacher, 57*(8): 38.

Weld, J. D. 1990. Making science. *The Science Teacher, 57*(8): 34.

Williams, R., and Sherwood, E. 1982. Activities in mathematics and science for young children using the school yard. *School Science and Mathematics, 82*(1): 76.

Yager, R. E., and Penick, J. E. 1985. (Eds). *Focus on excellence: Science in non-school settings* (vol. 2, no. 3). Washington, DC: National Science Teachers Association.

FIGURE 11–1   A view of the monitor for an interactive computer program on velocity

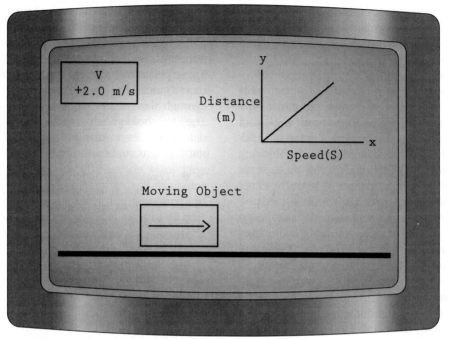

damental concepts and principles of genetics, especially when very little laboratory work and firsthand experiences are provided.

An interactive microcomputer program, for example, can instruct students on the effects of radiation on Drosophila (fruit fly). On the monitor, students are given opportunities to select the trait to be altered, the sex of the fly, and the amount of radiation to be given. After each trial, the results appear on the monitor. The learner can conduct repeated trials to select the appropriate level of radiation so that the organism is not rendered infertile or killed outright. From the program, students get a sense of what research biologists think about when they perform their work. As students work through the program, they can produce a variety of phenotypes, for example wingless, short body, and curly wing, all of which can be observed on the monitor. The students who use this type of program, along with classroom discussion and textbook reading, will likely gain greater understanding of sex chromosomes, sex linkage, somatic mutation, germ mutations, and other concepts than they would without this instruction.   ▲

Interactive Multimedia Systems

Computer and video technology have been combined to provide some of the most powerful curriculum resources that science teachers can have at their fingertips. With a videodisc, images can be viewed in full color on a large TV monitor. The videodisc can store thousands of images that can be accessed by videodisc players. Science teachers or students can select still pictures and view them as they would colored slides. They can also view pictures that show movement. Some of these multimedia systems have been referenced to many commercial textbooks series so that teachers can access the system with a barcode device.

There is a movement at the middle school level to promote multidisciplinary science programs. These programs also include themes that are useful for organizing science concepts and principles. Middle school science teachers need a variety of resources to instruct students in these programs. Consider for example a unit of study concerning ecosystems and the environment. Most likely the assigned textbook provides only definitions and explanations of common ideas related to this area such as niches, adaptation, competition, food webs, and biomes. Laboratory exercises would probably accompany the textbook, with additional hands-on activities from the teacher's resource files. However, an interactive multimedia system, such as the one described below, would enhance this unit of study.

▼ A videodisc program on ecology and the environment can take students to different parts of the world to illustrate simple and complex relationships among organisms. A video of the logging industry in British Columbia and the northwest United States can show that cutting down large trees affects certain bird populations. Furthermore, students can view the size of this deforestation and then discuss how it appears to be affecting other animal life. The scene can be changed to South America, where students can view the relationships between plants and vertebrates as well as the economic conditions of people who live in these areas. Since environmental factors and economic development are interrelated, this multimedia technology shows these interactions in a very powerful instructional format. The science teacher can team with a social studies teacher to reinforce students' learning outcomes in both disciplines. ▲

CD/ROM systems offer a very useful delivery system because they can store large amounts of pictorial and textual material on a 12-cm disk. The amount of information this system can house is literally encyclopedic—several hundred thousand pages of type-written text. Early versions of CD/ROM programs merely produced printed material on the screen. However, today's programs can provide pictures in vivid color and action along with textual material.

An example of a spectacular CD/ROM package that has been developed jointly by the National Geographic Society and IBM is called "Mammals: A Multimedia Encyclopedia" (Salpeter, 1990). The program contains sound, photographic images, text, and moving pictures. The user can select and view hundreds of different species of animals. For each animal, the program provides information about its habitat, diet, survival status, life span and reproductive patterns. This teaching resource supplies the user with the quality of animal study that has characterized the National Geographic Society's long tradition of excellence.

Multimedia presentations can benefit a large number of students at a given time, if they can see what is taking place. Several devices are available for classroom presentations that permit large groups to view multimedia presentations. For example large-screen monitors ranging in size from 26 to 35 inches provide excellent color images for student audiences. Video projectors can be used in large meeting rooms or auditoriums because they can project images directly onto large screens. Flat panel LCD displays can provide images produced from computers and lasers, which can be projected on a screen by a device similar to an overhead transparency machine. This device can be transported easily from classroom to classroom and used for large group instruction.

Telecommunication Networks

Electronic technology has made it possible for communications to take place between classrooms across the globe. Some schools are connected to telecommunication networks via satellite through computer and modem. This electronic system permits students, teachers, and administrators to communicate and share information with others living in different locations. Now it is possible for schools, planetariums, mu-

Electronic technology has made it possible for communications to take place between classrooms across the globe.

seums, educational centers, and other facilities to share programs and exchange ideas through this network.

▼   Middle school students studying earth science can communicate with other students around their state to gather on-the-spot data concerning the weather. They can construct regional weather maps by contacting students in other schools that are 50, 100, 200, or more miles away to determine weather conditions in those locations. Students can provide information about the temperature, cloud formations, wind direction, and wind speed that exist in their locations. The students can attempt to forecast the weather. They can provide students in the rest of the school with daily weather reports over the announcement system. Furthermore, the students can attempt to predict the next day's weather from the information they gather and can determine which factors seem to provide the most useful information for this purpose.   ▲

Telecommunications has made possible "distance learning" for many years, and some people living in remote places use it regularly. Distance learning is the application of electronic technology to help students receive instruction that originates from a distant location. The quality of these programs varies greatly. At one end of the

continuum are programs where students merely view an instructional presentation. On the other end are programs that are interactive and where the learner must prepare assignments and react to the instruction. The National Science Teachers Association (Harkness, 1991) has produced a position statement on criteria for high-quality distance learning in science education. A summary for this position follows:

1. *Interaction* should occur between the instructor and students so that critical thinking and feedback take place.
2. *Flexibility* should exist in the program so that it can be adjusted to a given group of learners and locality.
3. *Manipulative experiences* should accompany instruction so that it is hands-on as well as minds-on.
4. *Competent instructors* conducting the instruction must be able to interact with the intended audience.
5. A *variety of appropriate resources* must accompany the program to supplement it and reinforce learning.
6. *Appropriate technology* should be used that provides the best instructional delivery sys-

tem for the material to be learned, and not to use technology just for the sake of using technology.

7. *Evaluation* of the program and student learning must be part of the instructional package, and it should occur at several points during the instruction.

Obviously, a long-distance learning situation must go beyond individuals sitting passively and watching a video, regardless of how spectacular it is. Harkness (1991) cites the NSTA-sponsored JASON Project as a telecommunications program that merits science teachers' attention. To participate in the JASON Project, students and teachers meet at a museum on a designated day to accompany Robert Ballard on an undersea exploration. The students are expected to prepare beforehand for this experience using materials developed by NSTA. Other telecommunications programs to consider are National Geographic Society's "Kids Network" and NASA's "Spacelink" in which teachers and students can collaborate with scientists.

Virtual Reality Technology

Perhaps the most powerful instructional resource for science education is what is termed "virtual reality" or "virtual environments" or "virtual worlds." A virtual reality environment is created to be like a real environment but is not the real thing. Virtual reality technology goes beyond the common visual and auditory experiences from computer and video monitors; it includes perceptual and tactile interactions as well. By wearing a special helmet (or headgear) and gloves, the learner can experience a new world, a virtual world, different from anything experienced before. During the past decade, military and space programs have used this technology to train individuals in flying and space travel. The hardware for this technology exists at certain government-supported science and technology centers such as the Johnson Space Center and private corporations that develop and support high-tech industries. Today, software is in

the design stage for science educational purposes. However, this technology is still very expensive and this must change before it appears in classrooms.

SPECIFIC TYPES OF MICROCOMPUTER PROGRAMS

Microcomputer-Based Laboratories

Microcomputer-based laboratories (MBL) involve laboratory and firsthand experiences using the microcomputer to gather and display data. These laboratories may be used to quantify a variety of phenomena such as temperature, speed, length, force, heart rate, light intensity, and brain waves. MBL offer science teachers a powerful technology to enhance laboratory work, often making it more exciting and meaningful to students. Students can explore science phenomena in more accurate and precise ways than are possible by traditional laboratory methods—at least for certain experiences. As a laboratory tool, this technology permits students to be engaged actively in their laboratory work and to focus on the data almost immediately. "The students are not automated out of the experimental process but given powerful tools which may help them gain a 'feel' for the data: data become almost palpable as their links to sensory experience are clarified" (Lam, 1984–85, p. 1).

Unlike a simulated lab in which the computer is not employed to make actual measurements, MBL require students to manipulate equipment and to observe data measured by a sensor and then displayed on the screen. Consider, for example, a laboratory exercise to measure temperature. With a simulation, a student might see several objects on the screen, each having a different temperature. The temperatures are displayed, along with questions about the causes of these differences. However, with MBL programs students are required to take temperature measurements from real objects. They gather these data with a temperature-sensitive device called a thermistor. Thermistors that are properly interfaced to a computer are accurate instruments that can measure temperatures to within less than 1/10 of a degree Celsius. The student immediately observes the temperature readings and changes on the screen.

Many laboratory exercises can be carried out with the microcomputer. Some of these labs are already prepackaged and sold by software companies while others have been developed by science teachers. Any science teacher can learn to use this technology.

In order to use the computer for measuring and recording, one must obtain a device called a transducer or sensor to pick up the data from the environment. The transducer converts physical or chemical changes into electrical signals. Sensors are available to detect changes in pressure, sound, light, temperature, pH, etc. However, one must convert the electrical signal produced by the transducer from an analog signal to a digital signal.

Analog signals are smooth continuous electrical signals, similar to what one observes when drawing a line or a curve. Computers function mainly around digital impulses to process information. A digital signal is discrete or broken into small pulses. Analog signals entering the microcomputer from sensors are converted into digital signals before they are processed by the computer. The joy stick that is connected to the game paddle port of the computer, which many youngsters use when they play computer games, facilitates the digitizing process. Since the joy stick moves back and forth, this type of motion creates an analog signal that is fed into the computer at the game port connection, digitized inside the computer. Once the signal from the sensor is digitized, the data can be analyzed and displayed on the screen and calculations can be performed. In addition, the computer can transform data into graphs to help students see trends and anomalies.

The potential uses of MBL in science instruction cannot be overemphasized. Not only is the microcomputer an excellent measuring instru-

ment, but students are fascinated by it. The computer, and the graphics that it generates, gets students' attention and motivates them to participate in science laboratory work (Bross, 1986).

A science teacher might wonder how to conduct MBL if many microcomputers are not available. If one computer is available, pairs or small groups of students can use it on a rotating basis to conduct laboratory measurements while the other members of the class are engaging in other types of work. With three microcomputers, one can accommodate 24 students. For example, consider an MBL that uses the thermistor to study heating and cooling. Organize the students into six groups of four students each. Two groups of students can connect their thermistors to an adaptor box which in turn is connected to one computer. Each of these two groups can independently input the thermistor data into the one computer and observe it separately. If two groups of students can use one computer, then six groups of students can use three computers. Nevertheless, it would be easier to have six computers, one for each lab group. Many science teachers borrow five or six microcomputers from other rooms or from the school's microcomputer lab so that they will have enough for a special MBL they wish to conduct. Planning ahead can facilitate all of these arrangements.

The IBM Personal Science Laboratory (PSL) is a microcomputer-based laboratory designed for science instruction from middle school through college (IBM Educational Systems). This program provides its users with two options when electronic technology is employed to conduct laboratory work. Individuals can select experiments that are already designed, or they can create an experiment. The former is ready to go and comes with preset parameters. The latter allows the user to set the parameters for an experiment. Probes are available for collecting data on distance, temperature, light, pH, and time.

For the Apple Macintosh systems, Vernier Software has developed the Universal Interface. This is a flexible MBL program that is user friendly (Sneider & Barber, 1990). The software consists of simple pull-down menus. The data can be displayed in one, two, or four graphs. The probes that are available for investigating phenomena are as follows: photogates, voltage sensors, thermocouples, temperature probes, and radiation monitors. Users can also build a magnetic field sensor and pressure strain gauge to extend the capabilities of the program.

Simulations

Computer simulations permit science teachers to bring rich learning experiences into the classroom. Simulations illustrate real life or hypothetical situations that help the learner to visualize, in black and white or color, concepts and principles in "action" that would not be possible through lecture, laboratory experiences, demonstrations, or pictures. They often permit the learner to manipulate variables or parameters and then to observe the consequences of their choices. Simulations can bring into the classroom aspects of the world or universe that are too expensive, too dangerous, too difficult, or too slow or too fast in occurrence to be experienced firsthand (Tamir, 1985/86). Objects in space, molecular motion, radioactive material, electrical current, predator/prey relationships, and breeding are just some of the topics around which science simulations have been developed.

For example, a simulation of the heart can show the path of blood flow, highlighting the function of the chambers and valves. With the use of color graphics, oxygenated and deoxygenated blood is traced as it flows through the chambers and vessels of the heart. Other simulations can be used to illustrate the parts of the human circulatory system—its major arteries and veins, along with capillaries. Also available for science instruction are simulations of heart abnormalities, which are presented along with EKGs and their interpretations. Students can realize how technology is used to study the function and diseases of the heart—all of which

makes this area of study more relevant and interesting.

Simulations of weather fronts are also available for science instruction. These simulations show the movement of cold and warm fronts across regions and continents. They help the viewers locate where various types of weather patterns originate and how they produce rain and snow. These simulations make the study of meteorology more real and concrete for middle and junior high school students.

Simulations with various amounts of detail have been developed to illustrate the growth of plants. These programs show how plants develop from fertilization to seed germination, to the young plant, and to the mature plant. These simulations detail seed development, show tropisms, and illustrate how environmental factors affect plant growth.

Tutorial Programs

Tutorial programs go beyond drill and practice. They are used to teach concepts, skills, and new information. These programs, being highly interactive, require active participation from the learner. They often involve reading, solving problems, analyzing graphics, simulating laboratory experiments, and completing word problems. Although they are sometimes expensive, they can be very useful because they provide valuable learning experiences. Every science course has certain concepts and principles that many students fail to learn. For example, some students have trouble understanding density, genetic crosses, Ohm's law, balancing chemical equations, solving mole problems, and determining acceleration. Even after science teachers have invested considerable class time in attempting to teach these ideas to students, many students still fail to understand the concepts and principles.

The science teacher can provide tutorial programs for situations when students are not learning important concepts and principles.

These programs will carefully guide the learner through the steps necessary to develop the subconcepts that underlie the major learning outcomes. For example, a good tutorial program on balancing chemical equations might begin with writing chemical formulas, proceed to discussing chemical equations, continue with writing chemical equations, and end with balancing chemical equations.

Authors of good tutorial programs analyze the learning tasks involved in performing a given learning outcome. They break the learning into steps to increase the probability of the student performing the terminal task. These authors also include much practice after each step to promote understanding and retention. Branching is an important aspect of good tutorial programs because it provides appropriate remediation based on the student's incorrect answers.

Tutorial programs can be given to students to use during the school day or after school. They can be used in instruction if there is a microcomputer available. A display device can be placed on an overhead projector that permits many students to view the computer output on an AV screen or wall. An entire class can use tutorial programs in the microcomputer lab when many computers are available. Tutorial programs can be taken home by students who have computers in their home. They can also be used by homebound students.

Drill and Practice

The microcomputer can help students improve their performance in the science classroom. Many students who do not study or complete their homework and other assignments often do poorly on tests. These students fail to learn new vocabulary words and concepts that are found in their science textbooks or that are presented during a unit of study. They need to practice identifying, defining, writing key words, and solving word problems. Many students need repetition to achieve retention of the subject matter. Drill and practice microcomputer programs provide valuable practice for these students.

Drill and practice programs provide the learner with exercises that review subject matter. The list of commercial and teacher-prepared programs for this purpose is long. For example, in biology, practice programs are available for taxonomy, human anatomy, cell structure and function, protein synthesis, respiration, mitosis, digestion, parts of a flower, food chains, bacterial diseases, and more. Chemistry programs are available on various topics, such as significant figures, properties of matter, symbols of elements, naming compounds, balancing equations, and determining molecular formulas. A variety of programs is also available in other areas of science.

The next vignette describes a high school chemistry and physics program that uses microcomputers to help many become successful in these subjects.

▼ Mr. Moore serves as the master teacher for all of the chemistry and physics sections in a predominantly blue-collar community that has only one high school. He has organized these courses in a team teaching arrangement so that he lectures to a large group of students for two days in a row. On the third day, each large group of students is broken into subgroups of 30 individuals. Each subgroup of 30 students goes either to a science laboratory or to a computer laboratory. On the fourth day the groups switch, and those that were in the science lab the day before go to the computer lab, while those who were in the computer lab go to the science lab. On the fifth day, all students return to the large group instructional room for testing.

The computer lab utilizes microcomputer programs for drill and practice in which the word problems match those addressed in the lectures, in the laboratory exercises, and at the end of the textbook chapters. The programs were designed by Mr. Moore so that even the equipment and laboratory apparatus that appear on the color monitor are identical to those the students have already used. Students are expected to achieve 100% on the tests that they take at the end of each five-day instructional cycle. When students fall below 100% mastery, they are requested to engage in drill and practice problem solving using the computer lab at assigned times during the school day, before school or after school. There are *no* homework assignments. All work is completed in school. Mr. Moore has also produced microcomputer programs that generate many tests students can take to demonstrate their mastery of the subject matter.

▲

This instructional program is very traditional and is very similar to what college students experience in science and engineering courses. The computer is used in its simplest instructional format—drill and practice—as opposed to being used for simulations, microcomputer-based laboratories, or tutorials. Nevertheless, a large percentage of the students in this high school take chemistry and physics courses because they learn to master the basic concepts of these subjects and are successful in the course work. Many of the students go on to universities to study science and engineering. This teaching arrangement is possible because of the large number of drill and practice problems that Mr. Moore has developed for the computer lab as well as the large number of test questions that he has prepared for the tests that are to be given to the students.

Problem Solving

Computer-assisted problem solving experiences hold great promise for maximizing the potential of the computer as an instructional tool. Problem solving programs for the microcomputer go beyond computing numerical answers or producing verbal responses similar to those found in drill and practice, simulation, and tutorial programs. Good problem solving programs center on real situations that may have significant meaning to students similar to those described in Chapters 3 and 4. These learning experiences require considerable involvement and interaction before solutions are proposed and results are determined with the aid of the computer.

"The Voyage of the Mimi," for example, is a program that involves students in a great deal of study and analysis. It was developed by the Bank Street College Project in Science and Mathematics (1985). This technological age instructional program combines television video, microcomputer software, and printed material to engage students in the study of whales during a sea voyage. The study of whales provides students a real context in which they can study nature and be involved in showing them the human aspect of science. The program also shows students how mathematics and technology are applied to study and solve problems in daily life.

The instructional package described above consists of thirteen episodes of an adventure story on video, each lasting 15 minutes and paired with a 15-minute documentary. The students use printed guides and interactive software to explore selected concepts in depth. They also engage in LOGO activities (a special type of graphics program) and microcomputer-based laboratory simulations. Printed material and software that accompany the TV series stimulate problem solving by encouraging students to experiment with the elements of the natural world as observed through this instructional package.

Most problem solving programs are not as extensive as "The Voyage of the Mimi," which has been a program series on PBS. Nevertheless, there are many worthwhile problem solving programs for science teachers to preview for use in their teaching. One excellent microcomputer-based problem solving program can generate a great deal of student excitement in a science class.

USING COMPUTERS FOR RECORD KEEPING, TESTING, AND PLANNING

Record Keeping

A microcomputer can offer great assistance to a teacher who must keep daily records for many students. Entering and averaging grades, maintaining daily attendance records, and recording information about students can consume a great deal of time and be very tedious. Fortunately, many computer programs can be used to simplify these tasks and to save considerable time.

Most computer-assisted grading programs permit the teacher to list alphabetically the students in a given class along a left-hand column of the page, and allow the entry of grades across the top of the page, as one would view in a grade book. These programs also permit the teacher to weight the grades and to find numerical averages. For example, some science teachers count laboratory work one-third, daily work one-third, and tests and quizzes one-third. Others may count lab work twice as much as daily grades. Whichever way teachers desire to weight the grades, the program will compute the averages. Some science teachers have written their own programs so that they can produce a letter grade (see figure 11–2).

Teachers do not necessarily have to use a predeveloped software grading package. Spreadsheet programs can also be useful. Spreadsheets are commonly used in business for record keeping. They produce rows and columns, neatly displaying many data. Not only do spreadsheets permit freedom and flexibility, but they also save time. Fortunately, many individuals in local communities and schools can help a teacher get started in using the microcomputer for record keeping.

In addition to grading, the computer can generate seating charts to keep attendance. The seating chart is a matrix with squares that represent seats or desks in the classroom. When teachers place students' names in the appropriate spaces, they can take attendance easily while learning students' names.

Constructing Tests

Constructing science tests is a time-consuming task that can be facilitated with the microcomputer. One option is to use the machine's word processing capability to construct the test. Since the test is stored as a file, teachers can easily modify a test before administering it, if they so desire. They can easily make two versions of a multiple-choice test by taking the first version, modifying some of the stems and choices, and then saving it as a second version of the test. This procedure can be repeated several times to produce many versions of a given test.

Software and textbook companies have developed test generators for many science courses. These test packages are versatile and easy to use. They can produce many types of test items such as multiple choice, matching, true/false, and short answer. The programs also randomize test items, construct several test versions, and produce an answer key for each version of the test. Test generators that are designed for specific textbooks are keyed to the objectives for each chapter, and the user can select the objectives for the test items desired.

Preparation of Instruction

One of the most satisfying uses of the personal computer is for the preparation of instruction. Science teachers can use the computer's word processing and graphic capabilities to construct lesson plans, unit plans, paper and pencil exercises, lecture notes, laboratory exercises, assignments, word puzzles, and so forth in much less time than it would take to prepare these instructional materials with pencil or using a conventional typewriter since modifications of these written materials can be made in minutes. Corrections can be made without retyping the entire text. Since textual materials are stored in files, they can be saved and recalled for use at a future time. This saves a great deal of time retyping and modifying an exercise or a lesson plan.

Many school districts require teachers to prepare daily lesson plans and to submit plans to the principal or department head. This requirement provides both the administration and teachers with a record of what is expected of the students and the activities that will be used to help students achieve the objectives. These lesson plans always call for a list of instructional objectives, which can be prepared quickly with the computer because it permits the teachers to

FIGURE 11-2 An example of a gradebook for recording and averaging grades generated by a microcomputer program

CHEMISTRY I Pd. 3

STUDENT	LOCKER	TS0828	LB0902	LB0909	LB0901	TS0917	LB0924	TS0929	DG0105	TS0109	AVE	GRADE
1 AVOGADRO, AMEDEO	77	90	100	95	100	100	100	98	98	99	98	A
2 BECQUEREL, ANTOINE	23	80	89	90	90	82	92	87	52	70	82	C
3 CHADWICK, JAMES	93	88	84	90	95	93	97	83	100	99	91	B
4 DALTON, JOHN	59	90	100	95	90	71	92	78	76	77	85	C
5 EINSTEIN, ALBERT	59	92	91	100	95	84	95	93	90	74	90	B
6 FARADAY, MICHAEL	85	96	100	95	90	92	90	83	90	88	91	B
7 GAY-LUSSAC, JOSEPH	29	90	100	95	95	86	90	89	96	84	91	B
8 HEISENBERG, WERNER	69	94	100	90	95	74	97	69	90	0	75	D
9 JOULE, JAMES	29	84	100	100	100	73	97	59	64	70	82	C
10 LEWIS, GILBERT	69	84	89	95	100	56	98	61	64	64	77	D
11 MENDELEEV, DIMITRI	93	88	82	85	85	78	97	89	68	76	84	C
12 NEWTON, ISAAC	53	90	98	100	100	97	95	89	100	92	95	A
13 PASTEUR, LOUIS	85	80	100	50	50	69	95	78	90	68	75	D
14 RAMSAY, WILLIAM	53	86	98	90	100	75	100	89	76	89	89	B
15 SCHRODINGER, ERWIN	77	92	98	100	100	67	100	100	100	100	94	A

CLASS AVERAGE = 87

make changes and modifications on the screen. Some school districts require the objectives to be given to students on a handout so that students and their parents can study the learning outcomes for a given unit plan.

The Biological Sciences Study Committee (BSCS) has produced a staff development program to help science teachers use microcomputers in their curriculum. The program is called ENLIST Micros and its main objectives are to improve teachers' knowledge, attitude, and self-efficacy regarding the use of computers in instruction (Ellis, 1992). The ENLIST Micros workshop package is distributed by EME and covers the following topics: (a) the anatomy and use of microcomputer systems, peripheral devices, and trouble shooting; (b) science courseware such as tutorial, drill and practice, game, simulation, and data base; (c) the assessment of software and where to find it, (d) software utilities such as data bases, spreadsheets, graphing, data analysis, word processing, record keeping, test construction, and telecommunications; (e) the use of optical equipment; and (f) the linking of lab instruments to computers to build simple sensor interfaces.

SELECTING GOOD SOFTWARE

The quality of the instructional computer software is a major factor in determining the benefits students derive from them. The quality must be high—easy to use and instructive—otherwise the computer will make little contribution to science education. Many educational programs are available, but not all of them are effective for instruction. A number of programs appear to be useful when first examined, but a closer look will lead science teachers to conclude that they have little educational value. It is easy for commercial courseware developers to make colorful, flashy-looking programs that appear useful; however, in the final analysis these teaching aids have little instructional significance. Careful review and analysis of a microcomputer program is essential before it is used in science instruction.

Perhaps the first thing that science teachers should do before selecting a program is to reflect on the course they are teaching or will be teaching. They must identify key concepts, principles, laws, and science process skills that they wish students to master or what aspects of the nature of science and technology that they want students to explore. This process will give purpose to their search for quality software and will help them to incorporate computers into the curriculum as opposed to using them because "it seems to be the thing to do."

To examine a program, science teachers should read the author's description of the purpose. What is the program trying to teach? Some programs, for example, attempt to teach scientific laws. Chemistry programs instruct students on Boyle's and Charles's laws, which address gas pressure and temperature. Life science programs instruct students on plant growth and human anatomy. Other programs emphasize problem solving and the scientific method. Next, science teachers must decide on the type of instructional program they want to use such as drill and practice, tutorial, simulation, or microcomputer-based labs. Then, from the program description, determine for whom the courseware is intended. Usually the grade level is given for the program but often this information is too general. In addition to determining the merits of various important aspects of the software, a teacher should also take into consideration the cost of the courseware.

A teacher reviewing software should take into consideration the following factors:

- Ease of use
- Ease of learning
- Error handling
- Flexibility
- Documentation
- Accuracy of subject matter content
- Appropriateness of instructional approach to material

- Appropriateness of material for audience
- Adequacy of screen displays/prompts/instructions and so on
- Interest appeal
- Absence of features that restrict the utility of the program (Smith, 1985, p. 50)

Make sure a program is easy to use before purchasing it. Are the instructions easy to follow or does one have to spend a great deal of time figuring out what to do? Good programs must be user friendly or students will become frustrated and avoid using them.

Science teachers need to be concerned about content accuracy of microcomputer programs. They should become mindful of programs that contain factual errors and outdated information. They should also seek programs that use appropriate statistics and accurate graphs. Graphs are an important consideration because graphing is one of the process skills stressed in science education. Often computer programs use graphs that are attractive, yet difficult to read. Remember that not all microcomputer programs intended for science instruction have been developed or carefully reviewed by classroom teachers before they have been placed on the market. Individuals with programming background, who often take the major part in software development, may not know the type of graphing skills, for example, taught at a particular grade level. Also, these people are often not subject matter specialists.

Sethna (1985) suggests that an easy way to evaluate courseware is to develop a form and rate its important aspects. He recommends that users rate software in at least the following five areas: (a) ease of use, (b) documentation, (c) graphics, (d) content, and (e) instruction. Each area can be rated on a one-to-ten scale from low to high.

Sethna points out that perhaps the best people to assess the merits of science instructional software are the students. He suggests that students be given the program to work through. Some of the items that Sethna (1985, Appendix A) has given to students to evaluate science simulations are:

- The program was easy to use without help from the teacher.
- The program helped me learn what I was supposed to learn.
- The program was related to other work on the same topic.
- The meaning of special symbols used in the program could be displayed on the screen when needed.
- The program told me when I had control over waiting time or not.
- Helpful shortcuts were given when I was expected to run the program many times.

In conclusion, a microcomputer and the right software can assist teachers in enhancing their instructional programs. Nevertheless, science teachers must be keenly aware of the technical and the instructional quality of the courseware they wish to adopt. Technically, the courseware must be easy to use, and the screen display and graphics must be easy to view. Instructionally, the content of the program must be accurate and appropriate for the students. Good software can be selected only through systematic inquiry into what is available and careful evaluation of it for a given group of students.

SUGGESTED ACTIVITIES

1. Compile a list of individuals who can help you use the microcomputer and identify software for your personal and instructional use. Science teachers experienced in this area are a valuable resource. Some high school and college students are knowledgeable about microcomputers and software. Neighbors who work with computers can be of assistance.

2. Visit with science teachers who use the computer as a tool for record keeping and for instruction. List the many uses that these professionals have for the computer and the frequencies that they report for each purpose.
3. If you do not own a microcomputer, ask teachers who own them for their recommendations on which computer is most suitable for your needs.
4. Obtain catalogs and periodicals that review software. Identify programs that you might use in your instruction. Refer to the list of suggested resources at the end of this chapter.
5. Analyze microcomputer software that you might use in a science course. You may wish to use the rating scales discussed in this chapter.
6. If you have not used a microcomputer, arrange time to become familiar with one. Microcomputer labs are open to students on college and university campuses, and many public libraries have computer rooms for general use.
7. Take a course or a workshop to learn about computers.

BIBLIOGRAPHY

Bank Street College Project in Science and Mathematics. 1985. *The voyage of the Mimi: A teacher's guide.* New York: Holt, Rinehart, & Winston.

Bross, T. 1986. The microcomputer-based science laboratory. *Journal of Computers in Mathematics and Science Teaching,* 5(3): 16–18.

Eiser, L. 1991, March. Learning to save the environment. *Technology and Learning:* 18–26.

Ellis, J. D. (Ed.). 1989. *1988 AETS yearbook: Information technology and science education,* ERIC Clearinghouse for Science, Mathematics, and Environmental Education. Columbus, OH: The Ohio State University.

Ellis, J. D. 1992. Teacher development in advanced educational technology. *Journal of Science Education and Technology,* 1(1): 49–65.

Ganiel, U., and Idar, J. 1985, Spring. Student misconceptions in science—How can computers help? *Journal of Computers in Mathematics and Science Teaching:* 14–18.

Harkness, J. L. 1991, Feb./March. NSTA issues new position statement on distance learning criteria defined for high quality instruction in science education. *NSTA Reports:* 3–5.

Hirschfelt, J. M. 1987. A spread sheet for your gradebook. *The Science Teacher,* 54(6): 72.

Lam, T. 1984–85, Winter. Probing microcomputer-based laboratories. *Hands On! Microcomputers in Education—Innovations and Issues.* Cambridge, MA: Technical Education Research Centers.

McGreevey, M. W. 1991, March. Virtual reality and planetary exploration. In *Proceedings of the 29th AAS Gooddard Memorial Symposium,* Washington, DC.

Salpeter, J. 1990, Oct. Multimedia spreads its wings: New applications for the Amiga and PS/2. *Technology & Learning:* 40–44.

Sethna, G. H. 1985. Development of an instrument for evaluation of microcomputer-based simulation courseware for the high school physics classroom: Appendix A. Unpublished doctoral dissertation, University of Houston.

Sherwood, R. D. 1985, Summer. Computers in science education: An AETS position paper. *Journal of Computers in Mathematics and Science Teaching:* 17–20.

Smith, R. L. 1985, Spring. JCMST Guidelines for evaluation of software. *Journal of Computers in Mathematics and Science Teaching:* 50.

Sneider, C., and Barber, J. 1990. The new probeware: Science labs in a box. *Technology & Learning:* 32–39.

Stevens, D. J., Zech, L., and Katkanant, C. 1987, Spring. The classroom applications of an interactive videodisc high school science lesson. *Journal of Computing in Mathematics and Science Teaching:* 20–26.

Tamir, P. 1985/86, Winter. Current and potential uses of microcomputers in science education. *Journal of Computers in Mathematics and Science Teaching:* 18–28.

Vockell, E. L., and van Deusen, R. M. 1989. *The computer and higher order thinking skills.* Watsonville, CA: Mitchell Publishing.

Woerner, J. J., Rivers, R. H., and Vockell, E. L. 1991. *The computer in the science curriculum.* Watsonville, CA: Mitchell Publishing.

SUGGESTED RESOURCES FOR MICROCOMPUTER USERS

BSCS. ENLIST Micros, 830 North Tejon, Suite 405, Colorado Springs, CO 80903. Information on teacher and leadership training for using microcomputers in science teaching.

COMPress, A Division of Wadsworth, Inc. P. O. Box 102, Wentworth, NH 03282. A source for science course software.

CONDUIT. The University of Iowa, Oakdale Campus, Iowa City, Iowa 52242. See their catalog of educational software.

Electronic Learning. Scholastic, Inc., 730 Broadway, New York, NY 10003-9538. This magazine is published eight times a year and has many software reviews and useful information.

EME. Old Mill Pond Rd., P. O. Box 2805, Danbury, CT 06813-2805. See their catalog of science software and materials for ENLIST Micros training package produced by BSCS.

IBM Educational Systems. P. O. Box 2150, Atlanta, GA 30055. See their Personal Science Laboratory.

Journal of Computers in Mathematics and Science Teaching. The Association for the Advancement of Computing in Education, P. O. Box 2966, Charlottesville, VA 22902. A periodical devoted to the use of computer technology in mathematics and science teaching.

Optical Data Corporation. 30 Technology Dr., Warren, NJ 07059. See their catalog for science programs, especially "Windows on Science."

Preview! Cambridge Development Laboratory, Inc., Newton Lower Falls, MA 02162. A catalog of science courseware.

Project SERAPHIM. NSF Science Education, Department of Chemistry, Eastern Michigan University, Ypsilanti, MI 48197. See their catalog for descriptions of hundreds of inexpensive chemistry programs on floppy disk.

Sunburst Communications. 39 Washington Ave., Pleasantville, NY 10570.

The Computer Teacher. University of Oregon, 1787 Agate St., Eugene, OR 97403-1923.

The Science Teacher. National Science Teachers Association, 1742 Connecticut Ave., Washington, DC 20009. Refer to the "Idea Bank" and "Reviews" sections for information and ideas on science courseware.

Vernier Software. 2920 S. W. 89th Street, Portland, OR 97225. Get their catalog, *Science Software for the Classroom and Lab.*

Science reading materials that are selected and used properly can encourage and motivate students to carry out their reading assignments and can facilitate student achievement.

CHAPTER 12

Using Reading Materials in Science Instruction

Reading materials are some of the most important resources for instruction available to science teachers. Textbooks, laboratory manuals, newspapers, magazines, and nonfiction books should be part of the science curriculum. In spite of the fact that many students are poor readers and participate in more passive activity, such as watching television, there are many strategies that science teachers can use to engage students in reading and have them benefit from it. Reading can improve the scientific literacy of students and must be considered for teaching science through inquiry.

Overview

- Discuss the structure and function of typical middle and senior high school science textbooks.
- Review the textbook adoption process at the state and local levels.
- Examine many criteria and techniques that are useful when evaluating science textbooks.
- Present many strategies and techniques that can help students learn from science textbooks.
- Emphasize the importance of using newspapers, magazines, and nonfiction science books to stimulate interest in science, and to improve skills and knowledge.

Printed material remains the most widely used of all teaching aids in science instruction. Whether students are working in groups, pairs, or individually, they are regularly using science textbooks, paperbacks, popular science magazines, teacher handouts, laboratory manuals, and other printed matter. Furthermore, lectures and discussions often revolve around assigned textbook reading, laboratory exercises, and supplemental reading from reference books. In the laboratory, students must be able to read and interpret the directions associated with various types of laboratory exercises in order to carry them out successfully.

Many students have difficulty reading science textual materials and, consequently, cannot carry out their assignments effectively. Printed materials assigned by teachers are often written at grade levels not suitable for their students. It is therefore important that teachers evaluate reading materials to ensure they are at the appropriate reading level for their students. Even then, the teacher may still find that some students will have difficulty reading and understanding the material and will require special attention to surmount their reading problems. In such cases, some teachers select science materials at an eas-

ier reading level. Once students find that they can successfully read and comprehend the selected material, they often read more and eventually are able to handle their reading assignments. There are, however, students who have very serious reading problems that science teachers are not qualified to handle. These students must be given science reading materials that are suited for their reading levels.

Science textbooks as well as other printed material should contain more than definitions. Often science textbooks do no more than define terms and explain ideas. They fail to include the work of those who discovered ideas and those whose work led up to these discoveries. These texts leave out the creative process that goes into scientific inquiry and the thinking that reflects this enterprise. Science textbooks that present science as a body of knowledge, but rarely mention the investigations that produced this knowledge, do a disservice to the teachers and students who use them. Science textbooks must go beyond telling about and explaining ideas. They must illustrate how science is a way of thinking about the world in which we live.

Printed matter must be selected with great care to encourage and motivate students to com-

Many students have difficulty reading science materials. Science teachers should provide experiences to help students learn from their textbooks and other assigned readings.

plete reading assignments and to read independently. Reading materials that are selected and used properly can enhance student achievement and stimulate scientific interests. If the main ideas or the context in which the ideas are written is relevant and interesting, the students will concentrate and learn from their reading experiences. Nonfiction science books, newspaper articles, and journal articles are excellent resources for middle and secondary school students to read as part of their science curriculum.

THE SCIENCE TEXTBOOK

Science textbooks continue to be a major component of the curriculum at all grade levels of instruction. These teaching aids are widely and frequently used to support instruction and develop curricula. In many classrooms, they provide the majority of the instructional support beyond the teacher. Science textbooks contain much of the scientific information students receive (Mayer, 1983), which influences how they perceive the scientific enterprise.

Some science textbooks can be used to initiate inquiry and to suggest interesting investigations. The inquiry-oriented textbook poses questions that students can pursue, some of which have already been answered by others and some of which have not. The inquiry-oriented textbook can stimulate students to become active learners instead of individuals who passively absorb information. Students are challenged with problems that involve them in the collection and organization of data. Some science textbooks contain historical accounts of how laws and principles were developed from the thinking of the individuals who formed them. In addition, the tentative nature of science is often reflected in inquiry-oriented textbooks, illustrating how certain theories are modified and replaced as new information is acquired.

Unfortunately, many modern science textbooks are still encyclopedias of scientific information. These textbooks basically present science as an organized body of knowledge. Although many science textbook authors indicate that the discovery approach is stressed in the activity section of the book, often these texts provide answers to all questions and solve all problems. In many instances the details of all steps of an experiment are presented, and students are told what should be observed and what conclusions should be reached. These textbooks provide students with little or no opportunity to speculate, devise experimental methods, and draw their own conclusions. They emphasize the content of science and minimize the investigative thinking aspects of the scientific enterprise. These texts present the antithesis of science by precluding inquiry behavior.

Before selecting a textbook, the science teacher must consider a number of factors. A science course that is based on the conventional approach usually does not require special laboratory facilities or special materials and equipment. Instead, this type of course covers a great deal of subject matter. On the other hand, teaching science through inquiry involves more materials and equipment to accommodate student activities. Such an approach requires more class time than the traditional approach, limiting the breadth of the content covered. However, what is lost in terms of subject matter coverage may be offset by gains in understanding the nature of scientific inquiry and perhaps conceptual understanding.

Teachers must realize that textbooks are designed for a broad market. They must be usable in any state, school, and teaching situation. Textbooks cannot take advantage of the unique resources of the various communities, anticipate local events, or ensure relevance to any particular group of students. Teachers should also be aware that textbooks contain much more material than can be adequately and effectively taught in most science courses.

The range of students' abilities is generally very great in most science classes. This is certainly an important consideration when adopting and using textbooks. Able students, of course, can use the textbook without difficulty, whereas the less able need a great deal of help. Nevertheless, middle and secondary school science teachers can assist all students to profit from science textbooks if they learn certain strategies and techniques for their effective use.

In selecting a science textbook, science teachers must consider many aspects, including the purpose for the text, the supporting materials that accompany it, the content organization, readability, illustrations and pictures, and the end-of-chapter exercises. Along with these components, science textbooks should provide a reasonable balance of presenting science as a way of thinking, science as a way of investigating, science as a body of knowledge, and science and its interactions with technology and society. Recall that in earlier chapters these four themes have been discussed extensively.

Structure of the Textbook

The modern science textbook is a complex publication, packed with information and ideas. It usually comes in the form of a student's edition and a teacher's edition with annotations. The teacher's edition contains everything in the student's edition plus additional information and resources. The following discussion considers the basic features of both the student's and teacher's edition of a science textbook.

Student's Edition. The typical science textbook includes a preface, a table of contents, an introduction, sections or units divided into chapters, a glossary, appendixes, and an index. Science teachers should skim over these sections to begin their evaluation of a given textbook in order to gain an overview of the structure and content of the publication.

The preface gives a brief explanation of the approach used to present the subject matter. It usually indicates the role of the student in learning science and often stresses inquiry, discovery, and problem solving. The preface also explains the contents and sequence of units.

The table of contents is usually divided into sections or units, which are further divided into chapters. Each section or unit is written around a common theme or general topic. In some texts the authors recommend that the units be taught in sequence; in other texts the units can be taught in any order. Today's science textbooks are flexible, giving teachers a great deal of latitude in their use.

The science textbook invariably begins with an introductory section that usually defines science. The introduction usually presents and discusses the scientific method and scientific attitudes. This section stresses the importance of science to society and sets the stage for studying science.

Some textbooks begin each chapter with a list of instructional objectives. These objectives focus on the main ideas and important learning outcomes that can be expected from studying the contents of the chapter. Most chapters also begin with an introduction to motivate and interest students in what they are about to read.

The textbook chapters are divided into sections, which break down the main topics into smaller ideas. Most have questions listed throughout the sections to serve as checkpoints or self-checks. Some of these questions serve to stimulate further thought on the subject at hand. Illustrations, graphs, and pictures are prominent items in today's science texts. These are used to improve the readability and attractiveness of the text. Some textbooks highlight key phrases and key items in bold type. They also define key terms and stress important ideas in the margins. Chapters conclude with questions, problems, and ideas for further study. Some textbooks include laboratory activities within the text material.

The glossary, appendixes, and index are normally found at the end of the text. The glossary is an extremely helpful section that defines the key terms, technical words, and ideas, and can be an excellent instructional aid. Similarly, a good appendix should contain important information for the teacher as well as for the students. Many appendixes include formulas, units of measurement, tables of trigonometric functions or significant figures, scientific notation, graphs, and charts. A well-prepared index can be very useful in locating important words, principles, concepts, and ideas that are explained in the text.

Teacher's Edition.   Most publishers produce a useful teacher's edition to accompany the student's edition. The teacher's edition contains a teacher's manual located at the front of the student's text. The manual usually begins with a brief rationale of the textbook program. It provides unit and chapter overviews. Often, instructional objectives are listed with suggestions on how to determine when students have achieved them. Background information is given to provide the teacher with more knowledge of the topics in each chapter. Suggestions for demonstrations, along with tips for preparing laboratory exercises, are presented in the textbook chapters or in the laboratory manual. The teacher's edition also gives answers to questions and self-checks in the chapters, as well as to questions and problems at the end of chapters.

The student's text in the teacher's edition contains annotations, which provide useful background information, comments, questions to ask students, and answers to some questions. These may be located in the margins or elsewhere on the page. The marginal annotations are usually keyed to the main concepts being presented and are placed near them. The annotated questions and background information are valuable adjuncts for understanding the key ideas presented in the text, and, unfortunately, they usually do not appear in the student's textbook.

The general features just reviewed are some preliminary considerations for selecting a science text. A text with a thoroughly prepared teacher's manual can be very useful to inexperienced teachers because it can provide many helpful tips regarding the preparation of demon-

strations and laboratory work. The background information found in the teacher's manual can quickly fill in knowledge gaps that a teacher may have about a subject or a specific topic. The orientation of the textbook can be determined by a quick review of the teacher's manual and student's textbook. This overview will indicate the best use of the textbook.

Functions of Textbooks

Science textbooks serve many functions and are a valuable resource for teaching science. This is especially true for inexperienced teachers who have been assigned to teach in science areas for which they are not prepared or qualified. For example, many new teachers hired to teach science in senior high school are biology majors. These individuals may be assigned to teach only one or two periods of biology per day but several periods of physical science, for example. Many of these teachers are not well prepared to teach physics and chemistry because their strength and preparation are in biology. Thus, the textbook assigned for their courses becomes the curriculum. On the other hand, for experienced science teachers, the assigned textbook may be used for summary and review purposes or as a reference. Most experienced teachers have collected many printed resources over the years. For them, the textbook is merely one of many resources.

Most science textbooks are valuable curriculum resources and should be used to maximize learning. Textbooks are useful tools for teachers, students, and curriculum developers. Some parents take an interest in their children's textbooks and often complain to teachers and administrators when students are not given homework assignments in connection with the textbook.

Textbook for Summary and Review.   A science textbook is very useful for summary and review. Students, especially poor readers or slow learners, can benefit most from the textbook when they have been given a number of concrete experiences and activities before they read textbook assignments. The more experiences students are given with the concepts, principles, and terms associated with a given chapter, the more successful they will be in reading and learning the material in a given chapter. What students do before they are assigned reading material is as important as what they do during and after the reading assignment. Summary and review are effective when science teachers incorporate them into their lesson and unit plans, and use them after or in conjunction with laboratory activities, lectures, discussions, demonstrations, and field trips. They also work well when they are used in conjunction with a reading strategy.

Textbook as a Reference.   Science textbooks are useful as a reference; however, the information they contain is necessarily condensed and rather general. If the assigned textbook is to be used as a reference, it is best to encourage students to use other reference material as well, such as encyclopedias, specialized books, paperbacks, periodicals, and articles to supplement their reading. When the assigned text is used as a reference, it may help students gain an overview of a topic.

Textbook as a Course of Study.   In one sense, some science teachers use the textbook as their course of study and the chapter contents as their lecture material. These teachers often assign students to read a chapter in their textbook and require them to answer questions from the end of the chapter. They may also perform some of the activities suggested in the chapter and in the laboratory manual. This conventional approach, however, has its limitations. Working through the textbook, chapter by chapter, and covering its contents can be boring and detrimental to student's interest and attitudes toward science.

Some textbooks, however, have been written to accompany a particular science program and in effect constitute a course of study. When using such texts, teachers should make certain that the organization and sequence of the chapters in the

text is consistent with their own course objectives. The teacher should also recognize that the sequence in the textbook may or may not be one that students find logical or psychologically beneficial to their way of thinking.

Beginning teachers should not hesitate to use a textbook as a course of study during the first few years, especially if their course load is heavy and requires a number of different preparations. One of the most important aspects for using a science textbook as a course of study is the laboratory component. In some textbooks, the laboratory exercises are presented within the chapters; in others these exercises are found in a separately bound laboratory manual. These laboratory exercises can be valuable, concrete experiences during the course.

Textbook Adoption Process

Textbook selection for science programs throughout the United States is usually performed by teachers. Committees are formed for the purpose of identifying the most appropriate textbooks for a given state, a district, or school. In some states, the adoption process begins at the state level, whereas in other states this process begins at the district level.

In the states where textbook adoption begins at the state level, a committee of teachers is identified to select the textbooks in all of the subjects that are eligible for new textbooks. Such committees are composed of teachers in all subject areas, including mathematics, science, social studies, reading, and vocational education. Often, committees are not composed solely of science teachers to select science textbooks on the state level. Therefore, the selection of texts for all subjects in the state may be accomplished by a very broadly represented committee.

Textbook publishers often make presentations to the selection committee to promote their texts. Generally, the selection committee is charged with the responsibility of selecting a certain number of textbooks (which may range between five and seven) for a given subject. The competition is fierce among publishers to get their textbooks on the state adoption list. In addition, hearings are held for individuals or groups who wish to protest the adoption of any textbook. Special-interest groups who make their views known to the committee may have more influence on which textbooks get adopted or rejected than the silent majority of educators who rarely attend textbook adoption hearings.

The next level of adoption is the local school district. At this level a committee of teachers is identified to select the most appropriate text in a given subject. In an area where there is a statewide textbook adoption process, the local committee must select its science textbook, for example, from the list of several texts that the state committee has adopted. The local district can select a science text that is not on the state's adoption list, but funds will not be provided by the state to the district to purchase books not on the list. In states that do not have a state adoption procedure, the local committees can select any text they wish as long as it fits the district's guidelines. One guideline, for example, that has kept some innovative science programs from being implemented in the schools is the requirement that textbooks be hardbound. This has prevented many school districts and states from using modular or minicourse science textual material. Today, some states have made provisions to include multimedia computer technology science programs as a curriculum resource option.

Multiple Textbook Adoptions.   Textbooks are often written for special audiences, and such books can be helpful to the teacher trying to meet a wide range of needs in the classroom. Some schools make multiple adoptions possible, allowing the use of different texts at different times during the year for different groups of students. Although teachers may have difficulty keeping track of more than one set of textbooks, this technique does give the teacher flexibility in using the text as a teaching resource and for taking care of individual differences in the class-

room. Some school districts are adopting a combination of commercial textbooks and computer programs for courses. Instead of selecting one textbook for all students, they adopt one classroom set of textbooks and an interactive multimedia microcomputer program for a given course. This approach provides many up-to-date curriculum resources.

Textbook Evaluation

A teacher should strive for objectivity in evaluating textbooks. Merely looking through a textbook is not sufficient to make an evaluation. Even a careful reading may not leave teachers with an accurate impression of the book's usefulness. For instance, a teacher may favor a particular textbook despite serious defects because it contains an abundance of material dealing with the teacher's favorite topic or area of expertise. The teacher must also remember that purchasing textbooks is an expensive endeavor for a state, school district, or school. Therefore, it is imperative that science teachers be as discriminating as possible in this selection. They should avoid judging a book according to inappropriate criteria and must view the book as if they were the student, immature and untrained in science. The following aspects of a textbook should be considered carefully.

Content.   Of course, the content should be consistent with the goals and objectives of the present curriculum; consequently, changes in objectives can render textbooks useless. Thus, when selecting a textbook, keep in mind the broad ideas that have been discussed regarding the nature of science and technology. The content must reflect the scientific enterprise and the tentative aspects of scientific knowledge. It should deal with technology and include the practical scientific applications in everyday life. The content must be appropriate for the students' reading levels. Many science teachers complain that the adopted textbook is too difficult for their students; others say that it is too juvenile. Finally, the content must be accurate and up-to-date.

There is no question that the content of a science textbook is important. If this is true, science teachers must select content that reflects the nature of science. One procedure to accomplish this end has been developed by Chiappetta, Fillman, and Sethna (1991). The approach is based upon the ideas in his methods textbook and emphasizes a balance of themes that stress scientific literacy. The following sections describe these themes to evaluate science textbooks.

1. *Science as a body of knowledge.* Printed material emphasizing science as a body of knowledge presents, discusses, and asks the student to recall information, facts, concepts, principles, laws, theories, and so forth. It reflects the transmission of scientific knowledge where the student receives information. This theme typifies most textbooks and presents information to be learned by the reader. Textbook materials of this type:

(a) present facts, concepts, principles, and laws,
(b) present hypotheses, theories, and models, and
(c) ask students to recall knowledge or information.

2. *Science as a way of investigating.* Text that emphasizes investigation promotes thinking and doing by asking the student to "find out." It reflects the active aspect of inquiry and learning, which involves the student in the methods and processes of science, such as observing, measuring, classifying, inferring, recording data, making calculations, experimenting, comparing, and explaining. The instruction can include paper and pencil as well as hands-on activities. Written materials that reflect this theme:

(a) direct students to answer a question through the use of materials,
(b) encourage students to answer a question through the use of charts, tables, and the like,
(c) direct students to make a calculation,
(d) encourage students to reason out an answer, and
(e) engage students in thought experiments.

3. *Science as a way of thinking.* Text emphasizing science as a way of thinking illustrates how science in general, or a certain scientist in particular, went about "finding out." This aspect of the nature of science represents thinking, reasoning, and reflection. The reader is told how the scientific enterprise operates. Written materials that reflect this theme:

(a) describe how a scientist experimented,
(b) show the historical development of an idea,
(c) emphasize the empirical nature and objectivity of science,
(d) illustrate the use of assumptions,
(e) show how science proceeds by inductive and deductive reasoning,
(f) give cause and effect relationships,
(g) discuss evidence and proof, and
(h) emphasize how science is a discipline disposed to self-examination.

4. *Interaction of science, technology, and society.* STS material illustrates the impact of science on society, and the interaction of science and technology. It focuses upon the application of science and how technology helps or hinders humankind. In addition, it involves social issues and careers. However, the reader as the recipient of information does not have to carry out an investigation. Textbook materials that reflect the STS theme:

(a) describe the usefulness of science and technology to society,
(b) point out the negative effects of science and technology on society,
(c) discuss social issues related to science or technology, and
(d) address careers and jobs in scientific and technological fields.

Organization. At least two types of organization can be found in middle and secondary school science books. In one type, the organization of material is logical because it is developed in the way that a well-prepared person in science might organize it; this is the form usually seen in college texts and in high school physics and chemistry texts. In the other type, the organization is psychological because it is presented in the order that might be most meaningful to the students for whom the text is designed. Some general science and biology texts are organized this way.

Ideally, a textbook should be consistent with the organization of a course prepared by the teacher. If the book is to be used as a course of study, there should be a suitable selection of units, in an appropriate sequence. If the sequence is inappropriate, then the text should be flexible enough to allow teachers to rearrange units without losing the conceptual development of major themes. This will also allow the teacher to teach relevant units in the text when current science events of concern to the students are brought to the teacher's attention.

Teachers should also determine if the material within the chapters or units is well organized, which allows them to make assignments from the text with ease. Unorganized material will only confuse students and discourage them from reading the text and doing the homework assignments based on the textual material.

Readability, Style, and Vocabulary. Literary style has much to do with the readability of a book. Although style is difficult to judge, some points to look for are (1) length of sentences, (2) directness of sentences, (3) number of ideas per sentence, (4) use of lead sentences for paragraphs, (5) presence or absence of irrelevant thoughts, and (6) continuity of thought.

When evaluating a textbook, science teachers must decide whether the vocabulary is excessively difficult or inappropriate. They may want to refer to a standard word list designed for secondary school texts and other publications. The teacher should remember that new advances in

science may introduce words that are not included in some of the word lists.

Science teachers should also remember that a textbook that is easy to read may not be a suitable text for a given situation; readability is only one factor to be considered. Sometimes the text that is somewhat harder to read has other features that outweigh this point.

Quantitative methods can be used to predict the reading level of the material. Readability formulas, which are used to give indications of the difficulty readers will have with passages, have been developed by reading specialists. Readability formulas are also used to reveal the reading grade level of textbooks and other reading matter. Some formulas provide information concerning the words that cause difficulty; others are concerned with the length and complexity of sentences in the reading matter.

Formulas may overestimate the difficulty of reading material because they cannot take into account the specialized vocabulary background that the readers possess. This is especially true with science material. Even though these formulas can be used to some degree to predict reading difficulty, they cannot be a substitute for pretesting of the material. Reading formulas do not take into consideration the prior knowledge of the students, which is a critical factor in estimating readability and comprehension (Dishner et al., 1992). Unfortunately, publishing companies rarely pretest materials. In some cases, the authors may have used it in their own teaching and consequently revised it as reading difficulties arose.

Readability formulas are used as a quick way to assess the difficulty of textbooks. A simple formula that has been very popular is the Fry Readability Graph. This formula uses two variables—sentence and word difficulty—to determine readability. Because sentence length has been found to contribute to the structural complexity of a passage, the passages with long sentences are assessed to be more difficult. Similarly, the number of syllables per word has been found to contribute to the difficulty of words—simple words are usually monosyllabic and difficult words are polysyllabic. These two factors can be used together to determine the readability of a textbook.

Directions for using the Fry Readability Graph are as follows:

1. Select three 100-word passages from near the beginning, middle, and end of the book. Do not count proper nouns, initials, or numerals.
2. Count the total number of sentences in each 100-word passage (estimating to the nearest tenth of a sentence). Average these three numbers.
3. Count the total number of syllables in each 100-word sample. There is a syllable for each vowel sound; for example, cell (1), current (2), and voltmeter (3). Average the total number of syllables for the three samples.
4. Plot on the graph (figure 12–1) the average number of sentences per 100 words and the average number of syllables per 100 words. Most plot points fall near the heavy curved line. Perpendicular lines mark off approximate grade level areas.

Example readability calculation

	Sentences per 100 words	Syllables per 100 words
100-word sample, p. 30	9.5	112.0
100-word sample, p. 296	9.4	141.0
100-word sample, p. 437	7.8	174.0
Total	26.7	427.0
Mean (the sum divided by 3) =	8.9	142.3

The example shown here is a readability estimate of a middle school physical science textbook. When the average number of sentences per 100 words (8.9) and the number of syllables per 100 words (142.3) are plotted on the Fry graph, the readability estimate of the textbook is approximately at the upper sixth-grade level. Because this is an averaging technique, some of the text passages will be below the sixth-grade level and some will be above the sixth-grade level.

Science teachers are cautioned not to rely solely on readability graphs and formulas for determining the reading level of textbooks, because these procedures have their limitations (Singer & Donlan, 1980). Although sentence complexity and word difficulty seem to be the two most reliable factors that contribute to readability, they can be misleading. Researchers have not been able to tell the practitioner very much about syntax difficulties. It has been pointed out that students are more likely to comprehend written passages in which the syntax is similar to the syntax they use in their own oral language. Furthermore, when sentences are reduced in length to produce a lower readability index, they tend to fragment ideas and lower comprehension.

In the past, textbook authors and publishers have attempted to shorten longer sentences to improve readability, as estimated by a given formula. Although this rates a given sentence or passage as easier to read, it may really complicate the reader's ability to comprehend the passage (Vacca, 1981). The inferential burden of the reader may be increased when long sentences are artificially broken into short sentences because this serves to separate the intended meaning (Pearson, 1974–1975). Reading indices do not seem as important today in determining readability as they did many years ago. Remember, the inherent interest of the passages and the students' background play a large role in reading comprehension.

Vocabulary, and science vocabulary in particular, presents a problem in determining readability. Science vocabulary tends to be multisyllabic, which inflates the readability of a passage. In other words, science vocabulary tends to make the passage appear more difficult than

FIGURE 12–1   Graph for estimating readability (Courtesy Edward Fry, Rutgers University Reading Center, New Brunswick, NJ 08904.)

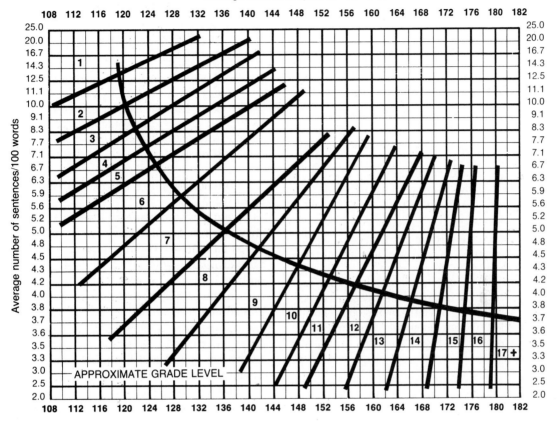

Average number of syllables/100 words

it is because the passage often consists of larger multisyllabic words. For example, *hydrocarbon, petroleum, distillation,* and *paraffin* all appear in certain middle and high school science textbooks. Although these words are multisyllabic technical terms, they may be familiar to many of the students and may not add as much difficulty to a written passage as a readability graph or formula might suggest.

If great accuracy is desired for determining the readability of a science textbook for a group of readers, then science teachers should involve the students in the process of determining the readability of a given text. Reading difficulty for a particular student depends as much on the student's reading ability, background, and interest in the material as on the reading level of the text.

The *cloze technique* uses a totally different procedure than the Fry Graph and readability formulas to determine reading difficulty. It presents a student with a passage from which words have been deleted and asks the student to supply the exact missing word. This tests the reader's ability to infer the missing words, which challenges one to determine syntactical and semantic relationships and to recall vocabulary words.

A suggested procedure to construct, administer, score, and interpret a cloze test is given by Vacca as follows:

1. Construction
   - Select a reading passage of approximately 275 words from material that students have not yet read but which will be assigned.
   - Leave the first sentence intact. Starting with the second sentence, select at random one of the first five words. Delete every fifth word thereafter, until a total of 50 words has been deleted. Retain the remaining sentence of the last deleted word.
   - Leave an underlined blank 15 spaces long for each deleted word, as the passage is typed on a ditto master.

2. Administration
   - Inform students that they are not to use their textbooks or to work together in completing the cloze passage.
   - Explain the task that students are to perform. Show how the cloze procedure works by providing several examples on the board.
   - Allow students as much time as they need to complete the cloze passage.

3. Scoring
   - Count as correct every *exact* word students supply. *Do not* count synonyms even though they may appear to be satisfactory. Counting synonyms will not change the scores appreciably, but it will cause unnecessary hassles and haggling with students. Accepting synonyms also affects the reliability of the performance criteria since they were established on exact word replacements.
   - Multiply the total number of exact word replacements by two in order to determine the student's cloze percentage score.
   - Record the cloze percentage scores on a single sheet of paper for each class. A teacher will now have, for each class, from one to three instructional groups that can form the basis for differentiated assignments.

4. Interpretation
   - A score of 40% to 60% indicates that the passage can be read with some competence by students. The material will challenge students if they are given some form of reading guidance.
   - A score of above 60% indicates that the passage can be read with a great deal of competence by students. They may be able to read the material on their own without reading guidance.
   - A score below 40% indicates that the passage will probably be too difficult for students. They will either need a great deal of background to benefit from the material or more suitable material. (Vacca, 1981, pp. 270–273)

A sample portion of a cloze test is given in figure 12–2.

In summary, problems of readability may be anticipated and avoided through techniques to assess the intrinsic nature of the material (such as readability graphs) or to indicate potential problems that individuals might have with a particular reading source (such as the cloze procedure). The unique vocabulary of science can also be a potential stumbling block to student understanding. Unfortunately, the various readability graphs and formulas, although easily applied, do not precisely assess this factor. Certainly, the daily increase in scientific knowledge brings with it an increasing science vocabulary. It is true that the meanings of many scientific words can be deduced by analyzing their component parts, but this is certainly not the case for a large portion of scientific terminology. Many terms do not result in meanings that are consistent with a simple combination of component parts (e.g., *entropy*). Furthermore, many scientific terms are not used in everyday conversation and are, therefore, unfamiliar to the reader. Many of these terms are so specific to science that they cannot even be found in some dictionaries (e.g., *clade* and *cladogenesis*). If a term is crucial to the meaning of a sentence and the student is unable to discern it, then she cannot be expected to understand the sentence.

FIGURE 12–2   A sample portion of a cloze test with answers (Adapted from W. D. Schraer and H. J. Stoltze, *Biology: The Study of Life* [2nd Ed.] [Newton, MA: CEBCO/Allyn & Bacon, 1987], pp. 249–250.)

The autonomic system consists entirely of motor neurons. Sensory information ____1____ this system is provided ____2____ the same sensory nerves ____3____ serve the somatic system. ____4____ in the autonomic system ____5____ in motor neurons in ____6____ brain or spinal cord. ____7____, the axons of these ____8____ do not extend to ____9____ organ involved. Instead, each ____10____ synapses with a second ____11____ neuron, which then carries ____12____ impulses to the muscle ____13____ gland. Some of the ____14____ bodies of these second ____15____ are located in ganglia ____16____ outside the brain and ____17____ cord. The ganglia, which ____18____ interconnected by nerves, form ____19____ chains alongside the spinal ____20____. Other ganglia are located ____21____ in the body. Some ____22____ them form large clusters ____23____ plexuses (plek-sus-sez).

Answers:

1. for	7. However	13. or	19. two
2. by	8. neurons	14. cell	20. column
3. that	9. the	15. neurons	21. elsewhere
4. impulses	10. axon	16. just	22. of
5. start	11. motor	17. spinal	23. called
6. the	12. the	18. are	

The following passage, which appears in a high school biology textbook, should illustrate this point.

The structures in the next ring, the *stamens,* and the ones in the center, the *pistils,* are directly involved in sexual reproduction. A stamen consists of an *anther,* which is a sporangium that contains the pollen *grains,* and a stalk that supports the *anther.* The pistil has three parts: a swollen base called the *ovary* that rests directly on the *receptacle,* a slender, stalk-like *style,* and a *stigma,* which is the tip of the pistil. The stigma is usually sticky or has hairs that help pollen grains stick to it. The ovary has one or more cavities that contain from one to several hundred *ovules.* An ovule contains the female *gametophyte.* After the egg is fertilized, the ovary develops into a fruit. The ovules become seeds. (Otto & Towle, 1985, p. 357)

The preceding passage is presented in a section on sexual reproduction in plants. The italicizing is ours, and its purpose is to indicate the specialized vocabulary used in scientific writing. Notice the high concentration of technical terms in the eight-sentence passage. The message should be obvious. The teacher must be sensitive to the technical vocabulary unique to the discipline of science and realize that students can learn only so many of these terms within a given chapter or unit. Science textbooks place too much emphasis on terminology and vocabulary, which results in students memorizing large amounts of information (Yager, 1983). The teacher's goal is to help develop student understanding of scientific principles.

Illustrations.   The quantity and quality of illustrations in a textbook must be considered.

Photographs should be clearly reproduced. Diagrams should be carefully made and attractive. Color in the illustrations adds eye appeal and when properly used has considerable teaching value. Some texts include colored illustrations of drawings and diagrams on transparent plastic sheets. Biology textbooks, for example, may include colored overlays of the digestive system, the circulatory system, and other parts of the human body.

Most books today are profusely illustrated. However, check to determine whether the illustrations have been well selected. The illustration should amplify the material printed on the same page, and the caption should tie in closely with the paragraphs around it. Photographs that have little or no relation to the text are "padding" and represent careless book production.

Teaching Aids.   Different parts of a textbook serve as valuable teaching aids. The table of contents and the index should be comprehensive. Some teachers like a glossary; if so, they should check for necessary technical and difficult words used in the book. The activities at the end of a chapter are important. If teachers wish to use them to address the individual differences among students, the range of difficulty and type of suggested activities should be great. If teachers wish to use them chiefly for summary, the activities should be closely related to the content of the chapter.

Mechanical Makeup and Appearance.   The following features are usually considered under the heading of mechanical makeup and appearance: attractiveness and size of the book, durability of the binding and paper, and legibility of print. The book should have an attractive overall appearance with an interesting cover design and binding. The size of the book may have unsuspected psychological implications. If the book is poorly proportioned—too long and narrow or too short and thick—it will be unappealing. If the book is excessively large, its mere encyclopedic bulk is likely to make the student feel that it will be dull and difficult reading. The size of textbooks has varied considerably during the past few years. Texts used to be large encyclopedias. Many of the better, newer texts are attempting to select content that is directed toward developing an understanding of the basic principles rather than being encyclopedias of information.

The length of line and the size and legibility of type are important mechanical factors that determine the ease and comfort with which the text may be read. If the book is rather wide, double-column pages will be more attractive and legible. Ample space should be left between lines to provide for ease in reading. The type should be sharp and printed on a good-quality paper.

Rating Science Textbooks.   Various checklists have been devised to make the selection of textbooks as objective and efficient as possible. Many of the checklists make use of the points that have been discussed in the previous sections of this chapter. These suggestions and others have been translated into a checklist that can be used to rate textbooks under consideration (figure 12–3). A point scale from 1 (unsatisfactory) to 5 (excellent) can be used to evaluate the worth of each textbook according to each criterion. Sometimes a major item is broken down into more specific components to improve the validity and reliability of the procedure. After the textbooks have been rated, their overall ratings can be compared, as can their ratings on specific criteria.

Other criteria can be used to assist in textbook evaluation along with those listed in figure 12–3. For example, one should determine whether there is a teacher's edition for the textbook and, if so, determine whether it provides background information and suggestions for laboratory activities and preparation and statements of goals and objectives for each chapter.

FIGURE 12–3   Science textbook rating system

Title of textbook _____   Publisher _____
_____   Year published _____

Author _____   Price _____

Give each item listed between 1 and 5 points based on the following categories:

1 point   Unsatisfactory
2 points   Poor
3 points   Satisfactory
4 points   Good
5 points   Excellent

The total number of possible points using this system is 290. A partial score for each category should first be determined. These partial scores may then be totalled to arrive at an overall score for each textbook under consideration.

Use the following chart to determine the overall quality of the textbook:

Scores between 261 and 290   Excellent
Scores between 232 and 260   Good
Scores between 203 and 231   Satisfactory
Scores between 174 and 202   Poor
Scores below 174   Unsatisfactory

Criteria	Unsatisfactory 1	Poor 2	Satisfactory 3	Good 4	Excellent 5
1. Content					
1. Content is up to date.					
2. Content is scientifically accurate.					
3. Content is appropriate for grade level of students.					
4. Content reflects the scientific enterprise.					
5. Content includes historical background and development of concepts and principles.					
6. Content includes moral and ethical implications of science.					
7. Content stresses the interaction of science, society, and technology.					
8. Content is relevant to students (e.g., rural or urban).					
9. Unit objectives are clearly stated.					
Partial score _____					
2. Organization					
1. The organization of topics or units fits the sequence of the syllabus to be used for the course.					
2. Organization is flexible, permitting variation in sequence.					
3. Material within the chapters is well organized.					
Partial score _____					

FIGURE 12–3 *continued*

Criteria	Unsatisfactory 1	Poor 2	Satisfactory 3	Good 4	Excellent 5
3. Reading level					
1. Reading level of text is appropriate for grade level of students.					
2. Technical words are kept to a minimum.					
3. Technical language is appropriate for grade level.					
4. Technical words are clearly explained when used.					
Partial score _____					
4. Understanding concepts and principles					
1. Concepts and principles are appropriate for grade level of students.					
Partial score _____					
5. Instructional approach					
1. Approach stresses science as inquiry.					
2. Content is presented at a variety of cognitive levels.					
3. Approach is suitable for a wide range of student abilities.					
Partial score _____					
6. Illustrations					
1. Illustrations are up to date.					
2. Photographs are clear and of good quality.					
3. Line drawings are well done and clearly executed.					
4. Illustrations are tied in with the content of the text.					
5. Captions for illustrations are well written and appropriate.					
6. Illustrations are useful in teaching.					
7. Illustrations are strategically placed within the text.					
Partial score _____					
7. End-of-chapter teaching aids					
1. Questions are well constructed and useful for review.					
2. Activities are suitable for a wide range of student abilities.					
3. Vocabulary lists are pertinent and useful to students.					
4. Suggested activities and projects are thought-provoking and challenging.					
Partial score _____					
8. Laboratory activities in text and/or accompanying manual					
1. Laboratory activities in text and/or accompanying manual are suitable for the cognitive capabilities of students at the grade level.					
2. Laboratory activities in text and/or accompanying manual involve skills that are within the manipulative capabilities of students at the grade level.					
3. Laboratory activities stress investigation.					

FIGURE 12–3 *continued*

Criteria	Unsatisfactory 1	Poor 2	Satisfactory 3	Good 4	Excellent 5
4. Laboratory activities are safe for students to perform.					
5. Facilities needed for laboratory activities are available.					
6. Equipment and apparatus are available for suggested laboratory activities.					
7. Activities are relevant to the content presented.					
8. Laboratory activities are included in text material.					
9. Laboratory activities are included in a separate laboratory manual.					
10. Laboratory activities can be conducted during a class period.					
11. Laboratory activities are identified in the text and correlated with the laboratory manual.					
Partial score _____					
9. Teacher aids					
1. Teacher's guide is available for text and is useful.					
2. Annotated edition is available for text and is useful.					
3. Unit tests are provided.					
4. Equipment list is available for laboratory program.					
5. Supply list is available for laboratory program.					
6. Student workbook is available.					
7. Library resource materials are suggested.					
Partial score _____					
10. Indexes and Glossaries					
1. Glossary is accurate and complete.					
2. Index is accurate and complete.					
3. Table of contents is accurate and complete.					
Partial score _____					
11. Physical makeup of text					
1. Textbook cover is attractive.					
2. Book is well constructed and durable.					
3. Textbook is not oversized or cumbersome.					
4. The print is attractive, and size is suitable for ease in reading.					
5. Page design is not cluttered.					
6. Paper is of good quality.					
Partial score _____					

Total score _____

HELPING STUDENTS LEARN FROM TEXTBOOKS AND SUPPLEMENTARY MATERIALS

Students' abilities to read and comprehend their textbooks, laboratory manuals, and other reading materials vary widely. With the mainstreaming of exceptional children into the regular classroom, a teacher can expect "an ability span equal to two-thirds the average chronological age of the students enrolled in a class" (Singer & Donlan, 1980, p. 35). This means that the achievement span for students enrolled in sixth-, eighth-, and tenth-grade science classes could be 7.3, 8.7, and 10 years, respectively (see table 12–1). In an eighth-grade earth science or physical science class, for example, students read between the fourth-grade level and twelfth-grade level.

In regular classrooms there are fast-powerful readers who read rapidly with high comprehension and fast-nonpowerful readers who read rapidly but do not comprehend difficult material. There are slow-powerful readers who read slowly with high comprehension and slow-non-powerful readers who read slowly and comprehend poorly. Poor reading ability reflects many different problems such as low IQ, bilingual background, learning disabilities, and dislike for reading. Poor readers usually demonstrate poor achievement in science courses. They are deficient in reading vocabulary, have short attention spans, tend to oversimplify, lack both confidence and abstract reasoning ability, and give up easily.

In spite of the poor reading ability exhibited by many students in our public schools, science teachers can help these students learn from their assigned textbook. The use of the textbook in instruction can improve thinking and reinforce learning in science. To accomplish this, science teachers must employ many of the reading strategies and techniques that reading and language arts teachers use to help students learn from text materials.

The degree to which students are prepared before they read printed material is directly related to their success in comprehending this information. Fortunately, many science activities can help prepare students to read their textbooks. For example, demonstrations can be used to illustrate concepts, principles, laws, and ideas that are presented in chapters. They can familiarize students with ideas that they will read about and provide an opportunity for students to verbalize and explain these ideas. Laboratory

TABLE 12–1   Sample for determining span of achievement*

A sixth-grade class

Average age of students	6	7	8	9	10	(11)	12	13	14	15	16	17	18	19	20
Grade level		1	2	3	4	5	(6)	7	8	9	10	11	12	13	14	15

|———————— 7.3 years ————————|

A 7.3-year span of achievement for sixth graders.

An eighth-grade class

Average age of students	6	7	8	9	10	11	12	(13)	14	15	16	17	18	19	20
Grade level	1	2	3	4	5	6	7	(8)	9	10	11	12	13	14	15

|———————— 8.7 years ————————|

An 8.7-year span of achievement for eighth graders.

A tenth-grade class

Average age of students	6	7	8	9	10	11	12	13	14	(15)	16	17	18	19	20	21
Grade level	1	2	3	4	5	6	7	8	9	(10)	11	12	13	14	15	16

|———————— 10 years ————————|

A 10-year span of achievement for tenth graders.

*The reading range of a class is two-thirds the chronological age (⅔ CA) of the students enrolled in a class.

Science teachers should provide reading activities to help students develop the necessary skills to comprehend their textbooks, laboratory manuals, and other reading materials.

activities and field trips can also give students concrete experiences with ideas that they encounter in the form of abstract symbols in their textbooks. Lectures and discussions are also excellent strategies for teachers to employ before giving students assignments from their textbooks. The rest of this chapter suggests specific reading strategies and techniques that teachers can use with science textbooks to improve student achievement in content areas, such as vocabulary guides, marginal glosses, a learning strategy called S4R, laboratory directions, practice in understanding symbols, diagram interpretation, graph plotting, and current articles and nonfiction science books of interest.

Vocabulary Guides

The large number of new words and terms encountered in science texts presents a problem to many science students. Often, there are many ideas and vocabulary terms to remember. Students need a great deal of help to incorporate them into their reading and speaking vocabulary. A vocabulary guide can assist students to learn

the science terms that they will encounter in a textbook chapter.

A teacher can prepare a vocabulary guide for students by going through a chapter in the science textbook and listing the important terms in the order that they appear. The vocabulary study guide shown in figure 12–4 gives an example for a physical science chapter on electricity. The students receive this list of the important terms. Each word is broken into syllables to help the student pronounce it correctly. A space is provided for the student to define the word. The learner is encouraged to use the glossary in the textbook and a dictionary to help define each word. In addition, it is helpful for the student to refer to the textbook chapter where the term is located.

Preparation and use of a vocabulary guide is only one important step in helping middle and senior high students learn science vocabulary. A teacher should encourage students to pronounce and define these terms as frequently as possible and urge them to use this vocabulary in classroom conversation and discussion. Verbalization is an important aspect of conceptualiza-

SUGGESTED ACTIVITIES

1. Evaluate two textbooks intended for a course in one of the science areas by applying the checklist in Figure 12–3. Determine which text you would use for a course of study based on your evaluation.

2. Read several sections of a science textbook and apply the Fry Readability Graph to judge the reading level of the text. Compare your assessment of the reading level from Fry results with your reading assessment of the material.

3. Select a science textbook and apply the cloze technique to determine the reading difficulty of a passage. Discuss your results with the members of your methods class.

4. Prepare a guide sheet that will help students with reading problems to read a chapter in their science textbook more effectively. Draw on the information discussed in this chapter.

5. Review several science textbooks and decide how suitable the textbooks are for (1) slow learners, (2) advanced students, (3) talented students, (4) honors science courses, (5) inquiry-based courses, and (6) traditional courses.

6. Examine several science textbooks in one or more of the science areas. List the special features that are unique to each textbook. Look for special sections such as questions for review, suggested readings, chapter summaries, vocabulary lists, and colored illustrations and make note of the quality of the diagrams.

7. Prepare a list of nonfiction science books that you would recommend for students enrolled in a science course that you expect to teach in the future or that you are teaching. How would you use these books in instruction?

BIBLIOGRAPHY

Bruner, J. S., Goodnow, J. J., and Austin, G. A. 1956. *A study of thinking.* New York: John Wiley & Sons.

Carter, B., and Abrahamson, R. F. 1990. *Nonfiction for young adults: From delight to wisdom.* Phoenix, AR: The Oryx Press.

Chiappetta, E. L., Fillman, D. A., and Sethna, G. H. 1991. A method of quantifying major themes of scientific literacy in science textbooks. *Journal of Research in Science Teaching, 28:* 713–725.

Chiappetta, E. L., Sethna, G. H., and Fillman, D. A. 1991. A quantitative analysis of high school chemistry textbooks for scientific literacy themes and expository learning aids. *Journal of Research in Science Teaching, 28:* 939–951.

Cho, H., Kahle, J. B., and Nordland, F. H. 1985. An investigation of high school biology textbooks as sources of misconceptions and difficulties in genetics and some suggestions for teaching. *Science Education, 69*(5): 707.

Dishner, E. K., Bean, T. W., Readence, J. E., and

Moore, D. W. 1992. *Reading in the content area.* Dubuque, IA: Kendall/Hunt.

Mayer, R. E. 1983. What have we learned about increasing the meaningfulness of science prose? *Science Education, 67:* 223–237.

Moyer, W. A. 1985. How Texas rewrote your textbooks. *The Science Teacher, 52*(9): 23.

Otto, J. E., and Towle, A. 1985. *Modern biology.* New York: Holt, Rinehart and Winston.

Pearson, P. D. 1974–1975. The effects of grammatical complexity on children's comprehension, recall, and conception of certain semantic relations. *Reading Research Quarterly, 7*(10): 155.

Pederson, J., Bonstetter, R. J., Korkill, A. J., and Glover, J. A. 1988. Learning chemistry from text: The effect of decision making. *Journal of Research in Science Teaching, 25*(1): 15–21.

Readence, J. E., Bean, T. W., and Baldwin, R. S. 1992. Content area reading: An integrated approach. Dubuque, IA: Kendall/Hunt.

Schraer, W. D., and Stoltze, H. J. 1987. *Biology: The study of life* (2nd ed.). Newton, MA: CEBCO/Allyn and Bacon.

Singer, H., and Donlan, D. 1980. *Reading and learning from texts*. Boston: Little, Brown and Co.

Stetson, E. G. 1982. Improving textbook learning with S4R: A strategy for teachers, not students. *Reading Horizons, 22*(2): 129.

Sutman, F. X., Allen, V. F., and Shoemaker, F. 1986. *Learning English through science.*

Washington, DC: National Science Teachers Association.

Vacca, R. T. 1981. *Content area reading.* Boston: Little, Brown and Company.

Wright, J. D., and Spiegel, D. L. 1984. How important is textbook readability to biology teachers? *The Science Teacher, 46*(4): 221.

Yager, R. E. 1983. The importance of terminology in teaching K–12 science. *Journal of Research in Science Teaching, 20*: 577–588.

room managers establish and carry out workable rules of behavior. These teachers give clear directions and deal quickly with inappropriate student behavior. Effective managers are well organized and pace their instruction so that activities and events flow smoothly from one activity to the other with minimal loss of time on task.

Sanford (1984) emphasizes that the best science teachers observed in her field studies required students to complete many tasks during each class period. In these teachers' laboratory classes, students could be observed busily working at hands-on activities. The teachers were frequently observed monitoring students' work. The best science teachers were those who were "task oriented, businesslike, and congenial" (p. 196), and ran orderly and productive classrooms.

IDENTIFYING CAUSES FOR STUDENT MISBEHAVIOR

As teachers work and associate with boys and girls, they will gain an insight into the causes for misbehavior. If Jane reads a book instead of performing the specific activity assigned to the class; Joe throws spit balls; Mary cheats on tests; Andrew horses around in the chemistry laboratory, causing safety problems; or Juan dozes off during the class period, there are reasons these students behave as they do. It is the teacher's responsibility to determine the basic reasons for the misbehavior before applying measures to correct it.

Misbehavior observed in students can usually be attributed to underlying conditions that are not necessarily obvious. A teacher should try to identify the deeper problems that are causing the misbehavior before attempting to deal with it. If only the symptoms are treated without knowing the conditions causing them, more undesirable behaviors may result, which may be more problematic than the behavior first observed.

Some Causes of Behavioral Problems

Students, as all human beings, interact with their environments, and the forces and pressures they encounter in and out of school can generate many of the misbehaviors exhibited in the classroom. The home environment, teachers, administrators, other students, and others not in the school setting can all influence student behavior. Additionally, factors such as academic aptitude and mental and physical health can have a great bearing on how a student behaves in a classroom.

Teachers should identify for themselves, as soon as possible, their own managerial styles. To accomplish this they should understand the managerial process and become familiar with the various models of classroom management. Teachers should familiarize themselves with classroom management models such as the authoritarian approach, the behavior modification approach, the instructional classroom management approach, and others to identify some of the strategies, techniques, and procedures that they feel can be used successfully in their own classrooms.

In addition to a basic knowledge of the models of classroom behavior, prospective and beginning teachers should also identify techniques and procedures that experienced teachers have used effectively in their own classrooms. Using them successfully will depend on whether they suit a particular teacher's style and personality and whether they are appropriate for the students, the classroom setting, the school setting, and the circumstances.

Effective classroom management is an important part of teaching. It involves a set of complex behaviors used to promote and maintain a proper environment in the classroom so that effective instruction can occur. Management and instruction go hand in hand; effective management is essential to doing a good job of instruction.

Home Background. Desirable and undesirable classroom behavior may be partially attrib-

Misbehavior observed in students can often be attributed to many factors. It is important that teachers identify the causes of behavioral problems before they attempt to deal with them.

uted to the home environment. Relationships with parents, siblings, and other relatives all influence students' conduct. Parents establish rules of conduct that are generally accepted by their children and have a direct influence on how they conduct themselves in a school setting.

Many of the problems exhibited in the classroom stem from standards of conduct acceptable in the home. Swearing may be an acceptable form of speaking at home but is certainly not acceptable in the classroom. Fighting may be an acceptable behavior at home, but is far from acceptable in school. Consequently, young people may be exhibiting behaviors that they consider normal and that reflect their parents' standards and rules of conduct when they act the way they do in the classroom.

The student who comes from a home environment with strict parents may react differently to the authority exhibited by a teacher. A student from such a home may resent authority at first and consequently resent the teacher. On the other hand, some students who come from a strict home atmosphere may be accustomed to authority figures and will readily accept them without question. Teachers, then, must learn how students are controlled at home before they can deal with the discipline problems that seem to stem from the home environment.

Students from homes where there is constant family friction and quarrels often develop into discipline problems. These students may be very emotionally involved in these situations and develop problem behaviors. The unsettled home environment is one variable that a teacher should consider when analyzing a student's behavior.

Often, the attitudes of parents toward education have a bearing on how students behave in school. Parents who do not consider an education important may communicate such feelings to their children, who, in turn, will have little or no respect for a high school education. On the other hand, students who come from homes where an education is considered very desirable may exhibit misbehaviors because their parents have established unreasonable expectations for their academic performance. Both situations may generate problems. In the first instance, the students may not be motivated to learn because they, too, regard the need for an education lightly. In the second instance, students who cannot meet their parents' high expectations may soon become frustrated, which often leads to low self-esteem, poor motivation, and behavioral problems.

Students who are allowed to stay out late at night or watch television until the early morning hours often become inattentive or fall asleep during the class period. They may have the aptitude to do the work but lack the energy they need to remain attentive. Other inattentive students may be involved in after school jobs, which require much energy and effort. These students may find it difficult to keep up with the work in class and to do the homework assignments. Teachers should identify the causes of inattentiveness. They will often find that these students can do the work without difficulty and are even interested in learning but lack the energy needed to carry out the assignments. A conference with the student during which the teacher emphasizes the need for rest may correct the problem. If it does not, then the parents of the student should be contacted.

Academic Ability. Students in a class have differing degrees of academic abilities, and some of the discipline problems the teacher encounters are due to the type of classwork the teacher provides. The teacher who focuses the classwork solely on the average students is not providing for the range of individual differences inherent in the typical classroom. The brighter students are usually not challenged and motivated in such classes and end up bored, irritable, and restless and exhibit contempt for the teacher, the school, and classmates. These students are potential discipline problems because they often become mischievous and interrupt classwork.

Students with low academic abilities may also exhibit behavioral problems if the teacher's lessons are presented at too high a level. These students become frustrated and their motivation suffers, possibly resulting in inappropriate behavior. It is the teacher's job to try to challenge all the students in a classroom. Each lesson must include portions that are at a level high enough to challenge those with high ability but at the same time offer the low-ability student a reasonable opportunity to master the material successfully. This is not an easy task, but the teacher must strive toward this end if he wishes to avoid behavioral problems associated with academic ability.

The Teacher. Teachers often reprimand students who are engaged in noisy activities such as talking, chewing gum, scuffling feet, and shuffling paper. These, however, are common behaviors among adolescents in a school setting. Many teachers consider these behaviors to be out of place and nonconforming. Such teachers often create discipline problems by insisting that their students be perfectly quiet and inactive in the classroom. Students in such a rigid environment end up irritable, restless, and threatened and eventually become behavioral problems. They must find an outlet for their energy. In general, it is best that the teacher not insist on an unreasonable degree of physical inactivity in the classroom if he wishes to avoid behavioral problems. Not all physical activity should be allowed, but common sense should be one's guide.

Teachers should also show respect for students and refrain from ridiculing them, both in front of their peers and privately. They should exercise control at all times and not derogate a student in public even when the situation appears to warrant it. Teachers who use humor at inappropriate times, sarcasm, or ridicule often cause discipline problems by backing students into a defensive posture in front of their peers.

Teachers who do not plan their lessons well, do not start their classes on time, deviate often from science lessons, and engage in useless talk and activity during a class period will soon lose the respect of their students and incite undesirable behaviors in them. In short, an atmosphere of disorder will promote disorderly behavior.

Many teachers find it difficult to understand students who come from socioeconomic backgrounds different from their own. Students, in turn, may have difficulty understanding the value system of the teacher. It is the teacher's responsibility to make the initial attempt at understanding the values of his students. If students sense that the teacher does not respect or ignores their deeply ingrained values, resentment will be felt and discipline problems will arise. Alternatively, students will respect a teacher who respects their beliefs and values. This, of course, will help diffuse any potential discipline problems.

Teachers can also cause discipline problems in the classroom because of their own personal problems. Teachers who have health, home, or work problems can allow them to affect their patience and relationships with students, which almost always leads to difficulties in the classroom. Teachers should always try to determine whether student misbehavior is due to the teacher's own problem. Once identified as such, the teacher should diligently work to eliminate the causes to avoid future undesirable situations.

The School Administration. The school administration is often responsible for problems that a teacher experiences in a classroom. Administrators who are unrealistic about the standards of conduct in a school setting, such as noise levels in the classroom or corridors, dress codes, and other codes of behavior, will cause students to resist and defy authority. Many causes of misbehavior can be attributed to unrealistic rules of conduct. Students in general will adhere to the established rules if they are within reason, even though they may be strict. Students will rebel when the administration constantly bombards them with new and unrealistic regulations.

Frequent interruptions over the intercom system made by school administrators can often generate problems. Students become uneasy, inattentive, and noisy as a consequence. Administrators should avoid frequent use of the intercom system. Interruptions should be restricted to certain times of the day unless emergencies occur.

Health-Related Reasons. Some students exhibit serious health defects such as poor eyesight, or heart, hearing, or respiratory problems. They also exhibit health problems associated with their teeth and with allergies and malnutrition.

Some illnesses are so severe that students are not able to exert the energy to work on assignments. Other conditions such as hyperactivity can cause restlessness and inattention. Hearing-impaired students who are required to wear hearing aids may be psychologically affected by the requirement and may avoid wearing the aid in school so as not to appear different from fellow classmates. As a result, these students may become inattentive and show a lack of interest.

A student who is inattentive in class may have a health problem; the inattention may be due to pain caused by a minor or severe illness or disability. It is important that teachers train themselves to be aware of such problems and refer these students to the school nurse, principal, and/or parents to avoid confrontations with these students.

Personality-Related Reasons. Many student discipline problems can be attributed to immaturity and poor judgment. They are also caused by insecurity, lack of recognition, lack of self-respect, lack of a sense of responsibility, and lack of self-control. The discipline problems caused by these situations may vary in magnitude. Students who talk constantly and listen occasionally are not considerate of others, lack maturity, and have no sense of responsibility to their fellow students or teachers. If students do not carry out assignments in and out of class, they lack maturity, motivation, and a sense of responsibility. These behaviors may seem minor, but when they are exhibited daily by the same students, they can create problems that eventually get out of control. These minor infractions should be dealt with as soon as possible to avoid more serious situations.

Some students exhibit misbehavior just to get attention and will go to any extreme to obtain attention from the teacher or classmates. Undoubtedly, these students are not getting the attention they require from their parents and others and will use the school setting to get it.

More serious problems such as stealing chemicals and equipment, cheating on tests, and talking back to the teacher stem from deep underlying causes, which a teacher must identify before dealing with them.

MODELS FOR CLASSROOM MANAGEMENT

There are a number of approaches that a teacher can use to successfully manage the classroom. Ultimately, the procedure to be used can be determined only after thoughtful consideration of the teacher's and students' characteristics. After all, one method of classroom management might be highly successful with one group of students and a dismal failure with another group. For example, an approach stressing permissiveness and student freedom would not meet with much success if the students were too immature to handle such freedom in a responsible manner. Alternatively, some teachers might not feel comfortable with a particular approach, and any attempts to use it will eventually fail. For example, a teacher whose personality is not authoritarian by nature would find it quite difficult to succeed in managing a classroom following the authoritarian management model. Because

the proper approach to classroom management is so highly contingent on both teacher and student characteristics, it is difficult to make any general prescriptions with respect to which model is "best." However, it is quite useful for the teacher at least to be familiar with his options. With this knowledge, the teacher is in a better position to make an informed decision as to which approach is appropriate for his classroom.

The list of classroom management approaches is too lengthy to mention in its entirety. However, three such approaches have proved to be widely accepted and successful. These techniques of classroom management are commonly known as the authoritarian, instructional, and behavior-modification approaches.

Instructional Approach

The basic assumption of the instructional approach to classroom management is that most problems can be prevented through well-planned instruction. Therefore, the teacher who uses this approach can prevent and solve managerial problems by preparing and implementing effective lessons. An effective lesson is one that is appropriate for the needs, interests, and abilities of students in the class. In addition, the well-planned and delivered lesson that is geared to the students in the class affords each student the chance to be a successful learner.

The instructional approach is predicated on the assumptions that well-planned and well-implemented lessons will (1) prevent management problems and (2) solve those that have already arisen. Although there is little evidence to support the latter assumption, the work of Kounin (1970) lends strong support to the former. Kounin found that the teacher's ability to prepare and conduct lessons that prevent boredom, misbehavior, and inattention is crucial to classroom management. Successful teachers presented well-prepared lessons that proceeded smoothly from one activity to the next with little waste of time or loss of focus. In addition, such teachers were quite adept at providing activities appropriate to their students' interests and abilities.

Conversely, Kounin found that teachers who experienced management problems were poorly prepared and therefore presented lessons that did not flow smoothly or maintain student interest. For example, the lack of "smoothness" inherent in the unprepared teacher's lessons took on many forms, such as:

1. sudden initiation of an activity with a statement or direction for which the class is unprepared
2. instances in which the teacher leaves one activity to start another and then returns to the original activity
3. premature termination of one activity to start another

In addition to this lack of lesson fluidity, unsuccessful teachers were found to spend inordinate amounts of time giving directions and explanations.

In summary, the instructional approach can be characterized by (1) well-planned and well-implemented instructional activities, (2) lessons that are appropriate for the abilities and interests of each student, (3) fluid movement from one class activity to the next, and (4) clear and explicit directions for each activity.

Authoritarian Approach

The authoritarian approach to classroom management is probably the most widely used method in today's public schools. In this approach the teacher's role is to establish and maintain classroom discipline by the use of a variety of managerial strategies. The teacher accepts the responsibility for regulating student behavior because the teacher takes complete charge of the classroom. More often than not, the student behavior is controlled by the enforcement of a specific set of rules of behavior.

Canter and Canter (1976a, 1976b) indicate that teachers have certain basic rights in order to control classroom behavior. They have the right to establish clear expectations, insist on correct or acceptable behavior from students, and, if needed, follow through with appropriate and reasonable consequences that have been established. They also emphasize that students must know their limits and that the teacher has the authority and right to set and enforce the limits.

The authoritarian approach is often considered to be synonymous with forcefulness and intimidation. However, the authoritarian approach need not degrade the student or involve harsh punishment. In its purest sense, the authoritarian approach to classroom management simply involves the establishment of a clearly stated set of rules and the consequences to be expected if the rules are not followed. The rules should serve to guide and limit student behavior. They tell the student what is and what is not acceptable. In short, they communicate to students what is expected of them. It is important that these rules be reasonable and enforceable and written in a clear, concise way so that all students understand what is and is not acceptable. In addition, the students should be well aware of the penalties that accompany breaking the rules. Charles (1985) suggests that students should take part in formulating class rules because students will believe that these rules are reasonable and thus will be more likely to adhere to them. "To be effective, those rules should be jointly formulated, reasonable, succinct, observable, enforceable, and enforced" (1985, p. 132).

Most advocates of the authoritarian approach do not support severe types of punishment. Such punishments have been shown to be relatively ineffective in the classroom. Alternatively, the terms *mild desist behaviors* and *corrective* are often employed to describe the type of punishments used in the authoritarian method. Such forms of mild punishment amount to the teacher informing students that they are misbehaving and that they should behave according to the rules that have been established. Such punishment is not intended to be a forceful or hostile way of handling an unacceptable behavior but, rather, a humane method to promote acceptable behavior. Such mild punishments, according to many researchers, are more effective than their harsh counterparts in promoting acceptable classroom behavior.

The major goal of the teacher who uses the authoritarian approach is to control student behavior. However, the authoritarian teacher does not force compliance, degrade the student, or use harsh forms of punishment. Instead, the teacher always acts in the best interests of the student. This approach is ultimately characterized by the establishment and enforcement of a clearly stated set of rules to guide student behavior and the use of mild forms of punishment as a consequence of noncompliance.

Teachers interested in learning more about the authoritarian classroom management approach should learn about the Canter and Canter (1976a, 1976b) assertive discipline approach. This model, based on a significant amount of research, has helped many teachers in maintaining classroom behavior. Assertive discipline helps the teacher take complete charge of his classroom in a nonthreatening but forceful way and, at the same time, interact with students calmly, patiently, and warmly. The approach is a positive one that permits teachers to maintain the best possible learning environment in a firm, nonhostile, and constructive way.

Behavior-Modification Approach

The major assumption involved in the behavior modification approach to classroom management is that *all* behavior, either acceptable or not acceptable, is learned by the student through interaction with her environment. Therefore, the role of the teacher in this particular approach is to control the students' environment carefully, which will in turn shape the students' behavior. Although the terminology may vary from practitioner to practitioner, there are four general categories of environmental contingencies that behaviorists believe will influence human behavior. These are: (1) positive reinforcement, (2)

negative reinforcement, (3) punishment I, and (4) punishment II. According to the behavior-modification approach, each student behavior that the teacher wishes to influence must be followed by one of these four consequences, which will in turn affect the frequency of that behavior by increasing or decreasing it.

Positive and negative reinforcements are used to increase the frequency of desired or appropriate student behavior. In the case of a positive reinforcement, a reward (e.g., praise, affection, or gold star) is given to the student immediately following the desired behavior. For example, if a student brings in a science-related article from the local newspaper, the teacher may respond by praising the extra effort in front of the class and mentioning that the reading of science-related literature other than their own textbooks will help them do better in science class. The introduction of this reward in response to the student's unsolicited behavior will increase the frequency of the behavior in the future.

A negative reinforcement will accomplish the same result as the positive variety, but it involves the removal of a particular punishment. For example, suppose a student rarely participates in class, but the teacher consistently calls on her anyway. Being called on without volunteering is very often aversive to students. In addition, the teacher consistently nags the student to participate more. One day, the student surprisingly volunteers to answer questions twice. In response to this the teacher stops nagging the student and calling on her when she does not volunteer. During the subsequent weeks, the student continues to participate in class with a frequency acceptable to the teacher. In this instance the *removal* of an undesirable stimulus was used to increase the frequency of a desired behavior.

Teachers using the behavior-modification approach can decrease the frequency of inappropriate behavior by using punishment. Punishment can be divided into two different categories, known simply as I and II. Punishment I involves the introduction of an aversive or undesirable stimulus, which is commonly called *punishment.* For example, if a student misbehaves in class, the teacher might respond by giving the student an extra work assignment during laboratory cleanup, or the teacher may simply reprimand her. In either instance, an aversive stimulus has been introduced that would decrease the inappropriate behavior in subsequent class meetings. The judicious use of punishment I has been found to be quite effective, but most educators prefer to use punishment II, because the former may carry negative effects with its use. Punishment II involves the removal of a reward or the withholding of an anticipated reward. If a student behaves inappropriately in class, the teacher may wish to exclude the student from participation in a field trip or class activity. In this instance the student is not allowed to participate in an activity that she enjoys, and thus it serves as a punishment. In the future, the student can be expected to eliminate or limit the frequency of inappropriate behavior in the hopes of being allowed to participate once again in desired class activities.

It is very important that the teacher carefully consider the individual student before using either reinforcements or punishments. Both these categories of environmental stimuli vary from student to student, and one student's reward might serve as another student's punishment. It is for this reason that behavior-modification techniques do not always succeed. However, if the teacher is careful to identify what is rewarding and what is aversive to a particular student, the research shows that this particular approach is extremely effective.

The timing and frequency of both reinforcement and punishment are important considerations in the behavior modification model (Buckley & Walker, 1970). If the teacher wants any student behavior to continue, the behavior must be rewarded or punished (whichever is appropriate) immediately after it happens. Acceptable behavior that is not reinforced immediately will most likely prevent the student from continuing the behavior, and behavior that is not punished at once may encourage the student to continue the undesirable behavior. The frequency of reinforcement is also an important consideration. During the early stages of promoting a particular student's behavior, the best approach is the use of continuous reinforcement. Quite simply, continuous reinforcement is used following every occurrence of the desired behavior. This approach helps the student to learn the desired behavior more quickly. However, once a desired behavior has been acquired, intermittent reinforcement has been found to be more effective than continuous reinforcement. Intermittent reinforcement can be applied on an interval schedule or a ratio schedule. With an interval schedule the teacher rewards the student after a certain amount of time has elapsed. The teacher who uses the interval schedule may choose to reward a student only once every class period regardless of the number of instances of the behavior during this time interval. In a ratio schedule the teacher may reward the student after a desired behavior has happened a specified number of times. For example, if a student has consistently behaved appropriately six times, then the teacher rewards the student for the consistent, appropriate behavior after the sixth occurrence. In general, a ratio schedule is useful for promoting repeated instances of desirable behavior.

Ratio and interval reinforcement schedules can be broken down into finer categories such as fixed or variable. In a fixed schedule the ratio or interval of reinforcement remains constant; in a variable schedule the ratio or interval of reinforcement varies. Ferster and Skinner (1957) studied each of the possible combinations of reinforcement schedules and found that various schedules had different effects on rate of responding, number of responses per reinforcement, number of responses made after reinforcement was discontinued, and so on. According to their research, fixed-ratio schedules resulted in frequent response rates but with less than desirable uniformity; variable-ratio schedules resulted in both stable and frequent response ratios with an added dimension of persistence. They found that behaviors reinforced with a variable-ratio schedule persisted long after reinforcement had been totally eliminated.

Thus, behavior reinforced with such a schedule is most easily maintained and hardest to eliminate. Fixed-interval schedules would occur only during the period of time immediately preceding the expected reward. The best example of this is a student studying for regularly scheduled tests. Rather than study continuously throughout the year, students study just before each test. With a fixed-interval reinforcement schedule, response rate is not stable; it is high immediately preceding the time for reinforcement and quite low at other times. Variable-interval schedules produce stable and uniform response rates. Furthermore, behaviors reinforced in this manner are difficult to extinguish.

Thus, it appears that variable-ratio or variable-interval schedules are the best for maintaining behaviors over time and should probably be used by the classroom teacher in place of either fixed-interval or fixed-ratio schedules.

The behavior-modification approach to classroom management centers on the effects of various environmental factors on student behaviors. These factors include positive reinforcement, negative reinforcement, punishment I, and punishment II. If a teacher wishes to increase the frequency of a particular student behavior, a positive or negative reinforcement is used. Alternatively, if the teacher wishes to decrease the frequency of a specific behavior, the use of a punishment (I or II) is recommended.

Other Management Approaches

The authoritarian, instructional, and behavior modification approaches to classroom management are only three of the options open to the classroom teacher. The intimidation approach is used by some teachers who attempt to control students through harsh forms of punishment such as severe reprimands, force, threats, and pressure. These practices are used to force the student to comply with the teacher's wishes. Most educators, however, feel that this approach should be used sparingly and only in instances of extreme misconduct. It should not be used for handling the normal types of discipline problems that occur in the classroom; that is, it should be used only in unusual situations. For example, it can be used when a student may be causing bodily harm to another student or creating a dangerous situation in the laboratory, such as throwing chemicals at other students or deliberately mixing chemicals that may cause an explosion.

The permissive classroom management approach allows students a great deal of freedom in the classroom. In essence, it advocates that students have the right and freedom to do what they want with little or no interference from the teacher. This approach, however, has very few supporters because it is believed that most students are not mature enough to handle such freedom. Many believe that the approach is difficult to implement in most school settings, because the amount of freedom allowed may prevent students from being productive learners and responsible individuals in classroom settings.

The socioemotional climate approach is derived from principles of clinical psychology and counseling. Those who have contributed to this model include Carl Rogers (1969), Haim Ginnot (1969), and William Glasser (1969). This approach stresses the importance of the positive effect that the application of logical consequences has on student behavior. It also argues against the use of punishment in helping students change their behaviors to those that are more desirable. In short, the teacher is expected to help the students understand the logical relationship between their behavior and its consequences.

Finally, the group process approach to classroom management, derived from group dynamics and social psychology theory, views a classroom as a social system in which group processes are the major factor. Because instruction is viewed as occurring in a group context, the nature and behavior of the classroom group are seen as crucial to effective learning. In this technique, the teacher is expected to foster the development and operation of an effective classroom group. Those interested in this approach should also refer to the works of Kounin (1970), Johnson and Bany (1970), and Schmuck and Schmuck (1979).

In conclusion, it should be remembered that each of these management approaches has its supporters. Different approaches will be successful for different teachers. It is up to the individual teacher to decide which approach is best for him and his students. However, the teacher may want to combine various aspects of the classroom management approaches to achieve the most appropriate model.

USEFUL SUGGESTIONS FOR GOOD CLASSROOM MANAGEMENT

Experienced teachers in general have the foresight to recognize situations that may lead to behavioral problems. They can analyze problems as they are developing or as they are encountered. Inexperienced teachers do not have this background and consequently cannot anticipate many of the problems that normally arise during the course of a day. Until they gain experience, they must rely on suggestions and techniques that have been tested and found successful in many classroom situations.

Many of the following suggestions and strategies for classroom management have been gleaned from various models already discussed. Other procedures presented have been used by successful teachers in many different situations and subject areas. The strategies, procedures, and suggestions are not guaranteed to work in all cases and therefore should be modified as needed to fit the teacher's style and circumstances.

Seek Student Cooperation to Maintain Control

Most students are basically responsible human beings and will react positively to teachers who show that they, too, are good, decent human beings. Teachers who have the respect of their students will find that their students will generally cooperate without hesitation. Teachers who seek student cooperation must show that they are firm and serious about their requests. They must display authority yet be proper and professional in the way they ask their students to cooperate.

Practice Good Planning

Beginning teachers must enter a classroom with the confidence of knowing what is to be done during each minute of the teaching day. Overplanning lessons can provide the confidence and ease needed to enter the classroom with authority. Plans, however, must be laid out in detail. It is good practice always to plan more work than can possibly be covered during the period. It is not uncommon for beginning teachers to underplan at first and find that they have nothing to do for more than 15 or 20 minutes at the end of the period. They stand before the class embarrassed and frustrated, not knowing what to do next.

Good planning involves overestimating what can be accomplished over a period of time. It also means that teachers should provide stimulating, interesting, and relevant experiences so that students are eager to learn. It is good practice to have an alternative plan available in case of emergency. Students will become uninterested and uninvolved when poor planning is evident. Problems such as inattention, increased talking, and class disruption may result. However, teachers who communicate organization and confidence will avoid many potential discipline problems.

Consider the School Calendar

There are certain times during the school year when students are apt to become behavioral problems. The day of an important basketball, soccer, or football game, the morning before an afternoon assembly involving an important entertainer, two days before the Thanksgiving recess, or the week before Christmas are periods when teachers should take steps to avoid potential problems. These are times when students must be kept involved and motivated. They should be given assignments that challenge and interest them and keep them active in order to compete with other interests associated with various events.

It is essential that teachers identify periods during the school year when other interests and enthusiasms compete with classwork, and plan suitable lessons and activities that will produce desirable student behaviors.

Organize Routine Activities

Many classroom activities are basically routine. Some of them can create confusion and noise if they are not handled properly and efficiently. Activities such as taking student attendance, passing out papers, organizing students into groups, and assigning student tasks can be considered routine but can also create problems unless they are handled efficiently. Teachers are asking for trouble when they spend 15 minutes taking attendance while their students are talking and engaged in mischievous activities. Teachers who spend unnecessary amounts of time passing out papers and collecting them create situations in which students keep themselves occupied by talking loudly and engaging in play. Many routine situations can cause a great deal of confusion unless simple rules are established beforehand. Simple rules for passing out papers and collecting them can avoid unnecessary misbehaviors.

Routines are also essential in the science laboratory. When students are asked to obtain materials and supplies from storage areas, they should not be allowed to go to these areas in large groups. They should be directed to secure supplies in an efficient way—that is, individually or in small groups. If students are allowed to go to secure supplies en masse, not only will discipline problems be created, but safety problems will also develop. Routines regarding safety procedures should be established and followed to avoid possibly serious safety problems.

Routines should be organized to carry out classroom work efficiently. With experience, teachers will determine the best procedures to follow to avoid unnecessary confusion, high noise levels, and mischievous activities.

Become Familiar with the School Routine

Familiarization with normal school routines can help one avoid a great deal of confusion and difficulties. Teachers should learn these routines as soon as possible so that they can handle them efficiently when they arise. If a routine is not known, mistakes can be made that can place both teachers and students in problematic positions. Many discipline problems can arise, for example, if the teacher does not know the procedures for issuing passes for students to leave the room or the building or when to send a student to the principal.

In general, school procedures are usually established for the following: (1) how to take attendance and keep records of attendance; (2) how to handle passes that permit students to leave the classroom or study hall; (3) how to deal with students who are tardy; (4) how students must check books from the science library or school library; (5) what to do when a student is injured in the science laboratory or if the student becomes ill during the class period; (6) how to take classes on an extended or short field trip; (7) when to send a student to see the school psychologist, principal, or guidance counselor; (8) what to do in case of an emergency involving a laboratory accident; and (9) what to do when disposing of dangerous chemicals.

Routines that must be followed by all teachers are generally pointed out early in the year and probably described in some written format by the administration. If the procedure is not known, teachers should find out *before* acting. Some routines, if not carried out as specified, jeopardize a teacher's position as well as create situations that can result in discipline problems.

Know the Students

It is good practice to learn students' names as quickly as possible. Teachers who demonstrate early that they know their students by name also show that they have a definite interest in them.

It is not easy to learn 30 or 40 names in a hurry, and if a science teacher has five classes, then he must learn 150 or more names. Seating charts are a help in learning names quickly and in associating names with individuals. It is advisable, at least at the beginning of the term, to ask students to use a particular seat. Once names have been learned, other seating arrangements can be made that may be more desirable.

In addition to knowing names, it is also advisable to know more details about students. Teachers should have information available regarding academic aptitude and achievement levels; health problems; family particulars such as number of siblings, occupation of father and mother, and status of the marriage; types of extracurricular activities in which the students are engaged or interested; handicaps such as hearing or visual impairments; and vocational plans.

Student files containing much of the information mentioned are usually available to teachers. Much information can be learned about students through individual conferences, observations made by the teacher in and out of class, and discussions with parents, guidance counselors, and other teachers.

Misbehavior in students can be attributed to causes that are unapparent, and knowledge of student background can often permit teachers to detect the underlying causes of misbehavior so that they can be dealt with effectively and efficiently.

Keep Contact with Students During a Lesson

Teachers should be aware of what is taking place in the classroom at all times. This means that there must be constant contact between the teacher and students while a lesson is being presented or during any other activity. Many teachers find that eye contact is a very effective way to maintain control. By using eye contact, the sensitive and alert teacher is able to spot behaviors in students that indicate a discipline problem is about to occur or has already developed. If the teacher recognizes the problem early, he can handle it effectively and with discretion. Sometimes a simple, long pause, combined with a stare in the direction of the situation, will curb the activity. Some teachers pause and then slowly walk toward the student or students who are misbehaving. These techniques can often prevent a problem from developing into something serious.

Maintaining contact with students during a lesson or activity can be accomplished only if teachers have complete confidence in what they teach and how they teach. Teachers who prepare their lessons well, know the content, know what to do next, and display interest in their lessons will have no difficulty in keeping contact with their students. Teachers who fumble through a lesson or other activity are so insecure about their teaching that they become oblivious to the behavior of their students. They do not sense how students are reacting to the lesson because they are so unsure of themselves. When students recognize a teacher's insecurity, they soon lose respect for the teacher and become inattentive and uninterested, causing behavioral problems.

Involve Students

The more involved students are during a lesson, the less likely teachers are to have serious discipline problems. The teacher should use a variety of strategies and activities to keep students actively involved during a class period. The involvement of students during a lesson, for example, "can take the form of asking questions, encouraging students to make comments about a lesson, allowing students to fulfill tutoring roles for those who need additional assistance in mastering a concept or skill and allowing students to work in large and small group settings" (Purvis & Leonard, 1985, p. 351). Getting students involved in various types of activities, such as demonstration work, leading discussions, delivering short lectures, presenting reports on projects, showing films and filmstrips, participating in games, and planning field trips, can take care of a variety of interests and abilities of students represented in a classroom. Many behavior problems occur when students are bored and have no involvement in what is taking place during the class period.

Practice Self-Control

Teachers should try to show self-control at all times. Teachers who have a tendency to explode over minor misbehaviors are not very good examples to students. Good and effective teachers are those who can control their emotions even when major problems arise. Teachers who are unable to control themselves and thus lose their tempers cannot deal with problems objectively, constructively, and professionally. These teachers in the long run become ineffective and lose student respect. Once students recognize that things are not under control, they are apt to create more problems that incite further misbehaviors.

It should be remembered that self-control is just as important for teachers as it is for students and that it is the teacher's responsibility to act civilized at all times. The best course to take when a discipline problem occurs is to under-react, avoid losing one's temper, and certainly not display any uncontrolled emotion. One should deal with the problem with concern, acting as an intelligent, rational, and mature adult.

Follow Good Testing Procedures

The administration of a test or quiz should involve a set routine. Students will learn the routine after several tests have been administered. In general, this involves procedures to use to avoid cheating, to use while papers are being distributed, and to follow when students need clarification of a question. Students should also learn the type of behavior they should exhibit while the test is in session. There should be a procedure for alerting students when the test is over and a routine to follow to collect papers at the end of the test.

Usually, teachers establish ways to avoid cheating. Some use alternate seating arrangements, others use two different forms of the same test, and some use both procedures at the same time. Usually, the teacher takes a position at the rear of the room after the tests have been passed out and the noise level has subsided. Once the teacher is positioned at the rear of the room, the students will avoid looking around at other students' papers for fear of being discovered by the teacher.

Students should know that the routine requires them to be quiet while papers are being handed out and that all books and notes should be placed under or in their desks. They should also be warned to avoid any activity that may give the appearance of cheating. Teachers should also be alert to see that no cheating occurs while questions are being asked and answered for clarification.

When teachers respond to students who require clarification of test items, they should learn to handle them without disturbing the rest of the class and to keep the noise level at a minimum. The amount of information that a student receives on asking a clarification question should be limited. Certainly, disclosure or hints of the correct answer to a question are unfair to other students and should be avoided.

Closing a test can be handled in a variety of ways. Once it is announced that the test is over, the students should immediately stop so that papers can be collected. Usually, to avoid confusion, it is desirable for the teacher to ask the students in each row to pass the papers forward to the first student in the row. Other routines can be used, but teachers should establish one and use it consistently so that students become used to the routine and know what to do once a test has been concluded.

Give Encouragement and Praise

Teachers devote a great deal of time reprimanding and disciplining students for their misbehaviors and often neglect rewarding them through verbal praise or a smile or a gesture for tasks well done.

Classroom management procedures should include techniques that acknowledge that

1. a class has performed well on a test
2. a student has answered a difficult and probing question well
3. a student has done outstanding work on a project or homework assignment
4. a student has mastered a particular skill in the laboratory
5. a class is doing well on a laboratory exercise
6. a class is well behaved when the teacher is out of the classroom

A number of instances afford teachers the opportunity to praise their students for appropriate behaviors. The acknowledgments should be made as frequently as the opportunities arise, and they should be made verbally, if possible. Praise and reward should not be reserved for special occasions but should be used as a matter of course. The frequent use of encouragements creates excellent learning environments in which students strive to receive rewards in one form or another as they achieve certain desirable levels of performance.

Techniques that teachers sometimes use to reward students include

1. giving students time to work on projects instead of devoting a full period to teacher-designated activities
2. permitting students to go to the library to ob-

tain more information about a topic that the students will present to the class
3. rewarding students by taking them on a field trip to a nearby industrial site or other facility
4. giving students an opportunity to work on interesting homework activities
5. permitting students to engage in education-oriented games
6. cancelling homework assignments because the students have already attained the desired level of performance on a topic, making further work repetitious and unnecessary

Teachers should give serious consideration to the use of praise and rewards. They should use them with discretion and only when students rightfully deserve the attention for outstanding work and performance. Teachers should always make certain that students fully understand the reasons for a reward, flattery, a pat on the back, or other indication of praise, and that it is given with sincerity.

Project Personality and Enthusiasm

A truly effective teacher is one who is able to accept students as they are—on bad days when they display their faults and lack of interest and on good days when they display their enthusiasm and interest, eager to work and learn. We all have good days and bad days, and as teachers we must consistently show enthusiasm and interest in students as well as the subject. Teachers who display their enthusiasm for students and their subjects will learn that students, too, will show enthusiasm and interest in their classwork.

The type of learning environment in the classroom is based on how the teacher interacts with the students. An ill-tempered teacher will only create an environment that is threatening and not conducive to learning. On the other hand, warm, friendly, and energetic teachers who understand their students and know their subjects will motivate students to behave in cooperative and meaningful ways.

Being a warm, understanding, and friendly teacher does not mean that one cannot display a firm, businesslike attitude. One can be interesting and humorous and still give the students a feeling that business is the order of the day. As a matter of fact, introducing occasional humor is a good tactic to use during a lesson or activity. Students learn that teachers are real, live human beings when they display humor.

The most important thing to remember, however, is that teachers must be consistent in the way they handle their students, so that the students know what to expect. Teachers who display inconsistencies by unpredictably reacting to student behavior with anger, laxness, rigidity, or apathy will only create problems because students have difficulty in adapting to these changing behaviors. On the other hand, teachers who are consistently calm and yet firm are less likely to have discipline problems.

Introduce Rules of Conduct

Rules can be used only when they save time and energy. Establishing rules for the sake of rules is poor practice, but those that help create a good atmosphere for teaching and learning are desirable. For example, rules may be established that state specifically when students are permitted to talk, how students should conduct themselves during a classroom recitation period, what students should do to maintain safety in the science laboratory, and what noise level is acceptable during a laboratory period or when the class is divided into discussion groups.

If rules are used, it is important that they be established as guidelines to control student behavior to establish a good learning environment. Rules should be kept as simple as possible, be direct, realistic, and understandable, and be enforced once established. They should be few in number. If the rule does not work for one reason or another, it is the teacher's business to discuss changes with students that will make the rule more realistic and enforceable. If it is not realistic, it should be abandoned. Arbitrary rules set up to create a stifling and unappealing environment will only induce undesirable behaviors in students.

Control Noise Levels

There are many situations in the science classroom that, by their nature, may produce relatively high noise levels. Question-and-answer sessions, for example, can degenerate into periods during which students yell, scream, or shout their answers, jump out of their seats, and move about, all indicating complete confusion and disorganization. The noise levels result in teachers shouting and yelling above student noise to maintain order. To avoid behaviors such as those mentioned, a question-and-answer session should be as well planned as a demonstration or other science activity. Good planning for this type of activity should include the procedures for maintaining a desirable learning environment. Students must be informed that chorus responses are not acceptable and that they will be ignored. Generally, chorus responses are generated by a small number of students who disregard the teacher's directions, even though they have been told they will be ignored when they answer in unison. Once teachers accept chorus responses and do not acknowledge students who raise their hands to respond to questions, they in effect are ignoring the majority of the class. Those students who follow the teacher's rules will soon become discouraged and refrain from future class participation. Furthermore, chorus responses do not give teachers the information they seek regarding the students' understanding of a topic or lesson or permit reinforcement and further explanations.

Allowing chorus responses can cause question-and-answer sessions to degenerate into chaos. Teachers may shout, yell, or slam books on the demonstration table to keep the noise level down. By this time the situation is out of control and almost impossible to handle.

Teachers should establish rules of conduct for question-and-answer sessions. The rules should state the conditions under which responses are accepted. Students are to raise their hands if they wish to respond to a question; the teacher may or may not acknowledge the student even if the hand is raised. The teacher may call on students who do not raise their hands. Other rules can be identified and put into effect as teachers gain experience and as the situations arise. Rather than imposing too many rules, one should be sure that the few rules established are clearly understood by all students and are enforceable.

Unusual noise levels also occur during laboratory sessions and during discussion periods. Again, rules of conduct for laboratory and discussion periods should indicate the appropriate and acceptable noise level. Students should have a chance to talk and interact during these sessions. They should not be inhibited from discussing procedures and their results when they are broken up into pairs or groups. However, they should be told the rules of the game. If they do not adhere to the rules, their misbehavior should be brought to their attention in the hope that it will be corrected. Certain signals can be used to control noise levels during these sessions—perhaps a bell, turning lights on and off—to signify to students that the noise level is too high. Teachers should avoid shouting or yelling to curb noise. This type of teacher behavior is not warranted if the students clearly understand what behaviors are expected of them at all times.

Discuss Misbehaviors with Students

It is sometimes desirable to ask students who continuously misbehave to remain after class or appear after school to discuss the reasons for their persistent disruptions. It is also appropriate to ask them to appear at other times when more time can be devoted to the problem. A meeting after school or during a study period can be scheduled if an after-class meeting is not desirable. During these discussions, teachers can find out the specific causes underlying the student's misbehavior and act accordingly to correct the situation.

Individual conferences are often very desirable to avoid student embarrassment and even teacher-student confrontations during class. Students will regard the situation seriously when the teacher asks for a private conference to discuss persistent misbehaviors. Conferences are particularly effective to handle problems that cannot be settled quickly and reasonably when they occur. During the conference, both the teacher and student can discuss the situation amicably and reasonably to correct the behavior.

Under certain circumstances, individual conferences are more effective if school officials are involved. Many students require this type of conference to realize the seriousness of the situation. In order to resolve certain types of misbehavior, Sauer and Chamberlain (1985) use the following types of conferences involving school authorities:

1. informal chat during which a school administrator talks to the offending student and tries to reach an agreement as to how the student is expected to behave
2. school conference that involves one or more school officials during which the student must agree to change her behavior
3. parent conference that involves the parent or legal guardian of the student along with the student and school officials

Conferences that involve school officials are generally very threatening to students and should be used as a last resort. Teachers can usually handle most behavioral problems effectively without the assistance of school administrators.

Change Seating and Isolate Students

Minor discipline problems can often be corrected by simply asking a student to take a seat away from a student or students with whom she interacted. Placing students who misbehave closer to the teacher where they can be monitored often does the job. Or placing them in areas where there are empty seats—away from other students—sometimes works as well.

Reseating students becomes a problem when all seats in the class are occupied and well-behaved students have to be displaced. The well-behaved student may react negatively toward the teacher, particularly if the student enjoyed the position in the classroom.

The technique works well on certain occasions when one or two students are involved. It is not a good procedure to use when many students are involved. Before locating a student in another area, make certain that the new seating arrangement does not create a more undesirable situation. The teacher should study the new seating arrangement carefully before acting and then determine if the misbehaving student can cause problems for students in the new location or whether the new location includes students who are also troublemakers.

Persistent misbehavior cannot often be corrected by new seating arrangements. Perhaps students may have to be isolated in another area of the science suite or even another room reserved by the administration for problem students. Persistent misbehavior that cannot be controlled in any reasonable way may have to be dealt with more severely. Students who are persistent problems and do not respond to corrective suggestion measures may require complete isolation. Students react differently to this technique. Some are ready to change immediately; others will not admit their faults and are not willing to cooperate and make desirable changes. The teacher has to be patient with this type of student who insists on innocence, because this situation may require counseling over a period of time to obtain the desired behaviors.

Know When and How to Use Punishment

The recommendations and suggestions made up to this point regarding classroom management can probably take care of the majority of classroom discipline problems. There are instances, however, when more drastic measures must be used to control behavior. When students continue to misbehave after the teacher has considered and tried all the previously discussed approaches, then it is probably time to use punitive measures.

Punitive measures can often have negative side effects; thus it is essential to use them only as a last resort. The teacher should also be sure to check the school policy regarding the use of punishment before proceeding. Usually schools have rules that teachers must follow when severe measures must be taken.

Punishment is anything (stimulus) that is perceived as adverse by an individual. As discussed earlier, what is punishment to one individual may not be punishment, and may possibly be rewarding, to another individual. Punishment is used to decrease the frequency of a particular behavior that may be considered disruptive, and possibly even dangerous, to the class or to the teacher.

Under certain conditions, punishment may be preferred to permanently suppress behavior. If the type of misbehavior is dangerous to the point that the student will be harmed or is harmful to others, severe punishment may be warranted. For example, suppose a student throws concentrated sulfuric acid out of an eyedropper in the direction of another student. This type of behavior should not be tolerated at all and must be stopped immediately. What should be done in such instances? Teachers should take the positive approach first but react to the misbehavior immediately. The student, as well as the class, should be shown what happens when sulfuric acid is placed on a piece of cloth. Once the student and classmates observe the reaction, they will probably think twice before repeating the incident. If this procedure does not work and the student insists on repeating the misbehavior or continues to perform dangerous acts in the laboratory, the teacher must take more severe measures. The student should be removed from the laboratory situation and isolated from other students during the period until the behavior is corrected. Repeated conferences with the student may also be in order. If again these avenues are not effective, the teacher should refer the student to the school psychologist (if there is one), principal, guidance counselor, or other administrative officer responsible for discipline. A parent conference is also desirable. Some might consider this approach to be harsh, but the unique nature of the laboratory compels the students to respect the rights of others. Squirting concentrated acid on a neighboring student is hardly analogous to shooting spitballs across the room.

Other types of behavior that warrant punishment include continually and persistently violating rules established to regulate noise levels during the laboratory or discussion periods; continually horsing around and disrupting classroom routines and procedures; deliberately destroying or damaging laboratory equipment, school property, and the property of others; misusing chemicals in the laboratory; preventing other students from conducting normal classroom activities; and cheating on a test. Punishments for the acts mentioned are certainly justified. Of course some of the misbehaviors listed are more serious than others, but they must all be dealt with in one way or another. When considering the use of punishment as opposed to a more subtle approach, one must weigh the overall effects of the behavior that is undesired. In instances where total class disruption and personal safety are involved, the teacher should give strong consideration to punishment.

For the punishments to be effective, the teacher must present them immediately after the misbehavior occurs. The punishment may vary in intensity depending on the act, and the teacher must weigh the degree of severity. Certainly, persistent talking is not as serious a violation as cheating on an examination. The two behaviors should be dealt with differently. The first act may require a reseating arrangement to take care of the problem. If this does not work,

then complete isolation from other students may be more effective. Persistent cheating may have to be dealt with more severely. The teacher may first give the students involved a warning and test them in an isolated area away from other students. If the behavior is observed a second time, the teacher may apply more direct measures. Some teachers go so far as to reprimand students in front of the other students who are taking the test. The reprimand is followed by tearing up the test papers before the whole class. This procedure, however, may have undesirable consequences. First, the whole class is disrupted, causing other students to become emotionally involved, which in turn may affect their test grades. Second, the students who are not receiving the punishment may consider the approach inappropriate and unfair, which may elicit further misbehaviors. The best way to handle the situation, if it is a repeat of a previous incident, is to take the test papers away from the students and discuss the incident with them at the end of the period. At that time, the teacher may choose to give the students involved a zero on the test or administer a makeup examination. Most likely the students will not repeat the act.

If one uses punishment carefully, making certain that it will do more good than harm, it can be effective. Punishment must also be administered so that it does not have an adverse effect on other students in the class. Often, students who are not experiencing the punishment will grow to dislike the teacher who continuously punishes students even for minor misbehaviors. Students will develop a disrespect for a teacher whom they perceive as unfair or unduly harsh. In short, the punishment should fit the crime.

The teacher should administer punishment only if another alternative is not available. If possible, he should issue a warning first. If the warning does not work and must be repeated many times with no significant improvement, the teacher should consider punishment. If this procedure does not work the first time, a warning followed by punishment should be repeated as necessary. This procedure, if it is repeated enough, will make the warning effective and the punishment unnecessary.

Another approach that teachers sometimes use, but that often backfires, is delivering punishment to the whole class for acts committed by one or two students. This action may curb the misbehaviors of the students involved, but the other students in the class may react negatively to the teacher and in turn probably cause further problems. On the other hand, the procedure may be effective if the teacher has a reputation of being a fair and understanding person who has the admiration and respect of the class. If the teacher does not have this reputation, most likely the whole class will react, causing more problems. In short, this approach can be highly successful if the teacher can create peer pressure against the misbehaving student or students. If such pressure is not established, the approach will not work.

"Time-out" is another kind of punishment that can be effective under certain circumstances, and it can take on many forms in a teaching situation. For example, the teacher can deprive students of working on a science project or participating in discussions, laboratory activities, field trips, science fairs, or school assemblies. Teachers do not regard these types of punishments to be as severe as those that involve undesirable and distasteful stimuli. The teacher is removing an activity that a student enjoys and replacing it with one that is unpleasurable. The student may be asked to go to a time-out room designated by the administration that is devoid of all external stimuli. Instead of going on a field trip, attending an assembly, or working on a project, students will be asked to go to the time-out room while the activity is taking place.

Sometimes time-out results in a reward for some students. Students may deliberately cause a problem to avoid doing work, resulting in a reward instead of a punishment. Teachers must use care when using this type of approach because it may be an effective discipline measure for certain students but not for others.

What about reasoning with a student? Is it effective? It may be with certain students. However, teachers who use reasoning as their only disciplinary technique may find it initially unsuccessful. When students misbehave and they have been warned and continue to misbehave, the teacher may try to reason with them. When reasoning alone does not take care of the problem, punishment should be used. If the punishment fails as well, another reasoning session with the student may be necessary. Reasoning with the student at this point may now be effective, since the student has now experienced punishment for the act and knows what will follow if the behavior does not change. The student also knows that the teacher means business.

What should the teacher do if reasoning and punishment fail and the student continues to be disruptive? First, the teacher should review the history of the situation and determine, if possible, whether the student is using the punishment to obtain attention. If this is the case, the teacher may try to ignore the misbehavior for a period of time and ask the other students in the class to do the same. Of course, if the behavior endangers others, it cannot be ignored, and the nonreceptive student should be removed immediately. In any instance when intelligently used punishment does not work, the teacher must consider whether the punishment was severe enough for the misbehavior.

The teacher should also determine if other students are positively reinforcing the student who is misbehaving. Sometimes the punishment is not severe enough and the positive reinforcement on the part of other classmates outweighs the punitive measures, so that the student continues to misbehave to gain the attention of the classmates and teacher.

Again, if the misbehavior is minor, the teacher should try to ignore it for a period of time and ask the students in the class to ignore it as well. Usually, if the student is not receiving needed attention, the behavior will not be repeated. Persistent inattention may often be the solution for students who continue to misbehave to elicit attention.

Teachers who have constant discipline problems and must frequently resort to punitive measures should examine their teaching situation. They should determine the causes of the misbehaviors. How often are the misbehaviors repeated? Are they teacher induced? That is to say, is the teacher doing a good job of teaching? Are the lessons stimulating, interesting, and relevant or boring, uninteresting, and irrelevant? Do the lessons provide for individual differences in the class, or are they geared for only the average student? Are a variety of activities used during the course of a science lesson, or does the teacher spend all the time asking students to read out of their textbooks or copy material written on the chalkboard into their notebooks?

A search for reasons for frequent discipline problems is certainly in order. Causes other than teacher-induced misbehaviors should be examined. The school situation may also be responsible for misbehaviors in the classroom. Students may know that they can get away with something if they know that the administration does not normally back up teachers' actions. The administrator who condones student misbehavior, serious or not, is only causing an atmosphere that can promote problems in the classroom. On the other hand, an administrator who is known to be firm, understanding, nonthreatening, and fair and who will not condone misbehavior regardless of degree, will create only a good teaching atmosphere in which teachers can work effectively and without fear of lack of support.

MANAGEMENT OF MIDDLE SCHOOL AND JUNIOR HIGH SCHOOL STUDENTS

Middle schools throughout the country have various organizational patterns. Some include Grades 6 through 8, while others include only two grades, such as Grades 7 and 8, 6 and 7, or 5 and 6. The junior high school pattern in most cases includes Grades 7 through 9. In the middle school, teachers will teach primarily preadolescent students. At the junior high school level, the teacher will deal with both adolescent and preadolescent students. The age span of students in the middle/junior high school grades varies depending on the school system organization.

Students at the middle school or junior high school level are difficult to manage unless the teacher has exceptional skills to maintain discipline. The students in this group are unique in many ways. In a particular class of middle/junior high school students, a teacher will encounter those who are mature and spontaneous and those who are immature and passive. They may be boisterous yet inhibited, and have attention spans shorter than those of senior high students. They may also be rebellious, often attempting to disregard parental or adult authority. They may be restless, lazy, and uncooperative. There are also social, physical, emotional, and intellectual differences not only within the age group, but also between boys and girls. Boys generally do not like to associate or be grouped with girls and vice versa.

Students between the ages of 12 and 14 undergo many physical and psychological changes that concern and perplex them. The teacher should be aware of these developmental changes in order to understand the students' behavior. The teacher will find that there are no stereotypes in this age group. Although a teacher will find some students who often want to be regarded as responsible individuals and who are capable of making their own decisions most of the time, in the same classroom there are those who want direction from the teacher all the time. Once the teacher realizes that students in this age group are unique in many ways and need to be understood, teaching them can be very rewarding and challenging.

Maintaining discipline in middle/junior high school classrooms requires a great deal of skill

FIGURE 13–1   Student characteristics and resulting curriculum and administrative needs that middle/junior high school science teachers must consider to improve their effectiveness and success at this grade level

---

FOUR IMPORTANT ASPECTS TO CONSIDER WHEN TEACHING SCIENCE TO PREADOLESCENTS
Intellectual Characteristics of Early Adolescents

- *Design instruction appropriate to students' concrete operational reasoning ability.*   Early adolescent students benefit from firsthand experiences that provide concrete exemplars of concepts and principles. The concrete stage of cognitive development limits the amount of abstract information students can process and find meaningful. Complex ideas and mathematical operations are difficult for many students to master.
- *Expect a short attention span.*   Students attend to a given learning task for a short period of time, then seem to be ready for a new experience.
- *Be aware that such students possess many interests.*   Students are motivated to engage in learning tasks when their interests are addressed. Their interest and knowledge schemas are tied together and form important mental structures to hook into during instruction.

Social-Emotional Characteristics

- *Don't be surprised when students display a variety of moods.*   Students may appear mature and orderly at one moment, followed by childish and disruptive behavior the next.
- *Take advantage of their tendency to seek peer acceptance.*   Students respond to and are attracted to their peers and friends. They spend considerable effort to gain peer approval.
- *Be prepared for students' tendencies to ignore and reject adult standards.*   Students want to gain their own independence and show their maturity, which can manifest itself in a rebellious and inconsiderate manner.

Science Curriculum

- *Provide a balanced curriculum.*   There should be a variety of learning outcomes. The first outcome should be to produce positive attitudes toward science, technology, and school science. In addition, the curriculum should develop mastery of basic skills and knowledge fundamental to science and technology.
- *Center upon relevant topics.*   The curriculum should address topics that are meaningful to the lives of students. Identify topics that have immediate interest and use science and technology to increase students' knowledge and understanding of these ideas.
- *Emphasize an inquiry-oriented mode of instruction.*   Use science process skills to help students process information and represent knowledge. Use problem solving and projects to provide relevant contexts for learning and to cause students to take more ownership in their work.
- *Use a variety of measures to evaluate learning.*   Assess many learning outcomes to help these students. Some do well on paper and pencil tests while others may do well on performance tests, especially students with limited proficiency in the English language.

Organizational Structure

- *Plan and organize the curriculum.*   Middle and secondary school science teachers must be given time to form their own curriculum so that they can provide for the unique needs of this population.
- *Provide for the integration and coordination of subjects.*   Teachers in all of the subject areas should be cognizant of each others' curriculum and reinforce the learning outcomes designated in their curriculum.
- *Use a variety of organizational patterns in the school.*   Multiage grouping, developmental grouping, homogeneous grouping, heterogeneous grouping, alternate scheduling, block scheduling, school-within-a-school, and so on must be considered to meet the needs of this age group. The faculty should be free to experiment with different organizational arrangements.

on the part of the teacher. Not all teachers are capable of handling this age group.

> Because of their unique physical development and behavior, preadolescent students have special needs that include a basic understanding by adults—especially by middle school teachers—of the physical and emotional changes. They particularly need warm affectionate teachers with a sense of humor who do not nag, condemn, or talk down to them. They also need opportunities for greater independence, and for assuming greater responsibility without excessive pressure. Finally, preadolescents need a sense of belonging and acceptance by peers. (Reinhartz & Beach,

Students in the middle school years need a variety of experiences and activities in order to have a productive learning environment. The teacher should be flexible enough so that students can pursue some of their own interests as well as the learning outcomes of the existing curriculum. The teacher should give students in this group opportunities to help them extend the cognitive skills they already possess, rather than spend all their time developing new skills. Because students in this group are constantly seeking independence, it is imperative that teachers recognize the necessity to provide activities that challenge their students in order to help them express themselves and to grow intellectually.

Middle/junior high school students behave best in an environment that promotes self-control. Teachers who are very rigid and control behavior through fear will not provide the atmosphere that will permit students to practice self-discipline. Students must be given a certain amount of liberty in the classroom in order for them to develop as responsible and reliable individuals who practice self-discipline. At the same time, by giving students a certain amount of latitude the teacher cannot allow behavior that will disrupt the learning environment. The students must recognize that the teacher is in control of the class and will not condone disruptive actions.

In conclusion, a teacher who provides a supportive environment that is friendly and promotes cooperation among students will foster proper classroom behavior. A variety of learning experiences that are intellectually demanding, which will allow students to use the mental skills they possess, will also help to develop self-disciplined, secure, and well-adjusted individuals. Figure 13–1 presents a set of characteristics of middle/junior high school students, as well as the curriculum and administrative considerations that should be used to plan an orderly and productive instructional environment in the middle/junior high schools.

SUGGESTED ACTIVITIES

1. Prepare a list of rules of behavior for a science class of your choice using the authoritarian approach as a model.
2. What set of rules would you give your students to maintain good control during a question-and-answer period?
3. Prepare a list of rules for behavior in an earth science laboratory, a chemistry laboratory, a biology laboratory, and a physics laboratory. Compare the lists. Discuss the differences with the members of your science methods class.
4. Suppose you are an inexperienced teacher and are taking over a science class that is known to be disruptive. What measures would you use to control the students when you first take over the class? What types of procedures would you use to maintain control throughout the term?
5. Nan is an eighth-grade student who comes to class late at least three times a week. How would you correct this behavior?
6. John is a seventh-grade student who cannot read the board without his glasses and he

will not wear them. How would you handle John?

7. Visit a science classroom and observe whether the students:
   a. seat themselves when the bell rings, and how they do it
   b. raise their hands before speaking
   c. hand in their assignment as the teacher directed
   d. clean up the laboratory area after doing the assigned exercise
   e. follow all safety rules during a laboratory period
   f. refrain from speaking while the teacher is attempting to teach a lesson

8. What rules were established by the teacher regarding each situation in question 7? What specific behaviors were expected by the teacher?

9. How would you deal with a student who waves an ignited Bunsen burner in the direction of other students?

10. What course of action would you take when a student ignores your directions and uses profanity in front of the class?

11. Mary constantly cheats on tests. What steps would you take to correct this misbehavior?

12. Suppose you are assigned to teach science to a group of sixth-grade students and you have prepared a series of 20 laboratory exercises that will require the use of a Bunsen burner, chemicals, a microscope, electrical equipment, plant material, insects, rocks, and glassware. How would you divide a class of 20 students in groups of three if half are boys and half are girls? What criteria would you use to group the students? What rules of conduct would you establish for such a laboratory program?

BIBLIOGRAPHY

Buckley, N. K., and Walker, H. M. 1970. *Modifying classroom behavior: A manual of procedures for classroom teachers.* Champaign, IL: Research Press.

Canter, L., and Canter, M. 1976a. *Assertive discipline.* Santa Monica, CA: Canter and Associates.

Canter, L., and Canter, M. 1976b. *Assertive discipline: A take-charge approach for today's educator.* Santa Monica, CA: Canter and Canter Associates.

Canter, L., and Canter, M. 1989. *Assertive discipline for secondary school educators: Inservice video package and leader's manual.* Santa Monica, CA: Lee Canter and Associates.

Charles, C. M. 1985. *Building classroom discipline.* New York: Longman.

Charles, C. M. 1992. *Building classroom discipline* (4th ed.) New York: Longman.

Farmer, W. A., Farrell, M. A., and Lehman, J. R. 1991. *Secondary science instruction* (pp 319–332). Providence, RI: Janson Publications, Inc.

Ferster, C. B., and Skinner, B. F. 1957. *Schedules of reinforcement.* New York: Appleton-Century-Crofts.

Gallagher, J. J., and Tobin, K. 1987. Teacher management and student engagement in high school science. *Science Education, 71*(4): 535.

Ginnot, H. 1969. *Between parent and teenager.* New York: Macmillan.

Ginnot, H. 1972. *Teacher and child.* New York: Macmillan.

Glasser, W. 1969. *Schools without failure.* New York: Harper & Row.

Glasser, W. 1990. *The quality of school: Managing students without coercion.* New York: Harper & Row.

Harrop, A. 1983. *Behaviour modification in the classroom.* Toronto: Hodder and Stroughton.

Hill, D. 1990. Order in the classroom. *Teacher Magazine, 1*(7): 70-77.

Johnson, L. V., and Bany, M. A. 1970. *Classroom management: Theory and skill training.* New York: Macmillan.

Jones, F. 1987. *Positive classroom discipline.* New York: McGraw Hill.

Kim, E. C., and Kellough, R. D. 1987. *A resource guide for secondary school teaching* (pp 295–317). New York: Macmillan.

Kounin, J. S. 1970. *Discipline and group management in classrooms.* New York: Holt, Rinehart and Winston.

MacDonald, R. E. 1991. *A handbook of basic skills and strategies for beginning teachers* (pp 168–185). White Plains, NY: Longman.

McDaniel, T. R. 1987. School discipline in perspective. *The Clearing House, 53*(2): 369.

McIntyre, T. 1989. *The behavior management handbook: Setting up effective behavior management systems.* Boston: Allyn & Bacon.

Medland, M. M., and Vitale, M. M. 1984. *Management of classrooms.* New York: Holt, Rinehart, and Winston.

Petty, K. 1988. Discipline in your classroom. *The Science Teacher, 55*(2): 34.

Purvis, J. S., and Leonard, R. 1985. Strategies for preventing behavioral incidents in the nation's secondary schools. *Clearing House, 58*(4): 349.

Reinhartz, J., and Beach, D. N. 1983. *Improving middle school instruction: A research-based self assessment system.* Washington, DC: National Education Association.

Rinne, C. H. 1984. *Attention: The fundamentals of classroom control.* Columbus, OH: Merrill.

Rogers, C. 1969. *Freedom to learn.* Columbus, OH: Merrill.

Sanford, J. P. 1984. Science classroom management and organization. In *1984 AETS Yearbook. Observing science classrooms: Observing science perspectives from research and practice.* Columbus, OH: ERIC Clearinghouse for Science, Mathematics and Environmental Education, Ohio State University.

Sanford, J. P. 1987. Management of science classroom tasks and effects on students' learning opportunities. *Journal of Research in Science Teaching, 24*(3): 249.

Sauer, R., and Chamberlain, D. 1985. Follow these six steps and learn to manage student discipline. *The American School Board Journal, 172*(1): 42.

Schmuck, R., and Schmuck, P. A. 1979. *Group processes in the classroom.* Dubuque, IA: Wm. C. Brown Group.

Sprick, R. S. 1985. *Discipline in the secondary classroom.* West Nyack, NY: Center for Applied Research in Education.

Walker, J. E., and Shea, T. M. 1991. *Behavior modification: A practical approach for educators* (5th ed). New York: Merrill/Macmillan.

Planning for Instruction

When planning science lessons, science teachers should use a variety of strategies to maintain student interest, such as lecture/discussion, hands-on activities, and demonstrations.

CHAPTER 14

Planning and Teaching Science Lessons

Planning and teaching are primary functions that all science teachers perform. The quality of planning and teaching is based on the competencies that teachers possess and determines what students learn. In many school districts, administrators periodically observe teaching to assess teacher effectiveness. The results are used for salary and tenure decisions. Preservice science teacher preparation programs devote considerable time to planning, practice teaching, and providing feedback to prospective teachers to develop their competence. Science teachers should plan their lessons to reflect the goals of science teaching and be able to translate these goals into clearly stated instructional objectives that are achieved through student-oriented activities.

Overview

- Present and critique examples of ineffective and effective science teaching practices.
- Reflect on the content of science lessons, especially on their relevance, interest, and potential to engage student thinking.
- Reflect on the methodology of a science lesson and discuss the instructional strategies and teaching skills that are directly related to science teaching.
- Discuss short-form, intermediate-form, and long-form lesson plans.
- Stress the merits of planning and teaching lessons, and of receiving feedback on these teaching functions.

HIGHLIGHTING EFFECTIVE AND INEFFECTIVE PRACTICES

Science teachers face a big challenge when they plan and teach science lessons because their instruction must reflect the dynamic nature of science, teaching, and learning. The task would be easy if they could transmit knowledge directly to students, and that knowledge would be learned, understood, and recalled with accuracy. The record is clear. The knowledge-presentation approach is not effective with most students in the middle and senior high schools. Science teachers must go beyond telling and teaching terms and facts. They must create learning environments that help students to understand science principles and theories, to explain scientific and technological ideas, to find personal meaning from the content and processes, and to apply this knowledge in their daily lives. In addition, science teachers must engage *all* students in instruction, not only those who seem disposed to science, and respond enthusiastically (McCormick & Noriega, 1986; Oakes, 1990).

Fortunately, a great deal of useful information is available to science teachers and educators that can highlight effective practices as well as those that interfere with students' learning (Tobin, Kahle, & Fraser, 1990). Effective science teachers use a variety of instructional strategies to help students build upon their prior knowledge and construct knowledge from sensory experiences. They use demonstrations, laboratory exercises, hands-on activities, discussions, and lectures, and implement these strategies in a thoughtful and thorough manner. Effective science teachers use questioning and wait time to engage students and to maintain their interest throughout their lessons. They encourage all students to participate in instruction. They also demonstrate the desire to help all students to learn a great deal about science and technology and to believe that science and technology are worthwhile enterprises in which they can participate.

Experienced teachers, as well as new science teachers, can benefit from practice in planning and teaching lessons and receiving feedback regarding these teaching functions. Planning, teaching, and feedback sessions can identify skills and perceptions about teaching science that need to be developed, reinforced, or altered. This aspect of pre- and in-service teacher education can be very useful to a science teacher's success in the classroom and growth in the profession.

The following two vignettes are presented and analyzed to highlight ineffective and effective science teaching practices.

▼ Mr. Stam teaches biology in an urban high school. He sits behind the desk as students file into the room and take their seats. After the bell, Mr. Stam takes roll and asks for yesterday's homework, which directed students to answer selected questions at the end of the textbook chapter just studied.

Today, Mr. Stam begins the chapter on the blood. He uses the overhead projector, writing on transparency material to present the lecture notes. Mr. Stam uses the assigned textbook as his lecture outline. He writes down key terms (e.g., plasma, red blood cells, white blood cells, and platelets) and provides a few phrases for each to highlight important information. The lecture moves at a rapid pace. Occasionally, Mr. Stam relates a personal note to embellish his lectures. For example, during today's session the students are told how he transported medical supplies and blood to the combat zones during the Vietnam War.

Once in a while Mr. Stam calls on a student to determine the student's knowledge of the content. He favors the males in the class, especially those whom he knows. For example, Mr. Stam frequently calls on Philip because Philip is on the cross country track team that Mr. Stam coaches. Philip is also a good student who has expressed the desire to become a doctor. Mr. Stam scowls at the girls because many of them are inattentive during his lectures. He warns that if they do not pay attention, they will fail his test. Mr. Stam is known for long, difficult tests. ▲

How would you evaluate Mr. Stam's teaching? From the vignette, one can conclude that Mr. Stam presents biology as a body of knowledge, using the lecture method as his primary instructional strategy. He believes that giving students large amounts of information is an appropriate way to teach. Therefore, Mr. Stam is an informer and a giver of information. He relies on the textbook and does not seem to be able to take advantage of the numerous resources and strategies available. In this situation, the teacher does all of the explaining and takes the primary responsibility for the conceptual organization of the subject matter. Given the teacher's philosophy and limited approach to instruction, the students are likely to find little meaning in the subject matter and to achieve little understanding of it.

Mr. Stam discourages the girls from taking part in class discussion. He even puts the girls down when they say something that is not entirely correct or when they are not paying attention to the lectures. On the occasions when Mr. Stam seems to include the girls in discussions, he does so with the more attractive females. Nevertheless, Mr. Stam favors his interactions with the boys, especially those he knows through athletics. He is a buddy to the boys with whom he is acquainted from the cross country teams that he coaches and the track teams he assists. An observer of these teaching practices can easily recognize that the teacher turns off many students in the classes by discouraging participation, especially from certain students such as females and males who do not take part in sports. Many science teachers are guilty of reinforcing sex roles and sexual stereotypes (Kahle, 1989), and this type of teacher behavior seems to contribute to the present nationwide situation where females are underrepresented in the sciences.

Mr. Stam rarely organizes students into small groups, except for laboratory work. He conducts one laboratory exercise per week. Given the enormous potential that cooperative grouping holds for increasing the interaction between students and the learning environment, it is unfortunate that Mr. Stam does not use this strategy. Just as unfortunate is the fact that he does not use the computer or conduct demonstrations. However, there are many science classrooms like this across the nation, where the presentation of information predominates and where many other useful teaching strategies and techniques are not implemented (Gallagher, 1989).

Let's look into the classroom of Mrs. Willis, another biology teacher just down the hall from Mr. Stam.

▼ Mrs. Willis stands at the classroom door and greets students as they enter class, taking attendance as students pass by. The students head straight to the lab bench areas to begin work immediately in their small groups. The class is studying blood and students are assigned to a given group based upon special interests.

Mrs. Willis forms groups to build on academic strengths and individual interests. The task for each group is to create an advertisement to promote the sale of blood and its constituents, for example, plasma, red blood cells, white blood cells, and platelets. Each group is expected to study the composition of blood and to make a presentation that would urge others to buy their product, at the same time demonstrating knowledge of this important body tissue. Students are encouraged to make posters, videotapes, and audiotapes, or to present skits. The students cooperate in their group functions, each carrying out a particular task and working diligently to prepare for their presentation to the other class members.

Mrs. Willis provides many resources for students to examine in order to learn about the composition of blood. On a table at the front of the room, she provides textbooks, pamphlets, and journal articles for students to study. A videodisc program is available that students can use on their own to view a program about the composition of blood. The classroom is a beehive of student activity.

Mrs. Willis stops the class 15 minutes before the end of the period and asks the students to take their seats. At this point, she determines each group's progress. She also checks for understanding of certain key concepts, calling on many students to demonstrate their knowledge and understanding. This is often accomplished by asking the students to construct a concept map or a concept circle. In addition, Mrs. Willis refers the students to the laboratory they conducted a few days before in order to reflect upon how well the information they gathered from the printed material and videodisc program confirms what they observed firsthand during the laboratory exercise.

▲

How would you assess the teaching effectiveness of Mrs. Willis? From the vignette, it is obvious that Mrs. Willis serves as a facilitator of instruction, helping students to seek, construct, and organize their own knowledge. She employs a variety of strategies and techniques to help students derive personal meaning from the study of biology, especially techniques that enhance their conceptual organization of the subject mat-

ter (Novak & Gowin, 1984). Mrs. Willis initiates the study of blood with a laboratory exercise that permits students to examine different blood cells under the microscope so that they can use this sensory data to build the notion that blood is a tissue composed of cells and other constituents. Often students carry the misconception that blood is really a liquid. A large percentage of class time is spent with students working in groups. This small-group work has improved the classroom climate as well as achievement. However, Mrs. Willis has fine-tuned the small-group work so that off-task behavior is at a minimum and students are productive during these sessions (Fraser, 1990).

Over the past 5 years the student body has changed dramatically. Now the school population is comprised of over 60% Asians, African-Americans, and Hispanics. As a result of this demographic change, the research she has read

When planning science lessons, teachers should use activities that allow students to work together in groups, in pairs, or alone.

regarding equity in the classroom, and her awareness of lack of participation in technical careers by females and minorities, Mrs. Willis has modified her instruction so that girls and minorities are active participants in the classroom. She is aware that females, for example, may be reluctant to take an active role in classroom discussions or to assert themselves in small-group activities. Mrs. Willis strives to help all of the students realize that biology is a meaningful discipline and that they can participate in careers in this field as well as other scientific and technological fields. By the end of the school year, many of the students realize that they may be suited to become nurses, pharmacists, opticians, and ecologists.

Mrs. Willis provides a great many resources for students so that they can seek out information in the classroom. She has a storehouse of textbooks, magazines, paperbacks, and pamphlets that students can read in order to find out information and to answer their own questions. She permits the students to use the computer, videodisc, and video equipment on their own. The confidence and trust that Mrs. Willis places in the students to use expensive equipment did not come easily, but she has found that students do not abuse equipment or the freedom that she has given them in the classroom because of the respect she has earned. In addition to the instructional resources mentioned above, a few laboratory stations are set up in the room for students to study and review important ideas through firsthand experiences.

REFLECTING ON THE CONTENT OF A SCIENCE LESSON

What should science teachers think about to prepare a science lesson? First, of course, they must determine what it is they want to teach. If a course syllabus is to be followed, part of the decision regarding what to teach has already been made. If science teachers are free to choose any topic, then they have a great deal of freedom to select ideas about which they are knowledgeable. Regardless of the situation, science teachers should carefully select the content and skills to be taught so that students will perceive them as meaningful and worthwhile. This approach to planning holds more potential to stimulate student interest and to motivate them to learn science than one that communicates the notion: "Learn this now because you will need to know it later."

Let's examine the relevance of a lesson plan of a prospective science teacher, Bill Cummings.

▼ Bill submitted to the student teacher supervisor a lesson plan on the endocrine system, which he was going to teach in about one week to a high school biology class as part of his field experience. The instructional objectives for the plan were as follows:

1. Define the terms endocrinology, hormone, and pituitary gland.
2. Name ten endocrine glands and identify their location in the body.
3. Describe the anatomy of the pituitary gland.
4. Explain and recognize the action of nine hormones produced by the pituitary gland. ▲

Although these objectives are clearly stated, they actually have little meaning for the students. They do not emphasize the importance of the endocrine system and how necessary these hormones are to maintain a healthy body. A better set of instructional objectives would stress the function of the endocrine system, the way hormones regulate physiological processes, and how abnormal growth and disease result when the system malfunctions.

When a science lesson is relevant, it will most likely gain and hold students' attention. Students will participate and respond favorably, if the information they receive and the activities they perform interest them. For these reasons, the student teacher supervisor met with Bill Cummings. The supervisor identified several aspects of the lesson plan that should be changed. One was the need to place greater emphasis on the function of the endocrine system and how its hormones affect the human body. She discussed the importance of male and female hormones and their contribution to athletic performance, relating how some athletes use drugs to increase the muscle mass and improve endurance. The supervisor pointed out the availability of many magazine articles and pictures of athletes who have used and abused anabolic steroids and how these teaching aids can be used in the introduction of the lesson to stimulate interest. She suggested that he address physical abnormalities stemming from the endocrine system that students might recognize, such as giantism and dwarfism—problems of hyperpituitarism and hypopituitarism respectively. She further recommended that the student teacher mention myxedema as a condition observed in some older people whose thyroid gland has become inactive, causing puffiness around the eyes, drooping facial features, loss of hair, and a general lack of vigor. The supervisor also mentioned that the lesson plan could include another abnormality known as testicular feminization. This syndrome results in a genetic male with female characteristics. After the discussion with the student teacher supervisor, Bill was able to develop a new lesson focus that would make his instruction more meaningful and interesting to students.

Prospective and novice teachers are often inclined to focus their instruction on facts and definitions, minimizing or omitting application and significance. This orientation should be reversed, if one believes that it is essential to make science meaningful. The science education profession recommends that science teachers modify students' beliefs so they perceive science as useful in their daily lives. This idea is especially important for the introductory lessons in a unit. Every time a science teacher plans a science lesson he or she should ask: "What knowledge is of most worth?" (Blosser, 1986, p. 608). This is not a trivial question.

Inspection of the instructional objectives for the prospective teacher's *original* lesson plan reveals that the learning outcomes reflect lower-

order thinking. The objectives are primarily at the knowledge level, typical of much instruction in many science classrooms. It is unfortunate, but too much science teaching stresses facts and definitions, when thinking and understanding should be the focus of science instruction (Tobin & Fraser, 1987). Exemplary science teachers work diligently to encourage students to think and to comprehend science course material. With this in mind, Bill, the prospective teacher, was requested to rewrite his instructional objectives to include learning outcomes at the application and analysis levels of Bloom's taxonomy.

Often novice and experienced science teachers believe that the first few lessons of a unit of study should "teach students the basics." These individuals believe that fundamental facts and concepts should be taught first so this information can be used to learn and understand the rest of the topic. Although this rationale sounds logical, these teachers do not realize that students already know something about the topic they are presenting. Students possess some knowledge related to most science topics, even though some of the knowledge may be shallow or incorrect. Therefore, science teachers should design introductory lessons with this assumption in mind and gradually develop the definitions, facts, concepts, and vocabulary words as the unit progresses. Sound curriculum planning must begin with what the learner already knows (Ausubel, 1963) and build and reform the knowledge with relevant contexts.

The amount of content covered during a science lesson is an important consideration for both middle school and senior high school instruction. Science teaching must stress thinking and understanding important scientific, technological, and mathematical concepts (American Association for the Advancement of Science, 1989). Lesson plans must reflect this ideal through activities that engage students in reasoning and seeking answers to questions. Lessons that cover a great deal of subject matter, and include many details and vocabulary, must be reexamined for their potential effectiveness.

In addition, a science lesson should reflect inquiry. An effective science lesson must stimulate student involvement, as evidenced by examining the instructional activities as well as the instructional objectives. Unfortunately, little active learning could be inferred from Bill's original lesson plan on the endocrine system. Consequently, he was referred to Chapters 3 and 4 in this textbook in order to reconstruct the instructional activities in a manner that would require students to observe, analyze, and explain. The prospective science teacher was asked to reflect on the instructional strategies discussed in this methods textbook. He was requested to determine how science process skills can be brought into the lesson to help students gather information and organize it. For example, pictures from athletic magazines and geriatric journals were suggested so that students could examine them for conditions brought about by too little or too much of a given hormone in the body. An inductive inquiry session that would encourage students to find patterns from exploratory activities was discussed. Strategies and techniques were considered that would help students build upon what they know, represent concepts in their minds, and organize ideas so that they have personal meaning. Frequently, science teachers conduct lessons that attempt to pour knowledge from written or spoken words into the minds of students (Tobin & Fraser, 1987). These teachers seem to overemphasize science as a body of knowledge and deemphasize science as a way of investigating and a way of thinking. They also demonstrate a lack of understanding as to how students learn best.

REFLECTING ON THE METHODOLOGY OF A SCIENCE LESSON

After teachers decide on what to teach, they should focus attention on how to teach it. Although this sequence is not sacred, some teachers begin their planning with the content and then address the methodology of the lesson. Effective science teachers use a variety of instructional strategies, teaching skills, and materials within a given lesson, especially those intended for the middle school (Rosenshine, 1990). As opposed to lecturing the whole period, these teachers may begin with a demonstration, move on to a brief lecture, conduct a short hands-on activity, and end with a review of major points. They ask questions that require students to think, and they encourage all students to participate in the lesson. This variety and transition from activity to activity is critical to the success of science courses.

In addition, when using a variety of instructional strategies within a given lesson, effective science teachers properly implement each strategy they plan to use. Whether they conduct a discussion session or a demonstration or a laboratory exercise, these teachers carry out each strategy well and thoroughly. The quality of their instruction is evidenced by the amount of student engagement, as reflected in the amount of student learning that occurs. All students, regardless of their intellectual ability, benefit from the implementation of good teaching strategies (Roadrangka & Yeany, 1985).

Instructional Strategies

An instructional strategy designates the way that a major segment or the entire lesson is approached. It is the general plan that will be used to achieve a given set of objectives. Some lessons are planned around the presentation of information, and thus the lecture strategy is used. Some lessons are planned around activities that require students to find out about ideas, and thus selected inquiry and discovery strategies are used. Some lessons are planned around the illustration of science principles or laws through the use of a demonstration. And, of course, some lessons are designed to use several instructional strategies.

Effective science teachers are those who use models to plan a lesson, selecting instructional strategies based on their potential to accomplish certain goals and objectives (Eggen & Kauchak, 1988). Science teachers have a variety of teaching strategies from which to choose to match appropriate teaching plans with various learning outcomes. If the appropriate selections are made and the plan is executed well, students will actively engage in the lesson and benefit greatly.

Let's review the instructional strategies discussed earlier in this textbook so that teachers can plan carefully for the inclusion of these approaches in a science lesson.

1. *Inquiry.* Inquiry is a broad strategy. Actually it is more than a strategy because it is fundamental to the scientific enterprise. Nevertheless, teaching science through inquiry suggests many strategies that can be used to construct knowledge, such as the use of science process skills, discrepant events, inductive and deductive methods, and problem solving.
2. *Demonstration.* Demonstrations illustrate and help to explain ideas through concrete means. They focus attention on key ideas and are an efficient means for guiding thinking and engaging participation.
3. *Laboratory work.* The laboratory involves students in hands-on or firsthand activities. This strategy can be approached in a variety of ways, such as using process skills, inductive or deductive methods, problem solving, and technical skills.
4. *Lecture.* Lectures involve the presentation of ideas and information, especially to large numbers of people. Lecturing is an efficient way to instruct a large group of students.
5. *Discussion.* Discussion permits students to express their views and to clarify their ideas. It can increase student involvement with instruction.
6. *Recitation.* The recitation session requires students to demonstrate their knowledge. It usually takes place toward the end of a lesson.
7. *Reading strategy.* Many reading strategies are available to involve students in examining the contents of a textbook. These strategies focus

students' attention on key words and concepts, and help them to internalize the information and review for tests.

Teaching Skills

Teaching skills are specific behaviors that teachers use to conduct lessons and implement instructional strategies. Specific skills are needed to introduce lessons, ask questions, give directions, provide feedback, and end lessons. Teaching skills must be developed to conduct an effective science lesson. These are the behaviors that promote student engagement during instruction.

In addition to student engagement, good teaching skills can help make learning more concrete. Exemplary science teachers use these skills frequently to give students concrete examples, ensuring comprehension of the subject matter (Treagust, 1987). For example, these teachers use expository and comparative advance organizers to help students derive meaning from explanations and to integrate what they are about to learn with what they already know. They constantly give students analogies so they can perceive abstract ideas that are common to the science curriculum. One might add, these teachers incorporate teaching aids such as diagrams, charts, and slides to help students visualize key concepts during their presentations.

The following list reviews the major teaching skills that a science teacher should consider when planning and teaching a science lesson.

1. *Set induction.* The set induction prepares the students for learning. It focuses attention on what will be taught and attempts to interest students in the lesson.
2. *Questioning.* Questions involve students in their learning by causing them to think and respond.
3. *Giving directions.* Giving directions communicates what is expected and directs students to proper and productive behavior.
4. *Interpersonal interaction and management.* Effective personal interaction between teacher

and students is necessary to establish a positive, productive learning environment. Accepting learner feelings and thoughts, giving corrective feedback, reinforcing student participation, and eliminating disruptive behavior are critical skills to use during science instruction.

5. *Closure.* The closure brings a lesson or teaching segment to an end. This act helps students to recite and review what has been presented, reinforcing main ideas and deriving additional meaning from the instruction.

CONSTRUCTING INSTRUCTIONAL OBJECTIVES

Instructional objectives are an integral part of most lesson plans. They focus instruction and describe what the learner should know and do as a result of a lesson, unit, or course of study. Some educators refer to instructional objectives as performance objectives while others call them behavioral objectives. In any case, most educators agree that instructional objectives should be stated in terms so specific that they can be measured or assessed. This section emphasizes the development of clearly stated objectives that can guide instruction and evaluation.

Instructional objectives can be placed into three categories or domains—cognitive, affective, or psychomotor. Objectives in the cognitive domain relate to intellectual abilities and skills such as recognition, recall, comprehension, and problem solving. Most of the objectives in science course work are written in the cognitive domain. Objectives in the affective domain relate to attitudes, interests, beliefs, and values. Objectives in this area are beginning to appear more frequently in science curricula because they relate to critical aspects of school learning. Objectives in the psychomotor domain reflect motor skills and hand-eye coordination. They occupy a special place in the teaching of science, especially in the laboratory.

A good instructional objective must describe a learning outcome that states what the student will be able to do, know, or believe as a result of the instruction. This must not be confused with an instructional activity, which indicates how the student will be taught. For example, watching a film on different types of sharks is an activity that should not be confused with that of naming different types of sharks or describing their behavior, which may be learned by viewing the film. One way science teachers can understand what is meant by a learning outcome is to determine what they want the students to be able to do, think, or know after they instruct them. The teacher should then write these outcomes in precise terms. The teacher should keep in mind several criteria when constructing objectives. A good instructional objective is one that is

1. student oriented
2. descriptive of a learning outcome
3. clear and understandable
4. observable (TenBrink, 1990, p. 75)

Some science teachers still write objectives that reflect what the teacher will do, rather than what the students are expected to learn. Somehow, teachers who construct these objectives have misunderstood the purpose of instructional objectives. These teachers describe what they believe they will be doing in the classroom, instead of describing what the students should be able to do as a result of the instruction. An example of an objective that might be considered teacher oriented instead of student oriented is as follows: "Explain the steps in making a parabolic reflecting surface." If this statement refers to what the teacher does and the teacher has no intention of assessing it, then the statement can be considered a teacher direction. The statement refers to the instruction that a science teacher might use to prepare students for a science project that involves the use of solar energy for heating objects. The student-oriented instructional objective for this activity might read: "Construct a solar cooker with a parabolic reflecting surface that will cook a hot dog."

Those who construct instructional objectives must be skilled in their work, similar to writers who wish to communicate well with their readers. Both groups achieve good communications by using verbs to specify action. Verbs give clarity and understanding to instructional objectives just as they give action and color to written and spoken communication. The proper use of a verb can convert a vague statement intended for use as an instructional objective into a clearly understood learning outcome. Examine each pair of the following statements and observe how the second statement of each pair is made clearer by the replacement of one or more action verbs:

Original	*Improved*
Learn the parts of a cell.	*Name* the parts of a cell.
Know Ohm's law.	*Use* Ohm's law to *calculate* voltage, resistance, and amperage in a circuit.
Understand weather charts.	*Interpret* weather charts.

Finally, good instructional objectives are observable; that is, they state learning outcomes that can be recognized. The actions specified by objectives are those that are overt, such as talking, writing, moving, and drawing. Some action verbs that are useful in stating objectives in behavioral terms are as follows:

construct	devise	state
demonstrate	make	graph
identify	measure	calculate
locate	estimate	predict
diagram	interpret	express
classify	infer	gather
show	hypothesize	

When teachers get into the habit of using appropriate action verbs to construct instructional objectives, they will avoid using vague terms and actions that are difficult to observe and measure. Some of these vague outcome terms include: know, understand, learn, appreciate, and familiarize. These terms are more appropriately used in goal statements associated with unit and course plans, not with instructional objectives associated with lesson plans.

Instructional objectives that are clearly written statements of observable student learning outcomes serve many important functions. They can define what the teacher should teach, because instructional objectives must reflect the goals and purpose of the lesson. They also serve as a guide to instruction. Activities are developed to ensure that the students achieve the objectives. Some teachers use objectives to focus attention on what the students are expected to learn in the lesson. Others use objectives as study questions. The objectives are also used to guide the evaluation process. The assessment procedures are based on the objectives and not the instruction.

Robert Mager (1984) provides a useful way to prepare and analyze instructional objectives. He indicates that clearly stated instructional objectives have three characteristics:

1. *Performance*—specifies the observable behavior that the learner is to exhibit.
2. *Condition*—specifies the conditions under which the learning outcomes will be assessed and what the learner will be given or denied.
3. *Criterion*—specifies the minimal level of acceptable performance that the learner will be expected to exhibit.

Using these three characteristics will likely result in clearly stated objectives. Furthermore, their use will help teachers specify objectives in the three educational domains and in the various aspects of scientific literacy. Although most objectives written by science teachers for daily lesson plans include only the performance component, these objectives can be clarified in many instances by including the condition and criterion components.

Performance Component

The first and most important characteristic of an instructional objective is the performance component, because it describes the behavior that the learner is expected to demonstrate as a result of the instruction. Action verbs are generally used to specify the performance component. Many of these verbs are listed in table 14–1. They can be used to prepare objectives in the three domains: cognitive, affective, and psychomotor. The following list gives nine examples of instructional objectives with the verb of the performance component underlined:

1. Label the structures of the heart.
2. Give the name of the chemical elements.
3. Analyze a compound.
4. Determine how well different materials stop radiant energy.
5. Identify the factors that influence the period of a pendulum.
6. Find the voltage of batteries.
7. Make a tapered glass tip.
8. Write a report on cloning in humans.
9. State your opinion on abortion.

All these objectives contain a performance term that indicates what the learner is expected to do. Verbs specifying the actions desired of the students are label, write, analyze, construct, and state. Although these objectives are concise, greater clarity can be gained by adding a conditional component to them.

Condition Component

The second characteristic that should be used to construct objectives is the condition component.

TABLE 14–1 Performance terms for writing instructional objectives in the three domains of educational goals

Cognitive	Affective	Psychomotor
Knowledge Define, describe, identify, label, list, locate, match, measure, name, select, state **Comprehension** Alter, classify, convert, distinguish, estimate, explain, extrapolate, generalize, infer, predict, summarize, translate **Application** Apply, calculate, compute, determine, predict, solve **Analysis** Analyze, differentiate, compare and contrast, relate **Synthesis** Arrange, compile, compose, construct, devise, design, reorganize, summarize, synthesize **Evaluation** Appraise, assess, conclude, evaluate, interpret, compare, criticize, explain, summarize, justify	**Receiving** Ask, attend, choose, describe, follow, listen, give, name, locate, select, reply **Responding** Answer, assist, discuss, do, help, perform, practice, read, recite, report, select, talk, watch, write **Valuing** Accept, argue, complete, commit, do, follow, explain, initiate, join, propose, report, study, work, write, differentiate **Organizing** Adhere, alter, argue, change, defend, modify, organize, relate, combine, explain, integrate **Characterizing** Act, confirm, display, propose, question, refute, solve, use, influence, perform, practice, verify, serve	Adjust, build, construct, calibrate, display, dismantle, dissect, focus, manipulate, measure, organize, prepare

Source: Based on information from B. S. Bloom (ed.). *Taxonomy of educational objectives. Handbook I. Cognitive domain* (New York: David McKay Co., Inc. 1956.) and B. S. Bloom, D. Krathwohl, and B. B. Masia, *Taxonomy of educational objectives Handbook II. Affective domain,* New York: David McKay Co., Inc., 1964.

This component gives the conditions under which the learner will be assessed or the circumstances under which the learner must perform. The condition component states what the learner will be given or denied or what he should have done to achieve a given objective. Inspection of the instructional objectives below, taken from the preceding list and given a condition component, will reveal that the objectives attain greater clarity than they had when only a performance component was used. Parentheses are placed around the condition components, and the performance components are underlined.

1. (Given a diagram of the heart), <u>label the structures of the heart</u>.
2. (With the aid of a periodic table of chemical elements), <u>give the names of the elements</u>.
3. (When given the structural formula for an unknown compound), <u>analyze the compound</u>.
4. (With black construction paper, white construction paper, aluminum foil, and glass), <u>determine how well different materials stop radiant energy</u>.
5. (Using two bobs of different masses and two strings of different lengths), <u>identify the factors that influence the period of a pendulum</u>.
6. (Using a voltmeter and a variety of good and bad batteries rated from 1.5 to 9 V), <u>find the voltage of the batteries</u>.
7. (Using a piece of glass tubing, a Bunsen burner, and a triangular file), <u>make a tapered glass tip</u>.
8. (Using information from at least three articles on cloning and from a discussion on this topic with four laypersons), <u>write a report on cloning in humans</u>.
9. (After studying the unit "Human Reproduction"), <u>state your opinion on abortion</u>.

The condition component adds clarity to an objective in specifying what is expected of the learner. For example, in the case of objective 2, by adding the condition "with the aid of a periodic table of chemical elements," one finds that the student does not have to memorize the names of more than 100 chemical elements but can instead refer to the periodic chart, which provides the symbols for all the elements. In the case of objective 6, by adding the condition "using a voltmeter and a variety of good and bad batteries rated from 1.5 to 9 V," the student can better realize the intent of the learning outcome, which appears to relate to the practical knowledge of science. Here the student is given an opportunity to learn how to determine the voltage of batteries with a voltmeter. The batteries specified in this objective are those used for electrical devices commonly found in the home, such as flashlights, cameras, and radios. Voltmeters are commonly used by electricians, garage mechanics, and electronic servicemen, and young people should become familiar with them. Nevertheless, without this condition, the student could completely miss the intent of this objective, because with only the performance component, "find the voltage of the batteries," the student could accomplish this objective simply by reading the voltage specifications printed on each battery.

Criterion Component

The third characteristic that should be used in the construction of an objective is the criterion. The criterion further defines the learning outcomes in behavioral terms. It gives the minimal acceptable level of performance that is expected. Speed, accuracy, quality, quantity, and reference to an established standard are used in the construction of the criterion component. A criterion component added to the conditional and performance components of an instructional objective gives greater clarity to that objective.

In the examples shown below, the criterion component is bracketed, the condition component is enclosed in parentheses, and the performance component is underlined.

1. (Given a diagram of the heart), <u>label the structures of the heart</u> [to include the chambers, valves, and vessels].
2. (With the aid of a periodic table of chemical elements), <u>give the names of the elements</u> [with 100% accuracy].

3. (When given the structural formula for an unknown compound), <u>analyze the compound</u>, [giving the names of its functional groups].
4. (With black construction paper, white construction paper, aluminum foil, and glass), <u>determine how well different materials stop radiant energy</u> [by presenting the data collected on this problem on a bar graph].
5. (Using two bobs of different masses and two strings of different lengths), <u>identify the factors that influence the period of a pendulum</u> [by demonstrating how each factor influences the period of the pendulum].
6. (With a voltmeter and a variety of good and bad batteries rated from 1.5 to 9 V), <u>find the voltage of the batteries</u> [with an accuracy of + or − .5 V].
7. (Using a piece of glass tubing, a Bunsen burner, and a triangular file), <u>make a tapered glass tip</u> [that can be used as an eyedropper].
8. (Using information from at least three articles on cloning and from a discussion on this topic with four laypersons), <u>write a report on cloning in humans</u> [that is at least three typed, double-spaced pages in length and either supports or condemns cloning].
9. (After studying the unit "Human Reproduction"), <u>state your opinion on abortion</u> [in one or two paragraphs].

The criterion components add greater understanding to the educational outcomes that are expected from the learner. For example, adding the phrase "with 100% accuracy" to objective 2 specifies that the chemistry students must learn all of the names of the chemical elements with complete accuracy. Adding the phrase "that can be used as an eyedropper" to objective 7 specifies that a certain quality must be achieved in making the tapered glass tip. If the taper is too narrow, liquids will not pass through, and if it is too wide, liquids will squirt out of the tip. Adding the phrase "that is at least three typed, double-spaced pages in length and either supports or condemns cloning" to objective 8 specifies both size and purpose for the report. The report must be a three-page position paper.

Instructional objectives do not necessarily have to be written with the three component parts in the order just described—condition, performance, and criterion. Often, these components are shuffled. An instructional objective is like any other written statement—the intent is to communicate. Therefore, teachers should constantly change and revise their objectives, using as many phrases as needed and using the order necessary to communicate the intended learnings.

DESIGNING A SCIENCE LESSON PLAN

There is no accepted format for designing a lesson plan. The sequence, number of elements, and amount of detail in a lesson plan vary considerably. Many school districts require teachers to use a short-form lesson plan outlining each day's lesson for one week on a single sheet of paper. The short form merely sketches what will occur during each lesson. An intermediate form of a lesson plan incorporates more detail than the short form. The intermediate form is usually one page long. The long-form lesson plan describes all aspects of each lesson in detail, including the handouts, worksheets, and tests. It contains many pages and gives great detail. Therefore, the long form can be used to analyze and reflect upon the potential effectiveness of an instructional plan. The long form is very useful for teachers who wish to develop and teach exemplary science lessons.

Short-Form Lesson Plan

The short-form lesson plan provides the minimum detail and number of elements. Usually, school districts require their teachers to sketch the teaching plans for each week. This gives the teacher, administrators, or a substitute teacher some idea of what should be taught on each school day. Figure 14–1 is an example of a short-form lesson plan for five class sessions of a unit on flowering plants. These lessons are intended

FIGURE 14–1 Daily lesson plan for one week of instruction (short form)

Instructional Objectives	Related Activities	Resource Materials	Teaching Techniques	Assignments
Monday State the uses of flowering plants	Students read and discuss "Parts of Plants Used for Food." Students identify flowers, stems, leaves, and roots that may be eaten as food. Allow students to taste different parts of a plant used for food.	Textbook: *Exploring Living Things* Handout no. 1 Various fruits, green vegetables, beans, etc.	Discussion, lecture, questions and answers	Students bring to class examples of edible seeds (e.g., vanilla, spices, coffee, chocolate)
Tuesday Describe the characteristics of flowering plants.	Students view 16 mm films: *Secrets of the Plant World, From Blossom to Fruit*, and *Wildflowers*. Class discussion related to films. Distribute copies of *Basic Characteristic Listings of Over 200 Garden Flowers* and discuss.	16 mm movie projector Films: M3334, M5077, 6395 from library Handout no. 2	Illustration, discussion, questions and answers	Students bring to class two simple flowers.
Wednesday Name, locate, and describe the parts of the flower.	Students perform laboratory activity, record observations and data, and answer related questions. Distribute copies of *Parts of a Flower* and complete activities 1–9 of the procedure.	Textbook: *Exploring Living Things* Handout no. 3 Live flowers, hand lens, forceps	Group work, demonstration, supervised study, questions and answers	Students complete worksheet on *Parts of a Flower* (handout no. 4).

FIGURE 14–1 *continued*

Instructional Objectives	Related Activities	Resource Materials	Teaching Techniques	Assignments
Thursday Describe the process of pollination and fertilization in flowering plants.	Students complete activities 10 and 11 of *Parts of a Flower*, recording observations and data and answering related questions.	Microscope, prepared slide of germinating pollen grains, glass slide	Group work	Students complete worksheet on *Pollination* (handout no. 5).
Friday Describe the characteristics of one flowering plant of California. Name the flowering plant of California by the common and scientific names.	Teams of two students select a wildflower of California and identify, locate, and describe the main parts. Construct a simple diagram of the flower and label it. Give the common name and proper scientific name.	Handout no. 6 Cloth material, glue, sewing thread, needle, and trimmings Various reference books	Illustration, group work, review	

for middle school life science students. Inspection of the example reveals that space is limited to describe the elements associated with the lesson. The elements contained in this particular short form are as follows: instructional objectives, related activities, resource materials, teaching techniques, and assignments. Because so little information is given in the short form, it is difficult to determine exactly what will take place and to critique it.

Intermediate-Form Lesson Plan

The intermediate-form lesson plan includes a moderate amount of detail about a lesson. Figure 14–2 shows an example of an intermediate-

form lesson plan concerning fog formation that is part of a junior high school earth science unit on precipitation. Analysis of this example reveals the advantages and disadvantages of the intermediate-form lesson plan. There are nine elements in this particular lesson plan. The time schedule gives the teacher an idea of what to do during each time segment of the class period. This will help the teacher gauge the length of each activity and realize how much can be accomplished during the period. There is plenty of room in this lesson plan to write the assignment, list the materials that are needed, and mention the references that will be used.

This type of plan has two major shortcomings. First, the objectives are usually abbreviated in

A technique that may work well is first to provide feedback on the effective aspects of the lesson. This approach sets the stage for a positive feedback session. Both the instructor and peers identify the skills and techniques that were executed well. These are listed on the chalkboard and on the feedback and evaluation form. The presenter is also encouraged to participate in this activity. Then suggestions for improvement can be given to the presenter, focusing on specific elements of the lesson and on skills that should be considered in future teaching.

SUGGESTED ACTIVITIES

1. Read the vignettes at the beginning of this chapter that contrast an ineffective with an effective high school biology teacher. Perform your own critique and compare it with those that follow the vignettes and those prepared by your peers.
2. Analyze Figure 14–3, the long-form lesson plan on static electricity designed for a middle school science course. Critique the lesson and compare your assessment with critiques prepared by your peers.
3. Design a lesson plan to be taught to a class of middle or high school students or to peers.
   a. Organize the content so that it will be relevant to the audience and identify strategies and skills that will actively engage the learners in the lesson.
   b. Carefully review this chapter to be sure the lesson plan is thoroughly prepared and will be perceived by the students to be a good or an exemplary science lesson.
   c. Give the lesson plan to an instructor or colleague for evaluation and feedback. Respond to suggestions by modifying the plan.
4. Prepare to present a science lesson.
   a. Gather all necessary equipment and materials.
   b. Practice the lesson in front of a peer, spouse, or friend, and determine perceptions regarding the potential effectiveness of the lesson.
   c. Present the lesson to the intended audience and ask for feedback on how well it was taught.
   d. Compare your lesson with lessons presented by others.
5. Obtain copies of lesson plan forms used in local schools. Compare and contrast them.
6. Obtain copies of the teaching evaluation forms used in local school districts. Study the categories and descriptors on these forms and compare them with the ideas and descriptors of effective teaching that have been discussed in your teacher education program.
7. Identify exemplary and effective science teachers in local school districts and arrange to visit their classrooms to observe their instruction. Use an observation form during your visits to focus your observations, if the science teacher is comfortable with this. Some teachers do not enjoy having an observer recording their teaching behavior or evaluating them, so discuss this with the teacher before you engage in the activity. Compare your observations with others who have completed this assignment.

BIBLIOGRAPHY

American Association for the Advancement of Science. 1989. *Science for all Americans: Project 2061.* Washington, DC.

Ausubel, D. P. 1963. *The psychology of meaningful verbal learning.* New York: Grune & Stratton.

Barufaldi, J. P. (Ed.). 1987. *Improving preservice/in-*

service science teacher education: Future perspectives, 1987 AETS Yearbook. Columbus, OH: ERIC Clearinghouse for Science, Mathematics, and Environmental Education.

Blosser, P. E. 1986. What research says: Improving science education. *School Science and Mathematics,* 86(7): 597–612.

Cruickshank, D. R., Holton, J., Fay, D., Williams, J., Kennedy, J., Meyers, B., and Hough, J. B. 1984. *Reflective teaching.* Bloomington, IN: Phi Delta Kappa.

Eggen, P. D., and Kauchak, D. P. 1988. *Strategies for teachers.* Englewood Cliffs, NJ: Prentice-Hall.

Fraser, B. J. 1990. Students' perceptions of their classroom. In K. Tobin, J. B. Kahle, and J. B. Fraser (Eds.), *Windows into science classrooms: Problems associated with higher-level cognitive learning.* New York: The Falmer Press.

Gallagher, J. J. 1989. Research on secondary school science teachers' practices, knowledge, and beliefs: A basis for restructuring. In M. L. Matyas, K. Tobin, and B. L. Fraser (Eds.), *Looking into windows: Qualitative research in science education.* Washington, DC: American Association for the Advancement of Science.

Heuvelen, A. V. 1986. *Physics: A general introduction.* Boston: Little, Brown, and Company.

Kahle, J. B. 1989. Teachers and students: Gender differences in science classrooms. In M. L. Matyas, K. Tobin, and B. L. Fraser. *Looking into windows: Qualitative research in science education.* Washington, DC: American Association for the Advancement of Science.

Mager, R. F. 1984. *Preparing instructional objectives.* Belmont, CA: Fearon Pitman.

Matyas, M. L., Tobin, K., and Fraser, B. J. (Eds). 1989. *Looking into windows: Qualitative research in science education.* Washington, DC: American Association for the Advancement of Science.

McCormick, T. E., and Noriega, T. 1986. Low versus high expectations: A review of teacher expectation effects on minority students. *Journal of Educational Equity and Leadership, 6:* 224–234.

Novak, J. D., and Gowin, D. B. 1984. *Learning how to learn.* New York: Cambridge University Press.

Oakes, J. 1990. *The underparticipation of women, minorities, and disabled persons in science.* Santa Monica, CA: The RAND Corporation.

Ornstein, A. C. 1990. *Strategies for effective teaching.* New York: Harper & Row.

Roadrangka, V., and Yeany, R. H. 1985. A study of the relationship among type and quality of implementation of science teaching strategy, student formal reasoning ability, and student engagement. *Journal of Research in Science Teaching, 22:* 743–760.

Rosenshine, B. 1990. On using many materials. In A. C. Ornstein, *Strategies for effective teaching* (p. 328). New York: Harper & Row.

Schon, D. A. 1987. *Educating the reflective practitioner.* San Francisco: Jossey-Bass.

TenBrink, T. D. 1990. Writing instructional objectives. In J. M. Cooper (Ed.), *Classroom teaching skills.* Lexington, MA: D. C. Heath.

Tobin, K., and Fraser, B. J. 1987. Introduction to the exemplary practices in science and mathematics education study. In K. Tobin and B. J. Fraser (Eds.), *Exemplary practices in science and mathematics education.* Perth, Western Australia: Curtin University of Technology, Science and Mathematics Education Centre.

Tobin, K., Kahle, J. B., and Fraser, B. J. 1990. *Windows into science classrooms: Problems associated with higher-level cognitive learning.* New York: The Falmer Press.

Treagust, D. F. 1987. Exemplary practices in high school biology classes. In K. Tobin and B. J. Fraser (Eds.), *Exemplary practices in science and mathematics education.* Perth, Western Australia: Curtin University of Technology, Science and Mathematics Education Centre.

When planning science units, a science teacher should include many hands-on activities in which students can participate. Laboratory work, field trips, and inquiry sessions can provide first-hand experiences for learning science concepts and principles.

CHAPTER 15

Planning Science Units

The advantages of planning are well recognized. Teachers can plan a total course, units of instruction, or daily lessons. The difference is merely the extent to which planning is applied. A unit plan is somewhat middle ground between a course and a lesson plan; consequently it is shorter than a course but larger than a daily lesson plan. When carefully planned to consider the students' interests and abilities, a unit plan can be very effective in producing desired learning outcomes in science programs. Science teachers should participate in the design and planning of their units of study in order to maximize their effectiveness.

Overview

- Present and discuss the components of a resource unit.
- Discuss and explain the components of a teaching unit.
- Present and analyze a science teaching unit.
- Provide suggestions for designing effective science units.
- Point out useful sources for constructing units of study.

A unit organizes the curriculum into a cohesive and meaningful instructional plan. Units of instruction break up a course of study into segments that are larger than lesson plans. A unit may consist of one or more topics. Each topic consists of facts, concepts, principles, and theories as well as skills. Perhaps the most efficient units are those of short rather than long duration.

Two types of unit plans can be used in science teaching—resource units and teaching units. *Resource units* are designed to identify a variety of resources that can be used to teach a particular topic. The resources can be drawn from many sources and organized in a variety of ways.

Teaching units are specifically designed to contain only those resources that are used for teaching a particular topic. They are carefully organized teaching plans that sequence content and experiences for students. Teaching units that are designed around relevant topics will probably stimulate student interest and help to motivate students to achieve the intended learning outcomes.

THE RESOURCE UNIT

The resource unit does not specify exactly what is to take place during instruction nor does it indicate the sequence of instruction. Instead, it provides a wide variety of activities and materials, which can be selected and used as needed. The resource unit is not a teaching plan per se, but a catalog of resources that a science teacher may use during unit or daily instruction.

The major advantage of the resource unit is its flexibility. Once teachers identify the instructional objectives for the unit, they may turn to this resource for suggested activities and instructional strategies. This flexibility permits the teacher to redirect an initial instructional plan to conform more with student abilities and interests as they are identified. For example, teachers may plan a lesson, teach it, and then find that it was not appropriate for the particular group of students. Then they may turn to the resource materials of the unit to identify more suitable student activities and instructional strategies to include in a new lesson plan. Another advantage of the resource unit is that it can be constantly updated as new activities and new teaching strategies are identified. Revising the resource unit requires no more than adding new activities and strategies as they are encountered and removing those that are no longer useful.

To use a resource unit effectively, a science teacher must identify the unit objectives, as well as the instructional objectives, and keep them in mind constantly. This will prevent the teacher from straying from the original specified objectives and timetable. Consequently, omission of important substantive content germane to the unit is avoided.

Format

The resource unit can consist of file folders identified by the type of resource material they contain. The filing system is used to organize the resources by category, such as laboratories, demonstrations, microcomputer programs, transparencies, films, readings, and projects. Each unit will have different categories identified to fit the unit and instructional objectives. For example, the unit entitled Unity and Diversity Among Living Things may have the categories listed in figure 15–1. A file folder would be created for each category and would contain items such as those listed.

File folders permit science teachers to update the unit material without difficulty. Teachers can remove the material as it becomes outdated and add new material at any time during the year as they encounter new activities, approaches, and information pertinent to the unit. The information that goes into file folders can be organized using microcomputers and word processing programs. The programs can store large amounts of information and organize it so that it is easily retrieved and modified.

FIGURE 15–1   Example of a filing system for a resource unit

Folder label	*Type of material in folder*
Unit objectives	Obtained from N.Y. State Biology Regents Syllabus.
Instructional objectives	Obtained from N.Y. State Biology Regents Syllabus.
Content outline and lecture notes	Outline as presented in N.Y. State Biology Regents Syllabus and revised by teacher. Lecture notes developed by the teacher.
Laboratory exercises and laboratory worksheets	Collection of specific exercises from commercial sources, as well as those developed by the teacher and/or other science teachers.
Demonstrations	Collection of references and books on biology demonstrations. Collection of teacher's demonstrations. Collection of demonstrations from other teachers. Collection of demonstrations from textbooks and other sources.
Films, film loops, and film strips	List of films available from Ward's Natural Science Establishment, Inc., Carolina Biological Supply House, local film library, Coronet Films, Athena Films, etc. Each film is reviewed and critiqued. A page devoted to each film is inserted in the file, which includes a synopsis of the film, as well as a guide sheet for use during the showing of the film.
2 × 2 inch slides	Reference to appropriate storage container, which houses 2 × 2 inch slides on the unit.
Transparencies	Reference to file of transparencies pertinent to the particular unit.
Materials and apparatus	List of needed materials and apparatus, including specific sources, costs, and reference to storage locations in the science area.
Books and articles	Includes references to textbooks, books, and articles in journals and magazines. Includes copies of selected articles from *Scientific American, Natural History, Discover, Omni, Science Digest,* etc.
Questions for oral discussions and question/recitation sessions	List of questions that can be used to guide discussions following lectures or discussion-question sessions. List of questions to ask students to review topics. Ideas for questions can be obtained from books, students, and supply house newsletters.
Sources for collecting live material	Includes local sources such as ponds, streams, and lakes and also commercial supply houses such as Ward's Natural Science Establishment, Inc. and Carolina Biological Supply.
Evaluation procedures	Collection of test items for quizzes and unit tests. (Refer to card file of test items). Includes psychomotor assessment of particular laboratory skills necessary to perform cytological and other laboratory work.
Provision for individual instruction	Collection of projects. Individualized learning activities derived from various sources such as professional journals (*The Science Teacher, American Biology Teacher,* etc.), teachers' manuals, those suggested by other teachers, and National Curriculum Programs and materials such as BSCS.

FIGURE 15–1 *continued*

Folder label	Type of material in folder
Career planning	List of books and articles on careers pertinent to the unit of study. (Obtain suggestions from guidance directors, university professors, local professional scientists in research laboratories, and local medical schools.)
Outside speakers	List of specialists in the local area, including research scientists and university professors.
Suggested field trips	Information about visits to microbiology laboratories at hospitals, medical schools, local universities, commercial biological laboratories, and also visits to local fields, streams, lakes, rivers, and woods.
Bulletin board material	Newspaper and magazine articles collected by students and teachers for display in the classroom.

THE TEACHING UNIT

A teaching unit is a highly structured plan, having a set of components that may vary from teacher to teacher. The terminology used to identify unit components may also vary. Some use "instructional activities" while others use "student activities." Some call "teaching strategies" "instructional strategies." Some call "instructional activities" "learning options." Although the terminology might be different, the meaning is the same. To avoid confusion, science teachers should use terminology with which they are comfortable and that can be understood by other teachers.

The following basic components are listed and defined for the unit plans discussed in this chapter:

1. Title
2. Unit objectives
3. Instruction objectives
4. Materials and equipment
5. Time frame
6. Instructional activities
7. Review
8. Evaluation and testing

1. *Title.* The title is usually a short phrase or a combination of words that captures the es-

sence of the unit. It can be direct and to the point or "catchy" and attention getting.

2. *Unit objectives.* These are generally broad (not specific) in scope and represent the expected student learning outcomes (concepts, principles, understandings) on unit completion. Such unit objectives are met by the achievement of the specific instructional objectives. For example, the unit objective for understanding the concept of homeostasis can be met by the achievement of a set of instructional objectives concerning positive and negative feedback systems. The unit objectives are, in essence, the content outline for a unit of study.

3. *Instructional objectives.* The instructional objectives are specific learning outcomes that students should be able to demonstrate at the end of the unit. They represent the knowledge and skills the students should possess as a result of instruction. Such objectives specify learning outcomes and acceptable levels of performance. They can also be used to evaluate instructional effectiveness. The instructional objectives indicate the specific content that is to be taught in the unit.

4. *Materials and equipment.* This is a list of materials and equipment needed for a particular unit that serves as a reminder to ensure that

such resources are available during the unit presentation.

5. *Time frame.* The time frame, or timetable, specifies the time allotted to the total unit and its individual activities.

6. *Instructional activities.* These are the learning activities in which the students will participate during the course of the unit. Instructional activities also include the specific teaching strategies the teacher will use during the course of instruction.

7. *Review.* Reviews can be used periodically during the course of instruction of the unit, but certainly before the unit test is administered.

8. *Evaluation and testing.* The specific evaluation plan of instruction is presented at this time. It should indicate the relative weight given to tests, projects, homework, laboratory work, and so forth. For example, if a system of total points is used for a final grade calculation, then the specific point value for each assessment component must be assigned. Although tests, quizzes, and other assessment instruments should be constructed prior to instruction, they can be revised after the content has been presented. Daily quizzes and other "flexible" types of assessment should be constructed while the unit is in progress.

Format

Once the instructional objectives have been stated, the teacher should develop a format suitable for use during the course of instruction. There are many types of formats that can be used. The example illustrated in table 15–1 shows a successful format which has vertical columns indicating the components of the unit—unit objectives, instructional objectives, instructional activities (student activities and teaching strategies), materials and equipment, resources, timetable, and evaluation.

The format outlines part of a unit adapted from a New York State Education Department biology syllabus entitled "Unity and Diversity Among Living Things." The unit objectives are as follows:

1. Define life in terms of the functions performed by living organisms.
2. Describe some of the schemes by which organisms are classified.
3. Identify the appropriate tools and techniques used for cell study.
4. Use some of the tools and techniques for cell study.
5. Recognize the role of the cell as the basic unit of structure and function of most living things.
6. Identify major biochemical compounds and some of the metabolic reactions in which these compounds are involved.

Table 15–1 presents that part of the unit concerned with meeting unit objective 5 from this list. Unit objective 5 is stated as follows:

Students should be able to recognize the role of the cell as the basic unit of structure and function of most living things.

To meet the overall broad unit objective 5, eight instructional objectives (learning outcomes) have been identified as follows:

Students should be able to:
1. Identify and distinguish among plant and animal cells.
2. Observe, identify, and explain the function(s) of each of the following cell structures: cell wall, cell membrane, cytoplasm, nucleus, chloroplasts, chromatin, nucleoli, endoplasmic reticulum, ribosome, mitochondria, Golgi complex, lysosome, vacuole, and centrosome.
3. Observe and describe the process of cyclosis.
4. Describe the contributions of each of the following scientists to the establishment of the cell theory: Leeuwenhoek, Hooke, Brown, Virchow, Schleiden, and Schwann.
5. Explain how cells are derived from preexisting cells.
6. Describe the process of mitosis and know its significance.

7. Describe the process of meiosis and explain its significance.
8. Explain the concepts of cell specialization and division of labor.

For purposes of illustration, the format in table 15–1 shows only three of the instructional objectives to meet unit objective 5. Associated with each instructional objective stated in the format are the instructional activities (student activities, teaching strategies), materials needed, resources required, timetable, and evaluation procedures (if any) to satisfy the objective.

The complete unit plan would include all the broad unit objectives specified for the unit, and all the required instructional objectives (specific learning outcomes) to meet each unit objective,

as well as the instructional activities, resources, timetable, and evaluation procedures.

AN EXAMPLE AND CRITIQUE OF A SCIENCE TEACHING UNIT

A detailed example of a teaching unit is given in this section to provide a concrete illustration of the form and structure of a unit of study. Each of the eight elements that have been recommended for inclusion in an effective science unit are included here to guide the critique of a unit developed by a science teacher, Miss Slone. Miss Slone developed this unit to introduce the study of oceanography, which is part of the earth science course she teaches.

TABLE 15–1 An example of a format for a teaching unit ("Unity and Diversity Among Living Things")

Unit objectives	Instructional objectives	Instructional activities	
		Student activities	Teaching strategies
5. Students should be able to recognize the role of the cell as the basic unit of structure and function of most living things.	Students should be able to 1. Identify plant and animal cells.	1. Prepare wet mount slides of onion skin stained with iodine. 2. Prepare wet mount slides of human epithelial cells from cheek stained with iodine or methylene blue. 3. Observe each under microscope. 4. Prepare diagrams of each and make comparisons.	1. Lecture on general structure of cell using overhead transparency diagrams of animal and plant cells. 2. Demonstrate wet mount slide making and staining techniques. 3. Show film strip, *Introduction to Animal Cells* (Ward's Natural Science Establishment, Inc.). Use 2 × 2 inch slides of electron microphotographs (Fisher Scientific Co.). 4. Discuss after lecture and laboratory the similarities and differences between plant and animal cells. 5. Show film *Cell, the Structural Unit of Life* (Coronet Films).

1. *Title.* The title "Mapping the Ocean's Floor" indicates that the unit relates to oceanography and emphasizes a specific segment of the topic—mapping the ocean's floor—which serves to make the study of oceanography concrete and interesting.

2. *Unit objectives.* The students will:

■ Learn how to construct a profile map of the ocean using the appropriate data.
■ Learn how the process of finding the ocean depths has evolved.
■ Understand the concepts related to marine geology to develop the background for further study of ocean geology.
■ Explore problems that have arisen because of pollution in the ocean.

■ Propose solutions to the problems of ocean pollution.
■ Use knowledge and skills needed to study the ocean; especially its chemistry, physical properties, and biological makeup.

The unit objectives present the scope of the unit on oceanography. They emphasize that the unit will include mapping the bottom of the ocean and discussing pollution. Miss Slone's approach to introducing her students to oceanography is to involve them in an activity that requires the application of science process skills and an activity that requires them to address a science and societal issue. Together, these aspects appear to provide a meaningful way to study marine geology in the middle school. The

Materials and equipment	Resources	Timetable	Evaluation

Laboratory:
    Onions
    Microscope slides
    Cover slips
    Iodine or methylene blue
    Microscope
    Eyedroppers
Lecture/discussion:
    Overhead transparencies of typical plant and animal cells
    2 × 2 inch slide projector
    2 × 2 inch slides of plant and animal cells
Film strip. *Introduction to Animal Cells* (Ward's Natural Science Establishment, Inc.)
    2 × 2 inch slides of electromicrographs (Fisher Scientific Co.)
Film. *Cell, the Structural Unit of Life* (Coronet Films)
	Modern Biology, pp. 42–51
	Prepared laboratory instructions and worksheet for observation of plant and animal cells (see file)
	H. Curtis *Biology,* pp. 76–139, 415
	Biology Resource Book, pp. 3–4
	Thorpe, *Cell Biology*
		Monday 9/10
		    Lecture and discussion on cells
		Tuesday 9/11
		    Laboratory preparation and examination of cells
		Wednesday 9/12
		    Review, discussion, lecture, and laboratory results.
			Examine laboratory worksheets.
			Evaluate diagrams of plant and animal cells.
			Evaluate written statements made by students on worksheet regarding the similarities and differences between plant and animal cells.

(*continued*)

TABLE 15–1 *continued*

Unit objectives	Instructional objectives	Instructional activities	
		Student activities	**Teaching strategies**
	2. Observe, identify, and explain the function of the following cell structures: Cell wall Cell membrane Cytoplasm Nucleus Chloroplasts Chromatin Nucleoli Endoplasmic reticulum Ribosome Mitochondria Golgi complex Lysosome Vacuole Centrosome	1. Using prepared slides of animal and plant cells, identify the following structures: Cell wall Cell membrane Nucleus Chromatin Nucleoli Vacuole 2. Using live material (protozoans and spirogyra), identify the structures listed in no. 1. 3. Using laboratory sheets with diagrams of cells, label the following additional structures: Endoplasmic reticulum Ribosomes Mitochondria Golgi complex Lysosome Centrosome 4. View laboratory demonstration of osmosis and diffusion using starch solution in plastic bag that is placed in beaker of iodine solution. Demonstration illustrates selective permeability of membranes. 5. Diagram and explain function of the cell structure identified in nos. 1 and 2 above.	1. Lecture/discussion of organelle function using charts and film, *Cells & Their Functions* (Athena Films, Inc.). 2. Lecture should explain how organelles perform necessary life functions. 3. Prepare demonstration on osmosis and diffusion.
	3. Observe and understand the process of cyclosis.	1. Prepare wet mounts of elodea. 2. Observe elodea leaves to see constant movement of cytoplasm. 3. Prepare wet mounts of amebas and observe formation of pseudopods, which results from cyclosis.	1. Lecture/discussion on cyclosis using 2 × 2 inch slides and overhead transparencies and charts. 2. Present film loop on cyclosis during lecture. 3. Demonstrate preparation of elodea and ameba wet mounts.

Materials and equipment	Resources	Timetable	Evaluation

Microscope
Slides and cover slips
Spirogyra culture
Protozoan culture
Methyl cellulose
Charts and diagrams of
  cells showing various
  organelles
Overhead transparencies
  of cell organelles
Plastic bags or dialysis
  tubing
Corn starch solution
100 ml beaker
Iodine solution
Rubber bands
	Modern Biology, pp.
	  42–61
	Laboratory guides,
	  accompanying
	  Modern Biology
	Biology Resource
	  Book, pp. 6–7
	Kimball, *Biology,* pp.
	  80–85
	Curtis, *Biology,* pp.
	  101–110
	Thorpe, *Cell Biology*
		Wednesday–Thursday,
		  9/12–9/13
		  Lecture/discussion of
		    organelle function
		Thursday 9/13
		  Demonstration on
		    diffusion
		Friday 9/14
		  Laboratory to observe
		    and identify cell
		    structures
		Monday 9/17
		  Discussion
			Examine laboratory
			  worksheets.
			Conduct recitation
			  Monday 9/17.

Elodea leaves
Amoeba culture
Microscope
Slides and cover slips
Film-loop (BSCS) on
  cyclosis
Film loop projector
2 × 2 inch slide projector
Charts, diagrams, and
  overhead transparencies
	Modern Biology, p. 186
	Biology Resource Book,
	  pp. 8–9
	Curtis, *Biology,* pp. 86–
	  87, 611
		Tuesday 9/18
		  Lecture/discussion on
		    cyclosis
		  Laboratory on cyclosis
		  Review of cyclosis at end
		    of class period
			Examine student
			  laboratory
			  worksheets.
			Give quiz (15 min.)
			  on cyclosis.

goals of this unit are specifically stated in the instructional objectives that follow.

3. *Instructional objectives.* At the completion of this unit, the students should be able to

A. Recognize the definitions of six out of eight of the key terms used in the field of geological oceanography.
B. Explain two methods used to determine ocean depth.
C. Calculate, given a set of sonar echo times, the corresponding depth of the ocean floor.
D. Identify five out of seven features of the ocean floor on a multiple-choice test.
E. Create a profile map of the ocean floor using sonar data and the depth formula.
F. Determine from the profile map the best location to initiate a research drilling operation.
G. After a classroom discussion on potential "ocean-killers" and a trip to the library to read at least one article on this topic, write a two-page report on how the ocean-killer problem can be solved and present the report to the class.

There are at least three aspects of instructional objectives that one should consider in examining Miss Slone's objectives. First, do the objectives reflect the goals of science teaching and the unit objectives for this unit? Second, are the objectives clearly and precisely stated? And third, are the objectives sequenced in a logical or psychological manner? Let us consider these aspects of instructional objectives.

Miss Slone's instructional objectives reflect many important goals of scientific literacy. Objectives A and D are learning outcomes for knowledge and information. Some of the terms to be learned in objectives A and D are *oceanographers, thermocline, deep zone, continental shelf, abyssal plains, ooze, guyots, bathyscaphe, trench,* and *fault.* These are basic terms in the study of oceanography.

Objectives B, C, and E focus on the technology and skills needed to study the ocean's floor.

To achieve these objectives the students learn about two methods that have been used to determine ocean depth: the sink-line method that uses a rope, and the echo-sounder method that uses a fathometer. The objectives further specify that the students will construct a profile map from sonar data, which will require that they use a mathematical formula to calculate the ocean's depth. The students will then analyze the data and the profile maps that they construct (objective E) to determine the best location for a research drilling operation (objective F).

Note that objective G addresses the science/societal topic of pollution. For this objective students gather and discuss information on pollutants being poured into the oceans. Here the students must carry out library research and make a presentation to the class. Undoubtedly, this will encourage them to clarify their own values regarding the importance of the ocean and who is polluting it.

The instructional objectives for this unit are clearly stated. Each of the objectives has a behavioral component (Mager, 1984) that specifies what the learner should be able to do as a result of participating in the unit of instruction. For example, in objective A the learner has to be able to recognize (identify) definitions for the key terms used in geological oceanography. In objective E the learner has to make a profile map of the ocean floor.

Miss Slone provided additional clarity and precision by including criterion components for many of the objectives. Objectives A and D indicate the level of acceptable responses expected. For example, in objective A the students are expected to recognize the definition of six out of eight of the key terms.

Miss Slone in some cases has even included a conditional component for her objectives, which indicates what will be given or withheld from the learner in the achievement of an objective. For example, in objective F the students determine the best location for a research drilling operation, using the profile map they created. In objective G the students write a report on ocean-killers only after discussing this topic in class and reading at least one article in the library on the topic. In other words, the teacher wants the students to acquire some background on the topic before they write about it, and not merely to write anything that comes to mind.

Another important aspect of instructional objectives is their psychological nature. Do the instructional objectives form a psychologically ordered set of learning outcomes that make sense to students? As pointed out in Chapter 3,

Robert Gagné's major contribution to educational psychology has been to recommend that teachers determine what it is the learner is supposed to be able to do at the end of the instruction and then determine the knowledge and skills the learner must acquire to demonstrate this terminal behavior. Following Gagné's (1977) recommendation, let us discuss the instructional objectives for this unit on oceanography.

Figure 15–2 presents a task analysis and learning hierarchy of the unit's instructional objec-

FIGURE 15–2   A task analysis and learning hierarchy of the unit objectives

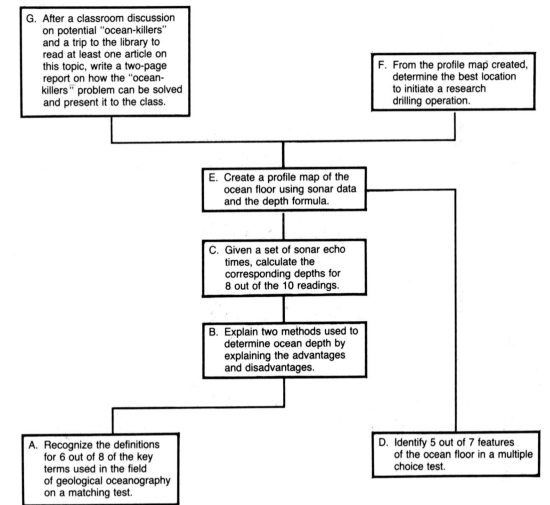

tives. Inspection of the figure shows that two instructional objectives, F and G, appear to be the major learning outcomes of this unit. Objective F concerns predicting where a research drilling operation should be situated. The accuracy of this prediction requires that the learner acquire certain knowledge of the ocean and its floor. The learned capabilities that lead up to making the prediction are as follows:

1. The learner must have developed a profile map from sonar data by calculating ocean depths at various points along the surface of the ocean (objective E).
2. To calculate ocean depths, the learner must know the formula for making depth calculations using sonar data—namely, the echo time for a sound signal traveling from the surface of the ocean to the bottom and back to the point of transmission. The knowledge and the skill to accomplish this task is identified in objective C.
3. Objective D addresses knowledge of concepts and terms related to the ocean floor and the construction of a profile map.
4. Objective B introduces the learner to the two methods that have been used to find ocean depths—the sink-line and the sonar echo method—to provide a background for determining ocean depths.
5. Objective A specifies at the beginning of the unit the key terms and ideas that are basic to this topic of oceanography.

Objective G concerns discussing pollution and ocean-killers, which relates to values and ethics. Achievement of objectives A, B, C, and D provide the background to make the topic concrete and the investigative experience necessary to discuss oceanography intelligently. Gagné (1977) also emphasized the importance of allowing the learner to acquire the appropriate concepts and rules for problem solving, without which the learner cannot bring to bear the proper learned capabilities to discover a meaningful solution to a given problem. We observe that Miss Slone appropriately placed the problem solving task of identifying a solution to polluting the ocean at the end of the instructional program.

4. Materials and equipment. These include: vocabulary study sheets, a filmstrip *The Open Sea* with a teacher-prepared study guide, a film entitled *Oceanography at Work* from Regional Educational Service Center, and a 16-mm projector.

Miss Slone listed the items above as a reminder to have them prepared and ready to go when the unit is taught. She photocopied vocabulary checkup sheets. She checked with Mr. Loucks, the other earth science teacher in the school, to be sure he was not using the filmstrip *The Open Sea* when her class was studying this unit. She ordered the 16-mm film, *Oceanography at Work,* several months in advance from the Regional Educational Service Center. The film was so interesting and informative that Miss Slone made certain to order it well in advance to ensure that it was available when needed.

5. Timetable.

Day 1:

Introduction to the unit.

Read chapter 10, pp. 236–257.

Homework: Do vocabulary checkup.

Day 2:

Go over vocabulary checkup sheet.

Present information on marine geology.

View filmstrip *The Open Sea;* do worksheet.

Day 3:

Discuss methods of ocean exploration.

Plot map of ocean floor topography using sonar data.

Homework: practice depth calculations.

Day 4:

View film *Oceanography at Work*.

Discuss ocean-killers.

Assign paper on ocean-killers.

Day 5:

Review for test.

Students can go to library to read about ocean-killers.

Day 6:

Give unit test.

Begin presentations on ocean-killers.

Day 7:

Complete presentations on ocean-killers.

At a glance, the timetable tells the teacher what should occur on each class day of the unit. It aids in planning and in improving the unit. If, for example, the students do poorly on one aspect of the unit, the teacher can analyze the amount of time and instruction devoted to this aspect and remedy the situation the next time she teaches it.

6. *Instructional activities.* Including all of the instructional activities for this unit is not possible in this presentation because of space limitations, but a few of these activities are presented and discussed. Consider the worksheet that goes with Activity 3, viewing the filmstrip *The Open Sea* (see figure 15–3). The worksheet focuses the students' attention on specific learning outcomes they are to achieve, such as the methods used to determine ocean depth and the features of the ocean's floor. With the worksheet, the students are more attentive and purposeful when they watch the filmstrip.

The fifth learning activity engages students in plotting a profile map of the ocean floor. Miss Slone passes out a topography map sketch of an ocean floor (figure 15–4). The students complete the profile map using the calculations they obtained from sonar depth information. She

gives them table 15–2, followed by an explanation of how to find the depth for each *point* (A, B, C, etc.) on the ocean, given the *time* it takes a sonar signal to be transmitted and received back from the ocean floor. The class discusses the formula $D = \frac{1}{2}t \times V$ for determining the distance sound travels in water, where V is the velocity of sound in water (approximately 5,000 feet per second).

7. *Review.* Miss Slone, along with many other science teachers, has realized the importance of setting aside a specific time during class toward the end of a unit to review the learning outcomes the students should have acquired. She prepares a dittoed sheet that requires the students to demonstrate their knowledge of the instructional objectives for the unit. Miss Slone requires them to work on the review sheet in class, and while they work she circulates among them to be sure they can demonstrate the objectives. This reinforces the learning that was intended for the unit, and in some cases provides a catch-up period for students who were absent or who lagged behind in the learning. She directs students who complete the review sheet and demonstrate mastery of the objectives to the library to read an article on ocean-killers in preparation for their report on this topic.

8. *Evaluation.* At the beginning of the unit, Miss Slone carefully explains the grading system and points that can be earned. The students seem to profit from this information. The grading system for this unit follows.

Evaluation component	*Points*
Unit test	36
Creation of a profile map	10
Determination of the best spot for research drilling operation	5
A two-page paper on a solution for ocean-killers	10
	—
Total points	61
Presentation of a solution for ocean-killers	A, B, C, or F

FIGURE 15–3   Worksheet for Activity 3, viewing the filmstrip *The Open Sea*

After viewing the filmstrip *The Open Sea* (Milliken Publishing Co.), answer the following questions:

1. What is bathymetry?

2. What method was first used to determine ocean depth? What was its disadvantage?

3. Another name for an echo sounder is _____ .
4. How does it work?

5. What are the three main sections of the ocean floor?

6. Name some features of the continental margin.

7. Name some features of the ocean basin floor.

8. What is the average height of a midocean ridge? _____
9. The ocean comprises _____% of the earth's surface.
10. The largest ocean is the _____ Ocean.

Inspection of the grading system shows that a student can receive a maximum of sixty-one points for the unit, plus a grade of A for the presentation. With five different components to the grading system, there are many ways for a student to earn points, such as taking a paper and pencil test, making a map, drawing a conclusion, and writing a paper. In addition, a student can earn a letter grade of A, B, C, or F by making an oral presentation.

Of great importance to the concept of evaluation is that the grading components must be re-lated directly to the objectives. The unit test, for example, assesses certain objectives in the following manner:

- Part A is a multiple-choice test with eight terms to be matched with their definitions specified in objective A, "Recognize the definitions for 6 out of 8 of the key terms used in the field of geological oceanography on a matching test."
- Part B is a short essay item that asks the student to explain two methods to determine

FIGURE 15–4   Ocean floor topography map

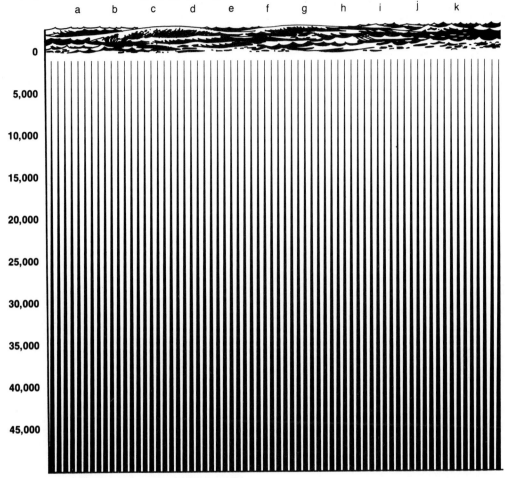

ocean depth; this corresponds with objective B.

■ Part C on the test asks the student to calculate the depth of the ocean floor at each of the listed positions, using the depth formula $D = \frac{1}{2}t \times V$, where t is given and $V = 5,000$ ft/sec. There are ten items in this section where a value for t is given. This part corresponds with objective C: "Given a set of sonar echo times, calculate the corresponding depths for 8 out of the 10 readings."

■ Part D on the test is a seven-item multiple choice test section that corresponds to objective D: "Identify 5 out of 7 features of the ocean floor on a multiple choice test."

Instructional objectives E, F, and G correspond to the other four evaluation components given. Thus, analysis of the evaluation component of this science unit indicates that the teacher matched the evaluation components with the instructional objectives, so that the stu-

TABLE 15–2 Ocean floor depth calculations*

Point	Time	1/2t × V	Depth
A	2	1 sec × 5000 ft/sec =	5,000 ft
B	3	1.5 sec × 5000 ft/sec =	7,500 ft
C	4	2 sec × 5000 ft/sec =	10,000 ft
D	5	2.5 sec × 5000 ft/sec =	12,500 ft
E	7	3.5 sec × 5000 ft/sec =	17,500 ft
F	7	3.5 sec × 5000 ft/sec =	17,500 ft
G	6	3 sec × 5000 ft/sec =	15,000 ft
H	6	3 sec × 5000 ft/sec =	15,000 ft
I	8	4 sec × 5000 ft/sec =	20,000 ft
J	10	5 sec × 5000 ft/sec =	25,000 ft
K	12	6 sec × 5000 ft/sec =	30,000 ft

*D = 1/2t × V

D, depth of water; t, time it takes the sonar signal to travel to the ocean floor and return; V, velocity of sound in water (approximately 5,000 ft/sec).

SUGGESTIONS FOR PLANNING SCIENCE UNITS

The unit plan is an effective way to organize instruction. Even though a great deal of time and effort is required for planning a science unit, the results are worth the effort. Some experimentation is necessary to determine the appropriate length and optimal amount of detail for a given unit plan. Consequently, when science teachers first begin to plan a unit, they should identify a title that communicates the central theme and the practical aspects of the unit. A meaningful title may attract students to a unit of study, and it

dents were tested and graded on what the unit was designed to teach them.

What can be said about the quality of this science unit, "Mapping the Ocean Floor"? For example, how would you rate the unit? Consider the following scale in making an evaluation of the unit:

Not observed	Ineffec-tive		Effective		Excep-tional
0	6	7	8	9	10

This unit is structurally sound and contains all of the essential components for an effective unit plan. The unit involves many aspects of scientific literacy, such as knowledge, investigation, technology, ethics, and values. It appears to focus on a practical aspect of oceanography—mapping the ocean floor—which should interest students in this topic. The unit appears to be of appropriate scope and length for eighth graders: it does not require the students to memorize an enormous number of new words and definitions. Furthermore, the unit enables the students to learn through various instructional strategies.

Unit planning should include activities that allow students to work alone, in pairs, and in groups.

will probably differ from the textbook chapter headings (Hoover, 1982). For example, a title such as "The Leaf: A Food Factory for the Living World" may have more meaning to secondary school students for a unit of study than the common textbook chapter title for this topic, "The Leaf and Its Function." Sometimes, however, attempts to make unit titles interesting can make them appear trivial and communicate very little meaning to the students.

When reflecting on the purpose of a unit, science teachers should think in terms of goals. These are broad and abstract statements of intent (Mager, 1975). Yes, there is certainly a place in science teaching to be general, and that place is when the teacher begins to plan a unit. At this point teachers should consider what it is they want the students to learn or, as Popham and Baker (1970, p. 7) recommended, ask the question: "What do I want my learners to become?" This is where teachers can take the central theme of the unit and think about it in terms of the goals of science teaching, asking themselves: What do I want to stress most? Do I want to stress knowledge, investigative skills, reasoning, problem solving, technology, values, or a combination of these?

A science teacher instructing a class of eighth graders who are poor readers would certainly have different expectations than a biology teacher instructing a class of tenth-grade honors students. The eighth-grade science teacher might stress basic science concepts and the application of these concepts to everyday life, emphasizing knowledge, investigation, and technology. The tenth-grade biology teacher might place more emphasis on students learning principles and theories than the middle school science teacher, so that the college-bound students will do well on entrance examinations and in college biology courses.

Unit plans vary, depending on the teacher, students, and subject matter. The teacher planning the unit should consider the goals of science teaching, the topic at hand, and the char-

acteristics of the students to formulate the objectives for the unit of study.

The instructional objectives of a unit specify the learning outcomes. Examination of the instructional objectives for a unit can tell a great deal about the unit as well as the developer. There are several aspects of the objectives that instructional planners must keep in mind when writing them, such as clarity, specificity, appropriateness, variety, and number. Clarity and specificity are discussed in Chapter 14. Insight into how appropriate the objectives are for a given group of students can be found in Chapter 3. In order to provide for a variety of learning outcomes, review chapters 4, 7, 8, and 10.

The instructional objectives must reflect the content of an instructional plan, and the content that is selected is an important factor in the success of the unit. Many science units address too much content. For example, one such plan examined was a biology unit intended for average students. The unit consisted of eighteen instructional objectives, and almost every objective was constructed at the knowledge level of Bloom's taxonomy. Many of the objectives for this unit began as follows: name, describe, list, identify, tell, and so forth. Analysis of these objectives indicated that the unit covered too much factual content and the students were required to memorize too many terms and definitions. This unit lacked learning outcomes that required the students to apply knowledge and analyze situations that related to the topics under study. The unit also lacked inquiry activities that permitted the students to construct their own knowledge; thus the massive body of information that the students were expected to memorize would probably become quickly forgotten because it would not be incorporated into their cognitive structures in a meaningful way.

On the opposite end of too much content is too little content. On occasion, science teachers produce unit plans that are too short and, therefore, too shallow. If teachers are developing a unit for students who generally demonstrate low levels of achievement in science, they should not necessarily list a small number of objectives for the unit. Units for these students may be more meaningful if they specify a variety of learning outcomes, such as those concerned with concepts, principles, the clarification of personal values, science process skills, and science projects.

Not enough can be said for developing science units that combine a variety of learning outcomes because they encourage students to apply their knowledge. Variety for the sake of variety may not be meaningful unless it ties together the concepts that are under study. One approach is to specify learning outcomes that require students to use what they are learning. Inquiry, problem solving, analysis, and critical thinking are important to emphasize in the terminal learning outcomes for a science unit.

Instructional activities make up the largest component of a science unit. They consume most of the instructional and planning time. Chances are that many individuals will plan a unit thinking about the instructional activities even before they think about goals and objectives. Instructional activities are so much more concrete than are many of the other elements of planning; consequently, they may be thought of first because of past experiences related to a topic.

Although we discussed the eight components of a teaching unit in a particular order, beginning with the title and ending with the evaluation, many instructional planners will not develop their units in this order and as systematically as the final product appears to indicate. Many planners will begin thinking about ideas and activities, and then abstract the goals and objectives from these ideas. Regardless of the way teachers go about the planning of a unit, they should include many student activities, so that rich learning experiences will be provided for the students and the instructional objectives will be achieved.

Inquiry sessions and laboratory activities can be used to stimulate students' thinking and to provide firsthand experiences for learning science concepts and principles. These activities help students participate in the scientific enterprise and construct their own knowledge. Demonstrations are useful for concretely representing abstract ideas and for introducing discrepant events that surprise students, causing them to question their own thinking. Field trips are also useful to stimulate curiosity and to provide concrete experiences. Science projects should also be used to stimulate investigations and to promote understanding of the subject matter.

Teaching strategies should also be considered in unit planning. For example, lecture, recitation, and discussion when carefully planned, are excellent instructional strategies that transmit information, clarify ideas, and increase meaning. These strategies should be included in science units. The use of films, posters, charts, and games should be considered along with the major instructional strategies. If there is an assigned textbook that has one or more chapters related to the unit, the teacher should consider using a reading strategy to take advantage of this instructional method to improve achievement. And, of course, the teacher should remember homework assignments.

Science teachers cannot use *all* of the instructional strategies and teaching aids mentioned in any one unit; nevertheless they should at least consider them in their planning and plan well those they selected for inclusion in a unit plan. There should be at least one instructional activity for each instructional objective. With at least a one-to-one correspondence between objectives and instruction, there should be an adequate instructional program. However, science teachers are encouraged to include additional activities that reinforce specific learning outcomes and tie the unit together.

Science teachers who use carefully planned units can play a role different from those who stay one day ahead of their students in their preparation. Unit planners can take on the role of managers in the learning environment. They can orchestrate instructional activities that serve students well, freeing themselves to observe, monitor, and facilitate student progress. Although these teachers will use the direct instructional delivery approach, they will often use strategies that allow students to interact directly with an instruction system such as a microcomputer or program workbook. Science teachers who plan poorly and end up delivering most of their instruction through lectures rarely get the opportunity to increase their students' interaction with other forms of instructional materials.

The final set of suggestions for planning science units concerns evaluation. The evaluation component must assess the stated objectives and students must be familiar with the objectives and the grading system that will be used to assess their achievement. Science teachers seem to agree that grades are big motivators for students. Students rarely want to do anything unless they know it will count toward the grade they will receive in the course.

Teachers should generally assess achievement at cognitive levels consistent with the stated objective. For example, if a given objective states that the student should be able to identify the definitions for eight terms related to mechanics, a matching or multiple-choice test that requires the students to match definitions with terms is appropriate. Testing students' understanding of these terms by giving a multiple-choice test that includes application, analysis, and evaluation-level questions is not really justified if the instructional objectives do not specify learning outcomes at these levels.

On occasion, a teacher may include a test question that attempts to assess the students' overall understanding of a unit, although no objective is specified for this. The teacher should reserve the right to ask such a question. Planning, instructing, and evaluating through the use of instructional objectives can be restrictive in some instances, and therefore the teachers should be careful that this approach does not inhibit flexibility. On one hand, teachers should convey as clearly as possible what they expect students to learn and evaluate them on these learning outcomes. Furthermore, they should be able to make some changes after the planning phase, so that they can include important learning outcomes and assessments that improve learning. However, students should not be penalized because the teacher is attempting to be flexible, or is inconsistent in planning and teaching.

In short, experienced teachers fully recognize the spontaneous component that exists in all instruction. Sometimes the unplanned, impromptu activity achieves the most valuable learning outcomes. Therefore, the teacher must always maintain a proper balance between essential unit planning and provision for instructional flexibility.

SOURCES OF MATERIALS FOR RESOURCE AND TEACHING UNITS

Resources for planning courses are dealt with in detail in Chapter 16, which includes a description of resources from the following:

- Existing local and state courses of study
- National curriculum projects
- Science textbooks
- Laboratory manuals
- Paperbacks
- Science teachers
- People and places in the community
- National professional organizations

This is only a partial list of sources or materials for resource units and teaching units. In addition to these, the teacher should also consider the following sources to obtain materials and other instructional aids:

- Science magazines such as *Science, Nature, Scientific American, Science News, The New Scientist, Science Digest, Discover,* and *Technology Review*
- Science magazines for secondary school students, such as *Current Science*
- Newspapers and other periodicals (e.g., "Science Times" section of *New York Times*)

TABLE 15–3 Sources for building a unit of study

Instructional materials and resources	Audiovisual aids	Brochures	Bulletin board materials	Career guidance/ideas	Charts	Current scientific knowledge	Demonstration ideas	Discussion ideas	Displays	Field trip ideas	Films	Filmstrips
Television and radio	X			X		X		X				
Testing services												
Technical science journals			X			X		X				
State and local curricula							X	X		X	X	X
Scientists and related professionals				X		X	X	X				
Science textbooks			X			X	X	X			X	
Science supply houses	X	X	X		X			X	X		X	X
Science education journals						X	X	X		X		
Science depts. of colleges and universities	X	X	X	X	X	X		X		X	X	X
Science teachers textbook manuals and guides	X							X		X		
Parent-Teacher Associations (PTA)			X							X		
Private industry	X	X	X	X	X	X		X		X	X	X
Popular science magazines			X	X	X	X		X				
Other science teachers	X	X	X	X	X	X	X			X	X	X
Newspapers and nonscientific magazines			X	X		X		X				
Nature centers and parks	X	X	X		X	X		X	X	X	X	X
National Science Teachers Association	X	X	X	X	X	X	X			X	X	X
National curriculum groups	X	X	X		X		X				X	X
Museums	X	X	X		X	X		X	X	X	X	X
Local science clubs	X	X	X	X	X				X	X	X	X
Local businesses	X	X	X	X						X	X	X
Laboratory manuals			X				X		X			
Guidance counselors		X	X	X							X	X
Government Printing Office		X	X	X	X	X		X				
Educators Guide for Free Films	X											
AAAS Science Film Catalogue	X										X	X

	C1	C2	C3	C4	C5	C6	C7	C8	C9	C10	C11	C12	C13	C14	C15	C16	C17
Group/individual project ideas	X	X	X	X	X	X	X	X	X	X	X	X	X	X	X		
Guest speakers			X													X	
Pamphlets	X	X	X	X	X	X	X	X	X							X	
Homework assignments/ supplemental student work ideas	X	X	X	X	X	X	X	X						X			
Instructional strategies		X	X			X							X				
Instructional objectives		X	X			X	X	X								X	
Laboratory materials		X	X					X	X	X							
Laboratory exercises	X		X	X	X	X	X	X	X	X	X	X	X	X	X	X	
Laboratory experiments	X		X	X	X	X	X	X	X	X	X	X	X	X	X	X	
Photographs	X		X	X	X	X	X	X	X	X	X	X	X	X	X	X	
Public exhibitions/ events		X	X	X		X	X	X	X	X	X		X			X	
Reference materials	X		X	X	X		X	X				X	X	X	X		
Science-related current events	X		X		X	X	X	X	X	X	X	X	X	X	X	X	
Games	X	X			X	X	X	X		X	X	X	X		X		
Science content			X	X	X	X	X	X	X	X	X	X			X		
Maps						X	X	X	X	X	X	X			X		
Technological applications of science	X	X	X	X	X	X	X	X	X	X	X	X	X	X	X	X	
Test questions		X	X	X				X			X						X

- Laboratory guides
- Teacher guides
- Laboratory kits and media packages from commercial supply houses
- Television programs such as "Nova" and the Discovery Channel
- Radio
- Museums
- Aquariums
- Planetariums
- Nature centers
- Public libraries
- University professors
- Scientists from industry

- Parent-teacher organizations
- Public relations directors for science- and technology-related industries
- Local utility companies
- Testing services
- National Science Teachers Association
- American Association for the Advancement of Science

This is only a partial list of sources that can be used in developing a unit of study. See table 15–3 for a more complete list of sources and types of instructional aids that can be obtained from these sources.

SUGGESTED ACTIVITIES

1. Develop a *resource unit* for a topic that you intend to teach in the future. List the unit objectives and the instructional objectives for the unit. Make a file of the materials you plan to collect for the unit, such as experiments, science projects, laboratory exercises, microcomputer programs, demonstrations, films, field trips, overhead transparencies, books, and other resources to meet the instructional objectives specified.

2. Develop a *teaching unit* for a science course in which you are interested. Use the format discussed in this chapter and give careful thought to the suggested components for an effective science teaching unit. In addition, consider the themes that have been recommended in this textbook to produce scientifically literate students, as well as those recommended by national committees on science education. Do the unit goals, unit objectives, and instructional objectives reflect the goals and objectives recommended for science teaching for the 1990s? Have you selected one or more topics that include the most relevant ideas that should be taught

to the students for which the unit is intended?

3. Analyze the instructional activities and review components for a science teaching unit that you have developed. Do the activities reflect *teaching science as inquiry* and *teaching science through inquiry* as these ideas were discussed in Chapter 4? Do the instructional activities help students to construct knowledge and achieve understanding of important learning outcomes? Do the activities encourage students or organize their knowledge with the aid of concept maps and concept circles? Do the activities provide enough recitation and review to result in a high level of mastery of all of the learning outcomes?

4. Obtain a teaching unit from an experienced science teacher. Use the questions provided above to critique the unit. Discuss the assessment of the unit with other members of the science methods class.

5. Identify or develop another format for a teaching unit. What components in your format differ from those suggested in this chapter? What are the advantages of your format?

BIBLIOGRAPHY

Cothron, J. H., Giese, R. N., and Rezba, R. J. 1989. *Students and research: Practical strategies for science classrooms and competition.* Dubuque, IA: Kendall-Hunt.

Gagné, R. M. 1977. *The conditions of learning.* New York: Holt, Rinehart and Winston.

Gunter, M. A., Estes, T. H., and Schwab, J. H. 1990. *Instruction: A models approach.* Boston: Allyn and Bacon.

Hassard, J. 1990. *Science experiences: Cooperative learning and the teaching of science.* Menlo Park, CA: Addison-Wesley.

Hoover, K. H. 1982. *The professional teacher's handbook.* Boston: Allyn & Bacon.

Lux, D. (Ed.). 1992. Science, technology, society: Opportunities. *Theory Into Practice, 31*(1): 2–83.

Mager, R. F. 1975. *Goal analysis.* Belmont, CA: Fearon Pitman.

Mager, R. F. 1984. *Preparing instructional objectives.* CA: Fearon Pitman.

Popham, W. J., and Baker, E. J. 1970. *Systematic instruction.* Englewood Cliffs, NJ: Prentice-Hall.

Posner, G. J., and Rudnitsky, A. N. 1986. *Course design: A guide to curriculum development for teachers.* NY: Longman.

in physics, chemistry, biology, and earth science. In addition, many have required courses in science for Grades 6 through 9. Normally, mandated courses of study can be organized in any order to fit the requirements of the local situation. In such instances, the unit sequence can be altered, but all the content is required.

Student Activities

The selection of content should lend itself to student-centered activities. Science teachers who prepare instruction with few or no firsthand activities generally relegate their students to listening to lectures and reading out of the textbook. Teachers should write units that incorporate many relevant, student-centered activities.

The selection of content based on availability of student-centered activities has a great deal of merit. The selection of material known to be teachable is advantageous, because it has already been tried by other teachers in their classes. An activity-based course of study benefits all types of students—slow, average, or even the academically gifted. One study has shown that active learning has a significant relationship to final grade. Students who are actively involved in carrying out activities, making observations, and discussing their results perform better than students who engage in more passive activities. This is quite a significant finding, because involvement is one variable over which both students and teachers have some control (Howe et al., 1983). Such courses also provide a great deal of flexibility. The teacher can plan individualized instruction for the academically gifted students and special activities for the slow learners, the disadvantaged, and those with disabilities.

Local Resources

When one is selecting content for a course, local resources should be an important consideration. Resources such as a local foundry, chemical plant, or paper mill that employs many of the students' parents might be important enough to consider as subjects for a unit of study in a given school. Schools located in or near a mountain range might develop a unit on the earth's history, glaciers, or igneous and metamorphic rocks. A school located near a lake, pond, river, or stream might use units on aquatic life. A school located in a city that has a science museum may promote units on animals, plants, rocks, and astronomy.

Science teachers should not overlook these resources when planning a course of study. They should make an inventory of these resources and bring it up to date periodically. The inventory can dictate, within limits, the content of a course of study, a unit, or even a lesson. Some of these resources may be important and interesting enough to warrant an entire unit.

Time Constraints

When science teachers are planning a course or a unit, they must take care to allow enough time to cover the amount of material planned for each unit in the course. If too much material is involved, the teacher may be forced to teach content for its own sake. Activities such as laboratory exercises, field work, and science projects that require additional effort would undoubtedly be neglected to save time. The value of such activities is well recognized, and they should not be neglected because of time constraints.

The teacher should plan a science course so that students can carry out experiments, make observations, and discuss results. Courses should be organized so that students can discuss topics of concern to them. Enough time should be allotted to initiate and solve problems and conduct discussions. These are all time-consuming activities, but they will give students a chance to explore their own special interests and develop a positive attitude toward science. The goals of science education are not met when a course or a unit is taught in a way that excludes student-centered activities. When time constraints are imposed, it is advisable for the teacher to plan the course units so that she can teach them in depth rather than superficially. When too much content is covered at the expense of student-centered activities, uncritical acceptance of principles and concepts is fostered, a situation that is certainly antithetical to the development of scientific thinking.

Structure of the Science Discipline

The structure of the science discipline is often used to determine course content. This approach shows how ideas are related and how they fit together in a logical and meaningful way. However, the teacher must take care when organizing a course around the structure of the discipline. Although the structure of the discipline often suggests an order in which the content should be taught, students may not be able to learn ideas, concepts, and principles in the particular order suggested. The discipline structure does not necessarily take into account what type of background a student needs to understand the ideas that are to be developed.

Knowledge of the psychology of learning is very important in selecting content based on the structure of a discipline. The science teacher should use the contributions of psychologists such as Ausubel, Bruner, Gagné, and Piaget when selecting and organizing a course of study based on the structure of a discipline.

ORGANIZING A COURSE OF STUDY

Various types of organization for science courses have been used through the years. Science courses have been organized around the structure of the discipline, around major topics, around major problems, and around social issues. There is probably no plan that is considered superior. This is left to the discretion of the individual teacher or teachers involved.

Flexibility is the most important characteristic of any organizational pattern. A course that mandates the content and instructional activities and allows no flexibility can stifle a teacher. Under such circumstances the teacher may not be able to take advantage of what may arise on a day-to-day basis. Science-based events may be impor-

tant and of interest to students but may have to be disregarded because of the time constraints and mandates. A mandated course may also restrict the teacher from taking care of the interests of the students. In spite of the inflexibility imposed, some teachers manage to find ways to adapt these types of science courses to the needs of their students. Other teachers who are given a free hand to adapt the course of study to the interest and abilities of the students will still follow the course outline systematically and rigidly.

Some courses of study are easier to use than others because of their type of organization. Courses and curricula composed of independent sections are the most flexible, permitting any order or sequence that best fits the teacher's requirements. Sequential courses are not as flexible, because the outcomes of one unit of the course may depend on the outcomes of another unit that has already been taught. The sequence of such courses is rather rigid and usually cannot be changed.

Flexibility is possible if teachers plan short, discrete, and independent units. Such units may be presented in any order that seems appropriate to the teacher for the particular teaching situation. Flexibility, for example, allows a teacher to take advantage of seasonal materials. If a catastrophe such as an earthquake or volcanic eruption occurs, the teacher should be permitted to teach a relevant unit in the course of study at the appropriate time. If such a unit does not exist in the outline of the course, the teacher should have the latitude to develop and teach the appropriate unit. Science courses should be organized to allow teachers the flexibility to teach units that are not specified in the course and to substitute these units for one or more specified originally in the course or curriculum.

Flexibility encourages teachers to take care of the special interests of students. Special units can be planned, project work fostered, and provisions for individualizing instruction can be made if flexibility exists. Logistically, teachers can organize a course of study so that they can teach 70 or 80% of the units specified in the course, with the remaining time allotted for unplanned or spontaneous units.

Organizing by Major Topics

It has been customary to plan a science course based on major topics that are in turn broken up into small and discrete units under each topic. For example, a course in physics may involve the topic Electricity and Electrical Energy. This topic can be broken up into discrete units entitled Electrical Charges, Electrical Circuits, and Magnetism and Electricity. A major topic heading for a course in earth science, The Earth's Crust, can include discrete units entitled The Nature of Matter, Minerals, Sedimentary Rocks, Igneous Rocks, Metamorphic Rocks, and Weathering.

The organization by major topics has been used successfully for many years. Organization into topics and then into smaller blocks of work or units permits the inclusion of problem solving situations, experimentation, field work, projects, and other types of student-centered activities. Technology and other applications of science can be incorporated with this type of organization.

This organization can be misused, however. The teacher can use this approach to cover a vast amount of subject matter through lectures, reading, and discussion, which avoids using activities such as experiments, problem solving, field trips, and demonstrations. Teachers can ignore the characteristics of the students—their abilities and backgrounds, how they learn science. They may also disregard the technological and commonplace applications of science.

However, topic organization can be very successful in meeting the goals of science education when each unit in the course of study has carefully identified objectives that

1. meet the needs of students in living today and in the future
2. take into account the mental and physical maturity of the students

3. permit the use of instructional activities that involve students in the scientific enterprise
4. help develop scientific literacy

Organizing by the Structure of the Discipline

A course organized around the structure of a science discipline may not necessarily provide the most desirable or appropriate way for students to learn the content of the discipline. The organization of the discipline does not necessarily take into account all that the students must learn to understand what science is, what scientists do, and how they do it. It may also disregard the interests of the students or their experiential backgrounds. The content is logically oriented from a scientist's point of view or from the point of view of a subject-matter specialist, but it may not be psychologically sequenced to suit the perceptual and cognitive orientations of the students.

When science teachers are planning a course organized around the structure of a discipline, it is very important to consider the characteristics of the learner. As already stated, the content organization of the course of study should consider the fields of cognition and learning. In addition, science teachers should consider what has been learned about the psychological and sociological nature of adolescents.

Organizing by Social Issues

A course that is organized around controversial topics, needs, and interests might involve types of experiences different from a course that is based on unit topics. A biology course may revolve around issues and problems of nutrition, health, genetic engineering, and ecology; a chemistry course might revolve around problems of acid rain, food and nutrition, water purification, and air quality; an earth science course might revolve around problems concerning environment, pollution, erosion, and subsidence. Courses planned around science and societal issues can be valuable, and through careful planning the content covered can be the same as that in more traditional approaches.

IDENTIFYING EVALUATION PROCEDURES

Evaluation procedures should be clearly stated and defined during the early stages of planning a science course. The basis for awarding student grades should be clearly indicated. The relative weights of weekly tests, unit tests, performance tests, written reports, laboratory reports, homework assignments, projects, and class participation should be established to arrive at report card grades. Other types of evaluation should also be indicated during the planning stage. These are used by teachers to assess the effectiveness of teaching and the understanding and skills of students. They are not used to determine a grade. Pretests, for example, are very useful in determining students' background and experience. They can help the teacher determine where to begin a unit or a lesson. Performance tests can be employed to assess whether students have the skills to perform certain types of laboratory tasks. Daily quizzes provide immediate feedback on student progress and their understanding of previous lessons. Question-and-answer periods of short duration provide immediate information regarding student understanding of the day's lesson. Evaluation procedures to be employed during a course should be scheduled at appropriate points before, during, and at the end of a unit or a course of study.

CALENDAR FOR A COURSE OF STUDY

When teachers are planning a course of study, whether it be for a semester or a year, they should set up a timetable designating the approximate amount of time to be devoted to each unit in the course. If flexibility in organization is

allowed by the district, this should be clearly understood. If certain units or instructional sequences are mandated by the district, it should be clearly indicated. If teachers are permitted by the district to select eight out of ten units, and three of them are mandated, this, too, should be stated.

A course based on 180 days of class meetings should allow for various types of activities such as laboratories, field trips, project work, and individualized instruction. Such activities are time consuming and should be carefully planned. Course organization must be well planned to ensure that enough time is available to employ the instructional strategies and to deal with topics in depth. The calendar should also take into account various activities that cut into class time. Assemblies, special field trips, band, athletic events, and fire drills are some examples of activities that can reduce the amount of teaching time.

After an overall course calendar has been determined, a complete and detailed timetable should be developed for each unit in the course. Such a timetable may indicate such details as when a film will be used, when a particular laboratory exercise will be conducted, or when a major piece of equipment will be required. This timetable could also indicate when a film should be ordered, when live materials should arrive from a laboratory supply house, or when a major piece of equipment is to be used by a particular teacher to avoid conflict of use by others who are teaching the same course.

Suggested dates for beginning and concluding units should be flexible to allow for conflicts and other interruptions (e.g., special activities associated with the unit, such as field trips). If flexibility is allowed with regard to the sequence of units, it is useful to designate the number of days that should be devoted to each unit of study in the course.

Dates for beginning laboratory activities, major examinations, field trips, and other activities should also be provided. If dates are not specif-

TABLE 16–1   A calendar for a course in chemistry

Unit number	Unit topics	Approx. number of days devoted to each unit
Orientation	Introduction to Chemistry: Its Relevance and Importance—Classroom and Laboratory Procedures and Expectations	3
Unit I	Scientific Measurement: Matter and Energy	13
Unit II	Atomic and Kinetic Theory: Atomic Mass and the Mole	12
Unit III	Chemical Formulas: Molecular Mass and the Mole	5
Unit IV	Chemical Equations and Reactions and Molar Relationships—Mass-Mass, Mass-Volume, Volume-Volume Problems	12
Unit V	Periodic Properties of the Elements	8
Unit VI	Chemical Bonding: Its Nature and Classification	13
Unit VII	Modern Atomic Theory: Bohr to Orbitals to Waves	10
Unit VIII	Solutions: Concentrations and Properties	10
Unit IX	Kinetics and Equilibrium: Enthalpy and Entropy	15
Unit X	Acids, Bases and Salts: Equilibrium and Titrations	12
Unit XI	Nuclear Chemistry: Radioactivity, Mass-Energy	10
Unit XII	Organic Chemistry: Nature, Nomenclature, and Reactions	12
Testing	One per unit, plus four to eight cumulative	18
Review:	For final examination	8
	TOTAL (approx. 35 weeks):	161

Industrial applications of chemical principles are integrated throughout.

SOURCE: Information provided by Mr. Ronald Morse, Chemistry Teacher, East Syracuse Minoa Central High School

ically indicated, the points at which the activities should occur during the course of the unit should be specified.

Course and unit calendars should also take the seasons into account so that live material can be collected in the field or ecological studies can be conducted in natural settings. Units in biology and earth science should be planned to take advantage of the weather and season. Table 16–1 shows a calendar for a high school chemistry course.

PLANNING A TEXTBOOK-BASED COURSE OF STUDY

In many instances, the assigned textbook becomes the basis for a course of study. The authors have spent a great deal of time in determining the content that is suitable for a wide range of teaching situations. They usually take into consideration the mandates of local and state departments of instruction, the mental ages of students, and the reading level of students of a particular age.

The content is usually organized in an effective way. Some authors organize the content to take advantage of seasonal materials so that teachers can plan appropriate field trips and use seasonal resources. Other authors use the structure of the particular discipline to determine the instructional organization. Textbooks in physical science are organized under major topic headings presented in a certain sequence, but the sequence is not necessarily required. Here, the teacher has the option of organizing the sequence of units to fit the teaching situation.

The textbook has a variety of features that add to its value as a course of study, especially for beginning teachers. Certain student activities are usually integrated into each chapter or unit. Additionally, there are many helpful teaching and study aids associated with each chapter or unit. The instructional objectives for each chapter or unit are usually stated in performance or behavioral terms. Other features include photographs, graphs, illustrations, tables, suggested experiments, demonstrations, field trips, problems to be solved, homework assignments, and other important suggestions.

Limitations of a Textbook as a Course

Textbooks are designed to sell to a broad market. The course suggested in a textbook is supposed to be useful in any part of the country, in any school, or in any type of situation. Consequently, the course cannot possibly fit a particular classroom situation in all respects. It cannot take advantage of local resources, the background of the teacher, or the background of the students. And, of course, it cannot anticipate national and local events.

Textbooks also contain more material than a teacher can possibly cover during a school year. Authors must present a vast quantity of material so that the textbook can be useful in various teaching situations throughout the country. This approach increases the possible sales of their textbooks.

Usually, a teacher must modify courses as presented in textbooks to fit the teaching situation. Material must be deleted, other material added, and the sequence of units must be rearranged. New material to take advantage of local resources must be added.

Textbooks are designed to answer the important questions. They solve most of the problems and, in many cases, detail all the steps of experiments and tell the students what they should observe and what conclusions they should draw. A teacher must be careful when using the textbook to ensure that students have an opportunity to pose questions, attack problems, make their own observations, and draw their own conclusions.

Adapting a Textbook to Fit a Local Situation

A course suggested by a textbook can be modified to fit the special needs of the teacher and students in a particular teaching situation. The teacher must do what is interesting and useful to students and must, therefore, revise the course to satisfy her needs as well as those of the students. The instructional objectives can be modified, added to, or deleted; content presented in the text can be reorganized to meet objectives. The content in one unit can be broken up and placed under other unit headings. Suggested activities can be deleted, revised, or augmented to meet the instructional objectives.

Textbook courses in physical science, earth science, and life science are usually planned for a wide range of students' interests and abilities. It may be necessary to examine the organization of these broad courses to determine if the sequence of units is desirable for the types of students to be taught. It may be necessary to arrange the sequences so that units covering content that is interesting and appealing to students are taught early in the sequence and given preference to those that are not interesting to the majority of students. For example, a unit on electricity and its uses might be taught early in a course in physical science in preference to one involving motion or forces.

Organizing a course so that some of the more interesting units are presented early in the year will make a great deal of difference regarding long-term acceptance of the course of study. Therefore, science teachers should intersperse less interesting units with the more interesting ones when organizing the sequence of the units.

Teachers should establish the criteria for deleting material when reviewing the textbook. Some material will be too difficult for most students; other material will be uninteresting and not useful. Some material may have been taught in earlier courses and therefore can be eliminated.

New units added to a textbook course of study can be based on the interests of students, the availability of materials, the availability of the facilities, local resources, philosophy of the school, local community pressures, and the needs of society in general.

RESOURCES FOR PLANNING COURSES OF STUDY

Many resources are available for identifying course content, instructional objectives and activities, materials and equipment, and many types of teaching aids. Among these are

1. existing courses of study prepared by individual science teachers
2. existing courses of study prepared by curriculum committees of local departments of instruction
3. existing courses of study prepared by curriculum committees of state education departments
4. courses and programs prepared by nationally recognized curriculum groups such as Chemistry Education for Public Understanding, Chemistry in the Community, Scope Sequence and Coordinated Science, and Project 2061
5. science textbooks, laboratory manuals, and paperbacks
6. experienced science teachers
7. community resources and national organizations

Local and State Courses of Study

Existing local and state science courses usually represent extensive work on the part of experienced teachers and educators. Groups of experienced teachers or groups of professionals— science educators, science teachers, learning psychologists, and scientists—may have devoted many years to developing these courses of study. Existing local and state courses are excellent resources that can be used to build a science course. They can provide ideas for content, laboratory work, demonstrations, field trips, individualized instruction, films, and other visuals. It is advisable for teachers to collect copies of existing courses for future reference. They can be obtained by writing to science teachers of a school district, state departments of instruction, or local boards of education.

National Curriculum Projects

The national curriculum projects mentioned earlier can provide teachers with many excellent suggestions and ideas for planning a course of study. The materials associated with most of these approaches include textbooks, laboratory manuals, student workbooks, teacher's manuals and guides, and many other teaching aids. These materials have excellent suggestions for laboratory activities, demonstrations, field trips, and projects. It is important that teachers select only materials from the approaches that are relevant to their students and of the appropriate level. Many teachers use materials from these approaches to supplement and enrich existing courses of study.

Science Textbooks, Laboratory Manuals, and Paperbacks

All science teachers would profit greatly by having a library of recent textbooks at their disposal. A basic library should include some college textbooks on science topics that relate to the teacher's subject areas. Textbooks used for the course and grade level taught are also useful as a resource. Each textbook presents a different viewpoint, many of which may be considered when planning a course of study. Many experienced science teachers have a library of textbooks and reference books in their classrooms. These can be examined to help decide what would be appropriate for one's personal library.

Laboratory manuals should also be part of a resource library because they are a rich source of interesting activities. Old laboratory manuals often contain activities that emphasize the practical aspects of science. The laboratory manuals that accompanied the national curriculum projects of the 1960s and 1970s are excellent sources for activities that emphasize the investigative aspect of science.

Paperbacks are sometimes good sources for ideas and information when planning a course of study. Some deal with a specific topic in great detail. Excellent paperbacks have been written on a variety of topics, such as human genetics, evolution, relativity, and thermodynamics. They can provide excellent background on various areas of interest to the science teacher.

Science Teachers

The experienced science teacher is an excellent source of ideas and information when planning a course. Individuals who have been teaching for some time usually have a storehouse of ideas that can help less experienced teachers strengthen their course. Most science teachers are willing to share their favorite laboratory activities, games, puzzles, teaching units, resource guides, paperbacks, films, and speakers. Many of them have developed course syllabi, new courses of study, and units of instruction and have field-tested experimental courses of study.

It is professionally acceptable to call on other teachers for help and suggestions. One can meet experienced and competent science teachers at professional science teachers' meetings. Science coordinators in school districts and principals of local junior and senior high schools can provide the names and telephone numbers of science teachers who can offer suggestions and ideas for planning science courses. Also, science instructors and professors at local colleges and universities are knowledgeable about these matters and can often be of assistance.

Community Resources

Teachers should make a file of the resources in the community that can enrich the science curriculum. There are numerous organizations, facilities, and individuals who can help enrich a science program, and most individuals and organizations are willing to help.

National and Local Organizations

Many national organizations have developed high-quality instructional materials that can be used in science courses and that are free to teachers. The American Cancer Society has produced a variety of teaching materials on cancer and diet. The American Heart Association has produced many instructional materials on health, exercise, and diet. The National Aeronautics and Space Administration has produced curriculum materials for the science classroom in astronomy, biology, chemistry, electronics, industrial arts, photography, and physics. These materials provide information, demonstrations, and experiments in space technology that relate to science. These are but a few examples of materials available to science teachers.

The following list can be used to start a resource file of community and national resources:

Agricultural Extension Service
Airports
American Automobile Association
American Cancer Society
American Heart Association
Arboretums and botanical gardens
Audubon Society
Colleges and universities
Fire departments
Fish and wildlife service
Flying clubs
Garden clubs
Geology clubs
Health departments
Hospitals
March of Dimes
Local and regional museums
National Aeronautics and Space Administration
Nature centers
Oceanography centers
Oil and petrochemical companies
Parks and wildlife departments
Photography clubs
Police departments (forensic laboratory)
Public libraries
Regional educational service centers
Science centers
Sierra Club
Steel mills
Technology clubs
Utility companies
Various local industries
Veterinary clinics

SUGGESTED ACTIVITIES

1. Construct a resource file that can be used to obtain ideas for designing a science course in which you are interested. The following resources should be included in this file: textbooks, laboratory manuals, paperbacks, science teachers, state and local boards of education, and national curriculum groups.

2. Start a resource library. Collect materials to include in the library. Request free materials from national and local organizations.

3. Examine a science course syllabus and determine the goals and objectives of instruction, and the evaluation procedures that are emphasized.

4. Analyze a science textbook as a course of study. Examine the textbook for flexibility, adaptability to your local situation, and the appropriate grade level for the intended audience. What changes, deletions, and reorganization should be made to fit a specific teaching situation?

5. Examine a state or local course of study. Determine the presence or absence of such features as seasonal arrangements and use of local resources. What changes do you believe should be made regarding the organization of the course to make it more effective for a particular teaching situation?

6. Make a file of local teaching resources for the school in which you will do your practice teaching. List the ideas for field trips and the outside authorities that may be called upon for assistance.

7. Evaluate a science course. How well does the course reflect the goals for developing a scientifically literate citizen?

BIBLIOGRAPHY

Biological Sciences Curriculum Study and Social Science Education Consortium. 1992. *Teaching about the history and nature of science and technology: A curriculum framework.* Colorado Springs, CO: Colorado College.

Bruner, J. S. 1966. *Towards a theory of instruction.* Cambridge, MA: Belknap Press.

Chiappetta, E. L., and Ramsey, J. M. 1991. Curricular models for science programs in Texas. *The Texas Science Teacher, 20*(4): 11–14.

Howe, A., Hall, V., Stanback, B., and Seidman, S. 1983. Pupil behaviors and interactions in desegregated urban junior high school activity-centered science classrooms. *Journal of Educational Psychology, 75*(1): 97.

James, R. K. 1981. Understanding why curriculum innovations succeed or fail. *School Science and Mathematics, 81*(6): 487.

National Science Teachers Association. 1982. *Science-technology-society: Science education for the 1980s* (an NSTA position statement). Washington, DC: The Association.

Posner, G. J., and Rudnitsky, A. N. 1986. *Course design: A guide to curriculum development for teachers,* New York: Longman.

U.S. Department of Education. 1991. *America 2000: An education strategy.* Washington, DC: The Department.

Assessment in Science Teaching

A good evaluation system must include an assessment of many different types of learning outcomes. Inquiry and manipulative skills should be part of the process.

CHAPTER 17

Evaluation in Science Teaching

Evaluation is the process that is concerned with making judgments of the value or worth or status of individuals, situations, or conditions, followed by decisions that affect those individuals, situations, or conditions. In education, student performance in the classroom is evaluated to determine progress and promotion. Teachers are evaluated by students and administrators to judge their effectiveness in the classroom. Science curricula are evaluated to determine if they are meeting the needs of students. Science facilities are often assessed to determine whether they are suitable for the type of instruction that is taking place or will take place.

Evaluation includes measurement and testing that provide the data to determine the status of individuals, conditions, or situations. The data collected will have to be interpreted in order to make judgments and decisions that will promote change.

Overview

- Discuss evaluation procedures in education.
- Give suggestions for determining student grades.
- Discuss the use of standardized tests in science teaching.
- Discuss the assessment of psychomotor objectives.
- Suggest ways to assess objectives in the affective domain.
- Suggest procedures that can be followed to assess teacher effectiveness in the classroom.
- Present ways that can be used to assess the science curriculum in middle and secondary schools.
- Offer suggestions for assessing the facilities and conditions for science instruction in the middle and secondary schools.

▼ Mrs. Hunt is the science coordinator of a small city school district. She is a very active, conscientious administrator who spends a great amount of time and effort attempting to develop a high-quality science program for the school district. For example, she involves the principals and the science teachers in each school to examine and evaluate certain facets of the science program in order to promote changes that will facilitate improvement. For example, she appoints two committees in each school—a committee to evaluate the science curriculum, and another to assess the science facilities and the safety and teaching conditions in the classrooms and laboratories. She and the principal attend all committee meetings to provide guidance. The committees are charged to make recommendations for improvement.

Mrs. Hunt is also involved in many other activities to promote improvement in the science program. In this regard, she frequently visits and observes beginning science teachers to guide them through their first year of teaching. She has found that the guidance they receive has helped in producing excellent results in terms of how successful the teachers are during their first year of teaching. In order to prepare the teachers for the visit Mrs. Hunt asks them to schedule a time during the work day when she can observe their performance in the classroom as well as have a conference immediately following the observation. She also provides them with an agenda, which consists of various items that will be addressed during the visit. She includes a list of tasks the teachers must complete to prepare for the visit, as follows:

1. Prepare a lesson on a new topic that involves students in a variety of learning activities including one hands-on activity. The lesson will be observed at the time of the visitation.
2. Prepare a frequency distribution of the test scores for a test administered to two different science classes.
3. Administer the instrument "The Science Teacher" (a questionnaire to be completed by students to assess science teacher performance in the classroom) to the class that will be observed during the visit and to another class that the teacher believes is difficult to handle. Summarize the results so they can be discussed during the conference.

4. Before the visit, make a safety inspection of the science classroom and laboratory and the storage area using the questionnaire "Science Safety Checklist" as a guide for the inspection. Summarize the results so they can be discussed during the conference.

Mrs. Hunt's approach to evaluating various facets of the science program is to involve the science teachers and administrators as much as possible in the evaluation process. In the case of evaluating science teachers, she relies heavily on teacher self-assessment in the beginning or formative phase of the process. For example, when she has a conference with the teacher after observing his or her classroom performance, she urges the teacher to determine whether the students were involved in the lesson by examining the information that the science teacher collected by performing the four assigned tasks above. She also asks the teacher to decide to what extent the students were involved in the lesson and to what extent the students were constructing their own knowledge. She allows the teacher to determine the procedures to follow if student involvement is low. Mrs. Hunt serves as a guide to help the teacher clarify certain perceptions and to determine the criteria that should be used to promote student involvement.

Since Mrs. Hunt is attempting to help new science teachers grow and develop during their first year of teaching, a great deal of formative evaluation is involved, which allows the beginning teachers to form their own judgments about their effectiveness. Mrs. Hunt provides very little negative feedback and rarely passes judgments unless she is urged to do so by the teacher or if the teacher's observations, assessments, perceptions, and decisions deviated significantly from her own. In the early stages of the evaluation of new teachers, Mrs. Hunt relies heavily on self-evaluation and improvement, whereas toward the end of the school year she uses summative evaluation procedures to make her own assessments and decisions relative to science teachers and program effectiveness.

During the beginning phase of the evaluation process, Mrs. Hunt tries not to compare science teachers, keeping other teacher references and comparisons out of the evaluation picture. She focuses primarily on criteria for making judgments. For example, when considering student involvement in a lesson, she urges the science teachers to determine whether 90% of the students are actually participating in the lesson, or, when the students' perceptions of their science teacher are analyzed, to determine if most (at least 85%) of the students like their science teacher. Or, in the case of the science safety inspection, she may ask the teachers when the fire extinguishers had been serviced last or if the caustic chemicals are stored properly.

Finally, Mrs. Hunt tries to rely on several elements in her evaluation procedures. She tries to look at a variety of factors when judging teacher effectiveness, such as planning lessons that involve students in a variety of learning activities, and taking into account such factors as classroom interaction, grading procedures, student grades and progress, students' perceptions of their teacher, and the extent to which the teacher takes safety precautions seriously. However, Mrs. Hunt does not overwhelm new teachers with too many assignments during the middle phase of the evaluation process. During the evaluation process, which is designed to assess the total science program in Grades 6 to 12, she employs many factors to determine how effective the science program is and what changes need to be made in order to produce a high-quality program in the school district. ▲

IMPORTANT CONCEPTS IN THE EVALUATION PROCESS

Evaluation requires that one makes a qualitative judgment on what is to be assessed. It is a process whereby judgments and decisions are made about individuals, situations, or products. Evaluation implies that a worth or value is placed on an individual, a product, or a situation and decisions will be made that will promote change in an individual, a situation, or a product.

Teachers are evaluated to determine their effectiveness in the classroom in order to justify merit, retention, or tenure. Courses and curricula are assessed to judge whether they are meeting the needs of students. Science classrooms and laboratory facilities are evaluated to determine their suitability for the type of instruction

that is taking place or will take place in the future.

In the classroom, students are evaluated to determine their status, progress, or promotion. But to make sound judgments and decisions about students, teachers first must decide what kind of information they need and how they will gather it. TenBrink (1990) suggests several techniques for gathering information that teachers can use to make judgments and decisions about students and about themselves as teachers. These include inquiry, observation, testing, and analysis. Inquiry involves asking students about their feelings, opinions, interests, likes, and dislikes. Observing students can give information about how well students perform certain skills, such as manipulating a microscope or bending a piece of glass tubing, or how well they behave in certain situations or interact with other students in a group or as a partner in the laboratory. Rating scales and checklists are often employed to record such observations. Paper and pencil tests can be used to measure achievement and cognitive outcomes. Analyses are made so the teacher can find out what is happening during the learning process. It might involve analyzing a test, a project, an assignment, or performance of certain types of psychomotor skills to determine where students are having problems. Analysis is done so that the teacher can help students as well as gather information about his own teaching.

These techniques have their advantages and disadvantages, and the teacher has to decide on the techniques or technique best suited to obtain the needed information. To carry out the evaluation process, TenBrink (1990, p. 339) suggests the following steps:

1. *Preparation.* Determine the kind of information that is needed as well as to decide how and when to obtain it.
2. *Gathering information.* Collect a variety of information as accurately as possible.
3. *Forming judgments.* Make judgments by comparing the collected information to selected criteria.

4. *Making decisions and preparing reports.* Record important findings followed with appropriate forms of action.

The steps mentioned above are considered in both the formative and summative phases of the evaluation process.

In summary, evaluation is one of the most sophisticated teaching functions, and, therefore, there is a definite need to understand the process and its concepts in the role of a teacher. Understanding the evaluation process means that teachers should not regard testing and evaluation as synonymous. Testing is only the measurement component of the evaluation process. However, the best way to measure achievement and cognitive outcomes is through testing. Teacher-made tests can reveal information about achievement of what has been taught. Standardized tests can measure the performance of a whole class or a student. They can also determine attitudes and changes in attitudes. Tests have a definite place in the evaluation process and the teacher can use them to collect certain kinds of information to make judgments and decisions about students and about teacher performance in the classroom. However, it must be emphasized that many techniques can be used in the evaluation process and testing is only one of them.

The discussion that follows will help explain the important concepts in evaluation. It explains the use of diagnostic evaluation, formative evaluation, and summative evaluation. It provides useful information necessary to determine the proper use of norm-referenced assessment and criterion-referenced assessment. Finally, the discussion deals with information useful to determine when to use evaluation techniques as opposed to only testing techniques.

Diagnostic Evaluation

Diagnostic evaluation can occur before the beginning of a unit of instruction or sometimes during the course of instruction, particularly when the teacher anticipates that students will have difficulty with the subject matter. Some teachers prepare and administer pretests to determine the students' level of knowledge and understanding of a particular science topic before beginning instruction on that topic. This information helps the teacher plan instruction to take into consideration the cognitive level at which students are functioning before starting a unit. Some teachers prefer to use standardized tests to collect information about students' knowledge and understanding of a science topic at the beginning of a course or, in some cases at the beginning of a unit. But in general, teachers usually prefer to use their own pretests to obtain needed information.

Diagnostic measurements can also be useful to collect information about a student's knowledge and understanding during the course of instruction. Teachers who find that students are functioning at a high level based on the information gathered can provide them with enrichment activities to take care of their needs. Students who are having difficulties and/or functioning at a low level can be given remedial work to help them overcome their problems.

Formative Evaluation

Formative evaluation is usually carried out during instruction in order to assess student progress and learning. The teacher can use the feedback to reinforce students in areas in which they are performing well and indicate to them where they can make improvements. Formative evaluations should be used throughout the entire instructional period to direct student learning and provide continuous feedback to shape students' behavior and learning.

Summative Evaluation

Summative evaluation is the final phase in an evaluation program. It usually assesses the overall situation or achievement. It often focuses on a comprehensive range of behaviors, skills, and knowledge. The most common example in education is the final grade the teacher gives to students in a course. This grade is usually determined by considering all the work for a given marking period, including grades on quizzes, weekly tests, unit examinations, laboratory work, and possibly grades earned on projects, oral reports, and homework. The grade permits the system to pass or fail students or determine whether some students will make the honor roll.

Norm-Referenced Measurements

Norm-referenced measurements are used to discriminate among individuals and often to categorize them. They rank and compare students in areas such as scholastic aptitude, language proficiency, and academic attainment. Norm-referenced data can be used to group students by ability and knowledge. The SAT (Scholastic Aptitude Test) and IQ test, for example, are norm-referenced instruments that provide scores that show students how they compare with others in a prescribed population. Standardized science achievement tests are also norm-referenced, and some can be used to predict how well a student will perform in a particular science area or a course.

Often, norm-referenced achievement tests are developed by drawing from a large and general body of subject matter, and the items are selected for their ability to discriminate among students. There are certain test items that high-scoring students will answer correctly, and low-scoring students will answer incorrectly. In common practice, many teachers do not adhere to this process for developing a test to be used for norm-referenced purposes, but they will use it for norm-referenced grading. For example, a science teacher who gives a 20-item laboratory practical and gives an *A* grade to the students who score at the top 10 percent of the score range has not developed a norm-referenced test but uses it as such to grade the students.

Norm-referenced tests have many purposes in science teaching. They can be used for screen-ing and tracking to place students in science classes for the gifted, average, and slow learners. They can be used to determine the effectiveness of a school's science program by comparing overall test results in science with norms established on the examinations of students in the state or in the nation. However, norm-referenced tests are not as ideal for monitoring student progress in a science course as are criterion-referenced tests.

Criterion-Referenced Measurements

Criterion-referenced measurements measure how well a particular student is progressing or achieving the objectives in a course. The measurements are used to assess knowledge and skills developed in specific areas. The test items are developed from a limited body of subject matter, and success on these items is judged against predetermined criteria or standards. Students are judged to be successful or unsuccessful based on their performance relative to criteria that have been stated for success or mastery. For example, a science teacher may indicate that students who receive at least 17 out of 20 possible points on a laboratory practical will receive an *S* (satisfactory) for the test, and those students who receive less than 17 points will receive a *U* (unsatisfactory).

Criterion referenced programs were popular in the 1960s when individualized instruction was being implemented in many schools throughout the country. These programs focused on student achievement of specific learning outcomes. One particular approach, mastery learning, allowed students to work at their own pace to master a body of science material. In these courses students were required to achieve a certain percentage of the objectives specified for a unit or block of work. The mastering of the material was determined after the students took a test in which they were required to receive a certain percentage, such as 70%, 80%, or another predetermined score, before going on to the next topic. If such competence was not achieved, students were required to continue studying the topic until mastery was achieved. Remedial instruction was also provided to help students achieve a level of proficiency. The teachers in these courses predetermined the standards for earning an A, B, C, D, or F before the students began the course. Therefore the teacher did not have to compare test results of students or grade them on a curve to arrive at a grade.

Self-Referenced Judgments

Teachers make self-referenced judgments by using information they have about a student and information they have accumulated from other sources about the same student. Progress over a given period of time is ascertained by giving a student a test before studying a unit, followed by a test at the end of the unit. Comparative scores will show how much the student has learned (or has not learned) over a given amount of instruction. Another type of self-referenced judgment that is commonly made is to compare how well a student performs a particular skill (for example, reading a vernier scale on a mercurial barometer) before and after instruction and practice.

Evaluation involves testing and standards. The standards are the criteria upon which performance and products are judged. Standards can be norm-referenced, criterion-referenced, and self-referenced. Precise measurements and carefully collected data are then used to make judgments, draw conclusions, and provide plans or a basis of discussion for future action. This should take place in formative and summative ways.

In conclusion, evaluation is a broad teaching function involving more than averaging test scores and assigning grades. A good evaluation system should place a value on many facets of a program, such as student achievement, instructional effectiveness, teacher effectiveness, and curriculum. An overall evaluation, which includes these components, should indicate the quality of a science program.

EVALUATION OF STUDENT OUTCOMES

Grades are important because they affect the lives of students. They are used as a basis for job recommendations, entrance to college, or a course of study. Grades can affect students' self-concept and their interest in schooling and science. Teachers take on a great deal of responsibility when they assign grades. Consequently, they must use the best means possible to provide equitable procedures to arrive at grades, which become a permanent part of science records.

When teachers evaluate students, they most often stress one aspect of evaluation—and neglect to obtain other useful information that they can use to determine a student's status at any time during instruction. For example, a pretest or standardized test can be used at the beginning of a course or unit of study, or a teacher-made pretest and posttest can be administered to determine student status over a period of time. If administered early enough in a course, tests that determine skill levels such as reading, writing, and mathematics can evaluate each student's ability to successfully complete the course based on their skill levels in those areas.

Other types of tests also provide students and teachers with information they need to improve themselves. Tests that measure psychomotor skills can help students learn their weaknesses in conducting aspects of laboratory work. Knowing students' weaknesses early enough, teachers can help them avoid a great deal of difficulty in the long run.

Standardized tests administered to students can be used for diagnostic purposes. They can help teachers determine aptitudes in science and predict success in a course. Used discriminately, they can give teachers very useful information about a particular student or a whole science class.

Determining Grades

The determination of grades is a serious responsibility. Grades, once entered in the records, have great impact on the lives of students. They represent degrees of success or failure. They are used to compare one student with another. They determine promotion and graduation and fitness for college or other advanced training. They are taken into account in awarding scholarships, and they are usually determined by one person—the classroom teacher.

A good evaluation system should include an assessment of many different types of learning outcomes. Knowledge and understanding of facts, concepts, laws, and theories should be part of this assessment. Inquiry and manipulative skills should also be part of the assessment process, as should motivation, interest, and attitudes. These outcomes can be measured using a variety of instruments and techniques, such as paper-and-pencil tests, laboratory practicals, laboratory reports, special reports, oral presentations, and project reports.

Types of Achievement.　The relative weights given to certain types of achievement in determining a report card or final grade depend on the philosophy of the teacher and the course objectives. If the teacher is process oriented, then he will give laboratory work, rather than content, greater weight. If the teacher is content oriented, then he will give more weight to acquiring information. Teachers who provide a balance of process and content in their teaching will probably give equal weight to both. The types of achievement include:

1. achievement that can be measured by paper-and-pencil tests (e.g., knowledge of words, definitions of principles, recognition of applications, and solutions of verbal problems)
2. achievement that can be measured by practical examinations (e.g., skills with equipment, tools, and procedures)
3. achievement that can be measured by assignments and extra credit work (e.g., inquiry skills, communication skills, organizational skills, interest, motivation, and attitude)

Keeping Records of Achievements.　Full records are essential for arriving at grades. Record forms should permit a variety of entries and notations as to the different types of achievements that students demonstrate. A portion of a record book adapted for science records is shown in figure 17–1. There are four major headings:

1. *Quizzes and unit tests.* Several spaces are reserved for the scores of short tests. A few spaces are reserved for the scores of longer tests and unit tests.
2. *Assigned work.* This section is treated as a checklist, with the exception that grades are given for work that surpasses the minimum.
3. *Laboratory and practical work.* Grades are entered for laboratory work that demonstrates cognitive and psychomotor learning.
4. *Extra-credit work.* The grades given for work that students undertake spontaneously and do largely on their own time are recorded. There may be no entries in this section for some students and only a few for others.

A record book is convenient to carry around and store. The records are compact and easily examined. However, there is usually not enough space in record books to specify the nature of the achievements indicated in points 3 and 4 above. Consequently, a busy teacher has difficulty in calling to mind precisely what each student has done; he is handicapped in arriving at final grades without having detailed records. Thus, some teachers prefer to keep their records on filing cards, because cards have more space than do record books, despite the problems of carrying and storing cards. The front of a card is used for test grades and notes on other achievements. The reverse side is used for notes on conferences and for observations of behavior.

FIGURE 17–1 A teacher's record book

Name	Quizzes						Unit test	Assignments			Laboratory work			Extra credit work			Average
Ackerman, Patrick	9	8	9	7	6	8	78	✓	✓	A	B	B	C	B			
Anderson, Richard	7	8	a	7	8	8	75	✓	B	✓	C	C	C				
Bentling, Nancy	10	9	10	10	10	10	95	A	A+	A+	A	A+	A	A+	A	A	
Brown, Roxanne	8	8	7	a	a	7	82	B	✓	B	A	B	A				
Calfiero, Ralph	4	6	3	6	6	3	40	✓		✓	D	D	0				
Case, Edna	7	9	7	7	7	6	80	✓	✓	✓	A	A	A	C	B	C	
DeMarte, Robert	10	10	10	8	8	10	100	A	B	B	B	B	A				
Doyle, Carl	6	7	7	10	6	6	45	✓	✓	B	C	A	C	C			
Drake, Faith	7	4	8	6	6	7	63	✓	✓	✓	D	D	A				
Hanrahan, Robert	9	8	7	9	9	4	85	A	✓	A	F	D	D				
Jewell, Vivian	8	8	8	9	8	9	94	✓	B	✓	A	B	A				
Kissel, David	6	3	6	6	2	a	48	✓	✓	✓	C	D	C	C	D	C	
Lindsley, Christine	10	7	10	8	9	10	95	✓	✓	✓	D	B	D				
McLaughlin, Joanne	5	6	5	8	6	7	64	✓	✓	✓	D	D	D	A+	A		
Nye, Kathryn	0	8	7	9	7	7	81	B	✓	B	C	C	B	C			
Olssner, Donald	a	a	a	a	9	9	95	a	✓	a	a	a	C	B			
Ruiz, Jane	7	8	8	4	8	8	82	✓	✓	✓	B	B	C	C		C	
Slocum, June	9	8	7	9	8	7	88	✓	✓	✓	B	B	B	C		C	
Smith, Michael	8	9	10	10	6	4	51	✓	✓	✓	A	C	D				
Williams, Louise	10	7	6	8	10	9	98	B	A	✓	A	C	A				

Assigning Grades. When detailed records are kept in the aforementioned fashion (see figure 17–1), final grades are determined by combining all the results of a student's work. Test grades are averaged, with long tests being given more weight than short tests. For example, if a test that ends a unit is three times as long as a typical quiz, perhaps the score on that test should have a value three times that of the quiz when it is averaged in with the other grades.

Grades for required work and for extra credit work are then averaged, proper weight being given according to the nature of the difficulties involved in each assignment. These results are combined with the test average to give a final grade.

There are a number of ways of handling the credit given for voluntary work. Some teachers include the grades with those on the required assignments. Other teachers work out a system for raising the final grade a certain number of points for each voluntary activity, the amount of increase depending on the grades earned in each case.

It is sometimes convenient to use numerical grades for tests and letter grades for other achievements, which requires the use of a conversion table. Each teacher should construct his own table to fit this method of scoring. A commonly used conversion table is shown here:

Score in percent	Letter grade	Significance of grade
90 to 100	A− to A+	Excellent
80 to 89	B− to B+	Good
70 to 79	C− to C+	Satisfactory
60 to 69	D− to D+	Poor
Below 60	F	Unsatisfactory

The method of amalgamating grades in determining final grades helps reduce some of the injustices that are bound to result when test grades alone are used. This process reserves the highest grades for the students who excel in tests and who exert themselves to do superior work. Students who pass tests easily are not penalized for failing to work, but they are denied superior grades unless they do more than the minimum. Students who have special aptitudes but lack all-around excellence can earn high grades by exploiting their special talents. Students who lack academic ability and find tests difficult can still earn satisfactory grades by sufficient hard work. Students benefit from grading systems based on predetermined criteria so that they are competing with themselves instead of their peers for grades.

STANDARDIZED TESTS IN SCIENCE

Standardized tests have been constructed and published by commercial testing bureaus that

1. measure general achievement in science
2. assess what students have learned in a science area at the end of a course
3. predict success in a course
4. measure aptitude in science

Standardized norm-referenced tests are best adopted for diagnostic purposes and not necessarily for use in the determination of a grade.

Factors to Consider in Selection

Before selecting a standardized or achievement test, teachers should determine whether the results of the test will give them the information they want. A good test should be both valid and reliable. A *valid* test is one that measures what the teacher wants it to measure, and this is determined by the content and objectives of the course. A number of factors determine whether the test is valid. The test items should reflect what has been taught; that is, the items should be closely tied in with objectives of a course or unit. The coverage of a topic on the test should be given the same consideration as the amount of emphasis that was given when the topic was being taught. If a standardized or teacher-made test in biology devotes 50 percent of the items to ecology and 50 percent of the items to genetics, this should be taken into account before the test is used. A teacher would not use such a test if he devoted 25 percent of instruction time to genetics and 75 percent to ecology. The items should be in proportion to the instruction given. Before using the test, the teacher should examine its contents to determine whether the test is balanced and reflects the teaching objectives. An item-by-item analysis is probably in order.

Another consideration in using a standardized test is its reliability. A *reliable* test is one that measures a student's knowledge and understanding of a topic in a consistent way. That is, if the test is given to the same student more than once on the same day, she should score the same each time. Reliability of a test depends on the quality of questions in the test. Poorly worded, tricky, or ambiguous questions decrease the reliability of the test.

A standardized test should also contain items that are reasonable to the student and geared to the level of difficulty that was presented during instruction and reflected in the instructional objectives.

Some achievement tests can be used to help determine different levels of achievement among students. Norm-referenced tests have this ability. A test that discriminates in this way can be used to separate students with different abilities or levels of achievement.

The items on a test should be written so that they are relevant or appropriate for the students who are taking the test. Items should be written so that they make sense to the student and make use of the student's background of experiences.

Uses and Advantages

Standardized tests may be used as follows in a science program:

1. *To determine a student's achievement levels.* This type of information can be used to determine where a student should be placed— a grade or section or an advanced course in science.

2. *To determine a student's growth and progress in a science area.* Generally, two forms of the same test are given at the beginning and end of a unit or at the end of a particular time period. Diagnosis can indicate whether growth is unusually slow or fast, so that special considerations can be prescribed.

3. *To determine a student's strengths and/or weaknesses in a science area.* A battery of science achievement tests taken by a student can reveal areas of achievement when compared with what is considered average for other students of the same age and grade. This type of information can be used to determine whether the student needs remedial work, or to guide her into special fields of science.

4. *To determine the relative standing of an entire science class.* The information obtained can guide the science teacher's design of a science program. A class that scores higher than the national norm may be able to do advanced work in a particular science area. A science class that scores lower than the national norm may be assigned to remedial work. Science courses can be designed to take care of both situations.

5. *To determine the areas of strengths and weaknesses of an entire class.* This information can be used to plan a balanced science program. It is useful to know which areas of science require special attention.

6. *To determine the progress of an entire science class.* Achievement tests that measure progress can be used to determine the effectiveness of the teacher's teaching strategies.

Limitations

Some teachers believe that the effectiveness of a particular learning situation can be measured by a standardized test. They often use the results of the tests to determine a final grade. In some cases these results are the sole determiners for a final grade.

Standardized tests should *not* be used as the sole determiners in making up final grades.

They probably should not be considered in the grading process at all except to reveal those students who know more than is indicated by the results of teacher-made tests. It must be emphasized that most standardized tests cannot fit a locally designed curriculum. And they are not designed to measure most of the desired outcomes of a science program in a particular school situation.

The weaknesses of standardized science tests only limit their usefulness; they do not make them useless. Administering them can help teachers ensure maximum benefits for their students.

PORTFOLIOS—THEIR ROLE IN EVALUATING STUDENT PERFORMANCE

▼ Mr. Prince required the students in his ninth-grade science class to maintain a science portfolio of their work during the year. He explained that the portfolio was to be a collection of their own science work and include reports of formal laboratory work done individually or with a partner or as a part of a group. He indicated that the students were expected to include all of their completed homework assignments, completed worksheets, tests, weekly quizzes, and the lecture notes they personally took during class. He also asked that they include the lecture notes he had prepared and presented to them in mimeograph form. He told the students that with each lecture he would often require the students to answer a list of questions based on the lecture. He asked that the responses be placed in their portfolios.

Mr. Prince explained other types of student work that could be included in their portfolios. For example, he said that he would often assign the students to read an article he selected from a newspaper or magazine, and would request them to prepare a report that included both a summary and a critique of the article. He indicated that these reports should also be placed in their portfolios. Summaries of other readings not assigned by the teacher could also be included. They were also to include reports of field trips, projects that involved scientific investigations, projects that were entered in science fairs, and assigned laboratory work.

After answering many questions about the type of material that would warrant inclusion in the student portfolio, Mr. Prince indicated that he would expect the students to design and organize their portfolios so that the material included could be easily found and identified. He mentioned that in the past students chose to separate the contents into files—such as quizzes and tests, laboratory work, projects work, lecture notes, and so on, and that each file also included a table of contents.

As the term progressed, a number of students became involved in many different types of activities that were not associated with regular classroom work. Some students volunteered to conduct their own science projects during and after school hours. Some projects engaged students in scientific investigations; others gave students firsthand experiences in collecting specimens of plants to study in the laboratory. The students who engaged in this project identified each plant collected by using a key that Mr. Prince provided. They also prepared a report that described the ecological conditions of the area where the plants were collected. This report was presented orally to the whole class and later placed in the students' portfolios. Two groups of students decided to conduct a survey to determine whether the people in the community believed that tight restrictions should be placed on certain industries responsible for industrial emissions that were polluting the area, knowing that such restrictions would probably cause the industries to move elsewhere and thus jeopardize the job market. Each student who participated in the survey wrote the final report and volunteered to participate in a group discussion before the class that dealt with the survey. This report was placed in the portfolios of students who conducted the survey. Another group of students designed a series of large safety posters illustrating the use of safety equipment such as fire extinguishers, fire blankets, and safety showers. Upon completion, the posters were placed in appropriate areas in the laboratory and classroom. The students had made preliminary sketches of the posters and these were placed in their science portfolios. Students also maintained a log of the daily activities that were conducted in science class. This was included in the portfolio. ▲

The preceding example describes the contents of a portfolio that includes original science work completed by students in one class during an academic year. An examination of the students' work showed considerable originality. The students included quizzes that had already been graded, as well as handouts and lecture notes that had been prepared by the teacher. Although this material was included in the students' portfolios, it was not considered original work and therefore was not used in the determination of a grade when the teacher evaluated the portfolios. The teacher considered only original work and important products completed by the students during the academic year.

What then is a portfolio? A portfolio, according to Collins (1991), is a form of authentic assessment. It "is a container of evidence of someone's skills, knowledge, and dispositions." The material in the portfolio is used to make judgments about the quality of performance of the individual who developed it.

Portfolios are used in different areas of endeavor. Artists develop portfolios of their art work. They select work that shows evidence of their ability as an artist and quality of their work. Photographers also produce portfolios of photographs they have taken. They usually include the pictures that best demonstrate the quality of their work. Artists and photographers do not include examples of their poorer work.

In science teaching, a portfolio should show the extent of student involvement in a science course. Portfolios are collections of work that provide evidence of student aptitude and competence. They also demonstrate a student's initiative, proficiency, skills, and capabilities.

Science teachers often require their students to maintain a science notebook. Some notebooks are collections of lecture notes copied by students from teacher-prepared material placed on the chalkboard. The notebooks are collected and graded by the teacher on the basis of neatness. This collection of material is not in any sense a portfolio. It does not contain evidence of original work produced by the students. Teachers who carry out this form of activity are not

very imaginative. They are really grading their own product. Others teachers require students to maintain a laboratory notebook, which is a collection reports of formal laboratory work and laboratory exercises conducted and completed by students. This collection of material is a portfolio of laboratory work and nothing else. It meets the requirement of original student work, but it does not show samples of other types of original work and products. In order for a collection of student work to be considered a portfolio, it should meet the following requirements:

1. A portfolio should contain products of *original* student work produced over a period of time. It can be a collection of original work produced over one term or longer.
2. The material in a portfolio can also include products that are not produced by the student. Handouts, study guides, lecture notes, review notes, worksheets, and laboratory notes provided by the teacher are examples of material that can also be part of a student's portfolio. These are documents that are evidence of the type of activities that have taken place during the course of science instruction over a period of time.
3. The collection of work in a portfolio should show different aspects of a student's abilities. It should demonstrate evidence of student aptitude and competence in one or more areas. It should also provide examples of a student's proficiency and capabilities and skills in one or more areas.
4. A portfolio should consist of materials that indicate that the student has mastered certain aspects of a course, for example writing reports, designing experiments, conducting experiments, conducting discussions, designing posters that include graphics which can be used in teaching science material, engaging in project work, or presenting a lecture on a science topic.
5. A portfolio should be evidence of student work and thus can be evaluated as such. The majority of material in a portfolio should be original student work and not the products of

science teachers or other individuals. It is evidence of what the student has accomplished and what the student knows.

MEASURING AFFECTIVE OUTCOMES

Most school science programs emphasize cognitive goals and neglect goals in the affective domain, which are concerned with attitudes, interests, and values. Only recently have secondary school science programs been interested in the identification of goals within the affective domain, because it is evident that student interests and attitudes play an important part in the learning environment and in the learning process. Science teachers and science educators believe that the affective domain should not be neglected even though outcomes are difficult to measure.

Taxonomy of Educational Objectives, Handbook II, by Krathwohl, Bloom, and Masia (1964) is concerned with the affective domain. The taxonomy is divided into five levels: receiving, responding, valuing, organizing, and characterization by a value or value complex. These terms and levels of taxonomy present problems when science teachers attempt to translate what they really mean. Also, clearcut boundaries within the domain cannot be exactly defined because they do not exist. How does one separate attitudes and adjustments, or attitudes and feelings, self-reliance and confidence, thoroughness and precision? Gronlund (1973), Nay and Crocker (1970), and Eiss and Harbeck (1969) have analyzed the affective domain and its levels. Their analyses are useful to science teachers who are interested in measuring affective outcomes and will aid them in understanding the five levels stated by Krathwohl, Bloom, and Masia.

Tests for Measuring Affective Outcomes

There are a number of ways to measure affective outcomes. One technique by Osgood, Succi, and Tannenbaum (1967) that has been extensively used involves a test consisting of a set of bipolar adjectives that are relevant to and describe a situation (e.g., classroom environment), an idea, and a school subject, such as science or mathematics. The adjectives or terms relate to the situation, idea, or object to which the student reacts. The adjectives used in the test are contrasting terms such as *bad/good, right/wrong, heavy/light,* and *fast/slow.* The terms are arranged in pairs (opposites), and the student checks a space along a continuum between the contrasting extremes that describes her perceptions. Tests of this type can be constructed to determine students' likes, dislikes, ability, and attitudes. The bipolar adjectives or descriptive phrases are listed in two columns, and the student is asked to respond to a collection of terms by checking or writing a number between the contrasting terms. Such a scale made up of bipolar terms is referred to as a *semantic differential instrument.*

A great deal of useful information can be derived from such tests. Assessment of student attitudes toward a particular science subject can be determined at the beginning of a science course. The same or a similar test can be given at the end of the course to determine if attitudes have appreciably changed. The same procedure can be used to determine what values a student has in the beginning of a course and how they have changed by the end of the course.

Example with Bipolar Adjectives.
Directions: In this test we want you to describe the feelings you have in biology class. There are no right or wrong answers. On the next page you will find the heading "BIOLOGY IS. . . ." The rest of the page shows pairs of words you can use to describe your feelings about biology. Each pair of words is on a scale such as this:

GOOD   1   2   3   4   5   6   7   BAD

You mark the number that best represents how you feel about biology. If you think biology is extremely good, you would circle number 1. If you think that biology is neither good nor bad, you would circle number 3 or 4. If you feel that biology is more bad than good, then circle number 6 or 7.

We are interested in first impressions, so work rapidly and do not go back and change any mark. Be sure to mark every item and circle only *one* number on each item.

BIOLOGY IS . . .

Good	1	2	3	4	5	6	7	Bad
Useful	1	2	3	4	5	6	7	Useless
Important	1	2	3	4	5	6	7	Unimportant
Interesting	1	2	3	4	5	6	7	Boring
Easy	1	2	3	4	5	6	7	Hard

Another set of adjectives can be used to measure a student's self-perception in a science class. The extremes in a test of this type might consist of ("In science class, I am . . .") *good/bad, interested/bored, involved/uninvolved, happy/sad, careful/careless, lazy/industrious.*

Science attitudes can be measured by using a scale developed by Renis Likert (1932). The test is easily constructed by using declarative statements in either positive or negative terms relevant to the area, subject, or topic being assessed. A five-point scale is used, and the student checks or circles a position on the continuum. The terms usually used are *strongly agree (SA), agree (A), undecided (U), disagree (D),* and *strongly disagree (SD).* Other symbols such as numbers can be used—5, 4, 3, 2, 1.

Example of Likert Scale.

Directions: Each of the statements on this questionnaire expresses a feeling or attitude toward science. You are asked to indicate on a five-point scale the extent of your agreement between the attitude expressed in each statement and your personal feelings. The five points are: strongly agree (SA), agree (A), uncertain (U), disagree (D), and strongly disagree (SD). Draw a circle around the letter or letters to indicate how closely you agree with the attitude expressed in each statement as it concerns you.

1. I am always under a   SA   A   U   D   SD
   terrible strain in sci-
   ence class.

2. Science is very inter-   SA   A   U   D   SD
   esting to me and I
   enjoy science
   classes.
3. Science is fun.   SA   A   U   D   SD
4. I have a good feeling   SA   A   U   D   SD
   toward science.
5. I really like science.   SA   A   U   D   SD
6. I have never liked   SA   A   U   D   SD
   science; it is the sub-
   ject I dislike most.
7. I am happier in sci-   SA   A   U   D   SD
   ence class than in
   any other class.
8. I feel more relaxed   SA   A   U   D   SD
   in science class than
   in any other class.

There are a number of ways to interpret results of such tests. One could compute a weighted mean for each statement or make a frequency distribution of the number of individuals checking each position for each item.

Direct Approaches to Determine Values and Attitudes

Direct questioning and personal discussion interviews are other methods often used to determine student attitudes and values. However, teachers must take care in the way they pose questions, because students have a tendency during interviews to respond in such a way as to please the teacher. Direct questions such as, "Do you like biology?" and "Why do you think biology is useful?" will produce responses that are intended to please a biology teacher. More honest replies can be obtained by asking questions such as, "What subjects do you feel secure about?" "What subjects do you study more than others?" "Do you like animals?" "Do you like to dissect a frog?" "Do you like plants?" "Does it make you nervous to perform an experiment in the science laboratory?" Such indirect questions can give teachers much information about a student. An excellent battery of questions can be constructed if teachers know their students well—particularly their hobbies, interests, and career choices. Questions can be constructed to help the teacher identify values without giving students an inkling of what he is trying to determine.

Student values, appreciations, and attitudes can also be determined through responses to essay questions such as, "Biology is a very interesting area of study. I like it more than any other subject I am taking. State why you agree or disagree with this statement." "Chemistry is boring and useless to me. Do you agree or disagree? State and explain your reasons." "State why you like or dislike performing an experiment in the physics laboratory."

Essay responses can help teachers determine the type of teaching environment they are providing for their students. They can also give teachers information about the effectiveness of their teaching and the attitudes of their students toward a subject or unit of study. They can help teachers determine how interested students are in the subject being taught. The feedback can help teachers improve their instruction and revise the science curriculum.

ASSESSING PSYCHOMOTOR OBJECTIVES

The laboratory is the place where students design and perform experiments, manipulate equipment, and use the processes of science— ask questions, formulate hypotheses, interpret data, and so on. It is the place where they use higher cognitive skills such as analysis and synthesis.

Laboratory outcomes are concerned not only with the cognitive and affective domains but also the psychomotor domain, the third major area of objectives categorized by Krathwohl, Bloom, and Masia (1964). The psychomotor domain is very relevant to science, because laboratory activities require students to perform such tasks as manipulating equipment, bending a glass rod, mixing solutions, preparing slides for observation under a microscope, and using a microscope. The list is almost endless.

Students' abilities and skills in the laboratory can be assessed through tests composed of a series of items related to a task or tasks. The observations shown in figure 17–2 can be made by the teacher to determine how well a student can manipulate, focus, and use a microscope. The teacher checks one of three categories as the observations are made.

Such checklists involving observations made by the teacher in a one-to-one situation are time consuming. However, checklists or scales of this type can be developed for different situations involving the assessment of laboratory skills. The degree to which a student can perform a given task or skill on such checklists can be determined by checking one of a set of alternatives, each of which represents a particular level of skill performance for a particular task or aspect of performance. For example, 1, 2, 3, 4, and 5 could represent (1) excellent, (2) good, (3) fair, (4) passable, and (5) inadequate.

Some teachers prefer to assess laboratory skills through laboratory performance tests or practicals. When they administer a practical they use a checklist similar to those involving direct observations but adapted to assess laboratory performance. The use of a checklist gives the teacher a more objective evaluation in assessing laboratory practical examinations. The teacher can measure psychomotor responses and other outcomes through laboratory practicals. The time factor in administering such tests should be considered. It is not unusual to devote several hours to set up and administer such tests to groups of students. The other considerations for administering laboratory practicals include availability of enough equipment for class members, availability of equipment that is in good condition, and a set of clear and concise directions that the students can follow.

Checklists can be structured to assess practical laboratory skills such as setting up apparatus to produce chlorine in the chemistry laboratory or setting up equipment to demonstrate the production of carbon dioxide by living organisms. Several models of such checklists are available in the literature. Those for the Biological Sci-

FIGURE 17–2   Example of a checklist to assess psychomotor objectives

E = excellent; A = adequate; I = inadequate

Gross body movements

1. Removes microscope from its case or space in the storage cabinet. Grasps the arm of the instrument with one hand and places the other hand under the base.   E   A   I
2. Sets the microscope down gently on the table with the arm toward student and stage away from student. The base should be a safe distance from the edge of the table.   E   A   I
3. Uses a piece of lens paper to wipe the lenses clean.   E   A   I
4. Clicks the lower power objective into viewing position.   E   A   I
5. Adjusts the diaphragm and mirror for the best light.   E   A   I
6. Places a prepared slide of human hair on the stage so that it is directly over the center of the stage opening.   E   A   I
7. Secures the slide in place with the stage clips.   E   A   I
8. Looks to the side of the microscope and slowly lowers the low-power objective by turning the coarse adjustment wheel until the objective almost touches the slide.   E   A   I
9. While looking through the eyepiece, with both eyes open, slowly turns the coarse adjustment so that the objective rises. The hair should become visible.   E   A   I
10. Brings the hair into sharp focus by turning the fine adjustment wheel.   E   A   I
11. Shows the properly focused slide to teacher.   E   A   I
12. Focuses the hair under high power and shows this properly focused slide to teacher.   E   A   I
13. Prepares to return microscope to storage area; turns the low power objective into viewing position and adjusts it approximately 1 cm above stage.   E   A   I
14. Returns the microscope, handling by the arm and base, to storage place.   E   A   I

Finely coordinated movements

The following observations are made by the teacher who judges how well a student can prepare and stain materials for observation under a microscope.

Preparation and staining of an onion cell wet mount slide

1. Rinses a microscope slide with water and wipes both sides with a clean, soft cloth   E   A   I
2. Rinses and dries a cover glass.   E   A   I
3. Cuts an onion lengthwise and removes a thick slice.   E   A   I
4. Peels the delicate tissue from the inner surface.   E   A   I
5. Uses a medicine dropper to place a drop of water in the center of slide.   E   A   I
6. Places a small section of onion tissue in the drop of water.   E   A   I
7. Lowers the cover glass over the onion skin.   E   A   I

8. Staining the specimen: adds a drop of iodine stain along the edge of the cover glass.   E   A   I
9. Places a small section of a paper towel on the opposite side of the cover glass. This will draw the stain across the slide by capillary action.   E   A   I

The psychomotor domain is very relevant in science teaching. Students' abilities and skills in the laboratory can be assessed by the science teacher in a one to one situation.

ences Curriculum Study program (BSCS) can be found in Tamir and Glassman (1971) and Lunetta and Tamir (1979). These models can be used not only to assess practical psychomotor skills but for other outcomes such as the ability to design experiments, make quantitative and qualitative observations, handle certain calculations, analyze and interpret, predict, formulate hypotheses, and apply techniques to new situations.

EVALUATING OTHER ASPECTS OF THE SCIENCE PROGRAM

Other important aspects of the science teaching program that should be assessed periodically in-

clude the science curriculum, space and physical facilities for teaching science, safety conditions in the science classroom and laboratory, the teaching and learning environment, and teacher effectiveness.

Evaluating the Science Curriculum

The middle/junior high school science curriculum and the high school science curriculum should be assessed to find out whether they are up to date and if they are serving the students in meeting their needs as future citizens. Important information can be obtained through a questionnaire that asks direct questions regarding the state of the present science curriculum. The instrument can be prepared by a committee of science teachers, principals, science supervisors, and possibly students, and then completed by individual science teachers and curriculum coordinators. Questionnaires can be constructed to include pertinent questions such as the following:

1. Does the curriculum meet the requirements to achieve scientific literacy?
2. Does the science curriculum give students opportunities to develop certain skills through laboratory experiences?
3. Does the science curriculum provide students with experiences to solve everyday problems?
4. Does the science curriculum give students a variety of learning experiences?
5. Do the science courses help students develop positive attitudes toward science?
6. Does the curriculum provide experiences that will help students solve everyday problems?

Other important questions can be included in the questionnaire regarding the flexibility of the present curriculum, the options for providing for individual differences, the teaching of science as inquiry, and opportunities for problem solving and making value judgments based on rational thinking and scientific evidence.

Assessing the Science Teaching Facilities and the Learning Environment

The classroom, the laboratory facilities, and other learning environments should be evaluated to determine if they are adequate for teaching middle/junior high school science or for teaching the various science courses offered in the high school science curriculum. Questionnaires prepared by a committee of science teachers, administrators, students, and possibly parents, which ask questions designed to give important information about the present state of the science teaching facilities, can be administered to all science faculty and possibly to students. The information collected and analyzed can be used to make recommendations for improving the existing science facilities. The following types of questions can provide useful information to determine the status of science facilities:

1. Do the science classrooms provide a good teaching and learning environment?
2. Are the science teaching areas well illuminated?
3. Are the science classrooms and laboratories overcrowded so that safety is a prime concern?
4. Are the laboratory facilities adequate for conducting experiments?
5. Are the laboratory tables heat- and chemical-resistant?
6. Are the laboratory areas equipped with electricity, gas, and water?
7. Are there adequate supplies and equipment to conduct the various laboratory exercises and demonstrations as recommended in the science curriculum?

Other questions can be posed, such as those dealing with flexibility of facilities to allow for a wide variety of activities in the classroom and laboratory, facilities for project work, facilities for group discussions, and facilities for experimenting.

Assessing the Safety Conditions in the Science Classroom and Laboratory

The safety conditions in the science classroom and laboratory should be investigated to determine what is needed to avoid potential dangers. Questionnaires designed by a committee of science teachers can provide useful information regarding the status of safety conditions. Answers to questions such as the following can provide this information:

1. Is there adequate lighting in the laboratory?
2. Is the laboratory equipped with fire blankets, fire extinguishers, and an eyewash fountain?
3. Is there proper ventilation for conducting certain laboratory exercises and demonstrations where ventilation is an important consideration?
4. Does the laboratory area have the approved number of exits?
5. Are students provided with laboratory aprons and goggles when needed?
6. Are first-aid kits maintained in the science classroom and laboratory?
7. Are the storage areas safe?
8. Are chemicals properly stored in the storage areas?
9. Are safety rules posted strategically in the laboratory area?

These and other pertinent questions can give a reasonable picture of safety conditions in the various science teaching areas so that recommendations for improvement can be made.

Evaluating Science Teacher Performance

Teachers should be interested in receiving feedback from students about their teaching effectiveness as well as how students perceive them as individuals and teachers. Questionnaires, to be completed by students, that include statements to which students react can generate much information about the teacher. The following types of questions can be asked:

1. How interested is the teacher in the students?
2. Do the students accept the teacher?
3. Does the teacher respect the students and do the students respect the teacher?
4. Does the teacher praise students when appropriate?
5. Does the teacher show enthusiasm for teaching?
6. Does the teacher give students encouragement when appropriate?

The students' perceptions of teacher effectiveness in the classroom can be judged by questions like these:

1. Is the science teacher well prepared to teach the science lesson?
2. Does the teacher provide a variety of learning activities when teaching a science lesson?
3. Does the science teacher give fair tests and examinations?
4. Are the tests based on course objectives?

Questions regarding classroom management, planning, teacher's personality, and other aspects of classroom performance can also be useful in evaluating teacher effectiveness. See Chapter 19 for further discussion regarding the evaluation of science teachers.

Many commercial evaluation instruments and guidelines have been published that can be used to examine science teaching practices in a school or school system. Some of these publications are excellent and can be used directly without modification while others need modification to be useful for a particular teaching situation. Some can also be used to provide ideas for developing evaluation instruments that will suit a particular situation.

The National Science Teachers Association (NSTA, 1989a, 1989b) has published two sets of modules that can be used to assess the science programs—one set consisting of questionnaires to assess the middle/junior high school science programs, the other set for assessing the high school science program. The guidelines for self-assessment of middle/junior high school science programs (NSTA, 1989b) include the following modules:

1. Our School's Science Curriculum
2. Our School's Science Teachers
3. Science Student-Teacher Interactions in Our School
4. Science Facilities, Safety, and Teaching Conditions in Our School
5. Middle/Junior High School Principal and Other Administrators Checklist
6. Parent questionnaire

The guidelines for self-assessment of high school science programs (1989a) consists of four modules as follows:

1. Our School's Science Curriculum
2. Our School's Science Teachers
3. Science Student-Teacher Interactions in Our School
4. Science Facilities, Safety, and Teaching Conditions in Our School

The guidelines were produced to encourage self-examination and discussion toward improving the local science program. The self-assessments are aimed at answering such questions as:

1. What does the nation, the community, and the school want the science program to accomplish?
2. How well does the existing school program measure up to these expectations?
3. What are the in-service needs of our school's teachers in resource materials, methods, content, and attitudes?
4. Are teachers qualified to teach what they are assigned to teach?
5. What are preservice needs of teachers in areas of methods, content, resource materials, and attitude? (NSTA, 1989a, 1989b)

The preface of the NSTA guidelines (NSTA, 1989a, 1989b) includes suggestions on how to conduct a self-assessment and procedures to follow after the administration has given permission to proceed with the evaluation. The documents are designed to be used by a task force or committee composed of concerned individuals including science teachers, students, principals, and parents. The team has the responsibility of organizing the assessment program, setting the assessment dates, collecting the completed questionnaires, and analyzing the data. After the final ratings have been made, plans are made to share the results with the administration and other concerned individuals, to determine paths of action for future developments of the various aspects of the science teaching programs.

The NSTA guidelines for evaluating science programs in the middle and junior high school and science programs in the high school are excellent to use as guides to develop one's own questionnaires, to use as is, or to use with modifications.

SUGGESTED ACTIVITIES

1. Develop a questionnaire using a Likert scale to determine certain attitudes, likes, and dislikes that students in a science class have about a particular science subject. Ask the members of your science methods class to critique the questionnaire and then alter it if required. Administer the questionnaire to a group of students who are taking a course in the science area you selected. What kinds of information were derived from your questionnaire? Discuss the results with members of your science methods class.

2. Develop a questionnaire to determine what values students have about a controversial subject such as abortion. Discuss and critique the questionnaire with other prospective science teachers. Administer the questionnaire to students who are taking a biology course. Summarize the information you derived from the questionnaire. Discuss the results with members of your science methods course.

3. Examine the evaluation system used by a middle school science teacher in one of the local schools and compare the system with one used by a high school science teacher. How do the systems differ? To what extent do the teachers use norm-referenced, criterion-referenced, and self-referenced evaluation? How would you suggest that the teachers involved improve their evaluation systems?

4. Study the grading system used by a middle school science teacher and compare it with the grading system used by a high school science teacher. How do the systems differ? To what extent does the middle school science teacher stress cognitive, affective, and psychomotor objectives in his grading system? To what extent does the high school science teacher use cognitive, affective, and psychomotor objectives when grading students? Compare the two grading systems and discuss the results with members of your science methods class.

5. List the positive and negative attitudes that a student may develop through the study of physics or another science. Develop a test that consists of bipolar terms or adjectives that are relevant to the subject area you have selected. Ask the members of your science methods class to critique the test and alter it if necessary. Administer the test to a group of students who are taking a course in the science area you have selected. Summarize the results of the test. What kinds of information did you learn by administering the test to a group of students? Do you think that the information is useful to a teacher who is teaching a science course in the area you have selected? Discuss the results with other members of your science methods course.

BIBLIOGRAPHY

American Association for the Advancement of Science. 1989. *Science for all Americans: A project 2061 report on literacy goals in science, mathematics and technology.* Washington, DC: American Association for the Advancement of Science.

Baron, J. B. 1991. Performance assessment: Blurring the edges of assessment, curriculum, and instruction. In G. Kulm and S. M. Malcom (Eds.), *Science assessment in the service of reform* (pp. 247–266). Washington, DC: American Association for the Advancement of Science.

Bloom, B. S. (Ed.). 1956. *Taxonomy of educational objectives. Handbook I. Cognitive domain.* New York: David McKay.

Champagne, A. B., Lovitts, B. E., and Calinger, B. J. 1990. *Assessment in the service of instruction.* Washington, DC: American Association for the Advancement of Science.

Collins, A. 1991. Portfolios for assessing student learning in science: A new name for a familiar idea? In G. Kulm and S. M. Malcom (Eds.), *Science assessment in the service of reform* (pp. 291–300). Washington, DC: American Association for the Advancement of Science.

Eiss, A. F., and Harbeck, M. B. 1969. *Behavioral objectives in the affective domain.* Washington, DC: National Education Publications.

Gronlund, N. E. 1973. *Preparing criterion-referenced tests for classroom instruction.* New York: Macmillan.

Hedges, W. D. 1966. *Testing and evaluation in sciences in the secondary school.* Belmont, CA: Wadsworth Publishing.

Hudson, L. 1991. National initiatives for assessing science education. In G. Kulm and S. M. Malcom (Eds.), *Science assessment in the service of reform* (pp. 107–125). Washington, DC: American Association for the Advancement of Science.

King, B. 1990. *A report on the assessment of students' portfolio entry.* Technical Report No. B4, Stanford University Teacher Assessment Project. Berkeley, CA: Stanford University.

Krathwohl, D. R., Bloom, B. S., and Masia, B. B. 1964. *Taxonomy of educational objectives. Handbook II. Affective domain.* New York: David McKay.

Kulm, G., and Malcom, S. M. (Eds.) 1991. *Science assessment in the service of reform.* Washington, DC: American Association for the Advancement of Science.

Likert, R. 1932. A technique for the measurement of attitudes. *Archives of Psychology, 22*(140): 1.

Lunetta, V. N., and Tamir, P. 1979. Matching lab activities with teaching goals. *The Science Teacher, 46*(5): 22.

Mullis, I. V., and Jenkins, L. B. 1988. *The science report card: Elements of risk and recovery.* Princeton, NJ: Educational Testing Service.

National Science Teachers Association. 1989a. *High school science programs: Guidelines for self-assessment.* Washington, DC: National Science Teachers Association. (Includes modules: Our School's Science Curriculum; Our School's Science Teachers; Science Student-Teacher Interactions in Our School; Science Facilities, Safety, and Teaching Conditions in Our School.)

National Science Teachers Association. 1989b. *Middle/junior high school science programs: Guidelines for self-assessment.* Washington, DC: National Science Teachers Association. (Includes modules: Our School's Science Curriculum; Our School's Science Teachers; Science Student-Teacher Interactions in Our School; Science Facilities, Safety, and Teaching Conditions in Our School.)

Nay, M. A., and Crocker, R. K. 1970. Science teaching and the affective attributes of scientists. *Science Education, 54*(1): 59.

Nimmer, D. N. 1982. The use of standardized achievement test batteries in the evaluation of curriculum changes in junior high school earth science. *Science Education, 66*(1): 45.

Osgood, C. E., Succi, G., and Tannenbaum, P. 1967. *The measurement of meaning.* Urbana: University of Illinois Press.

Schibeci, R. A. 1982. Measuring students' attitudes: Semantic differential or Likert instruments. *Science Education, 66*(4): 565.

Tamir, P., and Glassman, S. 1971, Feb. Laboratory test for BSCS students. *BSCS Newsletter, 42:* 9.

TenBrink, T. D. 1990. Evaluation. In J. M. Cooper (Ed.), *Classroom teaching skills* (4th ed.). Lexington, MA: D. C. Heath.

Worthen, B. R., and Sanders, J. R. 1987. *Educational evaluation.* New York: Longman.

Wall, J. 1981. *Compendium of standardized science tests.* Washington, DC: National Science Teachers Association.

Yager, R. E. 1987. Assess all five domains of science. *The Science Teacher, 54*(7): 33–37.

The physical conditions under which a test is given should be of primary concern. To assure maximum performance on a test, the temperature of the room should be comfortable, the lighting should be adequate and crowded conditions should be avoided.

CHAPTER 18

Constructing and Administering Science Tests

Science teachers must possess considerable expertise to construct valid and reliable tests that accurately assess student achievement and instructional effectiveness. This aspect of teaching requires one to use a variety of modes in order to determine many types of learning outcomes that form a complete assessment system. A good testing program must include many different paper-and-pencil items, even those that address manipulative and laboratory skills. Of no less importance is the teacher's ability to create a testing environment that is free from distractions and conducive to concentration. In addition, tests should be graded immediately in order to provide useful feedback to students.

Overview

- Stress the usefulness of Bloom's taxonomy in test construction.
- Present examples of test items at all levels of Bloom's taxonomy.
- Emphasize the importance of assessing higher levels of thinking using questions that involve science process skills.
- Describe the various types of test items and point out their advantages and disadvantages.
- Indicate points to consider to write good test items.
- Provide suggestions for administering tests under favorable conditions.
- Suggest useful procedures and techniques for scoring, grading, and analyzing tests.
- Discuss the purposes and construction of various types of tests.
- Provide procedures for writing unit tests.

QUESTIONS INVOLVING HIGHER LEVELS OF THINKING AND SCIENCE PROCESS SKILLS REASONING

The next four levels of Bloom's taxonomy (Level 3—application, Level 4—analysis, Level 5—synthesis, and Level 6—evaluation) are concerned with higher-level thinking. They involve such science process skills as examining variables, problem solving, critical thinking, hypothesis generation, hypothesis testing, and drawing conclusions. Tests that purport to evaluate science process skills are characterized predominantly by application, analysis, synthesis, and evaluation questions. They require considerable thought and imagination to develop and construct.

This area of testing is very important because a number of states are introducing statewide assessment programs. Such programs emphasize the testing of process reasoning skills in order to improve science education in that state and meet the reform goals advocated by the American Association for the Advancement of Science, *Science for All Americans,* Project 2061 (AAAS, 1989) and curriculum development programs supported by the National Science Foundation (NSF, 1989). Some of the assessment programs use Bloom's *Taxonomy of Educational Objectives* as a basis for their testing programs while others use the AAAS organization of process skills (AAAS, 1965). The AAAS organization of process skills provides a program of instruction for the primary and intermediate grades. The intermediate program builds on the competencies acquired in the primary grades and exposes students to the kinds of activities in which scientists are involved. The process skills for the intermediate grades include formulating hypotheses, making operational definitions, controlling and manipulating variables, experimenting, formulating models, and interpreting data.

The following examples using Bloom's taxonomy illustrate the type of test items that can be used to assess science process skill reasoning. Some questions can be constructed to assess only one skill at a time, while others can be more complex, involving the use of several process skills.

Level Three: Application

The application level requires that students produce solutions to problems. The student is required to identify the relevant information and rules to arrive at a solution. Application questions are often used in physics, chemistry, and biology. Words frequently used in application questions include *solve, which, use, classify, choose, how much.* and *what is.*

Example Application Questions.

1. In very cold conditions up to 80% of the heat made by the body can be lost through the surface of the head and neck. Which would be the best way to preserve the greatest amount of body heat if you were suddenly caught outside in below-zero weather and all you had to wear were boots, shorts, a T-shirt and light jacket?
   a. Wrap the jacket around the uncovered portion of your legs.
   b. Leave the jacket on as you usually wear it.
   c. Wrap the jacket around your head and neck.
   d. Use the jacket to keep the nearby air moving.

2. Complete the following reaction by selecting one of the formulas below that represents the product X.

Reaction:

$$\underset{H}{\overset{H}{}}\,C{=}C\,\underset{H}{\overset{H}{}} + Br_2 \rightarrow X$$

Product:

a.

$$H-\underset{\underset{H}{|}}{\overset{\overset{H}{|}}{C}}-\underset{\underset{Br}{|}}{\overset{\overset{H}{|}}{C}}-H$$

b.

$$H-\underset{\underset{H}{|}}{\overset{\overset{H}{|}}{C}}-\underset{\underset{H}{|}}{\overset{\overset{Br}{|}}{C}}-H$$

c. (structure with H—C—C—H, with H H above and Br Br below)

d. H—C≡C—H with Br Br below

3. Two disk magnets are arranged at rest on a frictionless horizontal surface as shown in the following diagram. When the string holding them together is cut, they move apart under a magnet force of repulsion. When the 1.0-kg disk reaches a speed of 3.0 m/sec, what is the speed of the 0.5-kg disk?

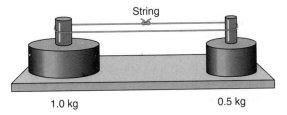

1.0 kg   0.5 kg

a. 1.0 m/sec
b. 0.50 m/sec
c. 3.0 m/sec
d. 6.0 m/sec

4. Davy found that the combining volumes of oxygen with 100 volumes of nitrogen are 49.5, 108.9, and 204.7. From these results Gay-Lussac concluded that the volume ratios are $2:1$, $1:1$, and $1:2$. What justification did Gay-Lussac have for such a conclusion?

a. The measured volumes are closer to these simple ratios than to any other.
b. Combination in definite proportions had already been established; simple volume ratios are, therefore, extracted.
c. The other gas reactions studied showed similar clustering of data around simple ratios.
d. Although subsequent measurements justified the conclusion, these data do not justify it.

Level Four: Analysis

The analysis level involves in-depth understanding of information. It requires that students separate an idea into parts or elements and demonstrate an understanding of the relationship of the parts to the whole. Words and phrases often used in analysis questions include *analyze, support, provide evidence, identify reasons, why,* and *provide conclusions.*

Example Analysis Questions.

1. Which of the following statements concerning mountain building is largely a hypothesis rather than the statement of a fact based on extensive observational evidence?
   a. Deposition of large quantities of erosional material has accumulated in geosynclines.
   b. Mountains have been produced by faulting on a gigantic scale.
   c. Heavy minerals have tended to sink toward the center of the earth, forcing the lighter segments upward.
   d. Mountains have been produced by the differential erosion of rocks.

2. The table below gives electrical resistance of five coils of wire at two temperatures. The resistance of each coil changes uniformly between 10° and 300° C. For each of the items written below the table, write in the letter for the coil that satisfies the specification given.

Coil	Resistance at 10° C (ohms)	Resistance at 300° C (ohms)
A	100.0	100.3
B	2.50	2.650
C	1.10	0.999
D	0.250	0.300
E	0.021	0.015

___ The coil whose resistance remains most nearly constant as the coil gets hot.

___ The coil whose resistance in the temperature range from 10° to 300° C is to be not greater than 2.5 ohms nor smaller than 0.25 ohms and to show as little variation in resistance as possible.

___ The coil whose resistance is most influenced by temperature.

___ The coil whose resistance is to be as low as possible and not decrease along with temperature.

Level Five: Synthesis

Questions on the synthesis level promote creativity. They require students to produce products, patterns, and ideas. At this level students are involved in analyzing phenomena and formulating hypotheses to explain them. Such questions permit students to devise ways to design experiments and test hypotheses. Students may be required to write a paper or a report in which ideas are synthesized or problems are solved.

Unfortunately, the teaching objectives of most teachers seldom include those that require creativity. Consequently, synthesis questions are seldom used in their testing programs. Comprehension, application, and analysis also involve higher-level thinking, but the tasks involved in answering those types of questions are not as unique and thought-provoking as those involved in answering synthesis questions. Synthesis questions require students to draw on many sources to put together a pattern or plan that has not been produced before.

Synthesis questions do not necessarily have one correct answer, but a variety of answers, all of which may be plausible. The directions used in synthesis questions often use words or phrases such as *write, predict, develop, design, synthesize, produce, solve, devise,* and *construct.*

Example Synthesis Questions.

1. Eight bean seeds were germinated, then divided into four groups of two seeds each. One group of two seeds was grown under red light, another under yellow light, another under blue light, and the fourth under ordinary white light. At the end of two weeks the growth of each group of plants was measured to see which group of plants had grown the most. Which option would *best* improve this experiment?
   a. Giving more water to the plants grown under red light.
   b. Increasing the number of seeds grown in *each* of the four groups.
   c. Growing just the plants under white light in sandy soil, but growing all others in humus soil
   d. Adding one more group of two seeds to the experiment and growing them under purple light.
2. Substance X is known to move through cell membranes. You suspect its movement is accomplished by simple diffusion. Design an experiment to test your hypothesis.
3. Dry gases X and Y react readily when mixed in a glass flask. However, if the flask is heated strongly and cooled just before the gases are introduced, no reaction takes place. Develop two hypotheses to explain these observations and state how you would test such hypotheses.
4. Now that you have studied the anatomy and physiology of animals and humans and the effects of radioactivity on tissues, describe the characteristics of a person who would be most likely to adapt to a radioactive environment.

Level Six: Evaluation

Evaluation, the sixth level of Bloom's taxonomy, concerns the most complex level of thinking. Questions at this level, like many synthesis questions, do not have one correct answer. They require students to make judgments about the value or merits of an idea, purpose, solution to a problem, procedure, method, or product. Students are required to make a value judgment and support it with their reasons for the evaluation, which demonstrates their knowledge and understanding of the topic. This requires them to use the other five levels of taxonomy to varying degrees.

Words or phrases often found in questions involving evaluation include *assess, decide, judge, argue, what is your opinion, appraise, do you agree or disagree, give an evaluation.*

Example Evaluation Questions.

1. Two hypotheses that account for the diversity of species on earth are (1) gradual accumu-

lation of mutations and (2) punctuated equilibrium. Which hypothesis do you favor and why?

*2. After they had blown through limewater, pupils concluded that their bodies give off carbon dioxide. However, one girl argued that since there is carbon dioxide in the air they inhale, the experiment proves nothing. The girl's argument
   a. is unscientific
   b. is invalid because there is too little carbon dioxide in the atmosphere to be considered
   c. shows that the pupils needed a control to check the effect of inhaled carbon dioxide on limewater
   d. ignores the reaction of isotopes of carbon dioxide produced during respiration

3. Sedimentary rocks can be distinguished from other classes on the basis of bedding, color, fossils, ease of breakage, and porosity. Which do you feel is the poorest criterion to use as a distinguishing characteristic? Defend your position.

The examples shown in figure 18–1 are more complex in that several process skills are involved. Questions 6 through 9, for example, present an experiment and ask the student to draw conclusions, determine sources of error in the experiment, and synthesize information. Questions 12 through 14 provide experimental data and ask the student to analyze information and draw conclusions. Questions 15 and 16 ask the student to draw conclusions from a report given by a missionary about a plant that an African witch doctor had used to cure him of an illness. The student is also asked to determine which procedure, from those listed, would be most appropriate to use in investigating the medicinal properties of the plant.

---

*From the New York State Education Department, Division of Educational Testing, Bureau of Test Development, *Teacher's Guide for the Biology Regents Examination,* The University of the State of New York, Albany, 1962.

TYPES OF TEST ITEMS

True-False Statements

True-false items allow students to judge a statement for accuracy. The statement is either right or wrong. True-false items can be used to determine how well students understand a concept, principle, or theory. Items can also be written to determine whether the students can make decisions, evaluate variables, interpret data, and understand the nature of science and the scientific enterprise.

Example True-False Statements.
Directions. The following statements are either true or false. In the blank to the left of each statement write the letter *T* if the statement is true or *F* if the statement is false.

   F   The chemical digestion of starch begins in the stomach.

   T   The liver secretes hormones that regulate the storage of glucose.

   F   The earthworm possesses an internal skeleton.

Advantages.   There are a number of advantages in using true-false items. They are easy to write and easy to grade. A great number of items can be answered in a short period of time, permitting a broad sampling of knowledge. Answers to the questions can be scored quickly, particularly if a key is used. The true-false test is an excellent tool to use to initiate discussions, for reviews, and for pretesting to determine the status of students at the beginning of a unit.

Limitations.   The limitations of true-false items should be considered when constructing tests. Teachers have a tendency to use true-false tests to evaluate factual information, forgetting that these tests usually have poor validity and low reliability. Students learn that certain key words in a statement can divulge whether the statement is true or false. Words such as *sometimes, often,* and *probably* are usually associated with true statements, whereas words such as *never, always,* and *none* are often associated

FIGURE 18–1   Process skill test items (From Biological Sciences Curriculum Study: Processes of science test, Form A, Colorado College, Colorado Springs, Colo., 1962.)

Questions 6-9 relate to the following experiment.

Suppose the following experiment is done for the purpose of showing that a plant growing in a bell jar decreases the amount of carbon dioxide, CO_2, and increases the amount of oxygen, O_2, in the bell jar.

A plant and a burning candle are placed beneath a bell jar as shown. The candle burns one minute and then goes out. Three days later the candle is ignited by means of the electrical apparatus and burns one minute before going out.

6. Before any conclusions could be drawn from this experiment, the experiment should be repeated with which one of the following changes?

(A) Use a larger bell jar

(B) Omit the candle

(C) Omit the plant

(D) Ignite the candle with a concentrated beam of light

7. The most probable source of error in this experiment would be

(A) impure materials in the candle.

(B) variations in the electrical energy supplied by the battery.

(C) the size of the bell jar.

(D) leaks between the bell jar and the platform.

8. Suppose it is definitely shown that a plant growing in a bell jar decreases the carbon dioxide, CO_2, and increases the oxygen, O_2, present. How is this finding related to the observation that plants restore the ability of air to support the burning of a candle?

(A) It adds information which contradicts the observation

(B) It adds information which fits with the observation

(C) It adds nothing to the observation

(D) It and the observation cannot both be correct

9. Which of the following questions is answered by the experiment?

(A) Do all parts of a plant take up CO_2 and give off O_2?

(B) Do plants require light to take up CO_2 and give off O_2?

(C) Do plants and burning candles have the same effect on air?

(D) Is the amount of CO_2 taken up by a plant greater than the amount of O_2 given off?

Questions 10-11 relate to rhythmical movements in plants.

10. Several plants known to exhibit rhythmical leaf movements were found to continue these movements even when the plants were grown under constant conditions of light, temperature, and humidity. Which one of the following conclusions (arranged in order of increasing generality) is the *most general* one supported by the facts presented?

(A) Rhythmical leaf movements of some plants do not depend upon variations in light, temperature, or humidity

(B) Rhythmical leaf movements of all plants do not depend upon variations in light, temperature, or humidity

(C) Rhythmical leaf movements do not depend upon variations in the environment

(D) Rhythmical movements in plants do not depend upon variations in the environment

11. If a biologist became interested in the problem of rhythmical leaf movements, which one of the following should he do in a careful investigation of the problem?

(A) Write to someone who is already working on this problem asking him to suggest the best experiments to do

(B) Read the published reports of biologists on the problem

(C) Talk to his friends to see what they know about the problem

(D) Look the subject up in an encyclopedia to find out more about it

FIGURE 18–1 *continued*

Questions 12-14 relate to the experimental data in the following table.

Several different plant parts were placed in sealed containers of equal volume. The amounts of CO_2 (carbon dioxide) used by the plant parts under different conditions were measured and recorded.

Container	Plant	Plant Part	Vol. of Plant Part (cubic centimeters)	Color of Light	Temp. (°F)	Time Elapsed (days)	CO_2 (carbon dioxide) Used (cubic centimeters)
1	Myrtle	Leaf	100	red	60	2	150
2	Myrtle	Leaf	100	red	80	2	200
3	Myrtle	Stem	100	blue	70	2	50
4	Oak	Root	100	blue	80	3	0
5	Oak	Leaf	100	orange	80	2	100
6	Oak	Leaf	100	orange	80	3	150

Assume that the experimental conditions not listed were identical in all six containers.

12. On the basis of the data in the table, one could properly compare the amount of CO_2 used in one day by

(A) myrtle leaves at 60° F and at 80° F.

(B) myrtle stems and myrtle leaves.

(C) oak leaves in orange light and in blue light.

(D) oak leaves at 60° F and at 80° F.

13. The experimental data indicate that the oak leaves used how much CO_2?

(A) More in orange light than in blue light

(B) More at 80° F than at 60° F

(C) More per day than did the myrtle leaves

(D) 50 cubic centimeters each day

14. Before it would be appropriate to compare the amount of CO_2 used in one day by an oak leaf and a myrtle leaf, it would be necessary to change the

(A) color of the light in container 1 to orange.

(B) temperature in container 1 to 80° F.

(C) temperature in container 2 to 60° F.

(D) color of the light in container 5 to red.

Questions 15-16 relate to the following report.

A missionary reported that the root of a plant much like the Rauwolfia plant had been used by an African witch doctor to cure him of a serious illness.

15. What is the most reasonable conclusion that can be drawn from this report?

(A) The plant is useless since witch doctors are not trained physicians

(B) The plant is useless because the missionary cannot judge how effective the plant was in curing his illness

(C) The plant may have been helpful since the missionary recovered after the witch doctor's treatment

(D) The plant was helpful because recent medical reports show that reserpine, a drug effective in lowering blood pressure, is extracted from Rauwolfia

16. Which of the following procedures would be most appropriate in a preliminary investigation of the medicinal properties of this plant?

(A) Administer portions of the plant to a group of human beings, with appropriate controls, and record the effects

(B) Select two groups of rats as experimental and control groups to test the effects of the plant

(C) Send purified samples of the drug to hundreds of physicians throughout the world to assure an adequate sample

(D) Cross the plant with Rauwolfia to determine how closely these two plants are related

Write the number of the correct symbol in the blank next to each element.

A	B
____ Lead	1. Si
____ Gold	2. Au
____ Silver	3. W
____ Tungsten	4. Ag
____ Iron	5. Hg
____ Mercury	6. I
____ Silicon	7. Fe
	8. Pb
	9. Ti

Directions. In the blank before each statement in column A, place the corresponding number of the appropriate word or phrase listed in column B.

A	B
____ A monosaccharide	1. Mitochondria
____ Contains RNA and protein synthesizing enzymes that make up DNA or RNA	2. Glucose
	3. Nucleus
	4. Ribosomes
	5. Chloroplasts
____ Centers of cellular respiration	6. Enzyme
	7. Nucleolus
____ Protein that acts as a catalyst	8. Nucleotides
____ Chemical process involved in formation of proteins and polysaccharides	9. Dehydration synthesis
____ Units that make up DNA or RNA	

Suggestions.   To write good matching questions, consider the following suggestions:

1. The items selected should include statements, words, or phrases that cover the same general topic or category. Related items should be used for a particular matching question instead of a hodgepodge of unrelated content. That is, chemical formulas should be related to chemical formulas, equations, or common names of substances, but names and dates of chemists should not be included in the same list of distractors. Each term in one column should be a plausible answer for those in a column to be matched.

2. Arrange a long list in some logical order, alphabetical or otherwise. If numbers are used, arrange them in ascending or descending order so that the students do not have to waste time searching for them.

3. The length of statements in a particular column should be somewhat consistent. List longer statements and phrases in the left column and the shorter ones in the right column. This helps reduce reading time and avoids having students read long statements or phrases more than once. It also permits rapid scanning of the list of shorter statements or phrases.

4. The number of answer choices should be greater than the number of statements to which they are to be matched.

5. The set of responses should be kept short—between six and twelve. Long lists are cumbersome for the student and require too much time for searching. Moreover, it may be difficult to find homogeneous items for more than twelve responses.

6. The directions should be short, concise, and clearly explained.

Advantages.   Matching-type questions are relatively easy to construct, particularly if they are to measure lower-level learning. The difficulty in constructing a matching-type exercise increases if one goes beyond lower-level learning. This type of question adds variety to a test, and teachers will find that students enjoy answering these more than multiple-choice, true-false, or completion questions.

Matching questions take up less space than a series of multiple-choice questions. Guessing is also kept to a minimum. A well-constructed matching item makes guessing difficult, particularly if the teacher has eliminated most of the clues and has included a sufficient number of items to match. Matching questions are easy to grade, and they permit a broad range of material to be tested.

Limitations. There are some limitations associated with matching questions. They are not especially effective for evaluating students' understanding of subject matter or assessing higher-level thinking. Multiple-choice items are more useful in this regard. Consequently, the matching type is generally used to test facts involving simple recall and recognition. Having too many items in a list can be confusing to the student since there is always a strong possibility that two or more associations can be made for any item.

Completion Items

Completion or fill-in-the-blank questions involve a statement with one or more blanks in which the student provides the information to make the statement complete and correct. Completion items usually require recall of information, unlike matching, true-false, and multiple-choice, all of which involve recognition. Items can also be written that require a student to perform certain calculations, the results of which go into the blank in the statement. Problems in physics, chemistry, and genetics that require calculations lend themselves to this type of item.

Example Completion Items.

Directions. Complete the missing information in each of the following statements to make the statement complete and true. Write the correct answer in the blank.

1. A gas that is explosive and lighter than air is _____

2. The product that results from the combustion of sulfur and oxygen is _____.

3. The chemical formula for potassium hydroxide is _____.

4. The number of different types of gametes that can be produced by a male guinea pig whose genotype is BbLlrrSs is _____.

5. The kinetic energy increase of an electron that has been accelerated to a potential difference of 20 million volts is _____. (Assume that $1 eV = 1.6 \times 10^{-19}$ J.)

Suggestions. To write good completion items, follow these suggestions:

1. Make all the blanks in the sentences the same length and place them at the end of the statement.
2. Avoid using more than two blanks per statement. More than two blanks may make the statement more ambiguous.
3. Avoid copying direct statements from a textbook or workbook because statements out of context may have a different meaning to the students than was intended.
4. Leave only key words out of the statement. It is best to limit the number of key words to one or possibly two. Avoid requiring long phrases.
5. Require specific responses when possible— dates, formulas, and results of computation.
6. Write directions clearly so that students understand where the answers should be placed (such as in blanks to the left of statement or on a separate answer sheet).
7. Make the answer blanks long enough for the answer.

Advantages. There are some advantages in using completion questions. They can be used to determine whether students understand basic rules and principles involving simple computational operations such as computing molar volume in chemistry. They can be used to recall dates, names, places, and other relatively unimportant facts. Completion statements permit testing for a wide range of material in a short period of time. The answers are also easy to grade.

Limitations. The limitations of the fill-in-the-blank items are serious and sometimes difficult to overcome. It is difficult to write unambiguous statements because in many instances several words may be correct responses even though they were not considered during the test construction. Normally, most of these items assess knowledge-level learning outcomes such as dates or names of places and people. These questions can rarely be used to test for upper levels of Bloom's taxonomy. In scoring fill-in-the-blank tests, the teacher must consider all possible responses to each item, which makes mechanical scoring difficult.

Essay and Long-Response Questions

Essay questions usually ask students to formulate responses in their own words. Two general types are the restricted type, which restricts the scope of responses, and the nonrestricted type, which gives students free rein to respond to the question.

In the restricted type the student either is given an outline that directs the response or is asked to stress certain points in the essay.

Example Restricted Essay Questions.
1. Discuss, compare, and contrast mitosis with meiosis. Point out in your discussion the basic differences between meiosis I and mitosis and between meiosis II and mitosis. Also, discuss the significance of each process. Include diagrams to illustrate and support your discussion.
2. List and describe 10 cell components. Discuss their roles in the life of the cell and state at least two unsolved problems concerning each component.
3. Explain in detail three methods for preparing a base and two methods for preparing an acid.

The unrestricted type of essay question gives the student free rein to discuss the topic with no direction.

Example Unrestricted Essay Question.
Discuss the statement: "The sun is the source of all energy on earth."

Long-Response Questions. Another variation is the long-response question, which allows the students to write a sentence, equation, formula, or short paragraph in response to a particular question or topic.

Examples.
1. Complete and balance the equations for each of the following chemical reactions:
    a. hydrochloric acid and sodium hydroxide
    b. nitric acid and potassium hydroxide
    c. hydrochloric acid and calcium hydroxide
    d. sulfuric acid and barium hydroxide
2. Write a definition for each of the following terms:
    a. work
    b. erg
    c. dyne

Suggestions. Consider the following suggestions when writing essay questions:

1. Use the open-ended question to measure the higher levels of Bloom's taxonomy—understanding, analysis, synthesis—outcomes that cannot be measured easily through objective-type items.
2. Use the restricted types when possible, because they are easier to score and they channel student responses.
3. Write questions so that students have sufficient time to complete the test.
4. While constructing the item, determine the criteria to be used to grade the response. The more specific the question, the more objectively it can be scored.
5. For essay questions that list the important points to be used in evaluating the response, determine a value for each point when constructing the essay question.
6. When grading long-response, unrestricted type questions, use the sorting method instead of a point system. In the sorting method, the teacher does not examine every sentence of every main idea, but instead judges the overall quality of the essay. The teacher then sorts the essays by placing the

best responses in one pile, the poorest in another, and the intermediate responses somewhere between the two extremes. A specific letter grade or a specific number of points can be given for each pile, representing the general quality of the responses in each particular pile.

Advantages. There are a number of advantages in using essay or long-answer questions. These types of questions take less time to write than many types of short-answer questions. However, great care must be taken in their construction. Various kinds of essay questions can be used to assess objectives that are not possible to evaluate by using short-answer questions. Essay questions are excellent to judge the organizational ability of students, as well as determine how well students are able to handle themselves at the higher levels of cognitive domain. These types of questions can be excellent pretest or posttest items to determine the level of student understanding before a unit is started or when a unit is completed. Essay questions can also be used to find out how well students understand a concept, theory, principle, or idea. They are good for detecting misconceptions.

Limitations. The limitations of discussion or essay questions are probably obvious. Essay questions are more difficult to score objectively because of the variety of responses given by students. This is especially true of long essay questions as opposed to the shorter, more restricted forms. The reliability of essay questions is lower than objective-type or short answer questions. To increase the reliability of essay questions it is often desirable to use the restricted type involving questions that are more specific.

LABORATORY PRACTICAL EXAMINATIONS

Laboratory practical examinations can be administered to assess whether students can carry out certain tasks in connection with laboratory work.

Different types of laboratory practicals can be developed for this purpose. Some of these skills are basically manipulative; others are cognitive and reflect the students' ability to identify or recognize.

Manipulative Tests

Manipulative tests assess students' ability to carry out certain procedures in science. The necessary apparatus, materials, and written directions to carry out the procedures are placed before the student. Students are expected to carry out tasks and demonstrate the end result. They may or may not be expected to explain the reason for the operations. They may be given a grade based on the end result only, or on each step of the operation, or on their explanations.

Manipulative tests can determine how well certain types of skills have been mastered. Normally, if large numbers of students are involved, the teacher has various setups distributed throughout the room. Each setup is numbered and the student is asked through written directions to use the setups in a particular way. It is not necessary to ask all the students to perform the same tasks. One student, for example, can be asked to measure the length of a steel rod located and numbered at one position in the classroom, and another student may measure the length of another rod located and numbered at another position in the classroom. The measurements are then recorded on an answer sheet with a corresponding numbered blank in which the student inserts his measurements.

Many variations can be used for manipulative skill assessment. Performance of a technique may be assessed on a one-to-one basis, where the students demonstrate their ability to bend a piece of glass tubing in the presence of the teacher. Or students may be asked to demonstrate, in the presence of the teacher, how they can handle and read a voltmeter.

Example Manipulative Tests.

1. A student is given a length of glass tubing, a gas burner, and igniting instrument and is

asked to make a right-angle bend in the tubing.

2. A student is given four miniature lamps, a dry cell, a small screwdriver, wire, and a switch, and is asked to connect the lamps in series and then in parallel.

3. Students are asked to examine four different cultures of protozoans. They are given several glass microscope slides, coverslips, and a microscope. They are asked to prepare the slides and identify the organisms in each culture using a key provided.

Advantages. Manipulative tests can assess the psychomotor domain of educational objectives. They can also give students who have difficulty verbalizing a chance to demonstrate their abilities and achievements. Manipulative tests, although difficult to administer, can measure many outcomes that reflect desirable goals of science programs. Furthermore, the teacher will find that students enjoy taking this type of test, for it adds variety to the testing program. A teacher with imagination and ingenuity can use manipulative tests to measure certain understandings and ways of thinking that cannot be measured using other types of testing procedures.

Limitations. Manipulative tests are time consuming to organize and administer, particularly to large groups of students. Such tests require materials and equipment for all students in order to be effective. This may be costly. If sufficient quantities are not available, the tests can be poorly organized and confusing to students.

Identification Tests

Identification tests involve giving students one or more unknown specimens to identify. In this testing procedure students are also provided with the necessary materials and instruments to help them make the identification.

Example Identification Tests.

1. Each student in the class is given 10 mineral specimens to identify, along with the materials needed for determining hardness and

streak. Each specimen is numbered. The students are asked to identify each specimen by using hardness and streak and place their responses in the corresponding numbered blanks on their answer sheets.

2. Each student is provided with six bottles, each containing a solution that is either acid, alkaline, or neutral. The student is provided with litmus paper to make the determination.

3. Each student is given four twigs and a twig key. The student is asked to use the key to identify the tree associated with each twig specimen.

Identification tests require careful organization. Caution must be taken to ensure that all students have proper materials to carry out the procedures. Care must also be taken that the quality of the material to be identified is the same for all students. Adequate time must be given for the test. Teachers might take the test themselves to judge how much time it takes for the identification and then decide how long the students might take to make the identifications. To be fair, the specimens must be unfamiliar to all students. Sometimes this may be difficult to arrange, since students will have had different exposures, depending on their background of experiences.

Another variation of an identification test is a recognition test, which requires the student to identify specimens or other science materials without the use of keys or other identification procedures. Familiar materials are presented to the students, who then write the name of each specimen on an answer sheet.

Example Recognition Tests.

1. Numbered jars containing plant specimens are placed in sequential order on the laboratory table. The following directions are provided:

   You are required to identify the 12 different plant specimens on this table. Each jar containing a specimen has a numbered label. The answer sheet provided is numbered from 1 to 12. Place the name of the plant in the space corresponding to the number on the jar. You will be given 1 minute at each position to provide the proper identification. At the end of the minute you will be asked to move to the next position. No lagging will be permitted.

2. There are 24 students in an earth science class. Twenty-four trays are distributed, each containing identical sets of 10 numbered rock specimens that the students have seen at other times. Each student is given one tray and is asked to identify the specimens, placing the responses on an answer sheet numbered from 1 to 10.

3. Ten large, mounted birds familiar to the students are displayed by the teacher, one at a time, before the entire class. The students write down the names of the birds in the order presented on an answer sheet with blanks numbered from 1 to 10.

4. Ten different pieces of chemical apparatus and glassware are placed at certain positions on the laboratory benches throughout the chemistry laboratory. Each piece is numbered, and the students are given 1 minute at each position to identify and write down the name of the item on an answer sheet with blanks numbered from 1 to 10.

Identification tests involving recognition have some value in biology and earth science because this type of test measures students' ability to recognize actual specimens and materials with which they have come in contact before. Identification of rock, bird, or insect specimens without keys may be advantageous during field work or other problem-setting situations.

In preparing identification tests, the teacher must make certain that all students have good specimens to examine and that the specimens are familiar to all students. If a rotating system is used, it must be set up so that it is not cumbersome. Enough space should be provided at each position so that students do not feel crowded. Teachers should make certain that each student has a turn at each position and also that enough time is allotted so that the students do not feel rushed. Teachers should take precautions against cheating.

Advantages.   Identification tests can be used to measure certain understandings and ways of thinking. They can be used to assess students' ability to carry out various kinds of identification procedures. In general, students enjoy taking this type of test because they can actually work with and handle materials and specimens during the testing procedure.

Limitations.   The limitations of identification tests are similar to those mentioned for manipulative tests. Setting up identification procedures for large groups of students may be very time consuming because the tests must be well organized to be effective. Adequate materials must be available for all students. Providing each student with the same number and quality of specimens or items may require a great deal of effort, time, and expense.

WRITING CLEAR TEST QUESTIONS

When your students turn in their test papers, their answers demonstrate their understanding of science. Right? Maybe not. Unless you have worded your test items so students understand what you are asking them, you may be challenging their reading ability rather than their grasp of scientific concepts. (Rakow & Gee, 1987, p. 28)

This statement has direct application to the construction of all types of test items, particularly those involving much verbiage, such as multiple-choice items.

The best questions are simple and direct. Each additional adjective and qualifying clause increases the complexity of a statement and the possibility for misinterpretation and ambiguity. The words used in the construction of a test item should be those that were used while teaching the topic. Generally, the use of short, familiar words minimizes reading problems associated with a test item. Long, complex sentences and prepositional phrases also can cause students to misinterpret a question.

The introduction of a new word in a lesson does not necessarily justify its use in a test item unless it has been used extensively during the course of instruction and has become part of the students' vocabulary. It is important for students to learn basic science vocabulary, but it is more important that questions be written to measure understanding rather than ability to use words.

As already discussed, multiple-choice questions should be written precisely and should avoid ambiguity to ensure that only one choice is correct. When writing multiple-choice questions, avoid wordiness. The items and the choice of responses should be in simple and clear language so that the poorest readers can understand what is intended by the question.

When writing matching exercises, there is always a strong possibility that two or more associations can be made for an item. Take care to avoid writing more than one association question for each item. A matching question can cause problems for the poor and/or slow reader if too many items are involved. Restrict the number of items to be identified or matched so as not to exceed ten or twelve in a set.

Completion questions must be carefully phrased to avoid ambiguity. Different interpretations of a particular question bring to mind a wide range of responses. For example:

1. Green plants produce _____.

This question may have several correct responses. A student who thinks in terms of photosynthesis will answer *food, carbohydrates, sugar,* or *starch.* A student who thinks along a different line may respond with *chlorophyll, cellulose, leaves, stems,* or *oxygen.* These responses are equally valid, and they do not represent errors in thinking on the part of the student. The fault is in the way the question is posed. All responses should be allowed because the question is ambiguous and poorly constructed. The question can be written more precisely to avoid these problems. For example:

1. Green plants produce their own food only in the structures of the plant that contain ____.

This question is well constructed because it solicits only one correct response and avoids problems of ambiguity.

Essay questions must be written so that students know exactly what is expected of them. They should not be broadly stated and open ended. Because this type of question requires students to organize ideas, those who are poor readers normally have difficulty expressing themselves and organizing information clearly. Essay questions should include an outline of the main points to be covered by the student. The question should also include instructions regarding the amount of detail required to answer the question satisfactorily.

For students who cannot organize information or write well, ask a series of short prose responses rather than one or two long prose responses. Generally, students with reading and writing difficulties will perform better when briefer answers are required.

Broad, open-ended essay questions solicit many varied responses, making the question difficult to grade objectively. A simple outline associated with the question, or the inclusion of specific areas to be covered, will guide the teacher when grading the question. Short prose responses provide a good sample of the content and the brief answers make the question easier to grade.

In summary, when writing test questions, consider the reading difficulties of students. Consider such factors such as vocabulary, sentence complexity, the ability of students to understand the question, and the use of diagrams and illustrations to make the questions clearer. Factors that may be helpful in rating test questions, particularly multiple-choice types, are listed in figure 18–2, "Test Item Readability Checklist" by Rakow and Gee (1987). This list can be used to rate a question to determine its suitability for a particular group of students. According to Rakow and Gee (1987), teachers who use the checklist

FIGURE 18–2 Test item readability checklist.
From "Test Science, Not Reading" by S. J. Rakow and T. C. Gee, 1987, *The Science Teacher, 54*(2), p. 28. Reprinted by permission.

Rate the questions using the following system:
5—Excellent
4—Good
3—Adequate
2—Poor
1—Unacceptable
NA—Not applicable

_____ 1. Students would likely have the experiences and prior knowledge necessary to understand what the question calls for.

_____ 2. The vocabulary is appropriate for the intended grade level.

_____ 3. Sentence complexity is appropriate for the intended grade level.

_____ 4. Definitions and examples are clear and understandable.

_____ 5. The required reasoning skills are appropriate for the students' cognitive level.

_____ 6. Relationships are made clear through precise, logical connectives.

_____ 7. Content within items is clearly organized.

_____ 8. Graphs, illustrations, and other graphic aids facilitate comprehension.

_____ 9. The questions are clearly framed.

_____ 10. The content of items is of interest to the intended audience.

2. The format of the test is organized, not confusing. Test items of the same type are grouped together, and the directions for answering the type of question are clear.

3. Test items are arranged so that the easiest questions can be answered first and the difficult questions later.

4. Test items are written in language familiar to students.

5. Technical words and scientific vocabulary used in test items have been presented during the course of instruction.

6. Tricky and absurd questions that would make the test less reliable have been avoided.

7. Several test items are used to test each objective to avoid an unreliable assessment of student achievement.

8. Each test item is valid and directly relates to the objective it is supposed to assess.

9. Teaching objectives are tested according to their importance. More questions are directed at important objectives and fewer aimed at those of less importance.

10. To make the test valid, all objectives are assessed. No objectives have been omitted.

11. All information and materials that the students need to complete the test are provided. Tables, reference charts, formulas, and other useful information are part of the test package.

12. Grammatical and spelling errors are eliminated.

13. Suggested length limits or time limits are clear, particularly on unrestricted essay questions.

to examine test items can reduce the reading difficulties that may be preventing students from demonstrating their science knowledge.

Carefully read and review the test before producing a final draft. Examine its contents to determine whether the following points apply:

1. Directions are clearly and concisely written in language familiar to the students. Avoid lengthy directions that take up time that could be used in answering questions.

After checking the test, the teacher should actually take the test and examine the questions for clarity and lack of ambiguity. By taking the test, the teacher can also judge its length and whether the students will have ample time to complete it during the allotted time. Shorten the test if necessary, but be careful not to omit questions that will ensure the proper assessment of the objectives.

ADMINISTERING TESTS

Just as constructing a well-written test is crucial, administering it properly is also important.

The best time to give a test is at the beginning of a class period. This will ensure that there is enough time available for all students to complete the test. It is also a favorable time because the students have moved about and relaxed during the interval between classes, placing them in the best situation to begin concentrating. A number of problems can be created when a test is administered toward the end of a period. Students are apt to be restless and nervous in anticipation of taking the test and will not concentrate during the period of time before the test is administered. Also, if the test is given toward the end of the period, teachers should be certain that there is enough time allotted to take it. This means that they must restrict the number of questions to ensure that the students have ample time to complete the test before the period ends.

The physical conditions under which a test is given should be of primary concern. Lighting should be adequate, and the temperature should be comfortable. The room should be neither hot nor cold, for both conditions may affect student performances on the test. It may be desirable to ventilate the room between class periods to provide the proper environment.

The test should be administered in an efficient and businesslike manner. The directions, if given orally, should be clear and concise. Written directions should also be clearly stated so that all students understand what is wanted. Any questions about the directions should be cleared up before the test begins.

Teachers should avoid interruptions during the test unless it is evident that an explanation is necessary to make the directions understanda-

A test should be administered in a professional and business-like manner. If proper seating arrangements are not made, cheating and other problems may arise during the course of the test.

ble. Such interruptions should be kept to a minimum. Errors in the wording of questions should be corrected at the same time directions are given. Students who have problems regarding legibility or possible misinterpretations can be helped through individual consultation.

Cheating is always a problem. The teacher must be alert to the problems that cause cheating and the conditions that promote it. Tensions build up when a great deal of emphasis is placed on the outcome of a particular test. The use of frequent tests, which gives the students an opportunity to practice their test-taking abilities, will build up their confidence so that they will be able to take more important examinations with ease and not resort to cheating.

Cheating also can be discouraged by using special seating arrangements. If possible, students should be seated in alternate rows, with empty rows in between. For major tests, some teachers prepare two forms of the test, using the same questions for both tests but scrambling the order of the questions. The two forms of the test are distributed so that every other row of students is given one of the forms.

Cheating also can be controlled and minimized if the teacher moves about the classroom and spends most of the time observing the students from the rear of the room. Teachers should avoid sitting at their desks, leaving the room, or doing chores that are not concerned with the administration of the test. The teacher who is not observing students during a test is encouraging them to cheat.

SCORING, GRADING, AND ANALYZING TESTS

Scoring, grading, and analyzing test results are important parts of the testing process.

Students benefit from receiving test results immediately, when they can still remember why they made certain responses. In some cases they can score their own papers. From an administra-

Grading a large number of test papers can be a burdensome task. The teacher who constructs a test without considering the time required for grading is not very efficient. Science teachers should find ways to simplify the grading task while constructing the test.

tive point of view, the tests that best lend themselves to scoring by students are those that call for single, positive responses, such as multiple choice, matching, and true-false.

Students should not grade tests with questions permitting alternate responses, such as completion and essay questions. If students are involved in grading such questions the teacher may need to make certain judgments regarding alternate acceptable answers.

Grading a large number of test papers can be a burdensome task. The teacher who constructs a test without considering the time required for grading is not very efficient. Teachers should look for ways to simplify the task while constructing the test.

Essay questions are not only difficult to grade but time consuming to read. A teacher's objectivity deteriorates after reading twenty or thirty essay responses. To improve objectivity, it is advisable to prepare a checklist of the points the students are expected to emphasize in their discussions. The teacher should use the checklist with discretion, remembering that students may not have thought of all the points and probably would have discussed them adequately had the points been brought to their attention.

Multiple-choice, fill-in-the-blank, and other short-answer tests can be scored easily by using special keys. For example, a matching-type question can be scored quickly if the student writes the response in a blank opposite the item with which it is to be matched.

The time for scoring multiple-choice examinations can be reduced by using an answer sheet, which requires the student to block out the symbol that corresponds to the correct choice given in the question. A key is made from an answer sheet by cutting or punching holes to correspond to the correct responses (figure 18–3). The key is used by placing it over the student's answer sheet so that correct answers can be seen through the holes in the key.

Mechanical scoring is becoming much more common in secondary schools, and in some schools the Scantron is available, which will score and provide item analyses instantaneously. It will provide difficulty or discrimination indexes for each item. This type of grading prevents the teacher from determining what types of errors or misconceptions the student may have had about a concept or principle, or the reasons why particular questions may have been missed by all students, or why all students answered some questions correctly. The procedures described later for analyzing tests will help in identifying the strengths and weaknesses of a given testing procedure.

Teachers must rely on their own judgment when determining the amount of credit that will be given for responding correctly to the various types of questions in a test. Usually, questions

FIGURE 18–3 A portion of an answer sheet from a multiple-choice test, correctly answered, and its key. The key is prepared by cutting out the correct choices from a blank answer sheet. When the key is placed over the student's answer sheet, the student's correct answers will show through the holes in the key.

Key   Answer
sheet

that require longer answers and originality and a high degree of thought will receive more credit than those requiring recall or recognition. Accordingly, the teacher might give one credit for multiple-choice questions and two credits for completion questions; a one-sentence response might be worth five credits, while a one- or two-paragraph essay could be worth ten credits or more. There is no rule of thumb for credit distribution. Teachers must be fair and sensible when determining the values for the various types of responses.

True-false tests and other either/or types of questions should be graded in such a way as to penalize a student for guessing. Theoretically, a student should be able to answer half of a set of true-false items correctly without reading the questions or without having any knowledge of the subject. Some teachers believe that a true-false item should not be given the same relative weight on a test as a multiple-choice item and that certain factors should be applied to reduce the weight of true-false items. For example, a correct answer might earn one credit, an incorrect answer might delete one credit, and no response (no-guess), zero credit. In general, it is probably best to use true-false questions to a moderate degree in testing.

Teachers assign credit using various possible combinations when grading a test. Some assign credits in units of 10 to simplify calculations. This is not necessarily the best or easiest procedure. Other combinations are probably more desirable. Test credits might be apportioned as follows:

Test Items (and Credits)	Possible Credits
15 multiple choice items (2 credits each)	30
6 completion questions (3 credits each)	18
3 one-sentence responses (5 credits each)	15
1 essay question (10 credits each)	10
Total	73

The test grade in this case can be calculated by taking the number of credits earned and dividing it by the total number of possible points. If a student earned 60 points on this test, the grade would be determined by dividing 73 into 60, or 82%. Another test with the following distribution can be handled in another way:

Test Items (and Credits)	Possible Credits
12 multiple-choice items (1 credit each)	12
5 completion items (2 credits each)	10
1 one-sentence-answer question (3 credits)	3
Total	25

The test grade in this case can be converted to a percentage by multiplying the number of

points earned by 4. Thus, a student who has earned 22 points will receive a grade of 88%.

Letter grades are appropriate to use when scoring essay questions. However, it is then necessary to convert points earned on short-answer items to a letter grade, or vice versa. The method often used to make this conversion is to give an A for anything between 90% and 100%, a B for anything between 80% and 89%, a C for anything between 70% and 79%, and so on.

The teacher may prefer to record the point scores on a test rather than a percentage or letter grade. Doing this has the advantage of using tests having different weights in determining a final grade. Because students sometimes do not understand how a grade is determined using point scores, it is good practice to convert scores to percentages or letter grades before papers are returned.

When grading tests that include both objective and essay responses, it is advisable for the teacher to score all of the short-answer questions first for *all* students before grading the long-response questions. This will make grading consistent. Long-response questions should be graded by reading all the responses each student has written for one particular question before grading the next essay question. When grading long-response questions, the teacher should have available a list of the points she believes should be emphasized in the response and the number of credits allowed for each point. Subjective judgments must be made when grading essay responses. Anything that teachers can devise to make the judgments more objective will make grading fairer and more consistent.

Teachers can present test grades to students in a number of ways. They can tell them the raw score earned on a test, percentage score, letter grade, or all three. When raw scores are given, the student wants to know how to translate the score to either a percentage or letter grade. The teacher may want to determine the cutoff scores to represent the letter grades. For example, take the maximum number of points, say, 50, and multiply this number by 0.9 to obtain the lower limit of the A range. The A range would thus be between 45 and 50 points. To obtain the lower limit of the B range, multiply 50 by 0.8, which is 40 points; anything between 40 and 44 points is the range for a B grade. The C range would be obtained by multiplying 50 by 0.7, which is 35, the lower limit of C, so that anything between 35 and 39 would be a C, and so on. The students can determine their letter grade by seeing in which category their score falls.

Criterion-Referenced Grading

In criterion-referenced grading, student achievement is judged relative to performance against an established set of criteria for grading, rather than performance relative to other students. The student is given the percentage score received on the test regardless of his relative rank with the rest of the class. For example, if the total test is worth 100 points, the student must earn 90 to 100 points to receive an A, 80 to 89 for a B, 70 to 79 for a C, and 60 to 69 for a D; an F would be received for scores below 60 points.

In most cases, these grades are recorded as is and not adjusted in any way. The teacher will allow as many As, Bs, Cs, Ds, and Fs as students earn. If the raw score for a test differs from 100 points, students should learn how to interpret their grades when they are given a raw score based on the total number of points. Students can easily accommodate to accepting raw scores for a test grade, provided the scores are easily interpreted.

Criterion-referenced grading is also used in mastery learning. In this case students are expected to attain a certain level of performance. If this level of performance is not met, they are given the chance to receive remedial help to reach the objectives. For example, if only 80% of the objectives are met as indicated by a grade of 80% on a test in which 90% is expected, then students are given remedial work to help attain the minimum of 90% when the next test is administered.

Norm-Referenced Grading

Grading on a Curve. Another procedure that some teachers use to grade test papers is to base the grades on a normal curve. In this grading procedure a student's grade depends on how well he has performed with respect to other members of the class. When the teacher uses this system, she has in mind a predetermined percentage of students who are to receive A, B, C, D, and F grades. The procedure assumes that students in a typical class can be categorized in a normal distribution.

The normal distribution curve is a mathematically defined theoretical distribution. The curve is symmetrical, with most of the scores in the center of the distribution and the number of scores decreasing as either extreme is approached. In a normal distribution curve, the mean, the median, and the mode are the same (figure 18–4).

When using a normal curve, the mean, standard deviation, and Z score are calculated, and grades of A, B, C, D and F can be determined based on the number of standard deviations the Z score is from the mean. (Refer to a standard textbook in statistics or tests and measurements for use of standard deviation and Z scores.)

In a normal distribution, the percentage of individuals who fall between different Z scores is constant. About 34% of the individuals will fall between the mean and +1 standard deviation, whereas 34% will fall between the mean and −1 standard deviation. About 14% fall between Z scores of +1 and +2 standard deviations above the mean, and between −1 and −2 standard deviations below the mean. Most of the scores will fall between +3 and −3 standard deviations above and below the mean in a normal curve. Assigning grades based strictly on this theoretical curve allows the teacher to assign 68% C grades, 14% B grades, 14% D grades, 2% A grades, and 2% F grades.

This type of grading system has been severely criticized, particularly when teachers use it to as-

FIGURE 18–4   The characteristics of the normal curve

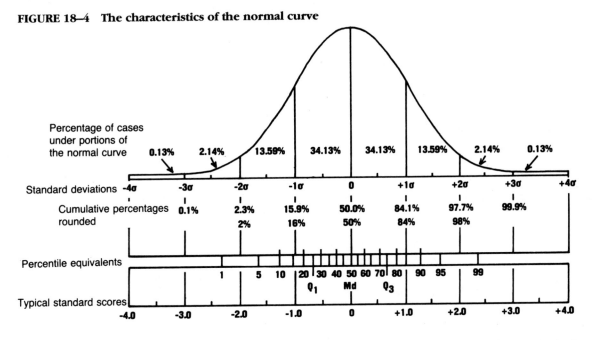

sess small groups of students. The size of classes in which the procedure is used is often too small to expect a normal distribution. Another criticism is that teachers can seldom produce tests that yield normally distributed scores. Where students are homogeneous, with similar aptitudes, a normal curve does not apply. This system of grading is also inappropriate when one is teaching by objectives and using criterion-referenced grading systems based on attainment of the objectives.

Teachers should use care in assigning grades based on a normal distribution. This method can be used for large groups, but is not necessarily recommended, and should certainly be avoided in small groups or when students are homogeneously grouped. It should not be used to cover up poor achievement resulting from inadequate instruction. The cause of the poor achievement should be found, and proper steps taken to rectify the situation. Poor achievement on a test certainly might be attributed to poor instruction, but the test itself may be the cause of the problem.

The Inspection Method.   Another method of grading that allows teachers flexibility in assigning grades is the inspection method. A frequency distribution is produced after the raw test scores have been determined. The grades are determined as follows:

1. Determine the median and give C grades to the large group of scores around it.
2. Determine the 25th and 75th percentiles to establish the approximate limits of the C group.
3. B grades are given to students above the C group, and Ds are given to those below the C group.

Certain characteristics of the test will be evident once the frequency distribution is set up. For example, if the test is too difficult, the scores will be skewed at the lower end of the curve; if the test is too easy, they will be collected at the upper end of the distribution. If the test does a reasonable job of assessing achievement, a distribution approximating a normal curve will be evident. If all students perform well, it is not nec-

FIGURE 18–5   Frequency distribution involving test scores for 40 students. A, B, C, D, and F grades are assigned by inspection

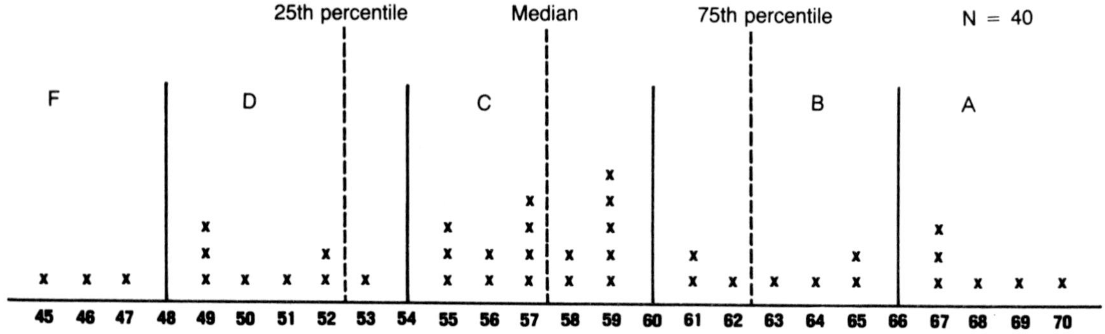

essary to assign F or even D grades. A grade of A may not be awarded if no one performs sufficiently well (figure 18–5).

In summary, grading, regardless of the system used, is always a problem for the teacher. Students are concerned about their grades and want to know how they will be graded. Students often feel that they are treated unfairly when they are given grades, so it is important that the teacher inform them of the system to be used and why the system is fair. TenBrink (1986) states, "Assigning grades has forever been a task teachers dislike. There seems to be no 'fair' way to do it and grading system seems to be subject to all kinds of interpretation problems."

CONSTRUCTING AND PLANNING TESTS

Tests are constructed for various purposes. Some are used for grading, others for diagnostic purposes. They can be given at the beginning or end of a unit, lesson, or course or at other times for different purposes.

Sometimes a single test may be used for more than one purpose. A diagnostic test may be given at the beginning of a lesson, unit, or course (pretest) to determine the level of student knowledge. The results of a test can tell the teacher where instruction should begin. The teacher may use this same diagnostic test at the end of a lesson, unit, or course to determine student progress (posttest).

Tests are also frequently administered to measure student achievement. Achievement can be measured by comparing student scores with predetermined criteria (criterion-referenced tests) or by comparing student scores with one another (norm-referenced tests). Criterion-referenced tests should be used in traditional approaches, as well as in individualized and mastery learning approaches. Norm-referenced tests are most commonly used in association with "grading on the curve."

Current emphasis in science teaching involves teaching process skills rather than teaching content for the sake of content. Tests can be constructed to assess these skills. Typically, these tests involve questions at the upper levels of Bloom's taxonomy—that is, application, analysis, synthesis, and evaluation.

Diagnostic Tests

Diagnostic tests, which are sometimes called *pretests,* are administered at the beginning of a course, unit, or lesson. The information obtained from administering such tests tells the teacher where instruction should begin.

Such tests are used to determine the extent to which students understand concepts and principles. They should not be given just to assess mastery of facts and technical vocabulary. Typically, objective or short answer questions are used in diagnostic tests because they permit the assessment of a wide range of material in a short period of time.

Careful construction of the pretest is essential. It is especially important to remember that students at the beginning of a course, unit, or lesson usually lack the technical vocabulary associated with a new topic. Tests that assess mastery of facts and vocabulary will not disclose the extent to which a student understands a concept or principle. A student could very well have an adequate understanding of the concept or principle, but a pretest based on facts and vocabulary would not disclose such information.

Pretests should be as short as possible to prevent students from being discouraged. Students who know little or nothing about a topic on a pretest could easily become frustrated and refuse to complete the test. A lengthy pretest consisting of many items that students cannot answer may also convince some students that they lack the background to understand the topic and that the topic is unusually difficult. Consequently, they may become discouraged and lack motivation to study the topic. On the other hand, students who have a considerable background may become bored if the test is too long and may consider the experience a waste of time.

Diagnostic tests should have a spiral organization. The simplest questions should be asked first and the more difficult questions relegated to the later portions of the test.

Figure 18–6 contains an example of a diagnostic test used to open a unit on the senses for a ninth-grade life sciences course. Notice the spiral organization of the items, beginning with topics commonly observed by most individuals and ending with information obtained from reference books or previous classroom experiences.

FIGURE 18–6 Pretest for ninth graders

The following statements are either true or false. Place a T in the blank to the left of the statement if the answer is true. If the answer is false, place an F in the blank to the left of the question.

____ 1. Most knowledge is gained through the eyes.
____ 2. Most people have either brown eyes or blue eyes.
____ 3. There is one eyelid for each eye.
____ 4. Tears wash dirt from the eyes.
____ 5. Only one eye at a time is used when looking at a nearby object.
____ 6. Sunglasses reduce the amount of light entering the eye.
____ 7. Objects can be seen more clearly by day than by night.
____ 8. The eyelids stay open for only a few seconds at a time.
____ 9. There are always tears in the eyes.
____ 10. Some eyeglasses make objects look larger.
____ 11. Colors of objects can be determined by starlight.
____ 12. A person can see clearly all objects in front of him without moving his eyes.
____ 13. Some eyeglasses help a person see objects that are far away more clearly.
____ 14. The black spot in the center of the eye becomes smaller in the dark.
____ 15. Both eyes always move in the same direction when they turn.
____ 16. The use of two eyes at once helps a person judge distance.
____ 17. Light enters the eye through the black spot in the center.
____ 18. There is a lens in the eye.
____ 19. The black spot in the center of the eye is called the pupil.
____ 20. The colored part of the eye is called the iris.

FIGURE 18–7   Mastery test for unit on The Compound Microscope and Biological Research

In order to indicate achievement of mastery of the material, you must correctly answer a minimum of 20 (85%) questions on this test of 24 questions.

1. Draw a diagram of the human cheek cell below: (The circle defines the field of view of the microscope.)

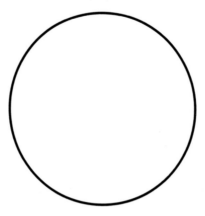

2. What is the approximate width of the above cell? (_____ microns)
3. Draw a diagram of the onion skin cell below. (The circle defines the field of view of the microscope.)

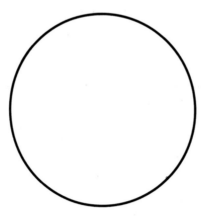

4. What is the approximate length of the onion skin cell? (_____ microns)
5. Which cells are more uniform in shape, plant cells or animal cells? (_____ cells)
6. Which cells that you have observed most closely resemble cork cells? (_____)
7. A micron is _____ of a millimeter.
8. List the steps you would use in computing the diameter of the high power field.

FIGURE 18–7   *continued*

Complete the following chart by making the appropriate computations. For example, column I shows a low-power diameter of 1.3 millimeters. Your task is to compute the equivalent low-power diameter (10X) in microns and place your result in space no. 9. Furthermore, you are also to calculate the high-power (43X) diameter for the low-power diameter (1.3) in microns and place your answer in space no. 10. Follow the same procedure using what is given in the various columns to compute what is required in other columns. Fill in all blanks.

I Low-power diameter in millimeters	II Low power (10X) diameter in microns	III High-power (43X) diameter in microns
1.3	9.	10.
11.	12.	350
2.4	13.	14.
3.6	15.	16.
5.0	17.	18.
19.	1,300	20.
1.0	21.	22.
7.0	23.	24.

Commercially developed standardized tests can be adapted and used for diagnostic purposes. The uses of such tests in the science program are discussed in Chapter 17.

Test for Mastery

Tests for mastery are used in self-paced or mastery learning approaches. In either approach, students are expected to achieve a predetermined level of mastery of a topic (the level of mastery is determined by the teacher).

Tests for mastery include test items derived directly from the instructional objectives (behavioral objectives) specified for the topic. Although the level of mastery required of the students varies among teachers, a criterion ranging from 80% to 90% is typically used by teachers before the student is permitted to go to the next topic;

some teachers even require 100% attainment of objectives.

Figure 18–7 shows a mastery test developed for a unit entitled The Compound Microscope and Biological Research. Behavioral objectives were listed for the unit. Upon completion of the unit, the student will be able to

1. compute the diameter (in microns) of the field of view of a microscope on both low and high power
2. compute the diameter, length, and width of various microorganisms and particles
3. determine the major structural differences between plant and animal cells
4. use the microscope to arrive at the size of microscopic organisms
5. use the microscope to determine the structure of a microscopic organism

The mastery test that was administered consisted of 24 questions. The students had to correctly answer a minimum of 20 questions, or attain a mastery level of at least 85%, before they could proceed to the next topic.

CONSTRUCTING A UNIT TEST

Before writing a unit test, list the objectives of the unit and then determine the type of questions needed to measure the attainment of each objective. Norm-referenced testing, which is the type of testing observed in most science classrooms, will probably involve questions primarily in the cognitive domain of Bloom's taxonomy. These can include questions that test recall of knowledge, problem solving, reasoning, attainment of concepts, principles, and critical thinking. Objectives that are categorized in the cognitive domain are arranged in six divisions: knowledge, comprehension, application, analysis, synthesis, and evaluation. Knowledge and comprehension are considered the first and second levels of learning in the taxonomy, and these are the ones that teachers test frequently. Of the other four divisions, the ability to apply knowledge is often tested, but analysis, synthesis, and evaluation are less often used by teachers when constructing unit tests. A unit test should contain questions for all levels of the cognitive domain whenever possible.

Unit objectives are met by the content and activities to which the students are exposed during the course of instruction. The unit objectives may be broad by nature and are therefore met by a wide range of instructional objectives, usually in behavioral form, which are more specific. If instructional objectives are written at various cognitive levels, the teacher is obligated to include questions in a unit test that assess a wide range of cognitive levels.

The number of questions in a unit test and the cognitive levels of the questions are significantly affected by time constraints. For example, synthesis and evaluation questions require more time to answer than simple recall or knowledge level questions. A test made up of questions that test recall and comprehension can include many more questions than one that is made up predominantly of analysis and synthesis questions. The teacher should attempt to strike a balance when constructing a unit test and not stress only recall and comprehension questions.

The number of questions used to test whether an objective has been satisfied will depend on the emphasis placed on the objective during instruction. Some objectives can be tested by one or two test items; others may require four or five.

A table of specifications such as the one illustrated in figure 18–8 shows how a semester test can be planned around the content. The content areas are listed in the left column; the various cognitive levels are named in the column heads at right. The teacher then determines from the lesson and unit plans the amount of time that has been devoted to the objective. Based on this time, the number of questions allotted to each objective can be determined. Other points to consider when constructing a unit test are

1. the amount of class time that can be devoted to the test
2. an estimate of the amount of time needed to answer recall and comprehension questions once they have been constructed
3. an estimate of the amount of time required to answer other types of questions at the synthesis, analysis, and evaluation levels
4. the total of the estimates, which determines whether enough time is available to complete the entire test. (Students should be given ample time to answer all questions. They should be given enough time to answer synthesis, analysis, and evaluation questions without feeling rushed.)

The next step is to start to write questions at all cognitive levels, placing each question on a three-by-five-inch card. The questions can be derived directly from the instructional objectives

FIGURE 18–8   Classification of questions for a final examination on cells and cell functions

Content	Knowledge	Comprehension	Application	Analysis	Synthesis	Evaluation
Life functions of living things	IIII	III	III	I		I
Classification of living things	卌 II	卌 II	II			
Structure of cell and functions	卌 II	III	III	II		
Major biochemical compounds	II	卌 I	II	I	I	
Metabolic reactions	卌 I	II	卌 II		I	I
Tools and techniques used for cell study	卌 I					

specified for the unit. Once the questions have been formulated for each objective, they should be classified in terms of cognitive levels. Using a chart similar to that shown in figure 18–8, tally the number of questions for each cognitive level for each content area. If there are too many questions in one category and not enough in another, adjust them for a more even distribution. Once the number of questions has been identified for each objective, check whether the number of questions for each objective is proportional to the importance of that objective and the amount of instructional time devoted to it.

Once the number of questions has been determined, plan the format of the test. Usually, short-answer questions involving recall and comprehension are presented in order of difficulty. Some teachers prefer to cluster questions around objectives rather than to randomly distribute them throughout the whole examination. Short-answer, multiple-choice, and true-false questions are usually presented early in the test, and more time-consuming essay questions are left for later. No specific rules govern the sequence of questions.

After the test has been constructed, estimate the amount of time required to answer all questions. If there are too many questions, reduce the number so students can take the test comfortably. Again, there is no rule governing the number of questions a teacher should ask on a test. Usually, about one minute or less should be given to answer objective questions. Five minutes or more should be estimated for analysis, synthesis, and evaluation items.

SUGGESTED ACTIVITIES

1. Make up several examples of different types of test items based on a set of instructional objectives for a unit of study. Discuss these with other members of your science methods

contributions are the most convenient way to exchange ideas with other science teachers.

Directions for submitting manuscripts to a professional journal can generally be found in any issue of the journal. Each journal has a specific format that must be followed, including special requirements for drawings, diagrams, and number and size of photographs. If specifics regarding format are not available in the journal, they can be obtained by requesting them from the journal editor.

OTHER ACTIVITIES FOR PROFESSIONAL GROWTH

Communication with Other Science Teachers

One of the most valuable sources of information and inspiration for beginning science teachers is the experienced teacher. Experienced teachers can be quite helpful in selecting objectives, methods, and materials, and in brainstorming problems that the beginning teacher has encountered in the classroom. A great deal of help can be obtained by asking the experienced teacher to critique and evaluate the teaching performance of the inexperienced one in an actual class situation or during a peer teaching session without students.

Even well-qualified teachers can benefit from contacts with fellow teachers, including those who do not teach science. Often teachers who are not in the field of science can help science teachers see the relationships of science to other disciplines. These contacts serve to better coordinate the various offerings included in the school curriculum.

Science teachers should not restrict their professional contacts to teachers within their own school; they should interact with teachers in other schools in their district and other school districts through visitations, conferences, or special meetings. Just as with professional meetings, such contacts are excellent ways to exchange teaching ideas and discuss mutual problems. A great deal of helpful information and insight can be gained through such contacts.

Travel

Science teachers can use travel to gather much information, experience, and material that will help them improve their teaching. A camera, preferably 35 mm, can be used to take photographs for use in the classroom. Advance planning is needed to obtain the greatest benefit from travel. Tourist bureaus, chambers of commerce, and travel agencies can provide information and literature about the places to be visited. Visits within the United States provide great opportunities to collect plant and animal specimens for biology teaching. Care must be taken, however, that no laws are broken when live materials are transported from state to state. Animals and plants can be studied in their natural habitats. Visits to museums give teachers many new ideas concerning exhibits and collections. Visits to local industries and places of geological importance and general scientific interest can be enhanced by taking photographs, obtaining samples, and collecting descriptive literature. A science teacher will acquire a great deal of information and material for enriching his teaching without detracting from personal enjoyment of the trip.

Summer Employment

Science teachers can contribute to their professional growth through science-related summer employment. Science teachers should take advantage of summer employment in industry if and when such employment becomes available. Such jobs enable teachers to gain information concerning the application of science in industry. Employment in laboratories, for example, can be fruitful in the development of skills and understandings of research methods for use in science instruction.

Science departments in colleges and universities often offer opportunities for science teachers to work as research assistants during the summer months. Working with university professors involved in research will help further develop a science teacher's laboratory and research skills. Such skills and the attitudes developed toward research can assist teachers in stimulating students enrolled in their science courses.

Hobbies

Teachers can enrich their science courses by having hobbies related to their teaching areas. For example, some science teachers are ardent entomologists or ornithologists. Some are excellent water biologists. It is not uncommon to find a chemistry teacher who is an avid bird watcher or a biology teacher who collects rocks. Physics teachers frequently are amateur astronomers or amateur radio operators. Such hobbies not only help enrich the courses, but also stimulate students to undertake science-related projects.

Research

Some science teachers have excellent in-depth knowledge in a science area, which qualifies them to do research. They may have advanced degrees in a science area or a great deal of experience in research through previous employment. Teachers with such backgrounds should be involved in their own research if time permits.

Teachers who think they do not have the space, equipment, or materials to do research should solicit the help of a nearby university. University professors will often cooperate with teachers on a research project by providing space and equipment. Teachers should not overlook this source of aid if they are at all interested in pursuing a research project.

Students who know that teachers are actively engaged in a research project may be stimulated to work on projects of their own. If teachers report their findings to the students, they may discover that many of them will become interested in participating in the research project.

High school teachers who publish their research in recognized journals receive recognition from their colleagues as well as their students. These teachers are often regarded as specialists and authorities in their science area. Members of the community will also recognize them as outstanding teachers and scientists. They are recognized as the teachers under whom parents would like their children to study.

Leaves of Absence

Science teachers should take advantage of leaves of absence or sabbatical leaves if these options are available to them. The teacher can use such extended leaves to update his science teaching methods and science content, or to work with a practicing scientist at a university on a research project, obtain an advanced degree to fulfill certification requirements, or acquire a background in a science area with which the teacher has had no previous experience. Generally, leaves are for a specified period of time such as one semester, one year, or longer, with full or half pay. The time period and pay have to be negotiated with the school authorities approximately a year in advance of the leave so they can take into account the financial aspects of the leave when they plan the school budget. They also need this lead time to recruit a replacement for the teacher taking the leave.

Keeping Up-to-Date Through Reading

A vast amount of valuable and pertinent literature is available to science teachers to keep abreast of new developments in science, technology, and science education. Teachers should devote time each week to reading for professional growth and development. This means that science teachers should either maintain a professional library or have one available to them in the school setting. Easy access to professional literature will encourage reading. Generally, if materials are not easily available, teachers will not read on a regular basis.

Science teachers need to have a professional library of science education journals and books on science teaching methods and content available to them. Books on the history and philosophy of science and books that contain collections of demonstrations and laboratory activities and other types of science activities can be very useful when planning units and courses of study. Curriculum materials including unit plans, courses of study, and syllabi prepared by state and local education departments and other science teachers should be collected for the professional library for future reference. Professional science education journals that are read on a regular basis can keep teachers up-to-date on recent developments in science teaching methods as well as provide suggestions for teaching activities. Articles in these journals are usually written by classroom science teachers and include many activities and suggestions for teaching science that have been tried in actual classrooms.

Periodicals that focus on recent developments in science and technology, such as *Scientific American, Science Digest, National Geographic, Discover, Omni,* and *Science* can keep teachers current in various areas of science and technology. Newspapers such as the *New York Times* and periodicals such as *Time* and *Newsweek* have regular features on various scientific topics. These valuable sources of information help enrich science lessons and science instruction in general.

Publishers are currently increasing the number of books on science-related topics. Science teachers should take the opportunity to benefit from reading such books in order to update themselves and increase their science background.

College textbooks are also excellent sources of information to help enrich the science background of the teacher. College science textbooks are revised periodically to keep the information current. College textbooks used during a teacher's college preparation are likely to be outdated and should not be used as a definitive source of science information.

Maintaining an up-to-date library can be a very expensive enterprise for the science teacher to undertake. If cost is a problem, then the school should be asked to maintain a professional library of books and journals that are useful in enriching and updating a science teacher's background in science and science education.

EVALUATING TEACHER PERFORMANCE IN THE CLASSROOM

Continuous evaluation of a teacher's classroom performance is essential for professional growth. The evaluation should consider teachers' performance in light of the goals they have set for themselves and the goals and instructional objectives that have been identified for a course of study.

Videotaping a science lesson is the simplest and most direct way to determine what is actually taking place in the classroom. Teachers can then assess their performance by viewing these tapes and asking themselves a number of questions such as, "How much of the class time do I dominate?" "Do I interact with all students?" "How do I react to an unexpected question or answer?" "How do I respond to an incorrect answer from students?" "At what cognitive level is my lesson presented?" "Are my questions thought-provoking?" "Do the answers to my questions indicate that the students are thinking?" "How much original thinking is done by the students?" These and other questions can be answered by analyzing a videotape recording subjectively.

Who should analyze such recordings? Self-analysis of the tape may not be the best way of obtaining information, because a teacher may overlook certain behaviors that, if pointed out and changed, would improve teaching performance. Using other teachers to evaluate teaching performance will solicit feedback from a different perspective. This type of feedback can very often give rise to constructive criticism and suggestions, which may help improve instruction. In short, the analysis of videotapes made by one's peers may point out important observations that might otherwise have gone unnoticed.

Suggestions offered up to this point for analysis of classroom behavior contain little or no structure and therefore can give only a general subjective feeling about teacher performance in the classroom.

Another way of obtaining information about teacher performance in a classroom is by asking students in a class to complete a questionnaire. The questionnaire is usually prepared by a teacher, a committee of teachers, or a committee of teachers and administrators. Typical questionnaires contain statements that are positively stated and associated with a scale (e.g., from one to five), which is used to rate the teacher's skill, performance, or personality traits. The students in the class are asked to complete the questionnaire with the assurance that their grades will not be affected by the results obtained. Students who complete the questionnaire are asked to remain anonymous. A sample questionnaire of this type can be found in figure 19–1.

An instrument called the "Teacher Credibility Questionnaire (TCQ)" (House & Lapan, 1978) is available and can be given to a class to determine a teacher's credibility. The results of the questionnaire can be compared with the data scores of 68 teachers whose students filled out the questionnaire. The questionnaire is very useful as a basis for discussion with students, supervisors, and other teachers.

Included in the National Science Teachers Association publications, *Guidelines for Self-Assessment of High School Science Programs* and *Guidelines for Self-Assessment of Middle/Junior High School Science Programs* (NSTA, 1989a, 1989b), are modules that can assist in the evaluation of science teachers. Each set, that is the set for high school science programs and the set for middle/junior high school science programs, contains questionnaires that can be used in the evaluation of science teachers. Module 3 of each series, entitled *Science Student-Teacher Interac-*

tions in Our School, can be used to assess the interpersonal relationship between teachers and students. Parts are devoted to quality of student-teacher relationships and teacher-student perceptions of teaching practices. They are both excellent instruments that can be used by the teacher and students to assess teacher quality and effectiveness. The instruments can be modified or used as guides to construct new questionnaires to meet each teacher's situation.

Module 2 in the high school series, entitled *Our School's Science Teachers,* is designed to evaluate (1) the teacher's science background and general education, (2) the teacher's professional education background, (3) the teacher's professional activities and development, (4) the teacher's contributions to the profession, (5) the teacher's attitudes, (6) student and teacher perceptions of teacher professionalism, and (7) teacher recruitment and selection policies.

Module 2 of the middle/junior high school series, also entitled *Our School's Science Teachers,* is designed to assess (1) the teacher's science background and general education, (2) the teacher's professional education background, (3) the teacher's professional activities and development, (4) the teacher's contribution to the profession, (5) the teacher's attitudes, (6) teacher recruitment and selection policies, and (7) school and community relations, as well as community involvement in the school program. This questionnaire also can be modified or used as is.

Information and feedback obtained through questionnaires, either teacher-made or published, that are not standardized can be the basis for discussion with other students, other science teachers, and administrators. The teacher can also make a self-analysis of performance by examining the results of the questionnaire.

It is essential that teachers make periodic assessments of their own performance in the classroom. This is certainly one way to obtain feedback to help improve instruction and evaluate performance based on instructional goals and objectives.

FIGURE 19–1 Example of a questionnaire to assess teacher effectiveness

The following statements concern the science teacher and what is happening during science lessons. Circle the appropriate number following each statement based on the scale below:

1 Almost always
2 Frequently
3 Sometimes
4 Once in a while
5 Not at all

Circle only one number for each statement. Example:

Science lessons are well organized. 1 2 3 4 ⑤

By circling number five, you believe that your science teacher does not have well-organized science lessons at any time. If you circled number one, you believe that your science teacher almost always has well-organized science lessons.

1. The teacher is interested in students.	1	2	3	4	5
2. The teacher knows the subject matter well.	1	2	3	4	5
3. The teacher is enthusiastic about the subject.	1	2	3	4	5
4. The teacher uses several types of activities during a class period.	1	2	3	4	5
5. The teacher often uses films, slides, and filmstrips with science lessons.	1	2	3	4	5
6. The teacher uses examples from everyday living whenever possible.	1	2	3	4	5
7. The teacher uses familiar objects to teach science lessons.	1	2	3	4	5
8. Lessons are well prepared and organized.	1	2	3	4	5
9. The teacher makes the objectives of a lesson clear.	1	2	3	4	5
10. The lessons are presented so that they are interesting.	1	2	3	4	5
11. The teacher makes the students think.	1	2	3	4	5
12. The teacher can explain concepts and principles so that they can be understood.	1	2	3	4	5
13. The teacher asks good questions.	1	2	3	4	5
14. The teacher makes good use of class participation.	1	2	3	4	5
15. The teacher uses the chalkboard frequently and writes all scientific words on the board as they are used.	1	2	3	4	5
16. The teacher gives the students a chance to do projects outside the classroom.	1	2	3	4	5
17. The teacher makes good and interesting homework assignments.	1	2	3	4	5
18. The teacher writes good, clear tests.	1	2	3	4	5
19. The tests are fair	1	2	3	4	5
20. The teacher uses the tests to help the students learn.	1	2	3	4	5
21. The teacher is available for extra help.	1	2	3	4	5
22. The laboratory is well organized.	1	2	3	4	5
23. The laboratory exercises are stimulating and interesting.	1	2	3	4	5
24. The laboratory exercises make students think.	1	2	3	4	5
25. The laboratory gives students a chance to ask questions and experiment.	1	2	3	4	5
26. The students usually give full attention to what is taking place in the science class.	1	2	3	4	5
27. The science class is usually exciting.	1	2	3	4	5
28. The students in the class are generally interested in what is being taught.	1	2	3	4	5
29. The science textbook for the course is written so that I can understand it.	1	2	3	4	5

FIGURE 19–1 *continued*

30. The science subject is interesting.	1	2	3	4	5
31. The teacher has a good voice and does not speak in a monotone.	1	2	3	4	5
32. The teachers and students in the class get along well.	1	2	3	4	5
33. The teacher is well liked by students in the class.	1	2	3	4	5
34. The teacher has good discipline.	1	2	3	4	5

The above questionnaire can be modified by adding the items below to make it useful for science supervisors to rate the science teachers. It would also be useful for self-evaluation.

1. The science teacher works well with other teachers.	1	2	3	4	5
2. The science teacher is involved with the total science program and not just the course he or she teaches.	1	2	3	4	5
3. The science teacher is concerned about the science curriculum and works with other teachers to improve it.	1	2	3	4	5
4. The science teacher is very professional and attends professional meetings.	1	2	3	4	5
5. The science teacher helps bring the community's attention to the school's science program.	1	2	3	4	5
6. The science teacher is a member of one or more professional organizations for science teachers.	1	2	3	4	5
7. The science teacher is involved in committee work associated with science curriculum improvement.	1	2	3	4	5
8. The science teacher reads professional journals, scientific publications, and current books on science to keep up to date.	1	2	3	4	5
9. The science teacher keeps up to date on new approaches to teach science, curriculum, teaching resources, and teaching aids.	1	2	3	4	5

MAINTAINING A RÉSUMÉ

It is advisable for science teachers to maintain a current record of professional activities that can be used for a résumé or curriculum vitae. This account of professional activities should include all aspects of a teacher's active career, such as college degrees earned; dates of degrees received, including major and minor areas of study; and a list of job positions, including places of employment, dates of employment, and the duties associated with the positions. In addition the teacher should include a report of participation in inservice courses and workshops along with the titles and descriptions of the workshops and inservice courses. If the teacher is continuing graduate study toward an advanced degree, it is advisable to indicate the reason for pursuing the degree and the progress that has been made toward the degree. Additional activities should be mentioned, such as attendance and participation at professional meetings with dates attended and reason for attendance, such as delivering a research paper or being a panel member. Membership in professional organizations and a description of the nature of participation in the professional organization should also be mentioned. Teachers should list their publications, including those that have been accepted by professional journals. Committee membership in and out of school along with duties associated with membership should also be indicated. Special honors received, such as awards by NSTA and other professional organizations, are important to mention.

The list of items to include in a résumé can be expanded or limited depending on the type and amount of information needed for a particular purpose. For example, those who request résumés often want an evaluation of classroom teaching. It is suggested that summaries of evaluation of teaching performance be available in case of need.

In conclusion, résumés should be updated on a regular basis. Often very important activities are forgotten if a continuous record is not maintained.

SUGGESTED ACTIVITIES

1. Join the National Science Teachers Association, your state science teachers organization, and the local science teachers organization. Dues for students are usually reduced. Carefully read the bylaws of each organization to determine its objectives. Attend a convention of one of these organizations if possible.

2. Obtain a recent copy of *The Science Teacher* and review the various sections to determine what is generally emphasized. Do the same for another journal specific to your area of expertise, such as the *American Biology Teacher* or the *Journal of Chemical Education*. What types of articles are emphasized in the journals you have selected? Discuss the results of your reviews with the members of your science methods class.

3. Examine several science periodicals, such as *Scientific American, Science, Discover, Omni,* and others that are found in public and school libraries. Determine the types of articles that are emphasized. Discuss the results of your inspections with other members of your science methods class.

4. Visit three schools in the immediate area and determine what inservice programs they offer for professional growth. Also determine other ways these schools foster professional growth (such as paying for graduate courses or en-

couraging attendance at meetings of national organizations).

5. Identify the statewide science teacher organization of your home state or the state in which your college or university is located. Obtain recent copies of their journals and newsletters. Determine how active the organizations are in that state by asking experienced science teachers about them. What are the advantages of being a member of such an organization?

6. Prepare a résumé including some of the items indicated in this chapter. Compare your résumé with those of other members of your science methods class. After discussion, rewrite your résumé to include items that will improve it.

7. Assume you are applying for a position as a middle/junior high school science teacher in a school located in a small rural community. The position has not been advertised but you know that one is available. The only experience you have had is student teaching. Draft a letter of application to the principal of the school. State your qualifications for the position. In writing such a letter of application, you will have to determine the important points you need to discuss in the letter, since you have no other guidelines. Compare your letter with those of other students.

BIBLIOGRAPHY

American Association for the Advancement of Science (AAAS). 1989. *Science for all Americans (AAAS Project 2061, overview report).* Washington, DC: Author.

Association for the Education of Teachers in Science. 1984. *1984 AETS yearbook—Observing science perspectives from research and practice.* Columbus, OH: ERIC Clearinghouse for Science and Mathematics and Environmental Education.

Bazler, J., and Simonis, D. 1989. Future science teachers of America: Strengthening professionalism in science education. *Journal of Science Teacher Education, 1*(2): 36–37.

Bieda, R., Gibbs, R., and Goldie, S. 1990. Collaborate with your collegiate colleagues. *The Science Teacher, 57*(1): 40–42.

Brinckerhoff, R. F. 1989. Resource centers serve the needs of school science teachers. *School Science and Mathematics, 89*(1): 12–18.

Evans, R. H. 1991. Engaging teachers in the science education literature, both as critical consumers and active contributors. *Journal of Science Teacher Education, 2*(3): 77.

Finson, K. D. 1989. Assessment of a pilot program for inservice teachers. *Science Education, 73*(4): 419–431.

House, E. R., and Lapan, S. D. 1978. *Survival in the classroom: Negotiating with kids, colleagues and bosses.* Boston: Allyn & Bacon.

Koballa, T. R., Jr. 1987. The professional reading patterns of Texas life science teachers. *School Science and Mathematics, 87*(2): 118–129.

Moore, K. D. 1987. Preparing science teachers: A proposal for excellence. *School Science and Mathematics, 87*(7): 545–557.

National Science Foundation (NSF). 1989. *Guide to programs, fiscal year 1989.* Washington, DC: Author.

National Science Teachers Association (NSTA). 1989a. *Guidelines for self-assessment of high school science programs* (modules 2 and 3). Washington, DC: Author.

NSTA. 1989b. *Guidelines for self-assessment of middle/junior high school science programs* (modules 2 and 3). Washington, DC: Author.

Rowland, P., and Stuessy, C. L. 1990. The effectiveness of mentor teachers providing basic science process skills inservice workshops. *School Science and Mathematics, 90*(3): 223–231.

Sanford, J. P. 1988. Learning on the job: Conditions for professional development of beginning science teachers. *Science Education, 72*(5): 615–624.

Showalter, V. M. (Ed.). 1984. *Conditions for good science teaching.* Washington, DC: National Science Teachers Association.

Spector, B. S. (Ed.). 1987. *Guide to inservice science teacher education: Research into practice. In 1986 AETS Yearbook.* Columbus, OH: The Ohio State University, SMEAC Information Research Center.

Sukow, W. W. 1990. Physical science workshops for teachers using interactive science exhibits. *School Science and Mathematics, 90*(1): 42–47.

Yager, R. E. 1988. Differences between most and least effective science teachers. *School Science and Mathematics, 88*(4): 301–307.

Index

ISBN 0-02-323551-9

9 780023 235511

90000>